Women in Engineering and Science

Series Editor
Jill S. Tietjen, Greenwood Village, CO, USA

The Springer Women in Engineering and Science series highlights women's accomplishments in these critical fields. The foundational volume in the series provides a broad overview of women's multi-faceted contributions to engineering over the last century. Each subsequent volume is dedicated to illuminating women's research and achievements in key, targeted areas of contemporary engineering and science endeavors.The goal for the series is to raise awareness of the pivotal work women are undertaking in areas of keen importance to our global community.

Monique Frize • Claire Deschênes • Ruby Heap

Karine Laporte
Collaborator

Women's Contribution to Science and Technology through ICWES Conferences

Springer

Monique Frize
Carleton University
Ottawa, ON, Canada

Claire Deschênes
Université Laval
Québec, QC, Canada

Ruby Heap
University of Ottawa
Ottawa, ON, Canada

Collaborator
Karine Laporte

ISSN 2509-6427 ISSN 2509-6435 (electronic)
Women in Engineering and Science
ISBN 978-3-031-33799-4 ISBN 978-3-031-33800-7 (eBook)
https://doi.org/10.1007/978-3-031-33800-7

This Springer imprint is published by the registered company Springer Nature Switzerland AG
The registered company address is: Gewerbestrasse 11, 6330 Cham, Switzerland

Paper in this product is recyclable.

Foreword by Ed. Jill S. Tietjen

The objective of the Springer Women in Engineering and Science series is to highlight women's accomplishments in science and engineering and raise awareness of the pivotal work that women are undertaking globally in areas of importance to the world. I agreed to be the series editor because of my now more than forty years of encouraging women to pursue scientific, technical, engineering, and mathematics (STEM) careers; of nominating women for awards to recognize their accomplishments; of encouraging professional development for women in the STEM fields; and of my personal mission of writing women into history and ensuring that their contributions to our quality of life and standard of living are not forgotten.

This volume on international engineering and scientific women documents the coming together of STEM women around the world since 1964 – at a time when women didn't even account for 1% of engineering degrees awarded annually, at least in the USA. It demonstrates the evolution of technology globally as well as the increasing contributions of women to STEM fields over that time. It is a vital part of the Springer Women in Engineering and Science series.

As I read the book, I was delighted to see mention of so many women that I am familiar with either through my writing or my nominations or personally, including through the Society of Women Engineers (SWE). Computer scientist Admiral Grace Murray Hopper is the woman I consider to be my catalyst for my advocacy work for women globally. She was my first successful nomination for the National Medal of Technology and my first successful nomination to the National Women's Hall of Fame. I received both of those honors for her: the first at the White House Rose Garden and the second, posthumously, in Seneca Falls, New York, the birthplace of women's rights in the USA. She has been included in several books within the Springer Women in Engineering and Science series and in others of my books as well. Although I never met her personally, I did speak with her on the telephone when I was preparing her National Medal of Technology nomination and after she was selected. The latter time, she gave me very specific instructions as to the citation she wanted in the press release!

"First Lady of Engineering" and a significant supporter of SWE, Lillian Moller Gilbreth was inducted posthumously into the National Women's Hall of Fame during the second induction ceremony that I attended in Seneca Falls. It was my pleasure to nominate her and to include her in several volumes in the Springer Women in Engineering and Science series and others of my books as well. I even got to meet two of her children among the famous Gilbreth twelve (immortalized in the book and movie *Cheaper by the Dozen*) – Bob and Ernestine!

I was very pleased to see significant mention of SWE co-founder and first President Beatrice Hicks whose posthumous nomination to the National Women's Hall of Fame was prepared by my collaborator Alexis Swoboda and whom I successfully nominated posthumously for an award from the New Jersey Inventor's Hall of Fame. Dr. Hicks has been included in my books – both in the Springer Women in Engineering and Science series and otherwise. I have written about solar pioneer Maria Telkes for several books in the Springer Women in Engineering and Science series as well as civil engineer Elsie Eaves. These women and quite a number of others in this volume grace the pages of two earlier books in the Springer Women in Engineering and Science series that I authored: one a history of women in engineering (*Engineering Women: Re-visioning Women's Scientific Achievements and Impacts*) and one a history of women in science (*Scientific Women: Re-visioning Women's Scientific Achievements and Impacts*). I saw friends in the book and many SWE Past Presidents (I am also a SWE Past President).

It is critical to the health of our world and to the world's economy to recognize the accomplishments of women and to ensure women's education in all fields of endeavor. I applaud the volume co-authors for rescuing the information on all of the International Conference of Women Engineers and Scientists, organizing it, and synthesizing it for this important book.

Series Editor, Springer Women in Engineering and Science Series Jill S. Tietjen
President and CEO, Technically Speaking, Inc.
Greenwood Village, CO, USA

Foreword by Gail Mattson

As a life-long advocate for women in science and engineering who has had the honor of being involved with the creation of the International Network of Women Engineers and Scientists (INWES) since the initial organizational meeting in 2001, I am very pleased to see this important historical record about the International Conference for Women in Engineering and Sciences (ICWES) and INWES appropriately recorded and published.

At the World Conference on Science for the twenty-first century, convened by the United Nations Educational, Scientific and Cultural Organization (UNESCO) and the International Council for Science in Budapest, Paragraph 90 was adopted encouraging special efforts to be made toward the establishment of an international network of women scientists and engineers. Through the support and encouragement of the Canadian Commission for UNESCO, and a successful grant from UNESCO, 20 women representing 10 countries and 8 organizations met in Ontario, Canada, May 2001, to explore the creation of such a network. I represented the Society of Women Engineers (SWE), USA, as the immediate Past National President. It was such a thrill to meet women scientist and engineers from around the globe who also had a passion for increasing the number and voice of women in STEM, as well as to also learn about this support from UNESCO.

After our first morning session, all agreed that we should look into how this new organization could build on the efforts of the existing international committee of like-minded women scientists and engineers who had been involved in the hosting of an International Conference of Women Engineers and Scientists (ICWES) every three years since 1964. The remaining sessions focused on how to structure, formally establish and operate this new international entity, and use all of our personal and organizational contacts to reach out to as many women in STEM around the globe to announce the formation of this new organization.

In July 2002, the ICWES-XII delegates supported the creation of INWES. The vote from representatives of 30 countries was unanimous. The following April 2003, INWES was incorporated as a non-profit corporation under the laws of Canada. INWES is governed by a board of directors consisting of directors representing organizations, universities/institutes, and individual memberships, with the mission "[t]o build a better future worldwide through full and effective participation of women and girls in all aspects of STEM." One of its primary programs is to ensure the hosting and expansion of an ICWES every three years, with more representation worldwide, multiple programs and workshops, and funding for participants from developing countries.

In April 2008, INWES became an official NGO partner of the operational type with UNESCO. This partnership involves a dynamic cooperation to help women and girls worldwide to have access to education, especially in Science and Engineering. Specifically, "[a]s a solution to tackle the issue of gender imbalances in STEM, UNESCO aims to support and promote networks of women scientists in various scientific domains and regions, such as, among others, [...] the International Network of Women Engineers and Scientists." Subsequently, INWES became a member of UN Women and in 2017 was granted the Economic and Social Council status to the United Nations. As a result, the priority actions for INWES are now aligned with the UN Sustainable Development Goals and are aimed at mainstreaming gender at all levels for:

- Goal 4: Ensure inclusive/equitable quality education and promote lifelong learning opportunities for all
- Goal 5: Achieve gender equality and empower all women and girls
- Goal 8: Promote inclusive and sustainable economic growth, employment and decent work for all

Since 2002, I have had the honor of supporting the growth and evolution of INWES to the worldwide organization today that now represents over 250,000 members from over 60 countries. Three of the important aspects of INWES that have contributed to this expansion were as follows:

1. The establishment of an interactive website and quarterly newsletters in which INWES members could share the success of their STEM activities
2. The establishment of regional networks where organizational members from different regions can recruit new members, help start up new Women in STEM organizations in countries where they do not exist, collaborate on similar issues and projects, and hold regional network meetings during the intervening two years between ICWES
3. Entering into Memorandum of Understanding with other international organizations such as the World Federation of Engineering Organizations and the World Federation of Scientific Workers to expand collaborative programs and our outreach

From initially representing SWE on the interim Board of Directors 2002–2003, I served has a Board Member and Treasurer (2003–2011), Conference Chair for ICWES-XVI in Los Angeles, CA (2014), VP Conferences (2015–2017), President (2018–2020), and now Immediate Past President (2021–2023).

Even though I have been very involved with INWES since the beginning, there were many details about the evolution of ICWES and INWES that I was not aware of or had forgotten. That is why I am so thankful for the authors to have spent the time and effort necessary to research and document these important international STEM organizations. I hope that you will enjoy reading the book as much as I did, and that many researchers and historians will dig further the material to reveal the Women in STEM contribution to science in the second part of the twentieth century to the beginning of the next one.

INWES Past President Gail G. Mattson
Senior ES&H Consultant, Longenecker & Associates, Inc.
Santa Fe, NM, USA

Acknowledgments

When we began writing this book, it soon became evident that we would need a fourth person to complete it. Karine Laporte, who recently obtained a PhD in Classics history, became an important contributor. She helped with several aspects of this book, such as organizing the ICWES database, preparing the Excel files with Claire, writing some of the Gender papers summaries with Ruby, finalizing the recuperation of ICWES-VIII content, participating in discussions, checking the references, formatting the text and making suggestions for corrections, etc. Karine, you were truly part of the team, we cannot thank you enough!

The authors would also like to thank CIWES-ICFIS for providing some financial support to hire Karine Laporte. Some of the support for her was provided by Claire Deschênes and Ruby Heap.

We wish to thank the Library at the University of Ottawa, especially Marie Noël and Marina Bokovay for their help in retrieving documents from the archives located at the university and for organizing the huge amount of information on ICWES, INWES, and CIWES-ICFIS. We acknowledge the important support provided by Troy Eller English (manager of the Society of Women Engineers archives) and the Institution of Engineering and Technology (IET) archives in the UK.

Thank you to Gail Mattson and Jill S. Tietjen for the two forewords.

For ICWES information, special thanks are due to Mahin Rahmani from Iran for lending us most of the ICWES proceedings, and to Suriya Thevar (Chennai, India) for digitizing the proceedings of ICWES-I to ICWES-XI (except VIII) in 2005 and organizing the information in a very helpful format. Thanks are also due to Akiko Tsugawa in Japan for her help in retrieving important information such as the list of participants at ICWES-XI. Thanks to Issié Gueye for providing the audio cassettes on ICWES-VIII held in Abidjan in Ivory Coast, Africa; and to François and Jennifer Aubry for translating the audio cassettes into text.

Finally, the authors wish to thank Springer, in particular Mary James, Jill S. Tietjen, and Pradheepa Vijay for the support and encouragement they provided to the authors during the writing of the book.

Introduction

Fifty-nine years have passed since the first International Conference of Women Engineers and Scientists was held in New York City in 1964. Since then, a group of women managed to ensure its continuity over a long period of time. Writing the story of the ICWES conference series is very much overdue, which is the goal of this book. Of course, making this happen needed serious research. Since several leaders of this international endeavor are deceased, it was very important to locate the proceedings of the conferences, whenever possible, and retrieve the notes of delegate meetings held at each of the conferences; these notes describe how the momentum was sustained between 1964 and 2002, established and revised guidelines on how the conferences should be organized and held a vote on the potential location of the next one. For this reason, in 2005, Claire Deschênes and Monique Frize decided to collect the proceedings from as many of the ICWES conferences as possible. We managed to find proceedings of most ICWES events and had the documents scanned and digitized. Although the idea to write the story of the ICWES conferences series was already in our mind in 2005, many years passed before the book could become a reality, due to the heavy load of academic activities of the potential authors. In 2020, we decided the time was right, as three of the authors were now retired.

It also made sense to bridge the ICWES story with the creation of INWES. Claire, Monique, and Gail Mattson were instrumental in creating the International Network of Women Engineers and Scientists (INWES), an organization incorporated in Canada in 2003, providing a solid foundation on which the ICWES conference series could continue to be held in various parts of the world. Part I of this book provides an overview of the scientific content from ICWES-I up to ICWES-XII in 2002, leaving for future research the analysis of the more recent ICWES conferences. Since the establishment of INWES during the ICWES-XII conference, there have been several ICWES conferences: ICWES-XIII in 2005, Seoul, South Korea; ICWES-XIV in 2008, Lille, France; ICWES-XV in 2011, Adelaide, Australia; ICWES-XVI in 2014, Los Angeles, USA; ICWES-XVII in 2017, New Delhi, India; ICWES-XVIII in 2021, in Warwick, UK, also held virtually, and ICWES-XIX in Aukland, New-Zealand.

Therefore, in addition to writing the history of the ICWES Conference series, the book describes, in Part II, how the informal committee of ICWES organizers was transformed into the International Network of Women Engineers and Scientists (INWES) as a not-for-profit corporation. This new organization was a direct outcome of the ICWES project.

The International Conference of Women Engineers and Scientists (ICWES) series was and still is unique in several respects. Most of the participants are women engineers and scientists presenting technical papers in their field of work or research. Each conference also has an important gender component, with presentations on women's experiences working or studying in these fields; or a description of outreach programs to encourage girls to consider such careers; or papers on various aspects of education for women and girls. Ruby Heap, a historian and co-author, has attended several ICWES conferences and added an important component to the program: a symposium on bridging women's studies and women in STEM (Science, Technology, Engineering, and Mathematics) fields; this began at ICWES-XII in Ottawa, Canada, and continued at each subsequent conference. Because women can present a technical paper, this can provide a justification to obtain travel support to attend the conferences for some of the women. But several women, especially in the early conferences, took vacation time to attend the conference and many paid their own travel costs and conference fees.

The analysis of the number of Science and Technology (S&T) and Gender activities may demonstrate how important preoccupations changed over time. This is an interesting aspect to consider in an analysis of the ICWES series. Another unique aspect of the conference series is that some of the organizers ended the conference by asking participants to develop a *resolution*, an important statement to be supported by all after the conference ended. The resolutions were sometimes based on the environment, such as at ICWES-III; others were based on ensuring a continuation of the conferences. The proceedings and notes of the ICWES series do not always record the business meeting, the resolution, or the participant and delegate list. This is unfortunate; however, when information exists, it is included in this book.

Sponsors of the conferences include industries, governments, politicians, universities, and individuals. The proportion of support from these sources and agencies varies by region. For several of these events, booths were sold to organizations and companies wishing to advertise their products or services. A particular aspect of conferences organized by the Society of Women Engineers (SWE) in the USA (1964, 1984, and 2014), include a career fair, where companies encourage students to meet with them and explore job opportunities. This aspect led to numerous booths rented by companies looking to hire women in these fields.

Each conference organizing committee offered either pre-conference tours or post-conference tours, or both, consisting of interesting tourist locations; attendees paid separately for these when they registered for the conference. All ICWES events also offered industrial tours during the conference, located in the geographical area. Usually there was also a program for accompanying spouses/partners. Finally, a unique component of each ICWES was that a representative from each country presented greetings; this began at ICWES-II in the UK and consisted of a few words from each person, delivered either at the welcome reception, or banquet, or at the opening ceremony. Women were also encouraged to wear traditional dress, which made a colourful and cultural evening. The social program was always well done, with tasty local foods, a friendly atmosphere, cultural demonstrations of traditional music and dancing or organized in a special museum. Many new friends were made at an ICWES conference!

The Internationalization of Scientific Conferences and Organizations

Without claiming to be exhaustive, the following review aims to provide elements of the historical context in which the ICWES Conferences Series appeared in 1964, and the creation of the INWES corporation which, since 2002, ensures their continuity.

It is interesting to note that international congresses emerged during the second half of the nineteenth century, owing to "a professionalization of knowledge, radically better and faster modes of transportation, and a proliferation of information conveyors" (Feuerhahn and Rabault-Reuerhahn 2010). Shifting existing local or regional meetings of experts coming together to discuss common research interests to an international scale, these conferences transformed scientific discourse by giving existing practices of writing and reading an unavoidable oral component and a much larger social dimension. This change also contributed to legitimizing further certain disciplines and ideas in the eyes of the wider scientific community. In addition to serious political incentives for countries hosting or participating in such events, international conferences were also a means to popularize a given discipline with a general audience, not to mention decision makers. Tied to contemporary issues by their very nature (that is, practical matters of hosting and funding, but also the fact of bringing people from various countries in one place), international congresses reflected ongoing social, geopolitical, economic, and cultural discussions.[1]

Relating to women, a number of international congresses date back to the nineteenth century or the early twentieth century as well. Table 1 lists a series of international conferences related to women; for example, the first International Congress of Women was held in Paris in 1878, during the third Paris World's Fair. This initiative was motivated by the need to bring together regional and national feminist movements campaigning for women's suffrage.

After WWII, more International conferences emerged. Well known among them, the United Nations organized four World Conferences on Women: Mexico City, 1975 (World Conference of the International Women's Year); Copenhagen, 1980

[1] On the internationalization of feminist movements from the second half of the nineteenth to the mid-twentieth century ("first-wave feminism"), see the two edited volumes Moynagh and Forestell (2012) and Forestell and Monaygh (2014).

Table 1 International conferences related to women

Name	Creation date and cities	Purpose
International Congress of Women	Paris, 1878	Regional or national feminist movements campaigning for women's suffrage ("International Congress of Women" n.d.)
International Alliance of Women (IAW)	Berlin, 1904	Feminists advocating for women's right to vote. Congresses every three years ("What is IAW – International Alliance of Women" n.d.)
International Congress of Working Women (ICWW)	Washington, DC, 1919	Bringing together labour feminists from all around the world ("International Congress of Working Women" n.d.)
International Socialist Women's Conferences	Stuttgart, 1907	Four conferences up to 1917, promoting the International Women's Day, helped launch the Socialist International Women in Stuttgart, and laid the groundwork for others ("International Socialist Women's Conferences" n.d.)
Communist Women's International	Moscow, 1920	Emerging from the Communist International (or Third International) ("Communist Women's International" n.d.)
Women's International Democratic Federation (WIDF)	Paris, 1945	Working on women's and children's rights, promoting anti-fascist and pacifist views. Consultative status with the Economic and Social Council of the United Nations. The International Women's Year was declared by the UN in 1975 at the request of WIDF ("Women's International Democratic Federation" n.d.)
International Conference for Women Leaders	Haifa and Jerusalem, 1961	Development in developing and industrialized countries. Issues identified by the United Nations ("International Conference for Women Leaders" n.d.)
International Conference of Women Scientists and Engineers (ICWES)	Washington, DC, 1964	
World Conferences on Women	Mexico, 1975	Four conferences organized by the UN (2. Copenhagen in 1980; 3. Nairobi in 1985; 4. Beijing in 1995) Issues specific to women's experiences. The International Women's Year was set in Mexico with a worldwide strategy
Congrès international des recherches féministes dans la francophonie (CIRFF)	Quebec, 1996	Every three years, gathering French-speaking scholars, students, activists, and public figures around themes dealing with gender studies and women's studies ("Congrès international des recherches féministes dans la francophonie" n.d.)
Gender Summits	Brussels, 2011	Platforms for dialogue between scientists, policy-makers, gender scholars, and stakeholders to examine how biological differences (sex) and socio-cultural differences (gender) impact the quality of research outcomes and to find solutions

(World Conference of the United Nations Decade for Women: Equality, Development and Peace); Nairobi, 1985 (World Conference to review and apprise the achievements of the United Nations Decade for Women: Equality, Development and Peace); Beijing 1995 (Fourth World Conference on Women). These conferences dealt with issues specific to women's experiences. The Mexico Conference was set during the International Women's Year and established a worldwide strategy to achieve the objectives of the International Women's Year. The Copenhagen Conference was held in order to review the advancement of women and gender equality, especially with regards to employment, health, and education, and proposed to further women's rights to inheritance, child custody, and citizenship. Following the Nairobi Conference, countries agreed on the "Nairobi Forward-Looking Strategies for the Advancement of Women," which listed "measures for achieving gender equality at the national level and for promoting women's participation in peace and development efforts" ("World Conferences on Women" n.d.). During the Beijing Conference, the "Beijing Declaration and the Platform for Action" was adopted by all countries as "an agenda for women's empowerment and [is] considered the key global policy document on gender equality" ("World Conferences on Women" n.d.). The fourth conference was followed by appraisals held every five years.

Focusing on scientific fields, various international[2] and regional[3] organizations emerged in the late nineteenth or early twentieth century. More specifically for women, as can be seen in Table 0.2, the Graduate Women International (GWI) was founded in 1919 in the wake of the First World War ("GWI History" n.d.), as well as the International Federation of University Women the same year, with a Canadian section called Canadian Federation of University Women (CFUW/FCFDU). The Medical Women's International Association was also established in 1919 an effort to promote and represent women physicians all around the world ("Foundation of MWIA" n.d.).

Regional or national organizations of women scientists and engineers were also created at the time, and since the 1970s, the number of organizations for – or about – women in S&T increased greatly. Dozens of local, regional, and national organizations of women in STEM fields emerged in many countries (Table 2).[4]

---

[2]To name but a few: the International Geographical Union (IGU, 1871), the International Association of Geomagnetism and Aeronomy (IAGA, 1873), the International Association of Geodesy (IAG, 1886), the International Astronomical Union (IAU, 1919), the International Union of Pure and Applied Chemistry (IUPAC, 1919), the International Association of Hydrological Sciences (IAHS, 1922), the International Union of Pure and Applied Physics (IUPAP, 1922), and later, the International Association for Hydro-Environment Engineering and Research (IAHR, 1935), the International Association for Vegetation Science (IAVS, 1939 as the International Phytosociological Society), and the Institute of Electrical and Electronics Engineers (IEEE, 1963).

[3]Such as the American Society of Mechanical Engineering (ASME, 1880), the American Association for the Advancement of Science (AAAS, 1848), the (Canadian) Association francophone pour le savoir (ACFAS, 1923).

[4]On women's scientific associations created during the second wave of feminism (from 1945), see Puaca 2014, pp. 127–168.

Table 2 International and regional organizations for women in science

Organization	Creation date	
Graduate Women International (GWI)	1919	("GWI History" n.d.)
International Federation of University Women	1919	
Medical Women's International Association	1919	To promote and represent women physicians all around the world ("Foundation of MWIA" n.d.)
National Association of Women Pharmacists	UK, 1905	
American Medical Women's Association (AMWA)	USA, 1915	
Society of Woman Geographers	USA, 1915	
The Women Engineering Society (WES)	UK, 1919	
Association amicale des femmes ingénieurs	France, 1920	Became the Cercle des femmes ingénieurs as part of AFFDU ("L'AFFDU vue de l'intérieur 1920–1995" 1997).
Association française des femmes diplômées des universités [AFFDU]	France, 1920	The French section of GWI
Graduate Women in Science (GWIS)	USA, 1921	
L'Association des femmes diplômées des universités de Québec (AFDU Québec)	Québec 1949	Founded in 1949, incorporated in 1959
Society of Women Engineers (SWE)[a]	USA, 1950	
Associazione Italiana Donne Ingegnere e Architette (AIDIA)	Italy, 1957	
Society of Japanese Women Scientists (SWJS)	Japan, 1958	
Association for Women in Science (AWIS)	USA, 1971	To advocate on behalf of women in STEM to achieve business growth, social change and innovation ("AWIS History" n.d.)
Organization for Women in Science for the Developing World (OWSD)	1987	Officially launched in 1993. Formerly known as the Third World Organization for Women in Science (TWOWS). A program unit of UNESCO, providing research training, career development and networking opportunities for women scientists throughout the developing world ("Organization for Women in Science for the Developing World" n.d.)
International Network of Women Engineers and Scientists (INWES)	2002	A network of organizations to build a better future worldwide through full and effective participation of women and girls in all aspects of Science, Technology, Engineering, and Mathematics (STEM) ("About INWES" n.d.)

[a]See Puaca 2014, pp. 65–84 on the creation and early history of SWE

As far as the authors know, the International Conference for Women Engineers and Scientists (ICWES), created in 1964, was the first international congress gathering women in science and engineering (S&G).[5]

To this day, ICWES has not been the focus of substantial research. Archival work has been conducted by the Walter P. Reuther Library at Wayne State University (USA) by collecting material from SWE members.[6] Emily Rees Koerner and Graeme Gooday, based at the University of Leeds, worked with Elizabeth Bruton from the Science Museum on a project called "Electrifying Women: Understanding the Long History of Women in Engineering." Rees Koerner and Goody held a webinar on April 8, 2022, featuring speakers working on various projects on the history of women in science and engineering, including INWES President Jung Sun Kim talking about "Two Decades of the International Network of Women Engineers and Scientists" and Monique Frize updating the audience about progress on this book.

Questions That Guided the Writing of This Book

Our goal is to produce the first in-depth analysis of ICWES Conferences, from their inauguration in 1964 until the establishment of INWES in 2002.

To achieve this goal, we were guided by the following key questions:

1. How did the organization of the conferences change over time, and who were the main actors involved in this process?
2. Could we detect concerns, emerging or recurrent, in the ICWES scientific programs, and how these concerns evolved over time?
3. Could we depict links between the ongoing ICWES Conferences and the growth and development of various feminist strands between 1964 and 2002?

Structure and Content of the Book

The book begins with two forewords: a first by Jill S. Tietjen, P.E., Series Editor of "Springer Women in Engineering and Science," who kindly agreed to include this book in the collection; a second by Gail Mattson, former president of SWE and INWES and cofounder of INWES.

Part I, "ICWES-I to ICWES-XII," presents quantitative and qualitative information on the first 12 conferences (1964 to 2002). It is organized in 13 chapters.

[5] Traditionally, "S&T" does not include medical fields.

[6] https://archives.wayne.edu/repositories/2/resources/1206/collection_organization. Accessed Feb 24 2023.

Chapter 1 presents the qualitative and quantitative database that was collected and examined to write this book and the description of the methodology the authors followed to write the book.

In each of the chapters from 2 to 13, the first component describes the organization, participation, and Business Meeting of Delegates. Extracts of important parts of opening keynote talks, other important presentations are also given for each ICWES. The second component of each chapter provides overviews of the conferences' Science and Technology (S&T) and the Gender content, along with notes on key historical contexts, continuity, and changing discourse. The book includes information that pertains to ICWES-I to ICWES-XII; the proceedings were analyzed, including the papers, posters, keynote speeches, activities for career development, panels, workshops on specific issues, etc. The authors intended to provide a glimpse of the rich content of the ICWES database and hoped that the text transmits some of their enthusiasm on the participation of women (and some men) scientists, for 38 years, in the second part of the twentieth century and the beginning of the twenty-first. Chapters close with impressions of some of the women who attended ICWES conferences at different periods (testimonials). We are fortunate to have been able to collect this information, even from women who attended the first ICWES in 1964! These testimonials clearly convey the unique experience in attending such a conference. Extracts or full texts are included in the book.

Part II, "INWES in ACTION!", concerns the set-up of INWES and events organized by INWES after its launching in 2002.

Chapter 14 recalls the first meeting held in 2001 in Merrickville, Canada, where expert women from many regions of the world brainstormed on how to create a permanent organization to ensure continuity to the conference series. The resolutions were adopted the following year at the Business Meeting for ICWES-XII in Ottawa. Chapter 15 provides the minutes of several important INWES meetings between 2002 and 2008. Chapters 16 and 17 look at the impact of this new organization through the two conferences following its creation: ICWES-XIII in Seoul, South Korea (August 26–29, 2005) and ICWES-XIV in Lille, France (July 15–18, 2008). Chapter 18 discusses the development of CIWES-ICFIS, an organization created with a charitable perspective in Canada.

The conclusion of this book presents some thoughts on how the authors believe the ICWES phenomenon started and how it managed to continue for nearly 60 years. It summarizes some enlightening discussions the authors had on the increasing participation and contribution of women in S&T. It also provides essential information for researchers and scientists who wish to continue digging and report on the extraordinary ICWES conferences.

A bibliography of the material mentioned in this book is listed at the end, along with eight appendices. Appendix 1 contains biographical information on some of the founders, leaders, and organizers of these events. Appendix 2 provides information on the evolution over time of the regions used by ICWES to divide the word. Appendix 3 details the participation by region at each conference ICWES-I to ICWES-XII. Appendices 4–6 include important papers taken from ICWES conference proceedings: Elizabeth Laverick's overview of ICWES-I to IX, Rita Levi

Montalcini's opening talk at ICWES-III, Laverick's notes on the organization of ICWES-IX, as a good example of how a successful ICWES conference was organized. Appendix 7 lists all INWES meetings between 2002 and 2008. Appendix 8 offers a testimonial by Dormer Ellis, a Canadian scientist who attended all the conferences ICWES-II to ICWES-XII, and ICWES-XVI.

We hope that readers will be inspired by the incredible journey started by a group of women leaders in 1964 and will wish to experience attending an ICWES sometime in the future!

Contents

Part I ICWES-I to ICWES-XII

1 Methodology in Creating and Analyzing the ICWES Database.... 3
Creating the ICWES Database ... 3
Qualitative and Quantitative Analysis of the Scientific Content of Each Conference ... 5
Component 1 in Chapters 2–13 ... 6
Component 2 in Chapters 2–13 ... 6
Quantitative Data ... 7

2 ICWES-I, Setting the Stage: ICWES-I, New York City, USA, June 15–21, 1964 ... 11
Society of Women Engineers (SWE): Host of ICWES-I ... 14
The Beginnings of SWE ... 14
Leaders and Organizers of ICWES-I ... 15
Organizations as Major Contributors to ICWES-I ... 15
ICWES-I, S&T Papers ... 16
ICWES-I, Gender Papers ... 21
Resolution Statement by the Delegates ... 25
Testimonials ... 25

3 ICWES-II, The Need to Continue: Cambridge, UK, July 1–9, 1967 ... 33
Participants ... 35
ICWES-II, S&T Papers ... 36
Links Between S&T, Scientists and Engineers, and Women ... 41
ICWES-II, Gender Papers ... 41
Business Meeting ... 45
Testimonials ... 50

4 ICWES-III, Turin, Italy, September 5–12, 1971 57
Technical Theme. "Planning for Progress," and a sociological theme, "Women's Professional and Family Duties" 57
Participants 57
ICWES-III, S&T Papers 61
ICWES-III, Gender Papers 68
Meeting of Delegates 72
Testimonials 75

5 ICWES-IV, Cracow, Poland, September 8–13, 1975 79
Technical Theme: "New Techniques in the Service of Mankind", and sociological theme: "Contemporary Sociological Problems of Women Engineers and Scientists 79
Executive Committee of the Conference 79
Participants 80
ICWES-IV, S&T Papers 81
A Glimpse of ICWES-IV's S&T Content 83
Women Scientists' and Engineers' Contributions to S&T 87
ICWES-IV, Gender Papers 88
Roshan J. Begum and Kamala Balaraman, "Contemporary Sociological Problems of Women Engineers and Scientists in India" 89
Halina Mortimer-Szymczak, "Feminine Occupational Work in a Socialist State. Selected Problems" 90
Danuta Mischke, "Women at the Higher Technical Schools in Cracow in the Thirty Years Period of Poland" 92
Delegates Meeting 93
Grants for WES Students to Attend ICWES-IV 93
Testimonials 94

6 ICWES-V, Rouen, France, September 4–8, 1978 99
Theme: "Technology and Freedom" 99
ICWES-V, S&T Papers 100
A Glimpse of ICWES-V's S&T Content 102
Women Scientists' and Engineers' Contributions to S&T 105
ICWES-V, Gender Papers 106
Jacqueline Nonon's Opening Lecture: "Technology and Freedom" 106
Nicole Bécarud's Closing Lecture 107
Commission 3, "Information of Women on Scientific and Industrial Careers and Their Training in These Careers" 108
Elizabeth Laverick, "The Education and Training of Professional Engineers in the United Kingdom and Its Influence on Women Engineers" 108
Anna Saengbangpla and Sasithorn Boon-Long, "Education & Information of Women Engineers in Thailand" 110
Commission 4, "Responsibilities of the Research Scientist" 112

Mary Jossy Nakhanda Okwakol, "Women's Role in Scientific and Technological Research in Uganda" 112
Business Meeting 114
Resolution Passed by Participants: Reported by Gillian Skinner...... 115
Testimonials 115

7 ICWES-VI, Bombay, India, September 8–12, 1981.............. 119
Theme: "Science, Technology and Society"...................... 119
"Indian Women Scientists Association" (IWSA) 120
"Inaugural Ceremony" 120
ICWES-VI, S&T Papers.. 121
A Glimpse of ICWES-VI's S&T Content......................... 122
S&T and Women in ICWES-VI 126
ICWES-VI, Gender Papers.................................... 126
Dormer Ellis, "The Entry of Women into Canada's Engineering Profession" 127
Somdatta Sinha, "Some Thoughts on the Present Science Education System and Its Relevance to Our Society" 128
Nicole Bécarud, "Successful Business Women Today in France" ... 129
Business Meeting 131
Notes from Isabel Hardwich, "Business Meeting"............... 131
Testimonials 131
Dormer Ellis 131
Bette A. "BK" Krenzer, "An International Conference: Impressions" 132
Gillian Skinner (WES Member), "Impressions of ICWES"........ 134
Maria Watkins (WES Member), "Impressions of India" 134
Emma C. Barth (SWE Pittsburgh Section), "Women Engineers of Other Nations: Their Career Access, Professional Concerns" 135

8 ICWES-VII, Washington, DC, USA, June 17–24, 1984........... 139
Theme: "Technology – An International Bridge".................. 139
Organizing Committee 139
ICWES-VII, S&T Papers 140
A Glimpse of ICWES-VII's S&T Content........................ 142
S&T and Women in ICWES-VII 146
ICWES-VII, Gender Papers 146
Jacqueline Juillard, "Looking Back Twenty Years Ago with a Foreign Eye"... 147
Susan M. Hobart, "Breaking Into Engineering"................. 149
Ila Ghose, "Women Engineers in India and Impact of Women's Polytechnics" 151
Business Meeting 153
Testimonials 154

9 ICWES-VIII, Abidjan, Ivory Coast, September 20–27, 1987 157
Theme: "Science, Technology and Development" 157
The Organization Responsible for ICWES-VIII: CIFISATS 157
The Opening Ceremony 158
Participants 159
The Program 159
Technical Program 159
ICWES-VIII, S&T Papers 160
ICWES-VIII, Gender Papers 161
Meeting of Delegates 162
On the Location of the Next Conference 163
Testimonials 164
  BK Krenzer, "ICWES VIII – A West African Experience" 164
Cicely Thompson and May Maple (UK), "Eighth International Conference of Women Engineers and Scientists Abidjan, Ivory Coast 20–27 September 1987: A Report" 166
May Maple, "Social Aspects – Random Jottings" 167

10 ICWES-IX, Warwick University, Coventry, UK, July 14–20, 1991 171
Theme: "Communication" 171
Jacqui Brookes Provides an Update on the Organizing of ICWES-IX 171
Jacqui Brookes Provided a Further Update on the Conference Organization 172
Elizabeth Laverick, "ICWES One to Nine" 172
Opening Ceremony 173
  The Welcome from the President of WES: Dorothy Hatfield 174
  Elizabeth Laverick, "Organising ICWES-IX" 175
"Final Thoughts" 175
ICWES-IX, S&T Papers 175
A Glimpse of ICWES-IX's S&T Content 176
S&T and Women in ICWES-IX 181
ICWES-IX, Gender Papers 182
  Monique Frize, "Women in Engineering in Canada" 184
  Margaret-Ann Armour, "WISEST—An Initiative at the University of Alberta" 186
  Lotte Algreen Reimer, "Women in Engineering in Zimbabwe" 188
Business Meeting 190
Testimonials 191
  Lee Arnold, "Reflections on ICWES" 191
"Conference Impressions" from Three Delegates (ICWES-IX, Conference Report, pp. 45–48) 192
  Jacqueline Juillard, Chambesy, Switzerland 192
  Mitsuko Kazuno, Miyama, Japan, "From Japan with Love" 192

Olivia Wellesley-Cole, Australia . . . 194
ICWES-IX Organizer, Elizabeth Laverick, on "The ICWES Phenomenon" . . . 194

11 ICWES-X, Budapest, Hungary, October 8–10, 1996 . . . 197
Theme: "Towards the Third Millenium: The Role of the Intellectual Potential in a New World" . . . 197
The Opening Ceremony . . . 197
ICWES-X, S&T Papers . . . 198
A Glimpse of ICWES-X's S&T Content . . . 199
New Trend in Natural Sciences and Technology . . . 199
Transfer of Knowledge . . . 200
ICWES-X, Gender Papers . . . 201
"Women in Georgia," Elizaveta Buadze . . . 203
Anita M. M. Liu and Lai Yin Ng, "Sex Culture of Female Students in Hong Kong" . . . 205
Ágnes Fényi, "Changes in the Role of Women During the [Past] 50 Years" . . . 208
Business Meeting . . . 210
Testimonials . . . 212
Lee Arnold, Mid-Atlantic Region MAL, "Special Report: Reflections on ICWES-X" . . . 212
Peggy Layne, "Report From ICWES-X, October 8–10, 1996, Budapest, Hungary" . . . 214
Elizabeth Laverick Reports on ICWES-X . . . 216
Elizabeth Rhodes, WES Member, "Destination Budapest—ICWES-X" . . . 217

12 ICWES-XI, Chiba, Japan, July 24–27, 1999 . . . 219
Theme: "Science and Technology for Global Ecology" . . . 219
Her Imperial Highness Princess Takamado, "Message" . . . 219
Michiko Ishii, Chairperson of ICWES-XI, "Welcome Address" . . . 220
Hiroyuki Yoshikawa, President, Science Council of Japan, "Welcome Address" . . . 221
Mitsuko Kazuno, President of the Society of Japanese Women Scientists, "Welcome Address" . . . 222
Susan Bird, Representative of ICWES and Former President, the Women's Engineers Society, "Message" . . . 222
Organization . . . 223
Cooperating Organizations . . . 223
Participants . . . 224
ICWES-XI, S&T Papers . . . 224
A Glimpse of ICWES-XI's S&T Content . . . 226
Invited and Plenary Lectures . . . 227
ICWES-XI Papers on Renewable Energy . . . 228
ICWES-XI Papers on Women in the Medical Field . . . 229

S&T and Women in ICWES-XI Oral Sessions 230
ICWES-XI, Gender Papers 232
Meihe Zhou, "Working Circumstances of China's Women Workers" .. 233
Susan Bullivant, "The Athena Project" 236
"Introduction" 236
Issues for the Sector 237
Issues Pertinent to the Project 238
Athena Project's Role and Programme 239
Components of and Findings from the Paper 240
Sumiko Sasagawa, "The Matthew Effect on Women Scientists: Case of Harriet Brooks" 241
"Introductive Description" 241
"Results and Discussion" 242
References (as Cited in ICWES-XI, Complete Proceedings, p. 543)... 243
Business Meeting 244
Declaration 245
Testimonials 245
Petra Godwin, Immediate Past President WES, "My Personal Experience" 247
Issié Yvonne Gueye, Ivory Coast 248
Joanna Maduka, Nigeria 249
Ila Ghose, Calcutta, India 249
Jacquie Carpenter, UK, "Renewable Resources and ICWES-XI" ... 250
Karla Solis Garcia, Peru 251
Dormer Ellis, Canada and ICC Representative 251
Monique Frize, Canada, ICWES-XII Chair 252

13 ICWES-XII: Ottawa, Canada, July 27–31, 2002 255
Theme: "Women in a Knowledge-Based Society" / "*Les femmes dans la société fondée sur le savoir*" 255
"Welcome" from the Conference Chair, Monique Frize 255
"Welcome" from the Program Chair Claire Deschênes 256
"Welcome" from CCWEST President Hiromi Matsui 257
ICWES-XII, S&T Papers 257
A Glimpse of ICWES-XII's S&T Content 259
ICWES-XII, Gender Papers 262
Mary Braithwaite, "An Assessment of the European Union Research Funding Programme 'Improving Human Potential and the Socio-economic Knowledge Base'" 269
Business Meeting 272
Agenda: ICWES Delegates Meeting, Tuesday, July 30, 2002 272
ICWES 12 Delegates Meeting Minutes, July 30, 2002, 2:30 p.m., Ottawa 273
Testimonials 280

Anne M. Perusek and Peggy Layne, "ICWES-XII: 'Women in a Knowledge-Based Society" ... 280
Jacky Crook and Dianne Patterson (WES Members), "ICWES-XII" ... 281
"INWES—Looking Forward to Korea 2005: Sue Bird Reports on Developments" ... 284
"INWES: Sue Bird Reports on the Latest Developments" ... 285

Part II INWES in ACTION!

14 Merrickville 2001, A First Meeting ... 289
Meeting to Discuss the Creation of a Transnational Organization for Women in STEM Fields, Held in Merrickville, Ontario, Canada, May 26–29, 2001 ... 289
Discussion of Structure and Funding ... 298
Characteristics and Role Desired for the New Organization ... 299

15 INWES Between 2002 and 2008 ... 309
2002 : The First Meeting of the INWES Interim Board (BM1) ... 309
Ottawa, July 31, 2002 ... 309
2003 : Interim Board Meeting 2 (BM2) ... 311
Paris, France, May 1–2, 2003 ... 311
2003 : Interim Board Meeting 3 (BM3) ... 317
Pai Chai University (Seo-gu, Daejeon, South Korea), November 13–16, 2003 ... 317
2004 : Interim Board Meeting 4 (BM4) ... 321
Nairobi, Kenya, August 15–19, 2004 ... 321
2005 : Interim Board Meeting 5 (BM5) ... 325
Ewha University, Seoul, Korea, August 24–25, 2005 ... 325
BM5 Minutes—Attachment 1 ... 340
BM5 Minutes—Attachment 2 ... 341
2005 : First General Members' Meeting (GM1) ... 342
(= AGM1) Seoul, South Korea (at ICWES-XIII), August 28, 2005, 3 p.m. to 5 p.m. ... 342
2005 : Board Meeting 6 (BM6) ... 345
Seoul, South Korea, August 29, 2005 ... 345
2006 : Board Meeting 7 (BM7) ... 347
Polytech Lille, Salle du conseil, Lille, France, May 23, 2006, 9 a.m. to 15:30 ... 347
2006 : General Assembly Meeting 2 (AGM2) ... 351
Held Online (Through Email), May 22–25, 2006 ... 351
2007 : A Challenging Year for INWES ... 354
2007 : Board Meeting 8 (BM8) ... 356
Wroclaw, Poland, June 29–30, July 1, 5–6, 2007 ... 356
2008 : General Assembly Meeting 3 (AGM3) ... 363

Held Online (Through Email), May 24–27, 2008. . . . 363
2008 : General Assembly Meeting 4 (AGM4) . . . 364
Polytech'Lille, Amphitheatre Appert, Lille, France, July 17, 2008, 1:30 p.m. . . . 364
2008 : Interim Board Meeting (BM9). . . . 372
Lille, France, July 12–15, 2008 . . . 372
2008 : Board Meeting 10 (BM10) . . . 385
Polytech'Lille, Pasteur Room, Lille, France, July 19, 2008, 9 a.m. . . 385

16 ICWES-XIII: Seoul, South Korea, August 26–29, 2005 . . . 391
Theme: "Women Engineers and Scientists: Main Force to Reshape the Future World" . . . 391
Organizers . . . 391
Sponsors of ICWES-XIII . . . 391
Participants . . . 392
Opening Ceremony . . . 392
Business Meeting . . . 393
Testimonials . . . 393
Asha Gopinathan. . . . 393
Sue Bird . . . 394

17 ICWES-XIV: Lille, France, July 15–18, 2008. . . . 397
Theme: "A Changing World: New Opportunities for Women Engineers and Scientists" . . . 397
Organizations Involved . . . 397
Description of the Conference (by the Chair, Monique Moutaud) . . . 397
Opening, by Monique Moutaud, Chair ICWES-XIV . . . 399
Welcome from Monique Frize, President of INWES . . . 400
Report from the Chair, Monique Moutaud . . . 400
Business Meeting . . . 401
Testimonials . . . 401
Pat Battams, "From the Editor..." . . . 401
Pat Battams, Ruth Carter, and Betty McCarthy, "Four Action Packed Days in Lille" . . . 402
Betty McCarthy, "Dear Editor", Writing About Her Impressions of ICWES-XIV . . . 405

18 Development of CIWES-ICFIS . . . 407
A Canadian Charitable Organisation to advance the Education and Success of Women in STEM . . . 407
Strategic Planning in 2019 . . . 410
Name Change . . . 412

Conclusion . . . 413

Appendix 1: Selected Biographies of ICWES Founders, Leaders, and Organizers 419

Appendix 2: The Evolution Over Time of the Regions Used by ICWES to Divide the World 425

Appendix 3: Participation at ICWES-I to ICWES-XII 427

Appendix 4: Elizabeth Laverick's Retrospective of the First Nine ICWES Conferences 429

Appendix 5: Prof. Rita Levi Montalcini's Opening Talk at ICWES-III 439

Appendix 6: Elizabeth Laverick, on the Organization of ICWES-IX 445

Appendix 7: All INWES Meetings Between 2002 and 2008 451

Appendix 8: Dormer Ellis 453

Bibliography 455

Index 465

Authors' Biographies

Monique Frize graduated with a Bachelor in Electrical Engineering at the University of Ottawa in 1966. She completed a Master of Philosophy and Diploma of Imperial College at Imperial College of Science and Technology in London, UK (1970), while holding an Athlone Fellowship from the British Council. In 1986, Monique completed a Master of Business Administration at Université de Moncton, and a doctorate at Erasmus University in Rotterdam, the Netherlands in 1989. Monique Frize was a clinical engineer at Notre-Dame Hospital (1971–1979) and director of Clinical Engineering Services Regions I & VII in Moncton (1979–1989). Monique was appointed as the first Northern Telecom/NSERC Women in Engineering Chairholder and full professor in Electrical Engineering at the University of New Brunswick, in Fredericton. In 1997, she was appointed as the NSERC/Nortel Women in Science and Engineering Chairholder for Ontario at Carleton University and the University of Ottawa. Monique was involved in dozens of committees and boards. She received five honorary doctorates and was appointed officer of the Order of Canada (1993) and Fellow CAE, EC, IEEE, IFMBE, CMBES. Monique is a distinguished professor at Carleton University, and professor emerita at the University of Ottawa. Monique is cofounder of INWES and of the INWES Education and Research Institute (now CIWES-ICFIS). Monique published six books that include: *The Bold and the Brave: A History of Women in Science and Engineering* (UOP, 2009); *Laura Bassi and Science in eighteenth Century Europe* (Springer, 2013); *A Woman in Engineering: Memoirs of a Trailblazer* (UOP, 2019). She published more than 250 papers in refereed journals and conference proceedings. Her research areas are clinical decision support systems, medical imaging, clinical engineering, and women in science and engineering.

Claire Deschênes graduated from Laval University in Mechanical Engineering in 1977. She obtained a PhD from the *Institut National Polytechnique de Grenoble* in France in 1990. Hired as the first woman professor in engineering at the Faculty of Sciences and Engineering of Laval University (Canada) in 1989, as specialist on hydraulic turbines. She founded the Hydraulic Machine Laboratory (LAMH) in

1994 and established the Consortium on hydraulic machines in 2007. The innovations made by Professor Deschênes and her research team have led to an increased understanding of the behavior of hydraulic turbines, and the measurements made in the laboratory have enabled the creation of databases, which make it possible to develop analysis methods both in academic and industrial environments. Claire Deschênes is also involved on the issue of women in science and engineering. She was the holder of one of the NSERC Chairs for women in science and engineering from 1997 to 2005. She is cofounder in 2002 of a French-speaking organization about women in STEM (AFFESTIM). She is also cofounder with Monique Frize and Gail Mattson of the NGO *International Network of Women Engineers and Scientists* (INWES) in 2003, and the charitable organization *Canadian Institute for Women in Engineering and Sciences* (CIWES-ICFIS) in 2008. Throughout her career, Claire Deschênes published many journals papers and collective editions. She retired in 2019 but remains active. She is director of the international multidisciplinary journal *Recherches féministes*, and president of AFFESTIM and CIWES-ICFIS. In 2022, we published with Louise Lafortune and Audrey Groleau the *Manifesto on Women in STEM – 50 Positive and Impactful Texts* (ISBN:9782897993825). Claire Deschênes has received numerous awards and distinctions for her career. Among these, she has been named Knight of the National Order of Quebec in 2021 and Member of the Order of Canada in 2019. She received two honorary doctorates, from the University of Sherbrooke in 2019 and from the University of Ottawa. She was awarded the *Prix du Québec Lionel-Boulet* in 2020 for her scientific career and for her contribution to the economic growth of the province of Quebec (first woman). In 2019, she lent her name to the *Claire-Deschênes Postdoctoral Fellowship Competition* of the Université de Sherbrooke. She received in 2015 the NSERC *Synergy Award for Innovation: Two and More Companies*, and the same year, she became Fellow of Engineers Canada.

Ruby Heap is Professor Emerita at the University of Ottawa's Department of History. During her career, Ruby committed herself to the development of Francophone and Anglophone scholarship and student training in the fields of educational history and of women's and gender history. She co-founded (1989) and co-edited the journal *Historical Studies in Education/Revue d'histoire de l'éducation* and was appointed in 1999 as Founding Director of the University of Ottawa's Institute of Women's Studies. She then served as associate dean at the Faculty of Graduate and Postdoctoral Studies and, more recently, as associate vice-president, research. During this period, Ruby also began researching the still largely unknown history of women in Canadian engineering. Through her involvement with the INWES Education and Research Institute (INWES-ERI), she also advocated the interdisciplinary study of women in STEM, and called for the creation of a national archive devoted to their history in Canada, a project that came to fruition in 2018.Ruby Heap is the recipient of the Professional Engineers of Ontario's President Award (2010), of the Order of Academic Palms from the French Republic (2016), and of the Royal Society of Canada's Ursula Franklin Prize in Gender Studies (2018).

Karine Laporte holds a Bachelor of Arts in Classics (Laval University) and a joint Master of Arts in Classical Philology (Laval University and Strasbourg University). She recently obtained a PhD in Classics at the Leiden University Centre for the Arts in Society (LUCAS), achieved with the support of the Fonds de recherche du Quebec – Société et Culture (FRQSC) and the Social Sciences and Humanities Research Council of Canada (SSHRC/CRSH). Her dissertation investigates the use of narrative patterns in the History of the Roman Empire written by Greek historian Herodian. Recent articles have explored further Herodian's method of composition and his place within third-century imperial literature. She works as a part-time lecturer at the Department of Literature, Theatre and Cinema at Laval University and as a research associate for the Groupe de Recherche sur l'Antiquité (GRANT) and the Institut d'Études Anciennes et Médiévales.

List of Tables

Table 1 International conferences related to women ... xiv
Table 2 International and regional organizations for women in science ... xvi

Table 1.1 An overview of ICWES-I to ICWES-XII ... 5
Table 1.2 ICWES-I to ICWES-XII content of proceedings ... 8

Table 2.1 Distribution of S&T sessions and papers in ICWES-I, Complete Proceedings ... 17
Table 2.2 Distribution of social and gender sessions and papers in ICWES-I, complete proceedings ... 22

Table 3.1 Distribution of S&T sessions and papers in ICWES-II, Complete Proceedings ... 38
Table 3.2 Distribution of social and gender sessions and papers in ICWES-II, complete proceedings ... 42

Table 4.1 Distribution of S&T sessions and papers in ICWES-III, complete proceedings ... 61
Table 4.2 Distribution of social and gender sessions and papers in ICWES-III, complete proceedings ... 68

Table 5.1 Distribution of S&T sessions and papers in ICWES-IV, Complete Proceedings[a] ... 83
Table 5.2 Distribution of social and gender sessions and papers in ICWES-IV, Complete Proceedings ... 89

Table 6.1 Distribution of S&T sessions and papers in ICWES-V, Complete Proceedings ... 101
Table 6.2 Distribution of Social and Gender Sessions and Papers in ICWES-V, Complete Proceedings ... 107

Table 7.1 Distribution of S&T sessions and papers in ICWES-VI, Complete Proceedings . . . 122
Table 7.2 Distribution of social and gender sessions and papers in ICWES-VI, Complete Proceedings . . . 127

Table 8.1 Distribution of S&T Sessions and Papers in ICWES-VII, Complete Proceedings . . . 141
Table 8.2 Supplemental S&T Papers in ICWES-VII, Complete Proceedings . . . 142
Table 8.3 Student competition at ICWES-VII . . . 143
Table 8.4 Distribution of social and gender sessions and papers in ICWES-VII, Complete Proceedings . . . 147

Table 9.1 Papers recovered in full from ICWES-VIII, recording . . . 161
Table 9.2 Distribution of social and gender sessions and papers in ICWES-VIII, recording . . . 161
Table 9.3 Distribution of social and gender sessions and papers in ICWES-VIII, recording . . . 162

Table 10.1 Distribution of S&T sessions and papers in ICWES-IX, Complete Proceedings . . . 176
Table 10.2 Distribution of S&T posters in ICWES-IX, Complete Proceedings . . . 177
Table 10.3 Distribution of social and gender sessions and papers in ICWES-IX, Complete Proceedings . . . 182
Table 10.4 Distribution of social and gender posters in ICWES-IX, Complete Proceedings . . . 183

Table 11.1 Distribution of S&T sessions and papers in ICWES-X, Complete Proceedings . . . 199
Table 11.2 Distribution of social and gender sessions and papers in ICWES-X, complete proceedings . . . 201

Table 12.1 Distribution of S&T sessions and papers in ICWES-XI, Complete Proceedings . . . 225
Table 12.2 Distribution of S&T posters in ICWES-XI, Complete Proceedings . . . 226
Table 12.3 Distribution of social and gender sessions and papers in ICWES-XI, Complete proceedings . . . 232
Table 12.4 Distribution of social and gender posters in ICWES-XI, Complete Proceedings . . . 232

Table 13.1 Distribution of S&T Sessions and Papers in ICWES-XII, Complete Proceedings . . . 259
Table 13.2 Distribution of S&T Posters in ICWES-XII, Complete Proceedings . . . 259

Table 13.3 Distribution of Social and Gender Papers in ICWES-XII, Complete Proceedings . . . 266
Table 13.4 Distribution of social and gender posters in ICWES-XII, Complete Proceedings . . . 266

Conclusion

Table 1 International Conference of Women Engineers and Scientists (ICWES) . . . 436

Abbreviations

AAAS	American Association for the Advancement of Science
AFFDU	Association Française des Femmes diplômées des Universités
AFFESTIM	Association de la Francophonie à propos des Femmes en Sciences, Technologies, Ingénierie et Mathématiques
AFFI	Association Française des Femmes Ingénieurs
AGM	Annual General Meeting
AIDIA	Associazione Italiana Donne Ingegnere e Architette (Italian Association of Women Engineers and Architects)
AITACI	Association des Ingénieurs, Techniciens, et Agronomes de la Côte d'Ivoire
APAGESTE	Association Promoting African Girls in Engineering, Science and Technology Education
APEC	Asia Pacific Economic Cooperation
ASCE	American Society of Civil Engineers
AWIS	Association for Women in Science
AWSE	African Women in Science and Engineering
BIEN	Basic Income Earth Network
BM	Board Meeting
CAETS	International Council of Academies of Engineering and Technological Science
CCWEST	Canadian Coalition of Women in Engineering, Science and Technology
CIDA	Canadian International Development Agency
CIFISATS	Cercle Ivoirien des Femmes Ingénieurs, Scientifiques, Architectes et Techniciennes Supérieures
CIHR	Canadian Institutes for Health Research
CIWES	Canadian Institute for Women in Engineering and Sciences
CNISF	Conseil National des Ingénieurs et Scientifiques de France (now IESF, Ingénieurs et Scientifiques de France)
CNRS	Centre National de la Recherche Scientifique
COP	Conferences of the Parties

CWIT	Center for Women in Information Technology
CWSE	Chairs for Women in Science and Engineering
DIB	Deutscher Ingenieurinnenbund (German Association of Women Engineers)
EU	European Union
EWB	Engineers Without Borders
FI	Association Française des Femmes Ingénieurs
GASAT	Girls and Science and Technology
GID-DB Gender	Institutions and Development Database
GM	General Meeting
ICC	International Continuing Committee
ICFIS	Institut Canadian pour les femmes en Ingénierie et en Sciences
ICRAF	International Centre for Research in Agroforestry
ICSU	International Council of Scientific Unions
ICT	Information and Communications Technology
ICWES	International Conference of Women Engineers and Scientists
IEEE	Institute of Electrical and Electronics Engineers
IET	Institution of Engineering and Technology
IFUW	International Federation of University Women
IIWE	International Institute for Women in Engineering
INWES	International Network of Women Engineers and Scientists
INWES-ERI	INWES Education and Research Institute (now CIWES-ICFIS)
IRSTG	International Report on Science, Technology and Gender
IWSA	Indian Women Scientists Association
K-12	indicates the range of years of publicly supported primary and secondary education (Canada and the USA)
KWSE	Korea Woman Scientists and Engineers
MAL	Members-at-Large (SWE)
MENA	Middle East and North Africa
MOU	Memorandum of Understanding
NFA	National Federation and Association
NGO	Non-Governmental Organization
NSERC	Natural Sciences and Engineering Research Council of Canada
NSF	National Sanitation Foundation
OECD	Organisation for Economic cooperation and Development
S&T	Science and Technology
SJWS	Society of Japanese Women Scientists
STEM	Science, Technology, Engineering, and Mathematics
STEMM	Science, Technology, Engineering, Mathematics, and Medicine
SWE	Society of Women Engineers
TWOWS	Third World Organization of Women in Science
UN	United Nations
UNDP	United Nations Development Programme
UNESCO	United Nations Educational, Scientific, and Cultural Organization

UNFCCC	United Nations Framework Convention on Climate Change
UNIFEM	Fonds de développement des Nations Unies pour la femme
UOP	University of Ottawa Press
USSR	Union of Soviet Socialist Republics
WEGES	Women Engineers' Group at the Engineers' Syndicate
WEPAN	Women in Engineering ProActive Network
WES	Women Engineering Society
WESTIP	Women Engineers, Scientists, and Technologists in Pakistan
WFEO	World Federation of Engineering Organizations
WISE	Women in Science and Engineering
WISEST	Women in Scholarship, Engineering, Science, and Technology
WITEC	European Association for Women in Science, Engineering, and Technology
WWII	World War II

Part I
ICWES-I to ICWES-XII

Chapter 1
Methodology in Creating and Analyzing the ICWES Database

Creating the ICWES Database

Finding the proceedings and reports for the ICWES conferences and having these scanned and digitized in 2005 were key steps in preserving the story of this unique project created by women engineers and scientists with great vision and leadership skills. The next essential part was to ensure that all the information was stored in a safe place; it will be available shortly for retrieval by historians and researchers at the Special Collection and Archives at the University of Ottawa Library. Most of the documents have been digitized, and a few remain in hard copy only.

The ICWES database contains the conference documents (programs, proceedings, reports, booklets, etc.) for ICWES-I to ICWES-XVII, as well as testimonials, bibliographies, letters, emails, transcriptions of testimonies and oral instructive sources, and some artifacts such as photos and, for ICWES-VIII, recorded tapes and their (partial) transcription into text. All ICWES events were a mix of presentations (scientific papers, keynotes, invited and plenary lectures), and different types of activities such as workshops, symposia, panels, round tables, poster presentations, student competition papers, business meetings, tours, and social activities.

Regarding the proceedings and conference material that were distributed to the participants, the collection was as follows. The Iranian Delegate, Mahin Rahmani, who had attended all 12 conferences (I to XII), was able to loan us almost all of the proceedings she had kept. The librarian Suriya Thevar, from Madras (now Chennai), India, kindly scanned in 2005 all proceedings and documents related to ICWES-I to ICWES-XI (except ICWES-VIII) and organized them in a dynamic database on a CD-ROM. This represents 990 papers and documents. Some of the original sources were sent to her by the SWE archivist and by a few participants. The authors of this book collected the more recent documents related to ICWES-XII to ICWES-XVII conferences, obtained from similar types of sources, sometimes in digital form, sometimes in hard copy. When the proceedings of a conference were contained in

M. Frize et al., *Women's Contribution to Science and Technology through ICWES Conferences*, Women in Engineering and Science,
https://doi.org/10.1007/978-3-031-33800-7_1

two or more separate documents, they have been combined, for the purpose of this book, into a single PDF file. The page numbers indicated in the present book refer to the pages of this combined PDF file.[1]

Although there were no proceedings for ICWES-VIII, some of the presentations given at this conference were recorded on 16 audio cassettes. Issié Yvonne Gueye, from Abidjan, Ivory Coast, made them available along with a few printed documents. These cassettes are now at the Library of Ottawa University. For the purpose of this book, they were digitized and merged into a single file that comes up to about 18.5 hours. This file was first processed through audio-to-text software and then manually corrected in certain places. The rendering quality of this transcription is variable because of background noise and very low volume. Nevertheless, it is fortunate that these precious cassettes are still available.

The ICWES database also contains information on the leaders and organizers of the different conferences, and on participants.[2] Each conference held a Delegate Business Meeting (one delegate per country attending), where one of the main goals was to identify volunteers to host the next conference. When several proposals were made, the delegates were asked to vote. These meetings also discussed how to ensure the continuation and success of the conference series and updated what they called the "guidelines" on how these conferences were to be organized. The approved resolutions would be action items for everyone to work on until the next ICWES. The minutes of these business meetings are available in the database for most of the conferences, but a few have been found missing, such as the meeting at ICWES-XI. When available, they have been included in the book.

Many reports and testimonials on the conferences can be found online in *The Woman Engineer*, the Women Engineering Society's (WES, UK) publication archived by the Institution of Engineering and Technology (IET) and WES, and in publications from the Society of Women Engineers (SWE, USA) (*U.S. Women Engineer*, which became *SWE Magazine*, and Members-at-Large (MAL) newsletters) and in the proceedings themselves; some parts have been reproduced in this book with permission. Monique Frize collected more testimonials by correspondence through emails, telephone calls, and letters between 2021 and 2023.

These testimonials constitute a rich source of autobiographical essays, which adds more complexity and diversity to existing historical narratives. This source thus helps to fill out gaps generated by the absence of primary documents pertaining to the still largely untold lives of women in various fields of science and technology.

The few biographies of ICWES founders, leaders, and organizers appearing in Appendix 1 were obtained through the same sources as above or through the Internet.

---

[1] For example, the paper "Irrigation and the High Dam in U.A.R" presented at ICWES-II and referenced as ICWES-II Complete Proceedings, pp. 192–199 in the present book is located at pp. 106–112 of the original copy of the proceedings' second volume.

[2] Some of the files located at the Library of the University of Ottawa contain confidential information on participants.

Qualitative and Quantitative Analysis of the Scientific Content of Each Conference

Table 1.1 first gives an overview of ICWES-I to ICWES-XII: dates, locations, main themes, and chair persons.

In Part I of this book (Chapters 2–13 concerning ICWES-I to ICWES-XII), quantitative and qualitative information provide an insight on some important aspects such as venues and the scientific content to help answer some of our initial questions presented in the introduction.

Table 1.1 An overview of ICWES-I to ICWES-XII

No	Year	Dates	Location	Main theme	Chair person
I	1964	June 15–21	New York, USA	Focus for the Future. Developing Engineering and Scientific Talent	Beatrice A. Hicks
II	1967	July 1–9	Cambridge, United Kingdom	The Application of Technology to World Food Problems/The Woman Professional Engineer	J. Cicely Thompson
III	1971	September 5–12	Turin, Italy	Planning for Progress. Women's Professional & Family Duties	Anne E. Amour
IV	1975	September 8–13	Cracow, Poland	New Techniques in the Service of Mankind/Contemporary Sociological Problems of Women Engineers and Scientists	Izabela Zachwatowicz
V	1978	September 4–8	Rouen, France	Technology and Freedom	Nicole Bécarud
VI	1981	September 8–12	Bombay, India	Science, Technology & Society	Sumati Bhide
VII	1984	June 17–24	Washington, DC, USA	Technology: An International Bridge	Naomi McAfee, Suzanne Jenniches, Pat Patterson
VIII	1987	September 20–27	Abidjan, Ivory Coast	Science, Technology, and Development	Marthe Achy Brou
IX	1991	July 14–20	Warwick, Coventry, United Kingdom	Communication	Elizabeth Laverick
X	1996	October 8–10	Budapest, Hungary	Toward the Third Millennium: The Role of the Intellectual Potential in a New World	Zsuzsa Szentgyörgyi
XI	1999	July 24–27	Chiba, Japan	Science and Technology for Global Ecology	Michiko Ishii
XII	2002	July 27–31	Ottawa, Canada	Women in a Knowledge-Based Society	Monique Frize

Component 1 in Chapters 2–13

The extent of regional representation was an important consideration to be included in this book for the selection of conference venues. The first part of each chapter of an ICWES conference indicates where it was held, who organized the conference, and provides the countries from which the participants came from.

The information on countries and regions included in this book relates to the 11 regions of the world defined after INWES was created. The current analysis includes Israel within the Middle East:

(i) North America
(ii) Latin America (incl. South and Central America, Caribbean, Mexico)
(iii) Western Europe (including Turkey)
(iv) Eastern Europe (including Russia)
(v) French-speaking Africa
(vi) English-speaking Africa
(vii) Middle East and North Africa
(viii) Central Asia (incl. India, Pakistan, Nepal)
(ix) Southeast Asia
(x) Far East Asia
(xi) United Kingdom, Australia and New Zealand

Appendix 2 provides more information on the evolution of these divisions over time since the beginning of ICWES.

Component 2 in Chapters 2–13

In the chapters covering ICWES-I to ICWES-XII, the scientific content has been examined in more depth. The scientific content has been treated separately for S&T and for gender. Each section was obtained in the following way:

For each ICWES, a brief report of keynotes or invited speeches is given. This is followed by descriptions of different topics in the scientific content of the conferences: historical context, issues scientists and engineers were working on, how they were performing their work, and research and development (technologies and methods), some resulting outputs, and hopes they had for a better future.

In each chapter covering an ICWES conference, there is a summary (in a table) of information such as sessions' name, number of papers per session expected according to the schedule of the conference, and the number of papers and abstracts that are present in the proceedings.

Quantitative Data

The table found in Appendix 3 summarizes quantitative information for the first 12 conferences: years, location, number of participants from the host country, number of nonlocal participants, number of participants from each of eleven current regions, and the number of countries represented at each ICWES. The total number of participants for the 12 conferences according to the registered participants' lists is 5056. This number of participants varies from 82 in Budapest (Hungry) to 1204 in Washington, DC.[3] The proportion of nonlocal to local participants is 30%. It is worth noting that there were at least 30 countries represented in more than two thirds of these conferences (8/12) and that, except for two of the conferences (ICWES-VI and ICWES-VII), all 11 regions were represented.

The papers, presentations, and other activities for the 12 conferences have been divided into two categories, gender and in science and technology (S&T). The classification into those categories logically follows the contribution content, rather than the background or discipline of the authors. The gender category includes papers, presentations, and activities related to gender and women's studies, feminism, outreach, and career development for women S&T. The S&T category includes papers, presentations, and activities that have not been selected for the gender category; they are related to natural and applied sciences, engineering, and sciences such as health, agriculture, history, geography social sciences, education, etc. Following these rules, for example, papers presenting an analysis to explain the low participation of women in STEM would be in the gender category, while a social science paper related to the impact of technology on—or by—women, would be in the S&T section.

Table 1.2 presents the final compilation of the presentations for each of the first 12 ICWES conferences, the number of papers in the proceedings (column 3), the number of papers with abstracts only (column 4), keynotes, invited papers and plenary papers (column 5), panels and round tables (column 6), and other activities (posters, student competition papers, and extra papers) (column 7). This summary was composed from the ICWES database' Excel files that list information taken from the programs and proceedings of each conference. These files contain information such as the number and titles of the papers, presentations, activities for each ICWES sessions, the number of subtopics by session (if any), the number of women, men, and unknown sex of the speakers, the ICWES region of the speakers and the pages where abstracts or papers can be found in the proceedings.

Table 1.2 is divided into gender and S&T contents. For the purpose of these tables, full papers (with or without abstracts) and papers with abstracts only have been counted separately. The panels, round tables, and workshops, which are usually presented in abstracts only, have been counted as a single activity, regardless of the number of participants they included.

---

[3] Note that the number of participants for ICWES-I in New York in 1964 and ICWES-VII in Washington, DC, in 1984 may be boosted, since these were joint SWE/ICWES conferences.

Table 1.2 ICWES-I to ICWES-XII content of proceedings

			Full papers in proceedings	Abstracts only in proceedings	Keynotes, invited papers, plenaries [1]	Workshops, symposiums	Panels, round tables	Others (posters, student competition, extra papers)	Total
ICWES			*Gender*						
I	1964	New York	0	1	0	1	0	0	2
II	1967	Cambridge	12	0	0	1	0	0	13
III	1971	Turin	22	2	1	0	0	0	25
IV	1975	Cracow	8	0	0	0	0	0	8
V	1978	Rouen	14	0	1	0	0	0	15
VI	1981	Bombay	9	0	0	0	0	0	9
VII	1984	Washington, DC	8	4	0	39	0	0	51
VIII	1987	Abidjan	0	0	0	0	0	0	0
IX	1991	Warwick	22	10	2	0	0	0	34
X	1996	Budapest	15	2	0	2	1	0	20
XI	1999	Chiba	17	3	5	0	0	18	43
XII	2002	Ottawa	137	0	3	6	14	14	174
Total			**264**	**22**	**12**	**49**	**15**	**32**	**394**
			Science and technology						
I	1964	New York	12	0	2	0	0	0	14
II	1967	Cambridge	21	0	1	0	0	0	22
III	1971	Turin	8	37	0	0	0	0	45
IV	1975	Cracow	36	0	0	0	0	0	36
V	1978	Rouen	21	5	0	0	0	0	26
VI	1981	Bombay	79	0	0	0	0	0	79

VII	1984	Washington, DC	46	14	0	0	0	22	82
VIII	1987	Abidjan	11	0	0	0	0	0	11
IX	1991	Warwick	47	12	6	0	0	15	80
X	1996	Budapest	24	0	3	0	0	0	27
XI	1999	Chiba	78	11	6	0	0	35	130
XII	2002	Ottawa	41	0	0	2	0	17	60
Total			**423**	**79**	**18**	**2**	**0**	**89**	**611**

Based on this table, we can see that the material contained in the proceedings includes 394 and 611 papers, presentations, and activities in the gender and S&T categories, respectively. The global ratio of gender full papers and abstracts is 36% (264+22 gender and 423+79 S&T). No specific trend in time regarding this ratio can be detected. However, most of the conferences presented more S&T papers than gender papers, showing the main interest of the participants for ICWES conferences. There are two exceptions: ICWES-III in Turin, with a percentage of 73% Gender (22 gender papers and 8 S&T papers), and ICWES-XII in Ottawa, with a percentage of 77% (141 gender papers and 41 S&T papers).

Chapter 2
ICWES-I, Setting the Stage: ICWES-I, New York City, USA, June 15–21, 1964

At a time when there were very few women professional engineers in most countries, it is surprising that an international conference was planned and held with great success, in New York, in 1964. The earliest data on engineering undergraduate degrees awarded to women in the United States was in 1966; the proportion of women awarded the degree that year was 0.4%.

It was an ambitious project, to bring women engineers and scientists from many countries around the world. Beatrice Hicks, the director of the First International Conference of Women Engineers and Scientists (ICWES-I) (see Appendix 1), invited participants to "consider the present and future opportunities in engineering and science and how we may encourage women to use and to enjoy these opportunities, and thereby to serve the progress of all peoples" (ICWES-I, Complete Proceedings, p. 18). The important factors that enabled this event to happen were the existence of the Women Engineering Society (WES), established in the United Kingdom in 1919, and the creation of the Society of Women Engineers (SWE) in the United States in 1950. These two organizations collaborated and reached out to hundreds of women from all regions of the world. Some of the international contacts were Members at Large (MAL) of SWE or of WES living in countries other than the United States or the United Kingdom.

It was opportune that the conference was held during the great World Fair and that the World Fair Organizers declared June 15, 1964, Women Engineers and Scientists Day. The World Fair had an important impact on the conference. Participants visited the exhibits on the first day, including the General Motors Futurama and Behind the Scenes, and the Second Mobil exhibit. The second day was devoted to industrial tours that included IBM's new Research Center, the Consolidated Edison Atomic Energy Plant, and the Bell Labs.

SWE held a conference every year in a different part of the United States, so it was decided that SWE's fourteenth conference in 1964 would be held in New York, and participants from around the world would be invited, thus making this a truly

M. Frize et al., *Women's Contribution to Science and Technology through ICWES Conferences*, Women in Engineering and Science,
https://doi.org/10.1007/978-3-031-33800-7_2

international event. The goal was also to attract participants from as many of the US states as possible. The report of the conference shows that there were 486 attendees, including 391 American attendees from 46 states and 95 participants from 34 other countries. Out-of-country attendance amounted to 19.5%. Participants came from as far places as Australia, Afghanistan, the Republic of China, Japan, South Korea, Pakistan, India, the Philippines, Uganda, and the USSR. Participants also came from Europe, South America, Mexico, the Middle East, Iran, and Israel. One remains impressed by the event's transglobal profile, with participants coming from regions and countries as far and as diverse (see Appendix 3).

Ruth Shafer, Chairman of the conference operations (see Appendix 1), gave an opening talk in which she invited participants to attend many of the sessions, briefly describing their content: the first session discussed the requirements in scientific and technological manpower; sessions II, III, and IV presented the types of talent needed in support of the several major areas of basic human needs; session V presented an overview of the status of the world's supply of such talent; session VI provided the opportunity for discussing how to develop scientific and engineering talent for the future (ICWES-I, Complete Proceedings, p. 14). The theme of the conference was: "Focus for the Future: Developing Engineering and Scientific Talent."

Aileen Cavanagh, President, Society of Women Engineers in 1964 (see Appendix 1), said in her opening remarks:

> "…this First International Conference of Women Engineers and Scientists has been an ambitious but much needed undertaking."
>
> "It has been needed because the place of women professionals in the fields of engineering and science has been debated without the benefit of facts for far too long, most often by people—including both men and women—who have never met a woman engineer or scientist…"
>
> "This Conference has been ambitious, as is any attempt to find rational perspective in a controversial issue. In nation after nation, poverty, selfishness, and ignorance have created severe economic problems. Today, determined to maintain world peace, the many nations of the world are striving to feed, clothe, house, educate, transport, and communicate with their peoples, no matter how large their numbers may become. While these nations are seeking responsible ways to govern their expanding populations without destroying family life, the world is trying to enter a new era of human understanding and international cooperation."
>
> "Its peoples seek to eradicate national prejudices, formed under past economic and social stress, by reciprocal communications and by sharing cultural and technical knowledge among the many nations. In this work to bind the world together, many engineers and scientists—men and women alike—are looking to the future needs of humanity; some are striving to conquer outer space and attempting to unlock the mysteries and vast resources of the oceans and the bowels of the earth, while others are working daily to nurture the hopeful spirit of life among the world's peoples." (ICWES-I, Complete Proceedings, p. 16)

Beatrice Hicks followed with welcome remarks: "This Conference has brought together men and women to exchange information and to create new patterns and new plans for the activities of women in engineering and science…We represent education, government, industry, professional guidance, personnel placement, and management, as well as engineering and science. But most important, we represent those people who recognize the need for progress and unselfishly give of their time and their means toward these goals" (ICWES-I, Complete Proceedings, p. 18).

Reading the wise words from the president of SWE and her colleagues, it is remarkable that not only did the women at the conference tackle such complex issues but that they shared experiences from so many countries and from all regions of the world. In a time when international mail was slow and international phone calls expensive, it is quite amazing that the organizers were able to reach and attract so many participants to this first ICWES from all over the world and from most of the American states.

To understand how the ICWES conference series began and how it managed to continue to this day, it is helpful to examine why and how SWE and WES were created. Lillian M. Gilbreth, in her keynote address, said:

> "In the beginning, before SWE was organized, the need for such a society was carefully weighed and its membership requirements clearly defined. The Certificate of Incorporation and the Bylaws of the Society are evidence of its standards. As a Technical Society, we accept the Code of Ethics and Canons of Ethics of the engineering profession. [...]"
>
> "Why women engineers? And why a Society of Women Engineers? Because the profession needs women, and because they need the opportunities for work that the profession affords. Because people with a common interest need the feeling of group solidarity that an organization gives. Because young women need knowledge of the opportunities in science and engineering, and help in being tested, educated, and trained, providing they have interest and the necessary aptitudes. Because many women, young and old, need the opportunities the profession gives." (ICWES-I, Complete Proceedings, p. 20)

In her talk, Gilbreth stated:

> "Our First International Conference gives us a timely opportunity to think and plan together—to recognize and evaluate our likenesses and differences. The future demands that we utilize both. If we are to work together effectively, we should recognize that, while each of us is different from every other person, our likenesses are of great importance. This applies both to skills and to satisfactions. The interdisciplinary research going on prove this."
>
> "We must come to know each other better, worldwide. This means better communication, definition of words to make sure they have the same meaning to all, and better interpretation and translation. This means more travel and exchange of experiences, more opportunities to know one another internationally, not only through books and all the libraries can bring us, but in our meeting, our workplaces, our social life, our homes."
>
> "One of our resources is what has been called 'management', which makes it plain that people everywhere have similar responsibilities. Everyone everywhere must learn (1) to manage himself, (2) to participate constructively in home and family life, (3) to be a good citizen, (4) to be a helpful volunteer, and (5) to earn one's way by doing a job that needs to be done. Our success in our engineering job is affected by our success in our other four jobs, and vice versa."
>
> "We have much to be thankful for, much to appreciate. Women engineers and scientists have welcomed and helped newcomers in the profession. Men engineers and scientists have helped us get education, training, jobs, promotion, and membership in organizations... The sharing of experiences in our First International Conference should help us to focus more effectively on the future and to go forward into it together, courageously and in the spirit of service." (ICWES-I, Complete Proceedings, p. 21)

Society of Women Engineers (SWE): Host of ICWES-I

SWE decided to combine its annual conference with a much larger, international context, succeeding to attract women in STEM (science, technology, engineering, and mathematics) fields from 34 countries, in addition to those from 46 States in the United States. It is evident from the documents pertaining to this first ICWES that several organizations collaborated with SWE to make it happen, among which were the International Federation of University Women (IFUW) and the Women Engineering Society (WES) in the United Kingdom. In fact, WES was to organize the second ICWES in Cambridge, July 1–9, 1967. A large delegation of WES members (22) attended ICWES-I, and it is easy to assume that discussions were held in New York between SWE and WES to encourage the latter to host ICWES-II.

The Beginnings of SWE

Prior to the formal creation of SWE in 1950, women student groups at Drexel Institute of Technology in Philadelphia, at Cooper Union (NY), and at City College New York began to hold meetings and networking activities. In May 1950, around 50 women, representing the districts of SWE (New York City, Philadelphia, Washington, DC, and Boston), held its first national meeting and elected its first president, Beatrice Hicks. The first official annual meeting was held in 1951, in New York City. In the early 1970s, SWE became involved in the activities of the larger women's movement. In 1972, representatives from women's scientific and technical committees and societies that included SWE formed an alliance to discuss equity for women in science and engineering. SWE's council endorsed ratification of the Equal Rights Amendment (ERA) and agreed not to hold national conventions in non-ERA-ratified states. Today, in 2023, SWE has over 40,000 students and professional members, including just over 4000 international members, and continues its mission as a nonprofit educational service organization ("Society of Women Engineers," under "History," n.d.).

From the information gathered on the ICWES conference series, it becomes clear that there was a group of women leaders who were determined that the first conference, in 1964, would be followed by many more in the future. They succeeded, as the conferences continue to be held every three to four years into the twenty-first century. How these women carried out the plan becomes very clear according to the notes of their business meeting at ICWES-II, which is discussed in Chap. 3. This was an ambitious project, but these women had a vision and the determination to make it happen.

Leaders and Organizers of ICWES-I

Here are a few words about the leaders at the first conference, which appeared in an IEEE publication in December 1963: "The New York Section of the Society of Women Engineers has announced the appointment of Dr. Beatrice A. Hicks (the first president of SWE in 1951) as Conference Director of the First International Conference of Women Engineers and Scientists to be held in June 1964 at the United Engineering Center in New York. Miss Ruth Shafer will be Operations Chairman. Dr. Hicks is president and director of Engineering, Newark Control Co. Dr. Lillian M. Gilbreth, President of Gilbreth, Inc., international industrial engineers, and management consultants, is the keynote for the conference. Ruth I. Shafer, chair of ICWES I" ("Women Engineers Plan International Conference," 1963, p. 745). Aileen Cavanagh was President of SWE (1963–1964) and spoke at the Opening of ICWES-I (on Hicks, Schafer and Cavanagh, see Appendix 1).

Beatrice Hicks introduced the keynote speaker, Lillian M. Gilbreth: "[a]s a world-renowned management consultant, she has spent her life in the studies, planning and activities that create the opportunities for progress. Born in Oakland, California, in 1878, and educated at the University of California and Brown University, Dr. Lillian Moller Gilbreth pioneered with her husband, Frank Gilbreth, in the field of motion study that became the profession of Industrial Engineering. Through the ensuing years, she continued to work for progress in the engineering management areas. In 1954 she was the first woman to receive the Washington Medal, awarded for her outstanding contributions to engineering and scientific management. Among her many honors she holds the Gantt Medal, the Wallace Clark Award, the Award of the National Institute of Social Science, and over twenty honorary degrees" (ICWES-I, Complete Proceedings, p. 18). Gilbreth's talk was titled: "Focus for the Future" (ICWES-I, Complete Proceedings, p. 18).

Organizations as Major Contributors to ICWES-I

In addition to SWE and the Women Engineering Society (WES), several organizations from other countries contributed to ICWES-I, at least in the sense of encouraging their members to attend. It is notable that several of these organizations hosted an ICWES conference later in the series. For example, the Associazione Italiana Donne Ingegnere e Architette (AIDIA), from Italy, organized ICWES-III in Turin, Italy, September 5–12, 1971; Anna E. Amour was the chair of ICWES-III and had attended both ICWES-I and ICWES-II. AIDIA was formally founded in 1957 by Emma Strada and engineer Anna E. Amour. The Society of Japanese Women Scientists sent seven members to ICWES-I and later hosted ICWES-XI in Japan in 1999. The Institution of Engineers (India), represented by Ila Ghose at ICWES-I, hosted, with the Indian Women Scientists Association (IWSA), ICWES-VI on September 8–12, 1981. Several chairs of sessions from ICWES-I reappear in future

ICWES events. For example, J. Cicely Thompson was the chair of ICWES-II, and Isabel Hardwich was its program chair and secretary. More details on this conference are found in Chap. 3.

It is interesting to note that the proceedings contain more than typewritten 300 pages.

ICWES-I, S&T Papers

The theme of the conference was "Focus for the Future—Engineering and Scientific Talent." What does it mean and how it was developed in ICWES-I from the science and technology (S&T) point of view is the aim of this section.

The 1960s may be considered the first productive generation after WWII. It brought unprecedented innovations that would change the world. Such innovations include memory chips and Dynamic Random-Access Memory, which increased the possibility to develop computers and satellites and changed many aspects of communications; however, these technologies were accessible only to a few scientists. Numerous smaller inventions appeared, such as audiocassettes, pocket calculators, Kevlar material, ATMs, etc. This was a new era of peace, and S&T innovations were useful for humankind and increased its well-being.

The ICWES-I conference brought some answers to the question raised by its theme. In her official opening, Aileen Cavanagh, president of SWE and chair of the conference, mentioned that the aims of the conference were "to look at the future technological needs in a peaceful word, define those areas of endeavor in which our scientifically inclined and inventive youth of today—boys and girls alike—can find their opportunities for service to their fellow men of tomorrow" (ICWES-I, Complete Proceedings, p. 17). According to the keynote speech of Lillian M. Gilbreth, a world-renowned management consultant and industrial engineer president of Gilbreth, Inc. in New Jersey, "What is the engineer expected to do? To 'utilize the resources of nature and of human nature for the benefit of mankind" (ICWES-I, Complete Proceedings, p. 20). These remarks and similar ones established the topics for the S&T portion of the conference.

In 1964, the population of the world was 3.3 billion. With an increase of 2% per year during that period, the population would have reached 8 billion by AD 2000, a huge increase that engineers and scientists wanted to tackle during ICWES-I.[1] Four basic needs for human beings were identified by the presenters: water, clothing, shelter, and food. To ensure success in providing these needs, transport of goods and resources, and energy supply were also important elements to consider. Each paper presented one aspect with a description of past and present states of development, challenges, and provided tips for potential solutions, innovations, or research subjects. Important questions were raised concerning the need to increase the standard

[1] In fact, in 2020, the world population was 7.79 billion.

Table 2.1 Distribution of S&T sessions and papers in ICWES-I, Complete Proceedings

ICWES-I S&T sessions	Session titles	# of papers expected	# of papers in proceedings
I	Keynote presentations: Spotlight on engineers and scientists (Isa Rischowski, M. H. Trytten)	2	2
II	Requirements on the Individual's standard of living	4	4
III	Focus on industrial and social needs	4	4
IV	Need of the World's standard of living	4	4
	Total	14	14

of living in underdeveloped countries,[2] as well as needs for the specialized training of engineers and scientists.

The ICWES-I conference had four S&T sessions, with 12 regular papers plus two keynote papers in the proceedings (Table 2.1). Although these sessions had no specific titles, the order of the conference presentations was very well thought through as it increased the level of specialization in the subjects gradually.

The first session considered the requirements for engineering and scientific manpower to meet the challenge of sustaining the expanding population of the world, and the technical services needed in support thereof. The following three sessions established the specific types of scientific and engineering talent needed, describing the state of the world in the major areas of human need, namely, food and water, clothing and shelter, heat and power, communications, and transportation.

Eight of the 18 authors (15 women and 3 men) were from the host country (USA), 4 from Western Europe, 1 from Eastern Europe, 2 from Far East Asia, and 3 from the United Kingdom, Australia, and New Zealand.

The following paragraphs give an overview of the rich scientific content of ICWES-I, aimed at providing a global perspective for humankind, sometimes with detailed scientific insights.

A Glimpse of ICWES-II's S&T Content

A good example of a scientific presentation was the "Planetary water Problem" (ICWES-I, Complete Proceedings, pp. 62–74) by John D. Isaacs, professor of Oceanography, Scripps Institution of Oceanography, University of California, where he worked on the topic of water, especially applied to agriculture. We know that the Nile, the Tigris, the Euphrates, and the Indus gave rise to civilizations. But today, humankind has populated almost the whole planet. Even if the total worldwide precipitation on land is theoretically sufficient to supply food for the human

[2] The authors are aware that the term "underdeveloped country" has a negative connotation and is no longer used. However, the expression "developing countries" was not used in the 1960s.

race by conventional agriculture, the natural distribution system is spotty. He also spoke of watering the desert and of the endangered rain forests. Agriculture is only one problem element related to water; fresh and clean water for other uses must also be addressed.

The author then proceeded to discuss new opportunities for technological solutions. For example, he mentioned how the natural distribution of water can be changed either through climate and weather control, or by the direct physical transport of water, agricultural techniques can be developed to be successful under a greater range of conditions, and water can be demineralized.

On the energy production question, J. Cicely Thompson, from the Nuclear Power Group Knutsford, England, in "Energy Supplies for an Expanding Demand" (ICWES-I, Complete Proceedings, pp. 113–142), summarized the main primary sources of energy in the 1960s as being hydroelectricity and fossil fuels (coal, oil, and natural gases), although minor sources such as drought animals, wood, and dung were already important in underdeveloped areas. According to Thompson, solar energy was already used for heating and the distillation of water in Israel and Japan, and geothermal steam was produced in some of the world's earthquake areas and successfully used for electricity generation in New Zealand and Italy. Although wind power was used on a small scale for drainage, irrigation, and other agricultural purposes, it was not yet economically feasible to use it on a larger scale to generate electricity. Nuclear energy was in its infancy. To tackle future energy needs, increasing the efficiency of the actual energy production was deemed to be useful. The author foresaw that commercial fuel cells would be available within the next decade. Photoelectric cells, combined with batteries, were seen as a promising source of reliable small power supplies.

Of course, technology may have a negative impact. For example, M. H. Trytten, director of the Office of scientific manpower, National Academy of Sciences (Washington, DC), in his keynote talk "Requirements for Engineering and Scientific Manpower," briefly describe problems brought to humanity and the Earth by technology: "On the negative side, we must learn to deal in a wholly new way with destructive elements growing in proportion to population and technological change, such as pollution of air and water, chemical and biological agents of destructive capability beyond their purpose" (ICWES-I, Complete Proceedings, p. 31).

Anneli Hattari, chief of laboratory, Printal Oy, Helsinki, Finland, in "Environment Pollution and Control" (ICWES-I, Complete Proceedings, pp. 143–160), described water and air pollution in great details. She wrote: "[...] the triumphal march of development has proceeded at the cost of pollution of our environment. Watercourses have to accept large amounts of industrial and residential wastewaters, oil, chemicals and other waste materials. Factories and dense population areas hurl huge amounts of smoke and fumes into the air. Nuclear tests pollute the atmosphere, soil, and water with radioactive fallout. Hundreds of thousands of tons of various chemical pesticides are scattered each year in fields, gardens, and forests, either as powdered or sprayed material or as injections. Diversified products, such as finely divided aerosols, are sprayed into the air. What have we done, and what must we do, to prevent the pollution of our environment?" (ICWES-I, Complete Proceedings, p. 143).

Hattari mentioned aerosol products (hair lacquers, insecticides and pesticides, deodorants, air fresheners, etc.), not yet as a concern about the ozone layer, but as being potentially harmful to health. She also was particularly concerned by the development of the nuclear field. On one hand, nuclear reactors in operation in different parts of the world produce atomic power and radioactive isotopes for peaceful purposes; on the other hand, radioactivity has become a frightening and menacing word since the day on which the first atomic bomb was dropped during WWII. Great powers have competed in an armaments race; hundreds of nuclear-explosion experiments have been carried out. To provide one negative effect among others, radioactive residues were found in milk during periods of experimental nuclear explosions.

Technological Transfer Between Developed and Developing Countries

Helping developing countries was presented during the conference as a new need for the future. It would be accomplished through the sharing of information and technologies: "There is no apparent solution except for the more advanced nations to assist in developing a technological base, which can support a working population" (M. H. Trytten, in "Requirements for Engineering and Scientific Manpower," ICWES-I, Complete Proceedings, pp. 22–33).

Ira Rischowski, VDI Elliott Process Automation Ltd. (Elliott Bros. Ltd.), London (England), discussed the participation of scientists and engineers in developing countries. "Some comments must be made about the technical assistance given to the developing countries. There are two main ways in which this aid is given, and in both engineers and scientists play an important part. There is first the direct aid in the form of plant and equipment sent to these countries. Almost always these will have to be accompanied by engineers and technicians to install them, commission them and service them. [...] It must be borne in mind that technology and science can only thrive in a congenial atmosphere. Where this technical climate does not exist, it has gradually to be built up. The engineers and scientists who go out to these countries in order to install plant and machinery must, therefore, be regarded as missionaries, whose aim at least in part must be the 'technisation' of the native population. The second way, and the more important one, is the direct training and education of the peoples living in the developing countries" (ICWES-I, Complete Proceedings, p. 55).

Young Sun Lee, a Lecturer at the Department of Textile Engineering, City College of Agriculture, Seoul, South Korea, in "Clothing—A Basic Requirement," added the point of view of developing countries: "[...] in the under-developed countries, engineers and scientists have to keep up with the advance of knowledge which has already been achieved in the developed countries and to find possible new fiber sources [for clothing] which are locally plentiful" (ICWES-I, Complete Proceedings, p. 86). This point of view was also taken up by John H.M.S. Miller, Deputy Director at the National Research & Technology Study, Dublin, Ireland, in "Management of our Natural Resources": "[…] the control of [natural resources] projects should as far as possible lies completely with the indigenous population although foreign expert advice may be essential. This is an education problem as well as one of development" (ICWES-I, Complete Proceedings, p. 212).

Links Between S&T, Scientists and Engineers, and Women

Many of the presenters discussed links between S&T and society, as well as man- and woman power to achieve these advancements, fitting the main conference theme.

One example among many others was in the talk of Ira Rischowski, an engineer working as a technical Writer at Elliott Process Automation Ltd. (Elliott Bros. Ltd.), London, England, titled "Technical Services Needed in Support of Engineers and Scientists." From her survey about electronic computers (analog or digital), she discussed their usefulness, as the newest tool available for engineers and scientists. "A computer cannot, fundamentally, solve any problem that could not be solved by a good engineer or scientist, given the necessary time and manpower. What the computer can do is to solve problems immeasurably faster and with greater accuracy. […] the computer is an immensely powerful tool that can give great relief from tedious work and enable accuracies to be achieved which could not be reached by means of 'hand calculations'" (ICWES-I, Complete Proceedings, p. 39).

Rischowski mentioned the ever-growing technical and scientific complexity of humankind societies, needing specialized engineers and scientists working together. She attempted to define what engineers and scientists were then and what they would be in the future; she provided several examples. She showed "that modern society industry and science are differentiated in both a horizontal and a vertical direction, and that the consequence of this division of labor is a total interdependence of all disciplines of science and technology, none of which can exist without all the others. The same applies to the hierarchy of skill and knowledge in each branch of science and technology" (ICWES-I, Complete Proceedings, p. 56).

Contributions of Women Scientists and Engineers to S&T

One of the underlying aims of ICWES-I was to show that women are making outstanding contributions to science and technology throughout the world. It is noteworthy to say, however, that women's specific contribution was not emphasized in the S&T section of ICWES-I. Most presenters spoke of women and men working together, or of the need for more women in S&T. However, Young Sun Lee, while presenting the need to develop new fibers for clothing and new method of production, mentioned: "Particularly to women engineers and scientists, processing of clothing is one of the stimulating fields because it is a natural interest of women" (ICWES-I, Complete Proceedings, p. 86).

ICWES-I, Gender Papers

As mentioned above, organizers of the 1964 ICWES conference did not hesitate to describe this inaugural meeting as a great success. Conceiving and offering an attractive program to the attendees was certainly a critical task for the conference leaders and organizers. Thinking about the chosen theme, "Focus for the Future: Developing Engineering and Scientific Talent," Aileen Cavanagh, then president of the Society of Women Engineers (SWE) and the meeting's official host, raised one fundamental question about women in STEM: Who wants more women engineers and scientists? (ICWES-I, Complete Proceedings, p. 17). According to SWE's president, many key sectors of society, in addition to women themselves, needed them, such as education, industry, and government, so the latter could "both develop strategic manpower reserve in critical categories and stabilize the costs of its engineering requirements" (ibid.).

This and several other issues and concerns related to the status of women in STEM, not only in the United States but around the world, were addressed in ICWES's inaugural program. It consisted of two components: one invited women scientists and engineers to present technical papers on their field of work, while the second proposed presentations focusing both on the struggles and successes of women engaged in, or aspiring to, careers in STEM. This section of the program can be related to the emergence of second-wave feminism in the early 1960s, to the broad-based campaign for women's rights, as well as to earlier and ongoing efforts to increase women's participation in the scientific fields.

As can be seen in Table 2.2, Session V covered the current status of women engineers and scientists, and information was published about Australia and New Zealand, Europe, Iran, Japan, North America, the Philippines, South and Central America, Syria, Turkey, and the USSR.

Session VI was in the form of a symposium on "Developing Engineering and Scientific Talent," at which various eminent people were invited to contribute statements as a basis for discussion. There is no report of the discussion as such, but a summary statement emerged from the conference that pointed out the many ways in which science and engineering were contributing to improved world living conditions, and the possibility of achieving the outstanding potential of the future by cooperation. Also emphasized was the inadequate supply of trained personnel to implement future world needs, thus highlighting the importance of attracting more women to these fields.

A look at the available quantitative data helps us draw a preliminary gender and transnational profile of those taking part in the various sessions. For example, we know that six speakers, originating from different parts of the world (Africa, Japan,

Table 2.2 Distribution of social and gender sessions and papers in ICWES-I, Complete Proceedings

Social and gender sessions	Session titles	# of papers in proceedings	# of reports or abstracts in proceedings
V	Report on the status of engineers and scientists	0	1 (session report)
VI	Symposium: Developing engineering and scientific talent	0	21
N/A	Special feature (contributions to session VI)		4

Australia, Italy, Guatemala, and the USA), took the stage for session V (chaired by Katharine Stinson, Past President of SWE). Of that number, five were women. In the conference proceedings, these reports are only available in a general summary produced by Naomi J. McAfee. The symposium held during Session VI (chaired by Lydia I. Pickup) attracted for its part 21 panelists, 11 women, and 10 men, most of whom came from the United States (14), with the remainder originating from Germany, Italy, the Philippines, Ireland, and the United Kingdom. A special feature section rounded off the symposium, with four women (all Americans) sharing an "inspirational personal account of how each was motivated to choose her career and what had been especially outstanding and satisfying in her work experience" (ICWES-I, Complete Proceedings, p. 269). The collected data thus indicates that the majority of the women participating in ICWES-I came from the host country, while the others made up a diversified group that encompassed both countries and continents.

Searching for and compiling impressions of conference attendees can help the researcher secure more detailed information on a session's participants, on the key themes that emerged and on the key concerns expressed throughout, as well as on the main recommendations tabled at the end of the session. The coverage of ICWES-I by the British journal *The Woman Engineer. The Journal of the Women's Engineering Society* is a good case in point. The 1964 fall issue included two reports on the conference, which were introduced by the president of WES, J. Cicely Thompson, who had herself attended the meeting as one of the British delegates. In her President's Message, entitled "International Engineering," Thompson observed that, to many delegates, "it was the spirit of friendship and unity which made the Conference unforgettable. We did not speak all the same tongue, but we all spoke the same language" (Thompson 1964, p. 1).

The first report was authored by engineer Veronica J. K. Milligan, who attended the conference as an international delegate from the United Kingdom. She noted that speaker after speaker, men as well as women, focused on the inadequate supply of trained personnel to implement future world needs. The possibility of cooperation in achieving the outstanding potential of the future throughout the world was then emphasized. A key theme that came up from these discussions was the critical role women could play, both as educators and as engineers and scientists. Thus, the training of more women in these fields emerged as one of the prime solutions to the problem outlined above.

Milligan's report indicates that this argument was strongly projected to the Conference participants by M. H. Trytten, the director of the Office of Scientific Manpower at the National Academy of Sciences in Washington, DC. Indeed, during Session I, in a lecture entitled "Requirements for engineering and scientific manpower" (ICWES-I, Complete Proceedings, pp. 22–33), Trytten spoke of the untapped reservoirs of talent in the United States, declaring firmly that women represented the greatest unused potential with regard to engineering and scientific talent. Trytten addressed the strong concerns expressed over women's persisting low employment as engineers. His view on this matter was an optimistic one, as he predicted that the changing needs of society would probably provide the impetus to improve the status of women, while the changing nature of jobs, such as the introduction of automation, would open up more opportunities for women as engineers.

Veronica Milligan also attended session V, during which a report on the current status of women engineers and scientists was presented and discussed. There was a strong interest in the data that was submitted on the proportion of women engineers across the world. For example, women were playing a more vital role in engineering in many of the so-called underdeveloped countries, despite their high unemployment figures. Countries such as Turkey, Syria, Japan, and the Philippines had higher percentages of women engineers, while the proportion was under 1% in most European countries, including France, Germany, Italy, and Great Britain. The situation in the United States was more problematic. While a far higher percentage of women engineers was quoted there, a recent survey had revealed that of the 7000 women who had reported themselves as engineers, many were actually below the level of professional engineers.

Not surprisingly, this session generated ample discussion on various issues related to women's professional work and the obstacles they face in their quest for a higher status. It was emphasized that women were able to compete with men on an equal basis when given the opportunity. They had, however, to overcome considerable difficulties, many linked to their marriage upon the completion of their education. There was a general agreement that various provisions needed to be made to allow women to interrupt their training or employment and be able to return to work without loss of status. In the case of women engineers more specifically, there was a firm call for a radical rethinking on the part of industry to utilize women engineers to the full.

The second report on the ICWES-I conference was submitted to the *Woman Engineer* by engineer Sarah C. Ferguson, another international Conference delegate from the United Kingdom (Ferguson 1964). Ferguson also attended session V. For her part, she noted that all the speakers present agreed that women were far more likely to be scientists than engineers in more "developed" countries. And since the need for engineers was less acutely emphasized, women were unfortunately channeled into more "feminine" pursuits.

Ferguson attended session VI as well. She noted that this symposium devoted to the theme of "Developing Engineering and Scientific Talent" was led by a large group of distinguished panelists who were asked to send their statements to the conference. There was a strong consensus regarding the need for more publicity on the opportunities offered to girls wishing to pursue a worthwhile career in engineering. More should also be done to encourage married women who wanted to work and retrain those who wished to return to work after a lapse of time. There were also references to the advent of automation, which tended to make the employment of married women easier by shortening the working week. One significant intervention was made by Constance E. Arregger, President of the British Federation of University Women (BFUW). She reminded the participants that only a small fraction of girls became scientists and a minute number engineers. There was here a "pool of ability" that was tapped to various degrees in different countries. There were also many "influencing factors" to consider when looking at the low numbers, such as parental attitudes, early concentration on homemaking, prejudices of headmistresses in favor of the arts and classics, job availability, and prejudice of employers. Finally, Arregger insisted on the complex problems linked to the employment of professional married women and of those women returning to paid work after an absence devoted to raising a family (ICWES-I, Complete Proceedings, pp. 238–239). In the end, all symposium panelists agreed that girls and women should be encouraged to study science or engineering, above all because there was a great need for their knowledge and talent.

This reference brings us back, once more, to the notion of "scientific manpower," which was linked to a strong concern about the insufficient supply of a talented workforce made of highly skilled engineers and scientists. As the conference proceedings demonstrate, a large number of ICWES-I participants, both men and women, expressed this concern and promoted, as a result, the view that women constituted the richest unused source of engineering skills. They rallied around Trytten, who had introduced his lecture with the following statement: "A large part of the greatest unused potential in our society in regard to scientific and engineering talent is that of women. There should be no doubt that the very fact of this meeting will contribute to a better understanding that women should not be ignored as a potential force in science and technology, and that engineering will become an increasingly accepted profession, open and hospitable to women" (ICWES-I, Complete Proceedings, p. 22). Significantly, Trytten went on to use the term "womanpower" in the rest of his lecture.

In the same vein, Sarah C. Ferguson ended her "Report on the New York Conference of Women Engineers and Scientists" with the following statement: "The vastly increasing populations, rapidly changing conditions, and the urgent need to conserve manpower make it obvious that new recruits are needed on an increasing scale. The Conference made it abundantly clear that there is a need and place for women to occupy important positions in the Scientific and Engineering Industries, complementary to and on an equal footing with their male colleagues. It is hoped that our Educational Authorities will take an urgent, keen, and interested look into the opportunities now available" (Ferguson 1964, p. 22).

Several themes transpire from the related "womanpower" papers included in the proceedings of ICWES-I. Some of the key ones include:

- The role of women's scientific and engineering organizations in the development and implementation of the "womanpower" strategy in the United States and other participating countries.

- The contribution of the burgeoning women's movement in the 1960s to the promotion of the "womanpower" strategy.
- The development of transnational alliances and collaborations to increase access and expand opportunities for women in engineering and science.
- Supporting and expanding the careers of professional women scientists and engineers: how to reconcile the family/work tensions to avoid the loss of skilled "womanpower" in STEM.

Resolution Statement by the Delegates

At the end of the conference, delegates were asked to approve a resolution that would be action items for everyone to work on until the next ICWES. This became a standard procedure at the end of some of the ICWES conferences. However, the documentation available does not always include this important feature. At ICWES-I, four action items were identified and approved by the delegates. They were:

1. To encourage each participant to report about the Conference in her home country or region.
2. To encourage women to increase their participation in the professional societies in their countries.
3. To encourage women to enter the fields (of engineering and science) and to improve their qualifications not only during their student days but throughout their professional life.
4. To maintain the central file of women engineers and scientists used for this Conference and enlarge it as much as possible. (ICWES-I, Complete Proceedings, p. 284).

> "So ended the First International Conference of Women Engineers and Scientists. It was attended by a delegation from the UK of some 22 engineers and scientists, who were so inspired that on their return they persuaded the Women's Engineering Society in the UK to host a second such Conference. This took place in the University City of Cambridge in 1967. The technical theme chosen was 'The Application of Technology to World Food Problems'." (Laverick, in ICWES-X, Complete Proceedings, p. 9)[3]

Testimonials

Carolyn Phillips:
Phillips, a past president of SWE, attended the first ICWES in 1964. She heard of the conference as a member of the SWE NY section where the idea started. She worked on some aspects of the program and wanted to meet attendees and hear from

[3] Elizabeth Laverick (WES member and prominent ICWES figure; see Appendix 1) presented a retrospective of the first nine ICWES conferences (1964–1991) in an opening talk at ICWES-X which took place in Budapest, Hungary, in 1996. The full paper, "ICWES One to Nine," can be found in Appendix 4, drawn from ICWES-X, Complete Proceedings, pp. 8–12.

them. She did not have financial support to attend, but she lived in New York, so expenses were not large. All the members of NY SWE were there. Her first impression of ICWES-I was an amazing gathering of hundreds of women from around the world, very adventuresome in 1964. Carolyn was on the hospitality committee and was quite busy picking folks at the airports.

> "With late planes and rerouting, we had crazy schedules; I also had to depart a day early due to a conflict with another meeting. I mingled with people from other countries, as that was one of the main purposes of being there. A few men attended; the organizers made a point of inviting select ones. The conference was stimulating, with differing options in some ways, but general agreement as to issues in others. What I liked the most about this first conference was the diversity of backgrounds and opinions and experiences of attendees. Issues with some parts of the conference were some boring or ill-prepared speakers, language issues, food. ICWES-I probably broadened my perspective as a young woman engineer. I wanted to attend ICWES-II and did. I made lasting connections: leaders in WES (UK) and Italy (Anna Amour)." (Email correspondence with Monique Frize, August 3, 2021)

Extract from an Unpublished Memoir of Margaret Pritchard, Engineer and SWE Member from Oregon, Relayed by Peggy Layne by an Email, March 25, 2021, to the Three Authors:

Margaret "Pritch" Pritchard, a well-known and beloved SWE member, has passed away. Peggy Layne says: "I remember her well from attending SWE annual conferences. She was quite a character" (email from Peggy Layne March 25, 2021).

Margaret Pritchard wrote about ICWES-I:

> "The International Congress for Women in Science and Engineering has an international emphasis. Women in the sciences and engineering from all over the world are invited to attend the conference. I was the West Coast Oregon Delegate at the first ICWES held in 1964 that SWE hosted. The conference planners asked me to run the 'sing song' entertainment for the evening. For that evening entertainment, we all had Indian headdresses on. We all sang songs together at first. Well, everything went fine until we started singing individually. Individual people would come up with a song they wanted to sing. When people quit volunteering, brilliant me here decided that all the women from the contiguous countries from Africa east to Iran (Middle East countries) could sing together. My friend Shirley from the SWE Seattle section was standing in the back of the room gesturing NO. Her dad worked in the foreign service. She knew I didn't have any idea what I was up to asking Israelis to sing with Iranians. Anyway, there was a Turkish woman who saved my bacon. She came up to me and said she knew a lullaby but didn't think anyone else knew it. She sang it and saved my bacon. By that time, Shirley got word to me to shut up. Then I regrouped and asked for a duet from the same nationality. That was so funny. I had no idea what I had done. Fortunately, we did not have a Middle East chorus or crisis."

Pritchard also wrote:

> "My dear darling friend, Dr. Maheene Ramane [= Mahin Rahmani], taught school at the University of Tehran. She is a very very lovely human being. She was also Secretary of Transportation for the Shah of Iran before he was deposed. After the Shah was deposed, she went and taught at the University of Tehran. The University of Tehran was its own enclave. They don't get involved in any of the politics that are going on in Tehran or outside campus. Both she and her husband were teachers there." (Email correspondence with the authors, March 25, 2021)

Extracts from "The President's Message: International Engineering," by J. Cicely Thompson, After ICWES-I:

"When the Conference opened, no one knew whether it would be a success or a failure, and there were no plans for a second, though, perhaps some hopes. However, success was quickly evident, together with the determination that so much international fellow-feeling should not be allowed to fade. The announcement that the Second International Conference would be in England in 1967 was immediately welcomed."

"The organizing of this Conference in addition to all the usual activities of the WES is going to mean a lot of hard work, but those of us who were at New York know that it will be worthwhile, and are encouraged by the enthusiasm already expressed by other members."

"Besides hard work, it is also going to mean raising additional funds to cover expenses that cannot be recovered from delegates' fees and to provide ready money for the next two and a half years. At the conclusion of the First Conference, over 500 dollars was raised for the second Conference and the wonderful amount of £170 was generously given for this fund during our Sheffield Conference by members."

"I feel very fortunate to be your President at the start of such an exciting project knowing that there is equal excitement and practical support forthcoming among members."

"In conclusion, I wish all members a busy and happy year." (Thompson 1964, p. 1)

Lee Arnold, "Reflections on ICWES":

"I was (and am) a college graduate in liberal arts with also a secretarial certificate. Why, after five years with IBM, I decided to go back to school for an engineering degree is a biography in itself and not necessary now. Suffice it to say that I did, and that led me to the Society of Women Engineers. I went to the first International Conference of Women Engineers and Scientists (ICWES) in New York in 1964 because (even at the age of 29) I was a starry-eyed student, and I wanted to hear Dr. Lillian Moller Gilbreth speak. So I cut a class and went to her keynote address at the opening ceremonies on Wednesday, and I was also able to attend the Student Day festivities on Saturday. That Conference generated a unique excitement which has been inherited and shared but never duplicated by any of the ICWES conferences that followed. I've always mourned that I could not afford to attend the entire event. It was women speaking up with a united force, which wasn't common in the early sixties—a unity not easily ignored. It was a brashly energetic organizational effort which flew full in the face of the times when women still had to do twice as much work to be considered half as good. Women had no widespread network and had never done this before. SWE's early leaders, Past Presidents Beatrice Hicks (1950–52) and Katherine Stinson (1953–55) were instrumental in this endeavor; many others—unsung heroes too numerous to list—added their efforts to the synergy." (Arnold 1996a, p. 7)

Anne M. Perusek, "'Focus on the Future': A Look at the First ICWES":

"ICWES-I was the first meeting of its kind to be attempted on a global scale, and it was an ambitious undertaking. Aileen Cavanagh was president of the then-14-year-old Society of Women Engineers, the conference host. Twenty years later, she reflected on the event in an article published in US Woman Engineer, the predecessor of SWE magazine. She effectively describes the backdrop of the event and its significance:"

"June 15, 1964. Will anyone who was there ever forget that day? 529 people from 39 nations found their way to the opening of a technical program called "Focus for the Future" at the United Engineering Center in New York City. In context, such a meeting seems rather ordinary, a technical meeting at the center for technology. But this was no ordinary meeting. As

best [we] know, it was the first international meeting of technically trained women and as such was unusual in 1964."

"Younger readers may not appreciate the setting: 1964 was the year of the passage of the law that prohibited discrimination in employment on account of gender in the United States. In those days, United States corporations divided their personnel organizations into male and female departments. The female personnel departments provided an excellent career opportunity for female personnel executives, but exempt women were often relegated to inferior job classifications, their opportunities limited by the expectation voiced by a corporate vice president when he said, albeit apologetically: "We don't expect our women to stay with us""

"[...] In the same article, Cavanagh recalls that in the fall of 1963, just about six months prior to the ICWES meeting, she spoke at a conference, appealing to the deans of United States engineering schools to reverse for men-only policy still in place in many of their institutions and admit women on the same basis as men in all their programs. The last of those institutions finally accepted women students in 1972. Despite this grim situation, the participants at ICWES did not dwell on such inequities. Instead, Cavanagh stated that:"

"The "Focus for the Future" program was technical in nature. It was built around the thought that all over the world the next decade would challenge and inspire engineers and scientists to ever more creative and imaginative work. The delegates saw themselves as involved in a tremendous program of international scientific progress. They lent their energies to the "what" and "how" of professional accomplishment rather than engaging in introspective discussion of women's place in industry or the rationalization of global sociological problems."

"The Society's first executive secretary, forerunner of the current position of executive director, was Winifred (Winnie) White. On the occasion of the 20th anniversary of the first ICWES, she shared her thoughts and filled in some of the details of the 'how' of that meeting. She described the genesis of the conference:"

"'Great oaks from little acorns grow' was never more apropos than it is to express the First International Conference of Women Engineers and Scientists. Our little acorn emerged from obscurity on Saturday, August 9, 1958. The board of directors was in session with Catherine Eiden presiding as president and several New York Section members listening in. From someone's hand a scrap of paper was passed to the other hands and the question of a SWE annual convention to be held in New York was raised. This was to include an international conference and be held at the time of the World's Fair. The following day, the board voted yes."

"With the momentum established White said that, 'From then on, the acorn grew rapidly and by 1964 was truly a great oak. Committees were formed; a grant was received from the National Science Foundation; many contributions were received from industry'. The logistics and networking required to put the conference together were challenges that were met with enthusiasm and dedication. White continues her narrative:"

"The boundless energy of SWE members started the ball rolling. The Belmont Plaza Hotel was selected, buses were hired, programs were printed, invitations were mailed all over the world. Without the assistance of women engineers in the United Kingdom, Italy, France and other countries the conference could not have taken place. Collecting the names of delegates around the world was the most difficult task, but within a week 40 names had been identified. The numbers later grew to nearly 2,000 women around the world."

"White tells us who chaired committees, who took responsibilities for what. Some of the names read like a 'who's who' of early SWE history. Of course, there were also people who worked quietly behind-the-scenes, whose efforts were every bit as critical, but weren't named. White recalls:"

"In 1959, Elaine Schapiro, New York Section chairman, appointed Dr. Beatrice Hicks chair of the steering committee, and Aileen Cavanagh, secretary. In 1963 Ruth Shafer became convention co-chair and took charge of the operations. Elsie Eaves was appointed public relations chair and Margaret Fox, chair of the technical program."

"Lydia Pickup has attended all but the most recent ICWES. She recalls that, as chair of the technical program, Margaret Fox (Foxy), 'cornered the speakers for the different panels, and worked with the government to get government funding to bring foreigners in.'

In that vein, Pickup also humorously recalls her experience chairing the closing panel session, which shed some light on social conventions of the day. As the panel drew to a close, the last presenter stated, 'You're all invited to come to Ireland. There are 700 engineers and only one woman engineer'."

"Pickup was seated by a live microphone, about to make her closing remarks. 'I'm in a suit,' she explains. 'I've got a hat on, my white gloves were there lying on my purse. I'm being very formal and precise, doing the best I can. And I popped into the microphone without one thought in my head, 'that's the best offer we've had all week'. The remark brought the house down, 'but, oh was I embarrassed. I could've crawled under the table'. Pickup believes that, aside from the technical sessions, the most valuable aspect of attending ICWES meetings has been 'all the wonderful people I've met'. White's synopsis of the first ICWES offers this final assessment:"

"Technical sessions discussed issues of transportation, energy supplies, communications, the environment and pollution control. Grace Murray Hopper received the Achievement Award and Lillian Moller Gilbreth was the keynote speaker. The conference was a resounding success, discussed for many years thereafter around the world. No wonder ICWES has survived until the present, and will, most probably, grow in the future." (Perusek 1998, pp. 14–15)

Veronica J. K. Milligan, "Impressions of the New York Conference":

"New York in June is not altogether the best of places to be; the temperature is around 90° F and there is a humidity that causes little beads of perspiration to trickle off the end of your nose. Most of the buildings are air-conditioned, but when you walk out into the street, the smell of hot, unpleasant air greets you. Moisture seems to be rising from the network of rivers on which New York is built. The best place to be on a hot evening is at one of New York's numerous cinemas which are efficiently air-conditioned. The pavements vibrate as the trains rattle through the underground. The 'subways' are only just below the surface as New York is built on rock. The large American cars race through the streets and taxi drivers dismiss you nonchalantly as they pass you by. Altogether New York can be a lonely city."

"How lucky I was to be a guest of The Society of Women Engineers! American hospitality in our case was very efficiently organized. Receptions were arranged by many of the large engineering firms. Private families arranged dinner parties. Tours were organized. We were never allowed a dull moment. One of the most impressive visits was to the World Fair [...] We saw the US Electric Power and Light Companies' 'tower of light'. A beam of light concentrated by reflectors from twelve searchlights adds up to 12,000 million candlepower. [...] The General Electric Company demonstrated a controlled nuclear fusion which was very spectacular [...] At General Motors' Futurama, we were whisked along on a conveyor-type car and given a series of glimpses into the future-underwater worlds of weird towns under the sea-miracles of the future such as underwater drilling for a substantial part of the world's oil supply and diamond mines on the floor of the ocean. It depicted the underwater world more certainly than space as the future development of man [...]"

"Another interesting visit was to the Con Edison's 275 000 kW atomic power plant at Indian Point. Twenty-four miles north of New York City, it went into operation in 1962. It supplements the Con Edison's system of generating stations that serve some eight million people in New York City and Westchester County [...] En route to this proud possession of the American Company, I was most surprised to see the condition of their distribution lines. Many times I have complained bitterly about the appearance of our overhead distribution and its destruction of scenic beauty, but by comparison with American distribution, ours is a picture... Poles straggle alongside the highways, poised at giddy angles and wires of all descriptions are festooned around them. To say they look like Christmas trees is an understatement. High voltage medium voltage, low voltage telephones, everything all draped on the same poles. Following our trip to Con Edison's Plant, we were entertained at the famous Patricia Murphy Candlelight Restaurant. After an excellent meal, informal entertainment was provided. We were all issued with Red Indian head-dresses and pipes of peace, and told we were to have camp-fire celebrations. All nations participated by singing their national folk songs. Even the gentleman delegate from Uganda sang a song. The British delegation reduced party to laughter by singing 'Clementine' with great gusto. Another of our conducted tours was around the United Nations Building. Situated between 42nd and 48th streets, the 18 acres were purchased by John D. Rockefeller Jr., in 1946, from the City of New York, and then gifted to the United Nations, a completely independent sovereignty maintaining its own police force and postal system. We were given a lecture on the history of the United Nations and it was pointed out that whereas originally women were engaged only in the Status of Women and the Social Committee, they were now actively engaged in many branches. Five official languages, English. French, Spanish, Russian and Chinese may be heard over headphones provided."

"An evening cruise was taken on the river, a very enjoyable evening. The city is almost entirely situated on islands."

"To me, the most interesting part of the visit was the Conference itself. More than 500 women engineers and scientists from thirty-five nations and all 50 states of the United States of America shared experiences, plans and aspirations during the conference. They showed that women are making outstanding contributions to science and technology throughout the world. Speaker after speaker pointed to the many ways in which science and engineering are contributing to improved world living conditions. The possibility of cooperation in achieving the outstanding potential of the future throughout the world was emphasized. Speakers, both men and women, mentioned the inadequate supply of trained personnel to implement future world needs and stressed the important part that women can play both as educators and as engineers and scientists." (Milligan 1964, pp. 11–12)

Sarah Ferguson, "Report on the New York Conference of Women Engineers and Scientists":

"During the course of the week, we were entertained at various luncheons and dinners. There was a Welcome Luncheon given by the SWE and the City of New York at which was read a message of welcome from the Mayor, Robert Wagner, in which he declared that the week June 15th to 21st was 'Women Engineering Week', repeated by several Governors and Mayors. I never found out what importance could be attached to this, as I saw no one selling flags in aid of SWE, but we rather felt that it was OUR week. We also were given a luncheon by the women's Program of New York, which concentrates on the training, re-training and rehabilitation of women. However, the luncheon which I enjoyed most was on the Saturday by the Students' section of the SWE Several students at various academic stages, spoke most enthusiastically of their work. Most of the speakers, as also many of the people I met, were at University, in one capacity or another, or in research. Many more graduates stay on to read for their Master's Degree and it also seems more common to switch Universities in the middle of a degree course. Many students work their way through college. This last made me very thankful for the British University Grants System. Other speakers were Lady Jeffreys, who gave an entertaining account of her own life in the engineering world, and Dr. Maria Telkes, born in Budapest, but now resident in the United States, a specialist on the use of Solar Energy."

"During the course of the evening banquet, the SWE achievement award was presented to Dr. Grace Hopper for her work on automatic computer programming. After the presentation, Miss Pauline Frederick, NBC correspondent at the United Nations, spoke on the work of the United Nations. She appealed for a change in individual and national attitudes; she emphasized that the day of gunboat mentality is over. The need today is not for armaments, but for conference tables and this is the main function of the United Nations. Miss Frederick ended by repeating her plea for whole-hearted support of the U.N., which could enable it to function far more efficiently."

"There was also a lighter side to the. Conference: the first day was World's Fair Day. In the afternoon, after a very pleasant picnic in 'Little Ole New York', one could wander or go by coach round the site."

"On Sunday, we had the last organized function Brunch. At this there was an auction in aid of SWE 'funds', and, due to the generosity of some of the International Delegates, particularly Signora Anna Amour, there were some beautiful and valuable objects auctioned, the proceeds of which were put towards the new International Conference in England in 1967. The final events of the Conference were a visit to the Neuberger Art Collection and a highly informal hot-dog party at the home of Miss Aileen Cavanagh, in Hanover. The collection is in an apartment in Park Avenue overlooking Central Park. The party was about an hour's ride out of New York and was yet another opportunity to meet delegates informally. In conclusion, I should like to say how grateful I am to the Caroline Haslett Trust for having appointed me as one of their delegates and of thus according me the privilege of attending this most interesting Conference in New York. The vastly increasing populations, rapidly changing conditions and the urgent need to conserve manpower make it obvious that new recruits are needed on an ever -increasing scale. The Conference did make it abundantly clear that there is a need and a place for women to occupy important positions in the Scientific and Engineering Industries complementary to and on an equal footing with their male colleagues. It is hoped that our Educational Authorities will take an urgent, keen and interested look into the opportunities now available." (Ferguson 1964, pp. 21–22)

Chapter 3
ICWES-II, The Need to Continue: Cambridge, UK, July 1–9, 1967

The second conference was organized and hosted by the Women Engineering Society (WES) in the United Kingdom and held at Newnham College at the University of Cambridge. Information on the Women Engineering Society (WES) was included in the proceedings of the conference (ICWES-II, Complete Proceedings). An extract is found below:

> "The Women's Engineering Society, founded in June 1919, is the oldest society of women engineers in the world, with members in Africa, America, Asia, Australia and Europe, as well as in the United Kingdom. All branches of professional engineering and allied fields are represented within the Society."
>
> "The two most important aims of the Society are to promote the study and practice of engineering among women, and to enable technical women to meet and correspond and to exchange ideas."
>
> "The Society has branches in London, Manchester, the Midlands, and South Wales, each with meetings and work visits. A conference, held alternately in London and the provinces, allows members of all branches to meet at least once each year. The conference, which usually lasts three days, includes technical papers given by eminent engineers or educationalists of both sexes, work visits, and at least one social function. [...]"
>
> "The Society's quarterly magazine 'The Woman Engineer' has an editor and business manager who give their services freely."
>
> "Most of the work of the WES is done on a voluntary basis by its members who give talks to schools and careers conventions, represent the Society on other organizations and, through Council and the committees, keep the Society active, dynamic and in the public eye. It was typical that about 20% of members were involved in some way in the organization of the national Conference" (ICWES-II, Complete Proceedings, p. 87).

M. Frize et al., *Women's Contribution to Science and Technology through ICWES Conferences*, Women in Engineering and Science,
https://doi.org/10.1007/978-3-031-33800-7_3

Extracts from "Second International Conference of Women Engineers and Scientists 1967," an Article on Preparations for ICWES-II:

"Over 8000 notices giving details of the second ICWES have been sent all over the world. We are grateful to The Society of Women Engineers in New York, who printed and mailed 4000 of these notices in the United States. Many of the additional mailing addresses have been supplied by our Overseas Conference Representatives, of whom we now have 27."

"Approaching 300 replies, from 29 countries, to the first general mailing have come in, and 60% of these, representing 24 different countries, confirm an expectance to attend. Some 60 people from 16 countries, have offered to present papers. So far only 16% of WES members have replied, but an appreciably larger percentage of members are already devoting time and energy to the organization of the Conference. The Executive Committee will welcome all volunteers. Funds collected now amount to over £1,600, but more is still needed."

"Our first Press Release, issued in August, has now been sent out in French and German in addition to English; SWE covered a tremendous list of journals in USA on our behalf, and the Overseas Conference Representatives have produced extensive lists. We are collecting Press cuttings. The main theme is the application of technology to world food problems; there will also be sociological sessions on the position of professional women engineers throughout the world. ("Second International Conference of Women Engineers and Scientists 1967," 1965, p. 13)

Bursaries Provided to Attend the Conference:

"The Caroline Haslett Memorial Trust fund has offered nine bursaries to assist women technologists from overseas to attend the Conference; these go to Mrs U. Angielska of Poland (one of our overseas Conference representatives), Miss A. Azmaz of Turkey, Miss K. Khubchandani of India (a Conference author), Miss D. Masson of France, Miss M. Muwanga of Uganda, Miss C.F. Novae of Romania, Miss V.S. Prabhavalkar of India, Miss P. Saengbangpla of Thailand[1] and Miss P. Sivaprakasapillai of Ceylon." (Neal 1967, p. 1)

Main Speakers at the Opening Ceremony at ICWES-II:

The welcome speech was given by the President of the WES, Rose Winslade; an opening speech was given by Lillian M. Gilbreth (USA); another opening speech was given by A. L. Armitage, vice-chancellor of the University of Cambridge.

"Miss Winslade is a member of the Society of Instrument Technology and was formerly Joint Manager of the Electronics Division of Research and Control Instruments Ltd in the Philips Group of Companies. [At the time of ICWES-II], she [was] Assistant Secretary at the Council of Engineering Institutions." (ICWES-II, Complete Proceedings, p. 18)

Winslade explained that the Conference was divided between two themes: first, "The Application of Technology to World Food Problems"; and second, "The Woman Professional Engineer." She mentioned that "the first of these themes presents a challenge to every engineer and scientist. A very wide range of engineering and scientific disciplines can and do contribute to the solution of food problems, but a great deal more could be done if the will and knowledge were more widespread. Engineers are generally in short supply and there are few women professional

[1] See Section "Testimonials" for Phulporn Saengbangpla's recollection of her first ICWES.

engineers. The second theme looked at women engineers' situation at the time and their future" (ICWES-II, Complete Proceedings, p. 19).

Armitage was the president of Queens' College and University Lecturer in Law. "He has been member and Chairman of Wages Councils and member of the Chairman's Panel of the Industrial Court, and he is also Justice of the Peace and Deputy Chairman of Quarter Sessions" (ICWES-II, Complete Proceedings, p. 18).

Participants

The participants were from 35 countries, including the United Kingdom. There were a total of 307 registrations, including 122 from the United Kingdom, 100 from the United States, and 85 from 33 other countries. The out-of-country proportion was 60.3% (see Appendix 3).

Six associations helping with the organization of the conference are noted in the proceedings; they are: Cercle d'Études des Femmes Ingénieurs de l'Association des Françaises diplômées de l'Université (France); Associazione Italiana Donna Ingegneri e Architetti (Italy); Society of Japanese Women Scientists; Women Chemical Engineers of the Philippines; Society of Women Engineers (USA); Women Engineers' Group at the Engineers' Syndicate in Cairo (United Arab Republic).

The executive committee was composed of J. Cicely Thompson as Chair; Hilda Blount as Honorary Treasurer; Molly Neal, Ira Rischowski, and Rose Winslade as Committee members; and Isabel H. Hardwich as Honorary Secretary.

> "The inaugural address was given by Dorothy Hollingsworth OBE, a chemist and Head of the Food Science Advice Branch of the Ministry of Agriculture, Fisheries and Food. Her address was titled 'Nutritional Goals in a World Context', which traced the growth of thought on the application of nutritional knowledge to the problem of feeding populations. Experiences in the UK since the beginning of this century were used to illustrate certain general principles. International action in matters of food started with the League of Nation's concept in the 1930s of 'the marriage of health and agriculture'. This was followed by the setting up in 1945 of the Food and Agricultural Organization of the United Nations for the purpose of raising nutritional levels and standards of living worldwide. Twenty years on and it was clear that the difficulties were even more daunting than anticipated, and a food crisis of staggering proportion was predicted by 1985 unless the rich nations drastically increased aid to poor nations for food and population control programs. This address provoked much discussion and formed a stimulating introduction to the ensuing sessions on: 'Enough for everyone'; 'The Use of Power in Agriculture'; 'Increasing Food Production'; and 'Future Trends'."

> "Again, there was a second theme: 'The Woman Professional Engineer' and Jo Webb, an American chemical engineer and author, gave the opening address titled 'Science Versus the Humanities – A Harmful Dichotomy.' She put forward the hypothesis that science and the humanities are not separate and unrelated but are subclasses of philosophy. She traced the division between the rational and the intuitive approaches to an understanding of life, through the ages to the present day. She expressed the view that changes in education to emphasize the importance of both modes of thinking were imperative, as was the need for,

and satisfaction to be found in, social responsibility combined with professional commitment. Jo Webb also reported on a survey, which she had carried out in which she invited 34 people from different countries to complete a questionnaire about women engineers. She pointed out that, professionally qualified women engineers were in a striking minority in every country in the world, but were one in three in the USSR, and one in one thousand in the United States. Her questionnaire was an attempt to collect information about women engineers on a common basis, something that the first ICWES had failed to do." (Laverick, in ICWES-X, Complete Proceedings, p. 9)

"ICWES-II was the first Conference to introduce 'The Bringing of Greetings Ceremony'. On the eve of the Conference, delegates were invited to attend in national costume, and each country delivered greetings to the host country. This was followed by a 'get to know you' party. Of course, the English are at a great disadvantage when it comes to national costume and rely on the Scottish and the Welsh to represent the United Kingdom. However, with countries such as Austria, Brazil, Ghana, India, Iran, and of course Japan, to mention just a few, it was a most colorful evening, and the UK delegates were put in the shade."

"ICWES-II vied with ICWES-I in providing a wide range of visits, some technical and some cultural (taking Jo Webb's comments on unification to heart). Based as we were in the City of Cambridge, using the facilities of the University, we were able to tour some of the historic colleges and enjoy such differing pleasures as an organ recital in King's College chapel, and a demonstration of Morris dancing – a ritual English folk dance symbolizing death and rebirth, performed by groups of white-clad men wearing bells and carrying sticks and handkerchiefs. The host Society also gave its *own* party in the historic town of Bury St. Edmunds where, in the Athenaeum, the eighteenth century assembly room where Dickens used to give readings, the delegates were 'treated' to a full-blown demonstration of the Scottish pipes, and haggis for supper. An additional attraction was the post Conference tour of the Irish Republic. All in all, the Conferences are a golden opportunity for delegates to learn more about the heritage of the host, as well as getting to know each other." (Laverick, in ICWES-X, Complete Proceedings, pp. 9–10)

ICWES-II, S&T Papers

The scientific and technical papers of ICWES-II were presented and discussed on July 3, 4, and 5 1967. They were published in volume 2 of the conference proceedings (ICWES-II, Complete Proceedings, pp. 88–433).

Some of the main scientific events worldwide on the scientific scene that year were as follows: Christian Barnard and a team performed the first successful human heart transplantation in South Africa; William Jason Morgan presented his theory of plate tectonics (published the following year); Syukuro Manabe and Richard T. Wetherald published a climate model predicting an increase in the Earth's global temperature of 2 °C if the concentration of CO_2 in the atmosphere was to double; the Chimpanzee Washoe began to learn American Sign Language; three American astronauts were killed in a fire during a plugs-out test for Apollo 1; and it was the beginning of the popular visionary Star Trek television series!

More down to earth and probably in phase with feminine preoccupations, for the ICWES-II conference, the S&T topics were exclusively related to feeding the human population in the context of population growth. In her paper titled

"Meteorology and the World's Food Supply" (ICWES-II, Complete Proceedings, pp. 255–268), G.M. DeSanto (senior scientist in the Laboratory of Applied Meteorology in the Technology Division of the GCA Corporation in Bedford, Massachusetts) presented by Aileen Cavanagh (USA) in the third session of the scientific portion of the conference, underlines this preoccupation:

> "The doubling of the world population expected by the end of the present century will clearly require a corresponding vast increase in world food production. According to United Nations estimates, more than 85% of the projected population increase will occur in Asia, Africa, and Latin America, where the current average daily diet is some 300 calories less than the minimum of 2300 calories considered essential for normal health and activity." (ICWES-II, Complete Proceedings, p. 256)

The opening lecture was given by the British scientist Dorothy F. Hollingsworth (1916–1994). Hollingsworth was a chemist and dietician. She worked at the Ministry of Food from 1941, where she studied nutritional aspects of the Wartime Food Survey. She became head of the Ministry's Nutrition Branch in 1949 and, at the time of the conference, was head of Food Science Advice Branch in the Ministry of Agriculture, Fisheries, and Food. In her presentation titled "Nutritional Goals in a World Context," among other elements, she reminded the audience that the growth of international action in matters of food stems from the concept of the "marriage of health and agriculture" proposed by the League of Nations in the thirties, now fostered by the Indicative World Plan of the Food and Agriculture Organization of the United Nations. To give a flavor of the rich discussion that followed her presentation, a participant expressed worries that the use of machines in agriculture would get people out of jobs. Also, an Indian delegate challenged Hollingsworth on the feasibility for every nation in the world to be self-sustaining in food production and on the presenter's assertion that the problem of the world's malnutrition is political rather than technological.

Hollingsworth answered as follows:

> "If I could answer the second question first: my personal view is that it is political and economic. The whole business of sending food from one place to another enters into world trade and it does seem to me that the economic and financial aspects create almost insuperable difficulties. If it was an ideal world and nobody had to buy anything from anybody else and prices need not be controlled, I think sufficient food could be produced now to feed everybody, indeed, if you added it all up together and divided by the number of people in the world, there is sufficient food. [...] The problem is one of distribution. Now your other question: '[...] No I do not think it should be. [...]. Take for example [England], it would be extremely difficult for this country to produce all its food, we import about half of it'." (ICWES-II, Complete Proceedings, p. 114)

ICWES-II brought varied answers to the numerous questions raised by the need to feed the growing population. Along with this issue, technological transfer between developed and developing countries was becoming a preoccupation. Such questions set the agenda for the rest of the conference. ICWES-II was organized in three S&T sessions, with a total of 21 papers distributed in 6 panels (Table 3.1).

Among the 22 authors (20 women, two men), 9 were from the United States (including the 2 men), 3 from South America, 3 from Western Europe, 2 from

Table 3.1 Distribution of S&T sessions and papers in ICWES-II, Complete Proceedings

ICWES-II S&T sessions	Session titles	# of papers expected	# of papers in proceedings
(1a)	Enough for everyone	3	3
(1b)	The application of technology: The use of power in agriculture	3	3
(2)	The application of technology: Increasing Food production	4	4
(3a)	The application of technology: Future trends	6	6
(3b)	The application of technology: Preservation	2	2
(3c)	The application of technology: Distribution	3	3
	Total	21	21

English Speaking Africa, 1 from the Middle East and North Africa (UAR Egypt), 2 from Central Asia, 1 from South-East Asia (the Philippines), and 1 from Far East Asia (Japan). Seven of the papers written by these authors were reported orally by American or European participants. The following paragraphs provide an overview of the papers and discussions.

In subtheme 1a, the countries of origin of the authors were Uruguay, Brazil, and Japan. While having the same broad preoccupation for feeding their populations, one is intrigued by the very different situations and challenges they face. In Uruguay, the focus was more oriented, at the time of the conference, on raising cattle for meat than on agriculture, and fish was only a supplement. Almost at the opposite, Japan depended on the sea and freshwater products to feed its population because the area of arable land necessary for pastures would be equivalent to twice the whole area available. Brazil was in an uncomfortable situation in 1967, not being able to properly feed its population due to, according to the author, demographic explosion, obsolete land structure, inadequate transport and warehousing, and food diet. However, the author expressed hopes that this famine would be overcome in the future.

Typical of some of the papers in the ICWES-II S&T sessions, the topic of the first presentation in subtheme 1.b called "The Application of Hydraulic Technology to World Food Production" (ICWES-II, Complete Proceedings, pp. 165–181), was written by Harold A. Lehman (Manager Engineering Services, Fluid Power Division, Plessey Incorporated, NY), and presented by Ann W. Fletcher (USA). His paper related different hydraulic machines useful in farm equipment and the use of hydraulic power that allows multiplying acreage output per man many folds. The second presentation (given by Anna E. Amour, Italy) explained that agricultural work is carried out in some areas of Italy in a more or less primitive way, while in other areas, agriculture is already fully mechanized. She advocated using electricity even to sustain traditional agriculture.

Elizabeth Laverick (UK) gave the last presentation of a paper written by Amina el-Hefny called "Irrigation and the High Dam in U.A.R" (ICWES-II, Complete Proceedings, pp. 192–199). Amina el-Hefny was a chemical engineer and head of the productivity measurement section at the Productivity and Vocational Training Department of the Ministry of Industry, in the United Arab Republic (U.A.R.). Her paper expresses hope for agricultural expansion through irrigation arising from the construction of the Aswan High Dam in UAR. Lake Nasser (created between 1958 and 1970) is fed by water from the Nile. This pharaonic enterprise led to the displacement of population and archaeological sites, but the expected benefits were large, in addition, of course, to the generation of electricity. Amina el-Hefny, more specifically in relation to *The Application of Technology: The Use of Power in Agriculture* theme, mentioned that the conversion from basin irrigation to perennial irrigation and the utilization of underground water should have positive effects on crops. We wonder if someone in 1967 expected that the regulating effect of the river floods would bring problems, such as reducing the amount of alluvium deposited to agricultural lands; today we know that this has occurred.

The second ICWES-II S&T session was on "The Application of Technology: Increasing Food Production." In this second session, the authors presented different avenues for increasing food production. Rowena O. Chester is one of them. From ICWES-II, Complete Proceedings, p. 204: Dr. O. Chester obtained a PhD for work on radiation damage to semiconductors at the University of Tennessee. In 1958 she joined Oak Ridge National Laboratory and is working on solid-state radiation detectors. In 1960, the study of agriculture became an avocation. She raises beef cattle and is a member of the American Angus Association and the Production Registry International. Her paper titled "The Use of Beef and Dairy Cattle to Increase the World's Food Supply" (ICWES-II, Complete Proceedings, pp. 204–213), presented by Emma C. Barth, describes how beef and dairy cattle can make a significant contribution to the world's food supply by producing high-quality protein from cellulose and nonprotein nitrogen compounds. She suggested using rumen microorganisms (the rumen is the first stomach of some ruminants) to transform substances that have initially little or no food value.

Letitia E. Obeng is a Ghanaian zoologist who received a PhD from the Liverpool School of Tropical Medicine. Two years before ICWES-II, she had been appointed as director of a new hydrobiological research institute to study the inland water system of Ghana, which includes the newly created three thousand square miles Volta Lake. In her paper "Man-Made Lakes as a Source of Freshwater Fish Production in Africa" (ICWES-II, Complete Proceedings, pp. 214–224), she shared positive and negative experiences on the use of dams for freshwater fisheries in developing countries. The implementation of dams is often intended for multiple purposes; for example, fisheries installed in the upstream lake would help to feed the population, and production of electricity would improve the country's economy. But the experience shows that not all projects are successful, because multiple problems may affect the quality of the water, and even the health of the population. She advocated that effective application of knowledge acquired from ecological studies and research are necessary for man-made lakes to contribute their share to the provision of enough food for all.

The third S&T session was on *The Application of Technology* and had three subthemes: *Future Trends*, *Preservation*, and *Distribution*. To cope with the fast-expanding populations, the application of modern technology for food production was explored in this session. This session certainly has, as it was the case for the preceding ones, great historical scientific value. There is no space to fully report the rich technical content in this book. Various research and projects were reported. To name a few: quality control for food canning; preservation of food using microorganisms or chemicals; techniques for transportation of food for long distances by sea boats, by refrigerated containers, or airplanes; and control of pests and diseases. The recorded discussions were also very rich in content.

Globally for ICWES-II, many of the papers report multidisciplinary work to tackle S&T issues. It is the case for a paper written by G. M. DeSanto and H.E. Cramer (respectively, senior scientist and director of the Laboratory of Applied Meteorology in the Technology Division of the GCA Corporation), presented by Aileen Cavanagh (USA). In "Meteorology and the World's Food Supply," they mention "Meteorology and Agricultural Food Production Meteorology and crop growth are inextricably linked through the energy exchange processes that occur at the air-earth interface and in the adjacent atmospheric surface layer" (ICWES-II, Complete Proceedings, p. 263).

Noteworthy, some discoveries raising great hopes that were presented at ICWES-II were later abandoned, as it often happens in research of innovations. Katharine Stinson, an Aeronautical Engineer, was president of the Society of Women Engineers from 1953 to 1955. After a long career at the Federal Aviation Agency, she was appointed in 1964 as technical assistant to the chief of the Engineering and Manufacturing Division. Stinson mentioned:

> "The use of aircraft in agriculture began as early as 1911 when experiments to combat insect pests were conducted, but it was not until World War II that the airplane came into general recognition as an agricultural tool. The rapid growth of aerial agriculture began in 1945. New and powerful insecticides such as DDT were developed during the war and techniques were evolved by the military services for effectively spraying insect ridden areas that had to be occupied by troops." (ICWES-II, Complete Proceedings, p. 334)

Of course, the international community learned that the use of DDT was to be banned, as early as the 1970s in most countries, due to its high environmental and health impact. Due to its high persistence, traces of it are still found in the soil today ("DDT," n.d.). If DDT is not used anymore, today spraying crop fields are performed by specially designed agricultural aircraft, that is, pilots attempting to fly just above the crops to reduce the drift of the sprayed material.

In "La production industrielle de protéines alimentaires à partir du Pétrole"[2] (ICWES-II, Complete Proceedings, pp. 312–331), C.Y. de Mayo, who had been an engineer and researcher at Société Française des Pétroles BP, reported the possibility

[2] "Industrial Production of Alimentary Proteins from Petrol" (authors' translation).

to use petroleum to achieve the production of valuable dietary protein. Petroleum contains paraffin that can be transformed into proteins by certain yeasts and micro-organisms. However, British Petroleum (BP) abandoned the idea in 1976 after having invested billions in development ("Vous souvenez-vous du 'steak de pétrole'?", Radiofrance 2019; Sullivan 1976).

Links Between S&T, Scientists and Engineers, and Women

The technical theme for the S&T sessions of ICWES-II was concentrated on the preoccupation with feeding the world population fully and justly. This could be seen as a feminine preoccupation. However, the issue of the participation of women scientists and engineers to tackle this problem was almost absent during these S&T sessions.

It was addressed once, specifically, when J. Cicely Thompson (UK) questioned Hollingsworth (UK), during the discussion after her opening presentation: "[...] could you elaborate a little bit more on what we as women engineers and women scientists can do to solve this very grave and serious problem?" (ICWES-II, Complete Proceedings, p. 111). Hollingsworth struggled to answer, mentioning that it is very hard to answer that question. She, however, provided examples where women were part of the solution for easing housework:

> "I was up in Leeds University at the beginning of this year, in their Department of Food Science and there was a lady from Korea studying their traditional food which is called kimshi, which is rather like sauerkraut. She was busy in the Department of Food Science making this food by traditional methods which she knew how to do, and finding out the science of it, so that she could go back home and, I suppose, put it into practice." (ICWES-II, Complete Proceedings, p. 111)

ICWES-II, Gender Papers

For the second ICWES conference, organizers chose to divide the event between two distinct themes: the first, already introduced, addressed "The Application of Technology to World Food Problems," while the second was dedicated to "The Woman Professional Engineer," with a focus on her current and future status throughout the world. Papers relating to the second theme appeared in the third volume in the original proceedings (ICWES-II, Complete Proceedings, pp. 434–636).

In her opening address, "Science vs. the Humanities: A Harmful Dichotomy," American chemical engineer Jo Webb expressed the importance of this issue; indeed, she had herself conducted a survey involving 34 women engineers from different countries to collect information on this group on a common basis, something, she concluded that the inaugural conference had omitted to do (Laverick, in ICWES-X, Complete Proceedings, pp. 9–10) (Table 3.2).

Table 3.2 Distribution of social and gender sessions and papers in ICWES-II, Complete Proceedings

Social and gender session titles	# of papers expected	# of papers in proceedings	# of abstracts without papers in proceedings
The woman professional engineer	12	12	0
	# of contributions (oral and written)	# of written contributions only	# of written contributions in addition to oral interventions
Discussion	25	1	2

A one-day conference was organized around the second theme. The major part was devoted to the presentation of formal papers (chaired by Lydia I. Pickup, President of SWE), followed by discussion sessions (chaired by Freda H. Gwilliam, Woman Educational Adviser to the Ministry of Overseas Development in the United Kingdom). For this theme at ICWES-II, 36% of the papers delivered dealt with gender and women's experiences (see again Table 1.1). Our collected data also confirms the Conference's international character. Those presenting papers on "The Woman Professional Engineer" (14 in total, 11 female, 2 men, 1 unknown) came from France (3), the United Kingdom (2), the United States (2), and Canada (2), with one presenter from Sweden, Germany, India, the Philippines, and the USSR. (A participant from the United Kingdom, Gwendolen Sergant, an active member of WES since she had joined the organization in 1945, presented the paper submitted by The Soviet Women's Committee which did not send speakers to the conference.) Worth noting as well is the low number of male presenters, with only two men participating, one from the United States, and the other from the United Kingdom. Similarly, of the 25 contributors who took part in the general discussion, almost all were women (22 to 3). A majority came from the United Kingdom (7), the United States (7), and Western Europe (7: 3 from France; 2 from Germany; one each from Switzerland and Turkey), with the remainder from Hungary (1), Nigeria (2), and India (1). Finally, from what can be gleaned from the conference proceedings, two participants in the discussion also provided written pieces, while another only contributed a written statement; two speakers took part in both sessions.

Overall, as was expected, the Conference proceedings include papers focusing on the role and status of women's professional engineers in their home country. Many presented results based on statistical analysis and on the conduct of surveys through questionnaires. This approach sought to provide new insights on topics such as the evolving socioeconomic conditions of women in professional engineering, the social and educational attitudes related to the employment of women's professional engineers in various countries such as Canada, Germany, Sweden, India, and the USSR, and the future role of women in science and engineering.

Some speakers entertained a fairly positive outlook on these issues and about the overall progress realized so far by women in science and technology in a session titled "The Women Professional Engineer." This first group included John B. Parish, an American professor of economics who had served on several "manpower" committees throughout the United States, who claimed that in the new environment of Big Science, the demand for both men and women would exceed supply for the first time, and that this would result in an increasing role for talented women scientists and engineers ("Future Role of Women in US Science and Engineering," ICWES-II, Complete Proceedings, pp. 553–573). Another speaker from the United States discussed how teacher education could become the "key to the science womanpower pool" by working to dispel the negative attitudes toward women in science which were inculcated unfortunately to children as early as kindergarten. In her view, this goal could be achieved by improving science education for elementary teachers. Through the early development of positive attitudes toward science among all children, the cultural bias against women "doing science" would hopefully be overcome (Gladys S. Kleinman, "The Key to the Science Womanpower Pool: Teacher Education," ICWES-II, Complete Proceedings, pp. 574–582).

Finally, on the international front, a speaker from India and another from the Philippines reported on the improving condition of women in science and engineering in their respective countries, with respect to their employability, wages, and working conditions. While there were still hurdles to overcome on these fronts, one could envisage gains in the coming years (K. K. K. Khubchanadi, "Social and Educational Attitudes to Women in Professional Engineering with Particular Reference to Indian Environments," ICWES-II, Complete Proceedings, pp. 494–508; Zenaida Gonzales Gordon, "The Filipino Woman Professional Engineer in the 60's," ICWES-II, Complete Proceedings, pp. 509–515).

The second group of participants shared a more skeptical view, focusing on the persisting obstacles facing women pursuing a career in science and engineering in their country, and on what future lay ahead for them. Speakers from France, Sweden, and Germany raised the issue of equality between the sexes in these professions, arguing that although the principle may by now be accepted and even guaranteed by law, equal rights had still not been fully achieved. Women continued to face considerable prejudice at work and in their larger entourage. They were still less paid than men for the same job, while positions of authority were basically out of reach for them. The difficulty in reconciling a woman's professional life with her responsibilities as a wife and mother was also a major hurdle, as she was expected to succeed on both fronts. A textile engineer from Germany insisted that the main reason for the very small number of active women engineers in her country was not a lack of technical abilities, but the "conventional attitude of society, which even today still prefers men to women in professional life and which grants the unemployed housewife the highest social recognition…" (H. Brocher, "Women Engineers Today in Germany," ICWES-II, Complete Proceedings, pp. 485–493).

A different perspective, however, was conveyed by the Soviet Women's Committee, in a paper on the "role played by women engineers and scientists in the USSR" (presented by Gwendolen Sergant; ICWES-II, Complete Proceedings, pp. 539–552). It discussed the increasing role and contribution of women engineers and scientists in the USSR during the last half-century, as the country was celebrating the fiftieth anniversary of the Great October Socialist Revolution in 1967. The past and current status of Soviet women scientists and engineers was reviewed with the help of general statistics relating to their employment, and of examples of successful careers pursued during the last 50 years by women in these fields. In the case of engineering, for example, 46% of the engineers and 71% of the technicians working in technology in 1963 were women. In 1965, the USSR had 6,646,000 scientific workers—a quarter of the world's scientific force—of whom 254,000 were women. How could we explain these impressive numbers? It all began with the Soviet Government's decision to quickly adopt a law granting women full equality with men and to adopt measures to achieve this goal in practical life. One such measure, which was strongly advocated by Lenin, the founder of the Soviet State, was to draw women into social labor by freeing them first from the drudgery of domestic labor. Creating the social and living conditions that would help both parents to successfully raise their children was considered an essential step to achieve women's full equality with men. As the Committee put it, the "[t]he Programme of the Communist Party of the Soviet Union envisages the total elimination of the remnants of the unequal position of women in domestic life and the provision of social and living conditions enabling women to combine happy motherhood with increasingly active and creative participation in social labour and social activities, and in scientific and artistic pursuits" (ICWES-II, Complete Proceedings, p. 551). For example, the Soviet State had taken steps to improve the professional training of women, setting apart additional funds to better prepare them for higher and secondary technical schools and reserving a percentage of vacancies for women in universities. As a result, "Soviet women had refuted by their work the erroneous views of opponents of women's equality who held them incapable of scientific work, particularly in the sphere of exact sciences" (ibid., p. 548).

Despite its absence, the Soviet Women's Committee's presentation during ICWES-II seemingly triggered the interest of the conference leaders and organizers. Jo Webb thus reported in her opening address that while professionally qualified women engineers were in a striking minority in every country in the world, they were one in three in the USSR and one in a thousand in the United States (Laverick, in ICWES-X, Complete Proceedings, pp. 9–10). The Soviet Union's obvious success at attracting women in engineering and science could thus raise questions and trigger discussions on the reasons behind the poor performance observed in other countries.

For example, in her "Post Conference Impressions" (see section "Testimonials" for full text), Rose Winslade, the president of the Women in Engineering Society (WES), and a member of the ICWES-II Conference Executive Committee, focused on the discussions related to the key issue of the equality between women and men. In her view, "that women of any nation, given the right education and social climate, can be the equal of men in intellectual achievement, was clearly demonstrated at this Conference." However, she observed that "during the discussions on whether or not women are wholly accepted in science and engineering, it became more and more clear that equality is often accepted in theory but not in practice." Winsdale then added: "it was disappointing that the Russian delegates had not, in the end, been able to attend or perhaps we might have got at least one difference of opinion on this matter" (Winslade 1967, p. 2).

Conclusion on Gender Papers

Winslade's observations bring us right back to the paper submitted by the Soviet Women's Committee, which contended that Russian women were fully equal in practice to their male counterparts in science, technology, and engineering, thanks to the support of the Soviet State. Both contributions should be discussed together, along with the other papers in the ICWES-II conference proceedings which address issues related to the "inequality of the sexes" in the 1960s. This could, for example, generate new and important insights into the emergence of second-wave feminism and its impact on the views and level of activism of women engineers and scientists in various countries, as full equality between the sexes in all spheres of society (including science and engineering) was one of the key objectives of the "women's liberation movement" which emerged during that turbulent and productive decade.

Business Meeting

At its business meeting on July 6, at 9 p.m., the main discussion was on the future of the conference series through the possibility of further international cooperation between women engineers and scientists from many countries. The Chair was J. Cicely Thompson, Chairman, Conference Executive Committee. Present were representatives of the six supporting societies listed above, as well as a representative from UNESCO, and one representative by country.

The following text is extracted from the minutes of the meeting (ICWES-II, Complete Proceedings, pp. 38–42); it explains clearly how the group planned to provide continuity for conferences to be organized and hosted every 3 years or so in various regions of the world:

Present

Representatives of Supporting Societies:
Le Cercle d'Études des Femmes Ingénieurs de l'Association des françaises diplômées de l'Université (Mrs M. Brylinski); Associazione Italiana Donna Ingegneri e Architetti (Miss A.E. Amour and Miss P. Ferrero); Society of Japanese Women Scientists (Miss K. Saruhashi and Miss M. Tanaka); Society of Women Engineers, New York (Miss L.I. Pickup and Mrs. O. Salembier); Women Engineering Society, London (Miss R. Winslade and Mrs. I.H. Hardwich).

One Delegate from Each Country:
Austria (Mrs A.M. Janovicz); Belgium (Miss P. Talpaert); Brazil (Miss S.M. Portella); Canada (Dr. M.E. Dormer Ellis); Ceylon (Miss P. Sivaprakasapillai); Czechoslovakia (Mrs J. Vejvodova); France (Mrs J. David); Germany (Mrs J. Knott-ter Meer); Hungary (Mrs Z. Szentgyorgyi); India (Mrs I. Ghose); Iran (Mrs M. Rahmani); Irish

Republic (Mrs K.E. Walshe); Japan (Miss D.U. Mizoguchi); Kenya (Miss E. Ricketts); Mexico (Mrs A. Peréz Lopéz); Netherlands (Mrs E. Jesse); Nigeria (Miss E. Adegbohungbe); Poland (Mrs U. Angielska); Portugal (Mrs A.V. de Seabra); South Africa (Miss P. Harding); Sweden (Mrs R. Wallin); Switzerland (Mrs J. Juillard); Thailand (Mrs P. Saengbangpla); Turkey (Mrs B. Özgüner): Uganda (Miss M. Muwanga); the United States (Miss I.F. French); Uruguay (Miss L. Chebatarott).

Officially Appointed Representatives of International Organisations:
Open Door International: Mrs. L.M. Guillerault-Daniel
UNESCO: Mr. W. Ryland Hill

People Who Had Expressed Great Personal Interest:
Dr. B.A. Hicks, Miss B. Rahm, Miss L.M. Maltby

In Attendance

The Conference Executive Committee:
Miss E.H. Blount, Mrs. M. Neal, Mrs. I. Rischowski, Miss R. Winslade, Mrs. I.H. Hardwich (Honorary Secretary).

Apologies for Absence:
Apologies for absence had been received from Mrs. L.E. Obeng (Ghana), Mrs. N. Abacumkin-Agnantopoulou (Greece), Dr. D.E. Gray (Hong Kong), Mrs. U.R. Ahmed (Pakistan), Mr. B.B. Ibrahim (Sierra Leone), Mrs. A. el-Hefny (U.A.R.), Mrs. L.E. Scarisbrick (International Council of Women).

The Chairman welcomed the delegates and explained that the Executive Committee had been moved to call the meeting by the worldwide enthusiasm for international cooperation between women engineers and scientists which had been displayed in the correspondence they had received. The Committee had already circulated some brief resolutions, but they did not want delegates to be bound by these. They felt this meeting should 'plant an acorn and let an oak tree grow'. [Authors' note: It can be assumed that the oak tree is a vision of the future international network: INWES!]

A lengthy and involved discussion ensued. Delegates helped each other with informal translations and the following points emerged:

A. Continuity of Conferences

1. There was a unanimous desire for the continuation of International Conference of Women Engineers and Scientists.
2. The Conference Agenda could be on the lines of the First and Second Conferences or it could be a discussion on and, preferably, a solution of some specific problem experienced by women engineers and scientists.
3. The duration of the Conference should vary with the agenda.

4. An international secretariat to organize the Conference was desirable but was prohibitive in cost. Each Conference would build upon its predecessors and self-help and enthusiastic volunteers could provide the necessary workers. However, it would be extremely difficult to organize a Conference if there were no already existent host organization. This did not have to be a women's organization.
5. A Conference Representative in each country was desirable. She should organize her own voluntary supporting and fund-raising sub-committee.
6. Three years was a reasonable gap between Conferences. A spacing of five years would tend to break up newly made contacts. An attempt should be made to hold the Conference outside the University terms and yet during a cheap travel period.
7. The Conference venue should move around the world.
8. Conference participation should be truly worldwide, and a great effort should be made to secure the attendance of delegates from all countries.
9. UNESCO might be able to offer financial assistance.

B. Other activities

1. A bulletin to disseminate information on women engineers and scientists was highly desirable, although it would probably cost a great deal.
2. A center to help keep the contact between supporting organizations was highly desirable but probably prohibitive in cost.

C. A Third Conference

It was agreed that the Third Conference should preferably be held in 1970. Each delegate present was anxious to do her utmost to assist in the preparations for the Conference and it was suggested that each should contact the representative of her country at UNESCO in the hope that it might provide financial assistance to the Third Conference. For a wide variety of reasons, no one present was prepared to offer to sponsor the 1970 Conference without further consultation with her home-country organization, so a small international committee was entrusted with the task of ensuring the continuity of the conferences. It was granted the power to co-opt further members and its basic membership was Ebun Adegbohungbe (Africa), Dorothy Mizoguchi (Asia), Bilge Ozguner (Europe), Angelina Perez Lopez (Latin America), Mahin Rahmani (Middle East), SWE President (North America) and Isabel Hardwich (Chairman). It was agreed that delegates to the Second Conference should be invited to make a small contribution to a fund to finance the work of this committee.

Other Business

1. To maintain contacts, Ilse Knott-ter Meer invited delegates to attend the meeting of the VDI in Germany in September 1967, where one session would be devoted to women in engineering. Berta Rahm invited delegates to join with the

International Union of Women Architects at their Second Conference to be held in Budapest in 1968.

2. Several delegates and groups of delegates thought they would be prepared to sponsor a conference in 1973.
3. The Chairman said the Executive Committee would send goodwill messages to absent friends in UAR and USSR. There being no further business, the Chairman thanked the delegates for attending, hoped she would meet them all again in 1970 and declared the meeting closed (ICWES-II, Complete Proceedings, pp. 38–42).

From Acorn to Oak Tree

This was a very interesting image and the notes of that meeting show the incredible vision these women had for what was needed in the future: a permanent secretariat, more steady funding, a newsletter to maintain contacts between participants and organizers. There was also a unanimous desire for the conferences to continue far into the future. In 2002, the International Network of Women Engineers and Scientists (INWES) was created, providing solutions for each of the early desires: a secretariat, more funding, a newsletter, and much more; this transformation of ICWES into INWES is described in Chap. 14.

Post-conference Impressions from the President of WES (Rose Winslade) on ICWES-II

> "This special issue of The Woman Engineer is to commemorate the Second International Conference of Women Engineers and Scientists. When the Editor asked me to contribute to this issue I thought that it would be an easy task to pick out one or two highlights. This has not been so easy because, looking back, I am amazed at the consistently high level of all events."

> "That the Conference was truly international is shown by those who have contributed their views to this issue. But much of what is published in this Journal concerns the social events with which most days ended and those of you who were not there might get the impression that these were the events with which the Conference was most concerned. More important, however, than the social events were the technological sessions, where the international contribution was even more obvious. Eminent women and some men from 34 countries and a number of international organizations brought both their scientific talents and their social conscience to join in the all-too-short discussions. A characteristic of the Conference was the high attendance at all lecture sessions."

> "From all the technical discussions on the theme 'Enough for Everyone' came the clear indication that while race, colour and creed were unimportant when discussing problems of worldwide interest, local food production and nutrition could be profoundly influenced by social, political and religious beliefs."

> "The technological sessions must, too, have proved a real lesson to people in the world of superabundance. Those countries in which there are grave shortages are not, as many Westerners believe, receiving aid passively and doing nothing to help themselves; they are in fact working with great determination and strongly held convictions of· the importance of the contribution that each educated individual can make and does, indeed, make to a

country's moral, economic and social progress. What emerged, too, was the interesting pattern that, whereas all countries suffer from a brain drain, women with brains tend to consider their education a part of their country's investment and seek employment in their own country. Consequently, in what we, in the West, have come to refer to as the developing countries, many very senior and responsible posts are held by women."

"What of these women? Are they people so wrapped up in their responsibilities that they can no longer enjoy themselves and laugh at themselves? Look at the pictures in this Journal and read their comments. I am sure that the full force of their personalities and their pleasure at being able to exchange views with others of similar interests will radiate through to you."

"That women of any nation, given the right education and social climate, can be the equal of men in intellectual achievement was clearly demonstrated at this Conference. But during discussions on whether or not women are wholly accepted in science and engineering it became more and more clear that equality is often accepted in theory but not in practice. It was disappointing that the Russian delegates had not, in the end, been able to attend or perhaps we might have got at least one difference of opinion on this matter."

"Some people have asked why, if women wish to be accepted in the professions on equal terms with men, was an international conference of women engineers and scientists organised. Wasn't this an implication that women themselves were erecting barriers? But it seems to most of us that, with women engineers in such a minority throughout the world, it is only by getting together on occasions such as these that they can make themselves heard and seen. All would agree that if the number of women in the engineering profession was comparable with the number of men then any special attempt to publicize themselves would be quite unnecessary."

"I would like to close by taking this opportunity to say a public thank you to all those members of the Society who worked so hard and so willingly' to bring this Conference to fruition. Also, I would like to express the gratitude of all workers to their families without whose sympathy and patience the work would often have been impossible." (Winslade 1967, p. 2)

Introducing the Bringing of Greetings (One Delegate per Country); this Custom Was Repeated at the Following ICWES Conferences:

"Perhaps the highlight of our Conference came early in the week when we met in Hall at Churchill College for the informal buffet supper on Sunday evening the 2nd July, followed by the ceremony of 'Bringing of Greetings'."

"Those of us who remembered vividly the novel ceremony which took place three years earlier in New York were prepared for an eventful evening, but this occasion seemed to have a freshness and spontaneity all of its own."

"Isabel Hardwich was in the Chair, looking wonderfully happy at the culmination of all the efforts to make the Conference really happen, and the Conference Committee were equally jubilant. The evening commenced with wine and chatter. The seating was quite informal so that delegates could move freely from table to table as they recognised old friends from the New York Conference or made new acquaintances, and over an hour passed before the first Greeting was given. Delegates had been asked to wear their national dress and many and varied were the charming costumes displayed." (Cridland 1967, p. 5)

Testimonials

Berta Rahm (Switzerland), "Need for a Third Conference":

"After the perfectly planned Conference (even including daily sunshine) in the beautiful City of Cambridge (with interesting lectures, discussions, films, receptions, excursions to technical, agricultural and historical places as well as an amusing introduction to the art of wearing a sari or kimono) all Conference delegates seemed to look forward to a third international meeting in three years' time."

"The brave organizing group of The Women's Engineering Society of Great Britain solved their task in such an excellent way, it is possible that no one will have the courage to follow."

"Cambridge was a real dream: old and new buildings on the river with their wonderful parks, lawns, flower gardens, and quiet courts, a variety of interesting colleges, a market hill. narrow shopping streets and old churches. How agreeable and fascinating it was to live in a college, to visit friends in their colleges, enjoy meals in both ancient and modern dining halls, and to be treated to lots of coffee, tea and cakes. How pleasant were the drives to old places with wonderful surroundings, such as Sawston Hall, Hatfield House and Woburn Abbey…"

"To choose a place after Cambridge would seem to be problem enough. Should there be no more international conferences because we have been so spoilt in every way? I have no doubts but that we should go on, even if there can never be a second place like Cambridge '67. Today there is a great need for cooperation. international understanding, exchange of knowledge and friendship. And as, especially for us women, this century is so different from the last and so much has still to be changed, we should remain in contact and even try to extend our contact to other countries. We can learn from each other, and help each other. An international conference might help in a country where the education and work of women engineers, scientists and architects is still regarded with curiosity, barely suppressed prejudice, or prohibited altogether."

"Here I mention my own country as an example – Switzerland, where the proportion of women students at universities and institutes of technology is one of the lowest in Europe; and where franchise for women in federal affairs (and in most cantons, even in cantonal and communal affairs) is unknown and refused. This seems to be strange, as Swiss universities opened their doors to women comparatively early. The University of Zurich has legally allowed women to study medicine, literature and science since 1867."

"One hundred years ago, in Geneva, Mrs Marie Goegg, a clever and courageous mother of three boys tried to found the first international association of women. She took up correspondence with women having the same aims in other countries: better education for girls; better paid work; emancipation and equal rights. She founded an international journal for women."

"Since then there have been many changes in the lives of women, but although so much has changed much remains to be done. Science and technology, international cooperation and agreements have helped women, and so too we have to offer our help. That is why I consider international conferences necessary, apart from the interest and pleasure of seeing new countries, making new friends and renewing old friendships. So let us hope a courageous and willing team can be found to organize a third conference. In the meantime our thoughts go back to charming and peaceful Cambridge and our thanks." (Rahm 1967, p. 38)

Lee Arnold, "Reflections on ICWES":

"The British picked up the enthusiasm, and Isabel Hardwich issued the famous cry, 'Come to England in 1971', and WES (Women's Engineering Society) made it good. I was just nearing the end of school and uncertain financially, but Ruth Shafer had become my mentor in SWE, and she talked me into it. I've been grateful ever since, and that was the beginning of the unbroken string of attendance at all nine ICWESes to date. It's a feat I share with only two other living people: J. Cicely Thompson of the UK, Chair of ICWES-II and a Member-at-Large of SWE (MAL) and Lydia I. Pickup of the USA (1968–1970 Society President and Fellow in the Pacific Northwest Section)." (Arnold 1996a, p. 8)

Dormer Ellis:

"Between 1967 and 2002, I attended *eleven consecutive* ICWES conferences. Also, in 2014, at the age of 89 years, I attended ICWES-XVI in Los Angeles."

"When I graduated in engineering in the 1940s, there was no organization of women engineers in Canada. In the 1950s, while teaching electrical technology at what is now Ryerson University, I became a Member-at-Large (MAL) of the Society of Women Engineers (SWE) in the United States. Although I knew that SWE's 1964 conference was to include women engineers from many countries, I was unable to attend. The dates conflicted with those of a Canadian research conference in Montreal where I was scheduled to represent the University of Toronto and present a paper. Therefore, I did not participate in ICWES-I."

"I heard about a second international conference of women engineers when the president of the Engineering Institute of Canada (EIC) asked me to go to Cambridge, England to represent Canada at the expense of EIC. I accepted this invitation. Going to England was not a novelty for me as, in the early 1950s, I had spent several years working as an engineer in Rugby, Warwickshire. The EIC paid my expenses for ICWES-II but the costs of attending eleven other ICWES conferences were personal expenses. I thoroughly enjoyed ICWES-II where I 'brought greetings from Canada' at the opening reception. My paper titled 'Factors Affecting Sex Differences in Aptitude for, Interest in, and Achievement in Mathematical and Scientific Subjects' was well received as it was research based. What I most enjoyed at ICWES-II was meeting interesting women from so many different countries."

"The eldest attendee was 92 years old. She was the mother in *Cheaper by the Dozen*! The youngest attendees were engineering undergraduates. I had the interesting experience of being presented to the Queen Mother, as were, one after the other, about a dozen others, each from a different country. This event was VERY formal. We were thoroughly briefed beforehand as to how we were to behave, what we could and could not say, etc. The Queen Mother told me that the Queen was in Canada at Expo '67 and asked if I had been there. I replied that I had and praised the British pavilion."

"During ICWES-II, a so-called Continuity Committee was set up, on which I was the North American member. The sole purpose of this committee was to ensure that another ICWES would take place. We accepted the offer of an association of Italian engineers and architects to host ICWES-III. Members of the Continuity Committee kept in touch and, when the date, venue, registration fee, etc. were finalized, we each conveyed the information to engineering organizations in our region. In my case, this meant informing the head office of the Society of Women Engineers and requesting each of the provincial organizations of professional engineers in Canada to publish the details of ICWES-III in their journals." (Letter to Monique Frize, March 15, 2021)

Carolyn Phillips:

"I did attend ICWES-II. One of the experiences I vividly remember is a very lovely but formal state dinner as part of the conference with fancy table settings with lots of silverware. Some of us were not sure what to use when so I finally realized I could watch one the British members sitting nearby and follow her lead. Worked well and I learned something new. I think it's those personal encounters with new things that one remembers more than the formal papers… the people and the experiences." (Email to Monique Frize, March 2021)

Letitia E. Obeng (Ghana), "Impressions of the Conference Banquet":

"The Second International Conference of Women Engineers and Scientists, like the previous pioneering meeting, was a great success. As with most international meetings, a good deal depends on organisation, split second timing of activities and their subsequent efficient execution. Efficient organisation requires imagination backed by a realistic approach to practicability and this in turn cannot be effective without energy, drive and personal interest of which there was much associated with the organisation of the Second Conference. The fact that most of the members of the organising committees had to come together from different parts of Britain and Ireland where they were engaged on their own official as well as private occupations makes the success which was achieved singularly pleasant."

"The same thoroughness and thoughtfulness which I had previously noticed to mark the preconference arrangements was obvious to me on arrival in Cambridge. I was soon to expect that this quality was common to all events associated with the Conference and for this the organisers deserve much praise. All arrangements including even the organisation of midday meals designed to give delegates a chance to see the Colleges of Cambridge typified the deep thought given to fine detail. It was not surprising therefore that the organisation of the Conference Banquet, the biggest social event, held on the night of the 5th of July at Churchill College, had the same noticeable quality."

"The banquet could perhaps be looked on as a half-way function which marked the end of the first section of the program on technology and its application to the food problems. Perhaps one may indulge in fancy and interpret the banquet to signify a physical assertion of the superb ability of women to blend a serious attitude to responsibility with an easy acceptance of social obligations. The change from the serious talks in the Chemistry Laboratory Hall to the gay atmosphere at the pre-banquet reception was a vivid demonstration of the adaptability of the human female to varying conditions."

"It was a very pleasing sight to find, grouped together, a large number of women from many countries, each radiating in her own way characters unique to her world, and, all together, jointly with each other by various modes and attitudes, producing a peculiar oneness in a highly pleasing and relaxed atmosphere. Traditional costumes of Africans and Indians are not so novel in Britain now, but the picturesque costumes from Mexico, Brazil. Uruguay, Finland and Germany alongside the graceful kimono-clad Japanese ladies provided a welcome international feeling."

"The display of age-old British tradition of the presentation of guests which preceded the pre-banquet reception may not have meant much to the British delegates, but I particularly noticed the effect it had on other delegates through casual remarks heard here and there among the people. The glamour of the rich display of colour which commingled in the reception hall, the glitter of jewellery, the gay cheerful laughter and the lightning flashes from the cameras contributed effectively to building a suitable introductory mood for the banquet. The atmosphere was cordial and as delegates and guests moved up the stairs under the benevolent gaze of Sir Winston Churchill, one could almost spot a twinkle of approval in his eye during the little time there was to stop and stare."

"The Banquet Hall as set, complete with College silver, rose to the occasion of a historic gathering of women from all over the world, perhaps the first it had seen, and perhaps the first it would see for a long time. At the High Table the impressive array of distinguished women personifying the essence of women's effort to contribute to the effective use of science and engineering, was most gratifying. I could not help but reflect, as I looked around, how pleasing it was to think that had the High Table been longer it would still have been possible to seat more able women engineers and scientists who had not been able to attend the Conference. To me, as a woman scientist from a developing country, this thought had inspiring associations. The presence, also at the High Table, of men from the world of Learning, Industry and Government brought into focus the essential appreciation of scientific training of women as a necessary contribution to activities aimed at improving the lot of mankind."

"This theme was reflected in all the toasts proposed and answered at the banquet. In spite of the wit and humour which characterised them and fitted the festive nature of the banquet, the speeches did not lose the serious theme of applying the know-how of science and technology to solving world problems. Alongside expressions of the appreciation of the efforts of women engineers and scientists in various spheres of work, there was a heartening assurance to women engineers that Industry would always endeavour to put them to profitable use. In all, the toasts stressed the importance of the contribution of women to engineering and progress, exhorted them to higher work and wished them more successes."

"The presence together of women engineers and scientists, the very young and the very old, the attention they paid to the brilliant and important banquet speeches flavoured here and there by snatches of Gaelic, Irish and Hindi, the amiable chatter that went with wine and fruit, breaking barriers such as only women can break, and the abundant goodwill that was established across the banquet tables as a result of a common interest of purpose, will go a long way beyond the objective of the banquet and accomplish much more that is good for the world than the Conference could even consciously have planned to achieve." (Obeng 1967, pp. 24–25)

Ragnhild Wallin, "Some Recollections of the Banquet from Sweden":

"In the evening of Wednesday the 5th July a selected number of guests of honour and about 300 delegates to ICWES were welcomed to Churchill College by the President of the Women's Engineering Society, Miss Rose Winslade. This evening she carried a graceful bunch of yellow roses instead of the Chairman's mallet which we had been accustomed to seeing her wield at meetings. Immediately, she and Miss Cicely Thompson, Chairman of the Conference Committee, put us all into the right mood for the banquet. The Mayor and Mayoress of Cambridge, Sir Charles Wheeler, Chairman of Associated Electrical Industries Limited, distinguished representatives of both University and Industry and not forgetting the Toastmaster, Mr. Quick, all seemed to appreciate the colourful costumes and festive appearance of the delegates. We had made use of ostentatious attributes from all over the world as bright silk headgear from Nigeria, colourful kimonos from Japan, abundant embroidered traditional dresses from Mexico, Iran and Switzerland, as sandals on bare feet, either silvered in latest Parisian style or worn with ageless dignity to gold-threaded Indian saris."

"While enjoying the meal we took the opportunity to get new viewpoints from our international friends, for example, what it means to be 'high' in university studies; being a Japanese scholar in the USA; or on how the increasing industrialization and emancipation of women has influenced the Scandinavian way of taking care of old people."

"When the strawberries were about to be finished the Queen's obedient servant, Miss Winslade, proposed the Toast to Her Majesty, and after some moments of standing reverence we were allowed to smoke! The gay relaxation prepared the way for the first speaker. Mrs Katherine Walshe from Ireland proposed a Toast to the City of Cambridge. In manner becoming to the venerable traditions of her subject and starting with 'Cead Mile Failte' or for the benefit of the very few who may possibly not be familiar with the Gaelic 'a hundred thousand welcomes', as you can see not only were the ancient Irish concerned with the problem 'Enough for Everyone' but they also had an appreciation of the metric system; Mrs. Walshe expressed our esteem and gratitude with typical Irish humour. We shall all take with us the memory of our unity in the City of Cambridge with the people of Cambridge reminding us of our duties as citizens of the World."

> "Toasts to the University of Cambridge, to Women Engineers and Scientists. and to Our Guests were brilliantly proposed and answered. Sir Charles Wheeler spoke of the encouragement his Company had given to women engineers and scientists and indeed, as the employer of Isabel Hardwich, it had generously provided the Conference Executive with Headquarters." (Wallin 1967, p. 26)

Katsuko Saruhashi (Japan), "Postscript from Japan":

> "The present time is an age of population explosion which is bringing about many serious and difficult problems for the lives of human beings. Among them the most serious question is the food problem. It was quite pertinent to the necessity of the time that the Second International Conference of Women Scientists and Engineers selected the food problem as the main theme to be discussed at the Conference. In the opening address, Miss Dorothy F. Hollingsworth historically surveyed world problems on nutrition and food and shed light on solving problems of the future. The food problem is a problem of the disparity between developed and developing countries. Without a solution of this problem, the international tension cannot be relaxed. Therefore, the food problem is also a problem for peace and one with which every woman in the world is deeply concerned."

> "It was my pleasure that the importance of studies on meteorology and oceanography in relation to the food problems was emphasized by several speakers. The solar radiation, atmosphere, land, ocean and fresh water are the natural bases of food production and we still know very little about how they are working together and how they can be changed favourably by man for the increase of common benefit of mankind. I strongly believe that the Conference achieved a great success.' One thing which made me rather disappointed was the absence of delegates from the USSR and some other important countries. I very much enjoyed a comfortable and clean room in Churchill College. It was my first and valuable experience of living in a famous English College."

> "I was also impressed to see how big an effort is being paid for the higher education of young people who will be leaders of the nation in many fields. I hope to learn also about the higher education of women in this country if I am able to visit England again. I thank you very much for the hospitality and friendship given to me by the Organizing Committee members." (Saruhashi 1967, p. 28)

Phulporn Saengbangpla (Thailand):

> "I first attended the ICWES conference in 1967; it was the Second International Women Engineering Society (ICWES-II) at Cambridge, England. At that time I was studying for

my Masters degree at the University of Manchester; if I were not studying there I wouldn't have known about it. Since then I have attended a few more conferences that followed.

My impression and experiences were,

1. Proud to be a representative of women engineers from Thailand and from Chulalongkorn University (at that time the only university that produced women engineers) and 38 years before TWEA [Thai Women Engineers Association] was established [in 2018]. I also felt pleased to be a part of a global group of Women Engineers.
2. Helping each other to self-develop in engineering works and application of engineering.
3. Exchanging views, knowledge and experiences concerning engineering education in various countries and professional careers.
4. Cultural exchange between the participants, because those attended were from different countries and different parts of the world, cultures and traditions are different.
5. All the papers presented made us better understand each other's way of culture.
6. The conferences were beneficial and enhanced our knowledge. There was cooperation in various fields and providing up-to-date information with each other."

"Apart from the points mentioned above, at the conferences the questions and answers session were very enlightening and enjoyable."

"Hoping that ICWES will continue to grow and be a pillar of support for women engineers worldwide" (email to Monique Frize, January 30, 2023).

Chapter 4
ICWES-III, Turin, Italy, September 5–12, 1971

Technical Theme, "Planning for Progress," and a sociological theme, "Women's Professional and Family Duties."

At the second conference, the delegates had expressed the hope that a group would volunteer to organize the third conference in 1970, three years after the event in Cambridge. However, organizing an international event of this kind is a huge undertaking. The Italian associations took the lead, and a very successful ICWES-III became a reality 4 years after ICWES-II, in 1971.

Participants

Several of the women who had attended the two initial conferences came to Turin. Examples are: Mahin Rahmani (Iran), Anna E. Amour (Secretary of ICWES-III), Ilse Knott-Ter Meer (Germany), Ila Ghose (India), Dorothy Mizoguchi (Japan), J. Cicely Thompson, May Maple, Elizabeth Laverick (United Kingdom), and several from the United States. There were 233 participants at ICWES-III, including 45 from Italy, 34 from the United Kingdom, 66 from the United States, and 88 from 29 other countries. The proportion of out-of-country attendance was 69% (see Appendix 2).

> "The technical theme of the Conference 'Planning for Progress' was based on the concept that planning requires a choice among the new ways and means attained by scientific and technological advancements to achieve real human progress."

> "The Conference took particular note of the concern of the members of the United Nations in the present state and future fate of the human environment, and appreciated especially the practical approach to the problem demonstrated by the setting-up of a special Secretariat to organize world consultation in Stockholm in 1972."

M. Frize et al., *Women's Contribution to Science and Technology through ICWES Conferences*, Women in Engineering and Science,
https://doi.org/10.1007/978-3-031-33800-7_4

"Believing in the need for a world approach to human environmental problems, the Third International Conference of Women Engineers and Scientists:

(a) associate itself with the aims and aspirations of the 1972 Stockholm Conference on the Human Environment;
(b) communicate its support to the UN Secretary-General;
(c) place at the disposal of the UN Secretariat the technological expertise of its members who are engineers and scientists from 32 member States;
(d) extend to the UN Secretary-General its good wishes for the success of the Stockholm Conference" (ICWES-III, Complete Proceedings, p. 21)."

These actions were decided in Turin, September 10, 1971, by delegates of ICWES-III.

"One of the major results of the Stockholm conference was the creation of the United Nations Environment Program (UNEP)":

"The participants adopted a series of principles for sound management of the environment including the Stockholm Declaration and Action Plan for the Human Environment and several resolutions."

"The Stockholm Declaration, which contained 26 principles, placed environmental issues at the forefront of international concerns and marked the start of a dialogue between industrialized and developing countries on the link between economic growth, the pollution of the air, water, and oceans and the well-being of people around the world." (United Nations Conference on the Human Environment, 5–16 June 1972, Stockholm, n.d.)

Organizations, People, and New Host Countries:

The inaugural session of ICWES-III took place in the historic hall of Palazzo Madama, where sat the First Senate of the Kingdom of Italy. The Mayor of Turin, Giovanni Porcellana, welcomed the delegates at this important international gathering of women dedicated to technical and social progress. Pietro Giulio Bosisio, President of the National Association of Italian Engineers and Architects (A.N.I.A.I.), then welcomed world colleagues. The National Secretary of AIDIA, Anna E. Amour (see Appendix 1), commemorated the first woman engineer of Italy, Emma Strada, the first president of AIDIA-(Associazione Italiana Donne Ingegneri e Architetti). Emma Strada had passed away in September 1970, less than a year before ICWES-III in Turin, Italy.

The inaugural address was given by Andrea Ferrari Toniolo, Professor of Electrical Communications at the University of Rome, who explained in five points the important links between technical progress and human progress. He ended his talk with "an invitation to further consider the subject presented, in the hope to give an answer at least to some of the questions, which should draw the attention of anyone, not only of women or of engineers" (ICWES-III, Complete Proceedings, p. 4).

Rita Levi Montalcini (director of the C.N.R. Laboratory of Cell Biology in Rome and Professor at the Department of Biology at Washington University in St. Louis, Missouri) spoke about the "Women Scientists and the Woman's Liberation Movement" (see Appendix 5 for full paper). She discussed "substantial contribution that women engineers and scientists can give in helping women to realize the full measure of their intellectual talent and dignity as human beings" (ICWES-III, Complete Proceedings, p. 4). Montalcini concluded: "A revolution needs not to resort to violence, but indignation, one of the most powerful and effective forces which move human actions, as proven by the tremendous impact that it played in exposing the infamy of slavery, is well appropriate in this early stage of the Women's Liberation Movement. Tomorrow, when women will have regained faith in themselves, and men will have recognized the advantage of their changed role, both will be able to work together in full harmony, to the great benefit of mankind, which—as all of you know, but as one is tempted all too often to forget—consists of men as well of women" (ICWES-III, Complete Proceedings, p. 13).

The opening ceremony was concluded with a reception offered by the City of Turin in the artistic rooms of Palazzo Madama, which houses the Museo Civico d'Arte Antica.

The inaugural address of FIAT to delegates was given by the vice president, Dr. Ing. Giovanni Nasi, at a lunch held in the Historic Museum, after a visit to the FIAT plant in Mirafiori. It is to be noted that there was also the presence of the representative of the Italian Government, Dr. Vincenzo Russo, Undersecretary to the Ministry of Public Works, who underlined that social improvement can only be attained through a joint action from men and women. He also argued that men need to recognize gender equality and accept parity for "family condition and civil society will advance on the way to human progress" (ICWES-III, Complete Proceedings, p. 4).

The above presentations can be found in more detail in the conference proceedings. It was interesting that the government representative tied human progress to a position of parity between women and men in 1971. Fifty years later, these seem to remain as nice words, but not so much action, not only in Italy, but in most countries in the world.

People Who Created AIDIA:

Anna E. Amour: Chair of ICWES-III, Anna E. Amour was present at ICWES-I and ICWES-II, and it seems very likely that she was one of the leaders who helped create the ICWES series (see Appendix 1).

Anna E. Amour presented the story of Emma Strada, the first woman graduating in engineering in Italy, who passed away a few months before ICWES-III:

> "The new century had just started when a beautiful young lady presented herself to the Secretary of the Royal Polytechnic School of Turin and asked to be enrolled in a course in engineering."
>
> "At the demand, the Secretary bewildered, went to consult the regulations but did not find any clause barring the entrance to a woman. Undoubtedly, no one of the founders of the School of Engineering had ever considered the possibility that a woman could enter the engineering career."
>
> "That was how Emma Strada was accepted in the School of Engineering and followed with enthusiasm and discipline the five years of study which until then had been the sole prerogative of men."
>
> "In those times Emma Strada went to school in a horse-drawn carriage through streets lit by gas lamps in which a car was a rare novelty. The electricity supply was at its beginnings and the electric light was installed in her home to celebrate her graduation 'cum laude' as Doctor in Civil Engineering."

"Her career developed at first in two ways, as an assistant lecturer in sanitary engineering and as a designer in her father's 'technical office of planning, construction and consulting'. She designed and directed the construction of civil engineering works, and of industrial plants not only in Turin but also in other regions of Italy."

"To be cited: a funicular railway from Catanzaro Città to Catanzaro Sala, partly running under the foundations of a group of preexistent buildings; a surface water tunnel in the copper mines of Ollomont; the excavation of a gold mine in the vicinity of Monte Rosa; and just before the second World War she studied the project to produce the liquigas from the waste products of refineries. The process is still utilized in a plant in operation at Marghera."

"Emma Strada's career developed however mainly in the field of civil engineering construction and ended during the second World War when she retired from the construction field and retained only the consulting activity in several industrial companies."

"Her example was soon followed by other young ladies who successfully attained the graduation in engineering. Some of them after a few years married and dedicated themselves to family duties, raising the children, then taking care of grandchildren and maybe also giving a hand to help a husband engineer in the exercise of the engineering profession, which for a man seems an obvious choice, while for a woman it still appears unusual if not strange."

"But Emma Strada was a rare woman exceptionally endowed with intellectual capacity and clarity of mind, which in her long-life experience developed in deep wisdom without lessening her youthful mind and her capability of looking with enthusiasm to life."

"She was fond of aerospace missions and would never have missed a TV broadcasting of astronauts on the moon."

"She was the first to consider the possibility of holding in Turin our Third International Conference of Women Engineers and Scientists; she examined estimates and proposals and encouraged with her approval the younger colleagues to agree to this initiative."

"'It is time to demonstrate that women also know how to make themselves useful to society', she said and then she added: 'even if I shall not be here anymore.'"

"And we are here starting this Third International Conference with the deep regret that she is not here anymore, youthful, elegant and wise as we remember her."

"We are here planning for progress, and not just the scientific and technological progress which has launched man on the moon or in the infinitesimal secrets of the atoms, but a real full progress that will help men and women to proceed along the way of civilization and to eliminate the dangers of destruction that we are facing because of some unbalanced scientific and technological achievements."

"Well, thinking to our theme, to plan for a real human progress, I went to the conclusion that, also under this point of view, Emma Strada gave us a great example."

"After the war, when she had retired from the engineering activity, she had reached a full maturity of mind that was sustained by her youthful strength."

"Our country was half destroyed and the way to recovery seemed almost impossible to be reascended and the dissensions incurable."

"In these hard days, Emma Strada was the cofounder of a group that assembled, beyond parties, all people faithful to an ideal of civil life, and accepted gladly a post of regional secretary, not to strive for wealth or power, but to make herself useful and to dedicate her experienced capacities to the monarchic idea, which she believed would have been better for her country, with passionate energy."

"Faithful, sincere, and disinterested, she fought for that cause, not for her own wealth or success, until the last months of her life."

"Maybe we do not always predict in which direction the force of events will push our future, because it is out of the human sight to estimate the full consequences of our scientific and technological progress, but undoubtedly our future will be better if we, men and women, will be willing to think about the common human aspects of our life and let our activity be inspired by the desire of dedicating our efforts to the human progress." (ICWES-III, Complete Proceedings, pp. 6–7)

ICWES-III, S&T Papers

The technical theme chosen for ICWES-III (Turin, Italy, 5–12 September 1971) was "Planning for Progress." Divided into six sessions, the S&T topics of ICWES-III were inspired by the world consultation in preparation of the United Nations Stockholm Conference on the Human Environment to be held in 1972. The conference secretary, Anna E. Amour, wrote on September 10, 1971, in the conference proceedings, as part of the final resolution:

"The technical theme of the Conference [...] was based on the concept that planning requires a choice among the new ways and means attained by scientific and technological advancements to achieve real human progress." (ICWES-III, Complete Proceedings, p. 21)

A total of 61 S&T presentations were listed in the conference table of content (ICWES-III, Complete Proceedings, pp. 40–46). However, only 8 full texts and 37 abstracts without papers are available, limiting the insight we can provide on the historical contribution of women in STEM at ICWES-III. The distribution is summarized in Table 4.1.

The 61 expected speakers were all women, among which 8 were from the host country, 20 were from the United States, 5 from Latin America, 6 from Western Europe, 5 from Eastern Europe, 2 from English-speaking Africa, 1 from Middle East and North Africa (incl. Israel), 3 from Central Asia (incl. India, Pakistan, Nepal), 1 from Southeast Asia, 1 from Far East Asia, and 9 from the United Kingdom.

Table 4.1 Distribution of S&T sessions and papers in ICWES-III, Complete Proceedings

ICWES-III S&T sessions	Session titles	# of papers expected	# of papers in proceedings	# of abstracts without papers (in proceedings)
T A	Energy sources	10	1	0
T B	Communication and transportation	14	3	8
T C	Computer technology	10	3	7
T D	Housing, town and community planning	14	1	11
T E	Industrial production	7	0	5
T F	Human engineering	6	0	6
	Total	61	8	37

A Glimpse of ICWES-III's S&T Content

Session T A was devoted to *Energy Sources*. The session Chair was J. Cicely Thompson (United Kingdom). The presentations reported on various aspects related to the following theme: hydroelectric development (in French), automation of nuclear power plants, thermal nuclear power plants, magnetohydrodynamic power generation, solar energy for water desalination, new power generation systems, carbohydrate physical properties measurement, water pollution, and pollution.

Unfortunately, all the summaries and papers of session TA presentations are lost but one, T A 7. This paper by Amour focused on "Solar Energy for Water Desalination" (ICWES-III, Complete Proceedings, pp. 52–59). She presented various "ways and methods of solar energy utilisation by conversion in thermal, electric and mechanical energy" (ibid, p. 52). She focused more specifically on explaining and schematizing an application for a desalination plant where a solar boiler would be used to feed heat.

Session T B, chaired by Ila Ghose (India), was devoted to *Communication and Transportation*. The subjects were either practical, such as normalization of new S&T terms in many languages, communication skills for readability, and communications between developed and developing countries, or they were technical, such as telephony, satellite communications, and satellite geodesy.

Concerning the last short paper called "Satellite and Geodesy" (ICWES-III, Complete Proceedings, pp. 69–70), the input of the author Suria Tatevian on space-geodesy is very interesting to read, knowing the importance of satellites in many applications today. Suria Tatevian was an engineer geodesist at the Research Associate of Astronomical Council of the National Academy of Sciences, Soviet Women's Comm., Moscow, USSR. According to her, simultaneous tracking of artificial satellites from several ground stations would be a very promising method to solve several practical problems. This well-informed scientist predicted that space-geodesy would allow, for example, to measure distances between tracking stations separated by oceans, and to document movements of the continents. The development of the world geodesic system took place between 1970 and 1990 ("Satellite Geodesy" n.d.).

Transport talks in session T B, echoing a theme already developed in ICWES-I seven years earlier, were related to practical concerns such as containers, heavy indivisible loads, rail traffic in Western Europe, and pipelines such as the Trans-Alaska pipeline. Elizabeth Laverick presented a paper called "Airport Instrumentation and Its Influence on air Transport of the Future." Technologies for the future were described during this session, such as the "L'aérotrain: transport moderne de l'avenir"[1] (Jeannine Plinate, France, ICWES-III, Complete Proceedings, p. 77), "The Channel Tunnel and Its Effect on Rail Traffic in Western Europe" (Hettie L. Bussell, UK, ICWES-III, Complete Proceedings, p. 73), and the "Concorde

[1] "The Aerotrain: Modern Transport of the Future" (authors' translation).

Supersonic Transport Aircraft" (Caroline Schofield, UK, ICWES-III, Complete Proceedings, pp. 76–84).

It is interesting to mention that these last three technologies suggested were actually developed. The Aerotrain was an experimental Tracked Air Cushion Vehicle developed in France from 1965 to 1977. As we know, the Channel Tunnel now connects Folkstone (England) with Coquelles (France). The construction of this railway tunnel began in 1988 and it opened in 1994 ("Channel Tunnel," n.d.). Regarding the second innovation, 20 copies of the Concorde aircraft were produced. The first commercial flights began in 1976 and ended 27 years later in 2003. The author, Caroline Schofield, concluded her abstract by writing: "The introduction of the Supersonic age opens up new possibilities in the drawing together of the continents. Concorde as its forerunner will be not only a unique combination of grace and power, but a practicable and profitable transport vehicle" (ICWES-III, Complete Proceedings, p. 84). The Concorde was certainly graceful and powerful. However, today we know that this great aerospace adventure was stopped due to high fuel consumption, the accident of Air France flight 4590 in July 2000 (113 death toll), and excessive noise ("Concorde," n.d.).

Session T C was devoted to *Computer Technology*. It is directly concerned with the development of computers and their applications. One could say, to present the historical context, that the computer expansion was still in infancy at the time of ICWES-III. Large processors shared by many individuals, called mainframe computers, were appearing in the industry, government agencies, and research centers. Programming was taught to future engineers using main frames and punch cards. Commercial microcomputers would only appear in 1977, although minicomputers were existing as single-user systems for devoted applications and students could acquire Hewlett-Packard personal programmable calculators.

At the time of ICWES-III therefore, in 1971, the authors (scientists and engineers) reported on the potential that computers would bring for different applications. They testified that computer systems were becoming more and more popular, not only because they are faster and have better memories than humans, but also because they can accomplish routine tasks or solve otherwise non-solvable complex mathematical systems.

One might remember as well that the prophetic Marshal McLuhan, professor at the University of Toronto, had already announced in 1962, concerning communications, the coming of the worldwide *global village*,[2] and its repercussions. The World Wide Web, as an example, was invented in 1989 by Sir Tim Berners-Lee, a British scientist working at CERN 18 years after ICWES-III. However, the 1960s and early 1970s were certainly exciting periods regarding communications and information services. Two contributions in the session T C were interested in information science. It is the case, for example, of a paper called "Computer-Based Information

[2] "Global village describes the phenomenon of the entire world becoming more interconnected as the result of the propagation of media technologies throughout the world. The term was coined by Canadian media theorist Marshall McLuhan in his books *The Gutenberg Galaxy: The Making of Typographic Man* (1962) and *Understanding Media* (1964)." ("Global Village", n.d.).

Services on the Computer and Information Sciences," by Margaret R. Fox, who was Chief at the Office of Computer Information, Center for Computer Sciences and Technology National Bureau of Standards, Washington, DC (ICWES-III, Complete Proceedings, pp. 92–99).

Fox wrote in her introduction: "Two facts have become increasingly evident. First computers are being applied in every facet of human endeavor [...]. Secondly, information is rapidly becoming the most significant of the five economic factors that affect the lives and livelihood of everyone in the world, in both the highly developed and the developing regions" (ICWES-III, Complete Proceedings, p. 92). In her conclusion, she predicted: "In summary, it is already evident that the integrated combination of computers and communication can and will have an important effect upon the fundamental structure of society" (ICWES-III, Complete Proceedings, p. 98).

Various applications were presented in session T C, such as computer-aided design and manufacturing, production control, radio communication systems, and highway research information services.

A contribution called "L'informatique au service de la production industrielle dans le domaine de l'optique,"[3] by Gisèle Hugues (ICWES-III, Complete Proceedings, p. 114), also stands out from the others because she speaks of a successful numerical analysis code. Hugues enforced that complex optical combinations could now be solved by computers, thanks to the expression of various fundamental physical laws and access to partial derivatives of aberrations with respect to fabrication parameters. She reported solving a system comprising 20–50 equations. Remarkably, optical lenses have been obtained following this numerical analysis in 1971: variable focal length, wide field, for aerial photography, and for the manufacture of integrated circuits. Today, numerical analysis is a common tool among scientists and engineers. Computers routinely solve 10^7 discrete equations.

Noteworthy in this session T C, Grace Murray Hopper presented an abstract called "Progress—Challenge to Computers" (ICWES-III, Complete Proceedings, p. 115). Born on December 9, 1906, in New York, Hopper graduated in 1928 from Vassar College, got an MS degree in 1930 from Yale, and a PhD in mathematics in 1934. This well-known pioneer had a long and fruitful career; she was a retired Commander, USNR, at the time of her presentation. It is said that the word "bug" is from her, when she and her colleagues found a moth stuck in one of the components of the Mark II computer at Harvard University in 1947.

Another of the ICWES-I themes, Session T D was devoted to *Housing, Town and Community Planning*. This preoccupation, raised by many authors, arises from urbanization caused by the effects of industrialization and population growth.

Only one full paper is available in this session: "Housing: Image of our Social Progress" (ICWES-III, Complete Proceedings, pp. 116–119). The author Estefania Chavez de Ortega was a Mexican Architect advisor in Urban Planning for the Governments of the States of Tabasco & Guanajuato, Mexico. She interestingly

[3] "Computing at the Service of Industrial Production in the Field of Optics" (authors' translation).

emphasized in her presentation that "the urban problems are symptoms, phenomena that reveal a confusion of the functions in the socio-cultural, economic and political structures of the communities" (ICWES-III, Complete Proceedings, p. 117). On the same subject, Izabela Kuninska-Zachwatowicz, a Polish architect at the State Design Office, and member of the League of Polish Women, mentioned in her abstract: "The Problems of Housing in the Aspect of the Modern Family Structure and the Technical Progress" (ICWES-III, Complete Proceedings, p. 120) that the "industrialization causes an unavoidable phenomenon of urbanization that means concentrated forms of human habitations together with their involved infrastructure, social and psychological consequences." She linked this situation to families living in these houses, in perpetual transformation in time.

Not limited to developing countries, the United States was also experiencing a housing crisis of its own at the time of the conference, as was discussed by Elsie Eaves (United States) in a paper called "Housing Headaches in the USA" (ICWES-III, Complete Proceedings, p. 121). Eaves was a pioneer civil engineer and author, consultant on housing costs to the National Commission on Urban Problems from 1967 to 1969. In her abstract, she emphasized the efforts made by various stakeholders to reduce the unit cost of housing. Patricia G. Burnham, a mechanical engineer from California, discussed on the housing industry in T D 5 "Required Factors for the Success of a Nationwide Factory Produced Housing Industry" (ICWES-III, Complete Proceedings, p. 122). Elizabeth Jacobson Rogers, an architect with the Facilities Engineering Construction Agency of the Department of Health, Education, and Welfare in California, focused on health, education, and welfare constructions for people in T D 6 "Health, Education and Welfare: Constructive Construction for People" (ICWES-III, Complete Proceedings, p. 123).

Susana Godoy Castrillon (Mexico) in T D 7 "Une solution possible au problème des taudis aggravés par la loi sur le blocage des loyers"[4] (ICWES-III, Complete Proceedings, p. 124), and Hande Suher (Turkey), in T D 9 "Housing and Resettlement Problems in Turkey" (ICWES-III, Complete Proceedings, p. 125), specifically spoke of the problems created by the apparition of slums following urbanization in developing countries and offered potential solutions.

Other authors spoke on specific solutions as well. It is the case of Liana Ferraris Zetto (Italy) on the utility of 3D models for the design and planning of constructions (ICWES-III, Complete Proceedings, p. 126), Mary Ellen Russell (United States) on using different branches of the geophysics as bases in community planning (ICWES-III, Complete Proceedings, p. 127), Judith L. Hamilton (United States, engineer geologist) on hydrogeological studies in urban planning for semiarid areas (ICWES-III, Complete Proceedings, p. 128), Wajeda J. Rob (East Pakistan, now Bangladesh) on the question of water availability and usage in open spaces near Dacca (ICWES-III, Complete Proceedings, 129), and, finally, Ayten Cetiner (Turkey) on the renewal of city centers provoked by the industrialization and urbanization phenomena in Turkey (ICWES-III, Complete Proceedings, p. 130).

[4] "Possible Solution to the Slum Problem Made Worse by the Rent Freeze Act" (authors' translation).

Session T E, on *Industrial Production*, was chaired by Lydwine E. Contesso from France. Topics such as digital instrumentation, the limits of measurement accuracy, and improving efficiency in the construction industry were discussed.

Mattie F. McFadden (United States) introduced innovations in industrial materials, finishes, and processes (T D 3 "The Role of Materials, Finishes and Processes in Industry in the 1970s"). Clairvoyantly, she underlined in her abstract some environmental issues that will arise with this industrialization: "The increased use of plastics and new developments in metals will be explored [in my paper] ... Whereas some materials contribute to our environmental problems and some cause challenges for waste disposal, treated with respect and properly handled, there are materials and processes to meet any challenge man could envision" (ICWES-III, Complete Proceedings, p. 133).

It is unfortunate that we cannot read the full paper presented by Bernarda Bartáková (CSSR) called "Industrialisation and Electrification of Tropical Regions as a Necessary Recondition for Living" (ICWES-III, Complete Proceedings, p. 135). This electrical engineer, graduated from Pilsen, worked for the development in various countries in East Asia at the time of the conference. She was the Scientific secretary of the Section of Technical Sciences at the Committee for the Complex Research of Developing Countries of the Czechoslovak Academy of Sciences. She is coauthor of the book *Tropicproofing Electrical Equipment*, published by Leonard Hill Books in 1967 and translated into several languages.

The final session T F was on *Human Engineering*. It was chaired by Dorothy Mizoguchi from Japan.

Martha E. Sloan (United States) presented an intriguing topic called "Population Modeling for the Engineer" (ICWES-III, Complete Proceedings, p. 139). She expressed a concern about the environmental consequences of engineering actions. According to the abstract, she intended to present mathematical analyses and simulation studies that can provide an insight into the best way to treat the environment. She also presented various models of population growth, ranging from the simple exponential model to complex systems models.

Martha E. Munzer (United States), a pioneer who graduated from MIT in 1922, advocated in "Durable Environment: Ecology and the Engineer" (ICWES-III, Complete Proceedings, p. 140), the necessity to build teams where ecologists, engineers, and other specialists would cooperate together to minimize the violation of fundamental ecological principles.

Four other abstracts spoke of safety, industrial hygiene, noise pollution, and women's health.

Women Scientists' and Engineers' Contributions to S&T

The proceedings of abstracts and papers provide little information on the authors' thoughts about the contribution of women to S&T, maybe because of the absence of many of the full papers. This is somewhat ironic since communications and information were important subject discussed in ICWES-III. However, here is a short report on four papers mentioning the role of women.

Suria Tatevian, in her paper "Satellite and Geodesy" (ICWES-III, Complete Proceedings, pp. 69–70), underlines implicitly the importance of having many women scientists to take part in space research like in the USSR (ibid., p. 70).

In the abstract for "Housing: Image of our Social Progress" (ICWES-III, Complete Proceedings, pp. 116–119), Estefania Chavez de Ortega would bring together professional women, housewives, and mothers to solve the housing crisis experienced at the time of ICWES-III:

> "It is crystal clear, as far as housing is concerned, that there is a possibility of professional women's active participation in contributing to this very important change, because the intention is to strengthen the creation of new and better perspectives of life. In this participation also, housewives and mothers should join, since they should—and are able to—spread out the knowledge of necessary requirements and technical solutions." (ICWES-III, Complete Proceedings, p. 119)

Gisèle Hugues is the only speaker to report in a technical session of her personal life (ICWES-III, Complete Proceedings, p. 114, "L'informatique au service de la production industrielle dans le domaine de l'optique," in French). She graduated from the French École Supérieure d'Optique and, in 1946–1947, was a researcher on spectrography at ONERA. In 1948, she abandoned her professional life for family reasons for a while, continuing nevertheless to participate in the research of her husband. In 1968, she resumed an active professional life at the CERCO Company (Centre de Recherche and Optical Calculations) of which her husband was the director. Interestingly, she mentions that she was already doing optical calculations on a computer in 1956.

It is in the very last abstract of the last technical session T F of ICWES-III that we find the only specific paper presenting what we would call today a feminist point of view. Anastasia Pavlovna Shitskova (USSR) gave a paper called "Hygiene of Woman's Labor and the Progress of Science and Technology" (ICWES-III, Complete Proceedings, p. 141). In her abstract, she poses the necessity of serious scientific investigations into the influence that new substances introduced into industry and agriculture have on women's health, and also the elimination of the unfavorable factors.

At the end of the conference, all the women (scientists and engineers) participants expressed, in the ICWES-III final resolution, a very womanly preoccupation toward the quality of human life that should address both the professional and family worlds:

> "Apart from any particular problem of their specialized field of activity, women engineers and scientists from 32 countries of the world agree that the quality of human life can be enhanced only be the responsible use of scientific and technological achievements, and that they will dedicate the efforts to this purpose, both in professional and in family life." (ICWES-III, Complete Proceedings, p. 20)

ICWES-III, Gender Papers

As we recall, the organizers of ICWES-III chose "Women's Professional's and Family Duties" as the Conference's main sociological theme. Subthemes were then retained to divide up the paper presentations into four "Sociological sessions":

- Session A: "Problems of Aptitudes or of Traditions?"
- Session B: "How to Integrate Alternatives".
- Session C: "Evolution of Women's Role".
- Session D: "Equal Opportunity: Deserved or Gift?"

As Table 4.2 shows, it was anticipated that a total of 29 papers would be presented under the four selected subthemes. Yet, only 22 full texts and 2 abstracts (without full papers) could be located in the ICWES-III conference proceedings. Looking at the program, a total of 32 speakers, all women, were expected to give papers. Only 2 participants came from the host country, while 9 came from the United States, 3 from the United Kingdom, 3 from Latin America (2 from Mexico, 1 from Argentina), 6 from Western Europe (2 France; 2 Germany; 1 Portugal; 1 Turkey); 4 from Eastern Europe (CSSR, Hungary, Poland, USSR), and 1 each from Canada, Nigeria, Iran, India, and Thailand. The first two sessions were chaired by Lydia I. Pickup (United States), while the other two were moderated by Lidia De Rita, a psychologist at Bari University.

Findings such as these confirm, without a doubt, the importance of ICWES-III proceedings for researchers interested in the establishment and growth of the organization. More specifically, the proceedings hold material that helps unravel a key chapter in ICWES's history, which consists of its evolving relationship with the women's movement since the inaugural conference in 1964.

In the case of ICWES-III, the meeting was held at a time when the "women's liberation movement" (WLM) had gained considerable ground during the

Table 4.2 Distribution of social and gender sessions and papers in ICWES-III, Complete Proceedings

Social and gender sessions	Session titles	# of papers expected	# of full papers in proceedings	# of abstracts without papers in proceedings	# of missing papers/abstracts in proceedings
S A	Problems of aptitudes or of tradition	8	6	1	1
S B	How to integrate alternatives	2	1	0	1
S C	Evolution of Women's role	11	8	0	3
S D	Equal opportunities: Deserved or gift?	8	7	1	0
	Total	29	22	2	5

mid-to-late 1960s, primarily in the industrialized nations of Western Europe and North America. Briefly, the movement was the result of an upsurge of interest in women's issues and in "women's status," and of women's growing involvement in a process described as "consciousness raising," which spurred the growth of feminist activism. The WLM's overall goal was to eradicate sexism and achieve women's full equality. Activists led, to this end, major campaigns to outlaw sex-based discrimination, suppress sexual harassment, and adopt "affirmative action" legislation and programs that were expected to oversee, protect, and reinforce the various laws already in place. Not surprisingly, a large number of these targeted education and the workplace, with the goal of achieving "equal opportunity" for women in these key sectors. Activists, thus, called for the increased admission and full participation of women in disciplines traditionally regarded as male, such as those in science, engineering, law, and medicine. Finally, the WLM was concerned with the increasing number of women who had to juggle family and career responsibilities on the professional and domestic fronts. Hence, the call for remedies to relieve busy women, such as the provision of childcare services and women's access to more flexible work schedules.

One can associate ICWES-III's choice of the theme and subthemes for the 1971 meeting with the WLM's reform agenda at the time. Both focused on issues related to the role of women and its evolution over time, the need for their increased training, equal opportunity for all women in all spheres, and the reconciliation of women's professional and family duties. These various elements were brought together by a common denominator, which consisted of a strongly articulated call for change.

As Elizabeth Laverick recalled some 30 years later at ICWES-X: "In the 'Sixties' with the increasing interest in non discrimination and equal opportunity in the Western World, conferences concerning the role of women in the scheme of things was not unusual, and the position of women in science and engineering, where women professionals had little visibility except in the USSR, was (and still is), a favourite subject for discussion and analysis" (Laverick, in ICWES-X, Complete Proceedings, p. 8).

As we noted above, the proceedings of ICWES-III include great source material that addresses, largely through the voices of the organizers and participants, the growth of feminism and of the women's movement during the so-called turbulent sixties. Submitted below are two notable examples of this material; one consists of an Address delivered at the Opening Ceremony by Rita Levi Montalcini (which was also submitted as a formal paper), while the other contribution consisted of a conference paper submitted by Evelyn Harrison.

Professor Rita Levi Montalcini, Opening Address, "Women Scientists and the Women's Liberation Movement"

At the Opening Ceremony of ICWES-III, Rita Levi Montalcini, director of the C.N.R. Laboratory of Cell Biology, in Rome, and a professor in the Department of Biology at Washington University, in St. Louis, Missouri, spoke to the audience about "Women Scientists and the Women's Liberation Movement" (ICWES-III, Complete Proceedings, pp. 10–13; see Appendix 5 for the full paper). Montalcini quickly observed that many viewed the movement as a "true revolution" (ICWES-III, Complete Proceedings, p. 12). It had started in the United States and spread from there to all other countries; it had then taken such a proportion and direction "as to disconcert even the liberally-minded persons of both sexes and to appear as repulsive to those who felt deeply attached to the century-old traditions of a patriarchally ruled society" (ibid.). In her view, a revolution did not need to resort to violence. Indignation was a more effective weapon to move human actions, as proven, she insisted, "by the tremendous impact it played in exposing the infamy of slavery." She thus felt that this indignation was "well appropriate in this early stage of the Women's Liberation Movement" (ibid., p. 13). At the same time, she wondered what would be the impact of the movement on the activity of that still too small number of young women who had selected a career considered to be a typical men's career such as that of engineer and scientist. She feared that the advantages that the movement hoped to obtain for women of the younger generation who had made this choice would be slim if today's scientists, "satisfied with the concessions which will certainly be granted in the near future and will make more academic positions for women, will lose interest in the situation prevailing in other fields, first of all, that of politics, which today as yesterday, and a thousand years ago, has been the undisputed man's privilege and field of activity" (ibid., p. 12). She herself had met many successful women scientists who had declared "their total indifference and almost annoyance toward this movement, stating that they had never suffered from discrimination, and in any way they saw no relationship between political activity and their own scientific activity" (ibid., p. 12).

Nonetheless, Montalcini ended her talk on an optimistic tone: "Tomorrow, when women will have regained faith in themselves, and men will have recognized the advantage of their changed role, both will be able to work together in full harmony, to the great benefit of mankind, which—as all of you know, but as one is tempted all too often to forget—consists of men as well as women" (ICWES-III, Complete Proceedings, p. 13).

Evelyn Harrison, "the Evolution of Women's Role in the United States"

One of the participants at ICWES-III, Evelyn Harrison, whose paper, "The Evolution of Women's Role in the United States" (ICWES-III, Complete Proceedings, pp. 195–200) can be found under Subtheme C, "Evolution of Women's Role," also focused her presentation on feminism and the Women's Liberation Movement. A graduate in civil engineering from the University of Maryland, Harrison was a successful member of the Federal Civil Service. Her interest in women's issues was clearly expressed through the high number of boards, commissions, and committees she had served on, including the Commission on the Status of Women, the District of Columbia Commission on the Status of Women, and the Women's Advisory Council on Poverty. She also worked for five years as director of the Federal

Women's program to provide equal opportunity for women in the Federal Government in hiring, promotion, and training (1964–1969). Harrison could add to her impressive CV the Federal Woman's Award for outstanding career women in government (ICWES-III, Complete Proceedings, p. 195).

In her presentation summary, Harrison confirmed that "today, there is a new wave of feminism sweeping the country, motivated in part by the climate of dissent and disaffection of youth and minority ethnic groups with the country's institutional and societal structure" (ICWES-III, Complete Proceedings, p. 195).

She then devoted the opening pages of her paper to a description and analysis of the American feminist movement: "Its voices are many; their ideologies differ. While the mass media has to some extent over publicized its more vocal activists, it has also underrated its wide range of support by both men and women throughout the country. Despite seemingly sharp ideological differences, the common thrust of all segments of the movement was geared toward major societal changes in the role of American women. Its objectives varied from the elimination of all educational and employment practices discriminatory in any way to women, to the attainment of equality of legal rights between both sexes under national law, to essentially a restructuring of the society and the role of men and women" (ICWES-III, Complete Proceedings, p. 196).

Harrison considered the "resurgence of the women's rights movement" as one of the key factors in the "greater public awareness of economic inequities and barriers to the full participation of women in American life" (ICWES-III, Complete Proceedings, p. 199). More specifically, the movement had generated the active consideration, in the US legislature, of the proposed Equal Rights Amendment to the Federal Constitution, which aimed to give equal rights to all persons under the law, regardless of sex. The proposed amendment had been approved in 1970 at the House of Representatives by an almost unanimous vote; however, it sadly failed to secure a vote in the Senate, mainly because of the substantial opposition of some women's and labor groups who feared the invalidation of existing legal protections.

The paper then proceeded to a discussion on the Women's Liberation Movement, which Harrison associated with the "new feminist movement" in the United States. She stated that the WLM had begun about two years ago, "around a number of university centers" (ICWES-III, Complete Proceedings, p. 200). She went on to explain that "its basic ideology is that the root problem facing society today is the general oppression of all women—their psychological subjugation to male chauvinism evidenced in economic, social, legal and cultural discrimination" (ibid.). In order for a woman to realize her full potential, insisted Harrison, "it is necessary that these barriers be eliminated, first from her own consciousness, and then from her environment" (ibid.).

Furthermore, she continued, the "doctrine which demands that [a] woman organize her life completely around the lives of others, neglecting her own needs and desires, is totally rejected." On the contrary, she "must acquire full awareness of the true nature of the traditional, subjugated role fostered upon her, reconstitute her identity, and then begin to reassess her environment in terms of her own humanness" (ICWES-III, Complete Proceedings, p. 200).

Harrison concluded by drawing a portrait of the current social, political, legal, and economic American landscape, which she felt would be conducive to the improvement of the status of American women: "In summary, the changing patterns of American women's lives today, together with the current dissent and disenchantment with social systems and institutions [...] is producing a climate, hopefully, for the eventual disappearance of the traditional passive, dependent, auxiliary role of women, a role no longer realistic in a viable society" (ICWES-III, Complete Proceedings, p. 200).

A preliminary review of the available material included in the 1971 ICWES-III proceedings provides compelling evidence of a strand of activism developing within the international community of women scientists and engineers, which could be linked to, and informed by, the "women's liberation movement" and by the resurgence of feminist interests in the overall status and role of women during the "the turbulent 1960s."

Meeting of Delegates

The meeting of delegates opened on Wednesday, September 8, 1971, at 10 p.m. The minutes are taken from the conference proceedings (ICWES-III, Complete Proceedings, pp. 22–23):

Present:
The Conference Secretary: Anna E. Amour
Chairman: Isabel Hardwich

Representatives:
Argentina, Tania P. de Cano (Centro Arg. de Ingenieros)
Austria, Meta Schummer, Sophia Machado Portella (Ass. Eng. Arquit. do Brasil)
Colombia, Blanca Molina de Kocsis
France, Sylvie Hoppilliard (A.E.P.F)
Germany, Rosemarie Berger (V.D.I.)
Holland, R.E.W. Kropveld
India, Ila Ghose (Inst. of Engineers)
Italy, Lidia Lanzi (AIDIA)
Japan, Dorothy Mizoguchi (Soc. Japanese Women Scient.)
Malaysia, K.H Ong
Mexico, Estefania Chavez de Ortega (Asoc. de Univers. Mexicanas)
Nigeria, Ebun Oni
Poland, Maria Grabczewska (PZITS)
Portugal, Anterna Valeriana de Seabra (Ordem dos Engenheiros)
Switzerland, Erna Hamburger (IFUW)
Tanzania, Margaret Kiame

Turkey, Bilge Ozguner
Uganda, Miriam Sebaggala
UK, Peggy Hodges (WES)
USA, Carolyn Phillips (SWE)
USSR, Suria Tatevian (Soviet Women's Committee)

Observers:
Argentina, Sonia Bevilacqua (Centro Arg. de Ingenieros)
France, Giséle Hugues (Cercle d'Études des Ingénieurs)
Mexico, Isabel Cisneros de Aguirre, Susana Gody Castrillon, Valeria Prieto, Edith Nunez Tovar
UK, John Hardwich, Beatrice Hicks
USA, Lydia Pickup, SWE

Preamble:
Dr. Amour opened the meeting and introduced Dr. Luigi Richieri, who on behalf of the Order of Engineers of the Province of Turin welcomed delegates.
Dr. Amour called on Mrs. Hardwich, Chairman of the International Committee which was set up at the end of the Second Conference, to take the Chair.
Telegrams and messages of greetings were read.
Prof. Hamburger presented the greetings of the IFUW, announced its new board and the hope that many countries will have a group of women engineers to deal with technical matters.

Future Conferences:
It was unanimously agreed that the International Conference of Women Engineers and Scientists must continue. Four offers to organize a future Conference were received.

(a) The President and the Secretary General of the Polish Association of Building Engineers and Technicians, PZITB, conveyed their intent to organize the Fourth International Conference in Poland.
(b) The President of the Institution of Engineers (India) offered to extend every assistance and cooperation for a Conference to be held in India in a near future.
(c) Mrs. Ozguner offered to organize the Fourth Conference in Turkey.
(d) The delegates from Mexico offered to organize the Fifth Conference in Mexico with the assurance that they had the backing of the local Engineering Institutions.

Gratitude was expressed for all the offers and much discussion ensued.

Fourth Conference:
On the majority vote, in a paper ballot, allowing one vote per country, the invitation to hold the next Conference in Poland was accepted. The preferred date was 1975, in order not to clash with the meeting of the International Federation of University Women in 1974 in Tokyo.

Mrs. Grabczewska expressed her pleasure at the decision and said that the PZITB would be delighted to welcome delegates to Warsaw in 1975.

Invitation from Japan:
Miss Mizoguchi regretted that the Society of Japanese Women Scientists was unable to organize the Fourth Conference but extended a warm welcome to all who planned to attend the IFUW 1974 Conference and join her Society for a meeting.

Organization of Future Conferences:
The content and organization of technological sessions gave rise to an animate discussion; different opinions were expressed, some of them as follows.

Dr. Oni supported Prof. Jackson's plea for the provision of working groups at the next Conference; she suggested to have smaller sessions to discuss in greater details specialized subjects and asked to set up a Committee to prepare a workable format for the planning of future sessional meetings.

Dr. Amour said that the acceptance of almost all the papers offered was a deliberate choice aiming to make the Conference a demonstration of the talent of women engineers and scientists and their capacity to play an active role for human progress.

Mrs. Prieto appreciated the great number of papers as an opportunity to know how other countries resolve problems.

Lydia Pickup added that some companies would not let their employees attend if the themes of the Conference were not of general interest.

Dr. Hicks proposed to read the summaries of the papers in general sessions, then to discuss them in separate sessions for working groups to develop individual subjects.

Dr. Kropveld would have preferred 10% of the time allowed for the presentation of papers (not too specialized) and 90% for discussion.

Mrs. Hardwich remarked that the three Conferences were quite different from each other and suggested that an attempt should be made to report back the results of the discussions, so that some definite conclusions could be drawn from the Conference.

Expressions of Thanks to the Organizers of Third Conference:
The meeting ended at 23.00 h with a hearty vote of thanks to Dr. Amour and AIDIA for their good work in organizing the Third Conference" (ICWES-III, Complete Proceedings, pp. 22–23).

An interesting aspect of ICWES-III was the development of a resolution by the participants. After debates and discussions, the conference secretary, Anna E. Amour, reported final agreement on the following resolution:

> "Aside from any specific problems relating to their specialized field of activity, women engineers and scientists from 32 countries of the world agreed that the quality of human life can be enhanced only by the responsible use of scientific and technological achievements, and that they will dedicate their efforts to this purpose, both in professional and in family life" (ICWES-III, Complete Proceedings, p. 20).

Here we see major actions being taken by the women engineers and scientists from many countries:

1. Offering their support for an environmental conference to be held in Sweden. The 1972 United Nations Conference on the Environment in Stockholm was the first world conference to make the environment a major issue.
2. To be conscious of the responsible use of scientific and technological achievements in both their professional and family life.

As this third ICWES concluded, it became clear that the women leaders who created and led the ICC (International Continuing Committee) were determined to ensure that an ICWES would happen every three to four years, hosted and organized by a local Association or Society. The format was now becoming more established and acted as a template for future events. In fact, as we examine the next ICWES conferences until the twenty-first century, the pattern was similar and proved to be highly successful in attracting many women in STEM fields from all around the world to participate, with a number of Societies offering to host the next ICWES.

Message from the President of WES (Peggy L. Hodges) on ICWES-III:

> "As I take office as your new president and look back over the past two years: the outstanding event which comes to my mind is the Third International Conference of Women Engineers and Scientists which was held in Turin in September. You will read much about the conference in the pages of this journal, but I would like to, underline the feelings of all those members of the Society who were able to participate. We particularly enjoyed meeting women engineers from other countries all over the world and discussing with them personal experiences and views on the position of women engineers in their respective countries. It was especially exciting to renew friendships which were made during the [ICWES-II] Conference in Cambridge. This made the efforts of the many members involved in making the Cambridge conference so successful doubly worthwhile; [ICWES-II] exists not only in its own right but as one of a series of meetings which we hope will go on indefinitely into the future. We can all look forward to the next link in the chain in the visit to Poland in 1975" (Hodges 1971, p. 1).

Testimonials

Lee Arnold, "Reflections on ICWES":

> "A multi-layered concept of communication -the personal and the technological—has grown on me throughout a series of Conferences. In 1971, at ICWES-III in Turin, Italy, I was sadly at a loss. Work demands had prevented perusal of guidebooks or phrase books, and I depended upon my travelling companions, Mary Munger, Dot Peabody, and Del Coyle of SWE-Hartford. Lack of language severely limited my understanding of session contents; however, I did come away with the beginnings of a long friendship with the Conference Chair, Anna Amour" (Arnold 1996a, p. 9).

Dormer Ellis:

> "I found ICWES-III, held in Turin, Italy, just as interesting and exciting as its predecessor. I brought greetings from Canada at the opening reception and also gave a paper titled, 'In Canada, Must Engineers be Men?', which outlined the very slow growth in the number of women practicing engineering in Canada. ICWES-III attracted 275 delegates from 35 coun-

tries. Industrial tours were an integral part of the programme and small groups of women engineers were given technical tours of FIAT and OLIVETTI. Many of the participants at the Turin conference had been in Cambridge four years before and there were joyful reunions" (Letter to Monique Frize, March 2021).

Rosemary West, "An Opinion of the Purpose and Achievements of ICWES-III":

"This conference was attended by delegates from more than 30 nations from all over the world: nothing unusual for an international congress except that nearly all these delegates were women, highly qualified in some branch of science or technology. We met both formally and informally with many opportunities to discuss the world's technological development, the contributions made by women in particular and the problems, some of them remarkably similar facing the nations which can be broadly grouped into under-developed or developing and, dare I say it, over-developed. Also, it was inevitable that we should discuss the part played by women in the scientific life of their country, contrasting the comparatively large part played by women in the Soviet Union, Poland and Czechoslovakia with the pitifully small part in the rest of the world. So many speakers mentioned that their country's society had once been more or less matriarchal but that women's position had then degenerated over a period of centuries and was once more improving, that it was tempting to extrapolate this improvement into the future and hope that one day real equality throughout all strata of life would be so much taken for granted as to be no longer a subject needing discussion."

"Actual progress in this direction since our last international conference in Cambridge appears to be small, however. This congress has amply demonstrated that some women can make it to the top. The properties of the brain, it would appear, are not sex-linked, and yet from many sets of statistics there emerges a picture of female brains being employed below their capacities in the less important jobs, while their physical abilities are employed to the full in mundane· but necessary functions like satisfying hungry appetites and keeping homes dean and tidy. The question arises then: are not many women quite content to serve mankind in whatever capacity is necessary? And should more of them be encouraged into scientific disciplines, unless they have special aptitude, when at the present time in the United States, Canada, and the United Kingdom there is a surplus of engineers and scientists, making it harder for young graduates of either sex to find suitable employment? It seems to me that the answer must lie in more consideration of each person as an individual and setting aside past generalities about what is man's work and what is woman's work so that they each have a chance to fulfil themselves in any chosen career."

"There can be little doubt that the true purpose of this congress is the exchange of scientific knowledge on a global scale. The spoken word makes more impact than the written word and an illustrated talk followed by immediate discussion makes more impact still. A congress is, so far, one of the best ways of effective communication of this sort, notwithstanding the simultaneous translations needed to overcome the language barrier and the complexity of arrangements needed to assemble so widely scattered a population. Parallel research is being initiated in various parts of the world and needs the benefit of being well communicated to be most effective. Furthermore, ideas and techniques developed in one particular discipline can find applications in other disciplines as, for example, the many uses to which computer technology has already been put. A broadly based scientific conference such as ours provides an excellent opportunity for good ideas to spread and be useful elsewhere."

"All this has been made possible by the sponsorship and support of a number of organizations, many of which are Italian, and to all of them we are greatly indebted."

"The delightfully unhurried Italian approach to me and the excellence of their cuisine contributed a great deal to the pleasure and general sense of well-being of delegates. Unfortunately, time is not elastic and there were difficulties in finding enough time for the adequate presentation of a hundred or so papers without, some of them having to be unduly compressed and thereby losing some of their impact. A possible solution to shortage of time was the suggestions that some of the technological sessions be held simultaneously in nearby lecture theatres."

"To complement the lecture sessions, we were entertained by a number of local manufacturers and invited to visit their plant. One of the highlights of our stay in Turin was undoubtedly the visit to the Fiat Mirafiori car factory, one of 12 Fiat factories in the environs of Turin and the most excellent lunch they gave us afterwards in the automobile museum. It was novel to me and very interesting to travel through machine shops and past assembly lines without exertion, in coaches especially provided for that purpose."

"The Alps are always interesting, but more so on this occasion because of our visit to the hydroelectric generating plant of ENEL, in a tunnel into the Italian side of the mountains at Venalzio; followed, after a superlative lunch, by a visit to the largely man-made lake of Moncenisio in the French Alps from which the water power is derived."

"It was delightful too, to be taken back in the past by a visit to Turin's seventeenth century baroque Teatro Carignano, still decorated in the original style of red and gold, and hear music by Vivaldi much as it might have sounded three centuries ago" (West 1971, pp. 22–23).

Chapter 5
ICWES-IV, Cracow, Poland, September 8–13, 1975

Technical Theme, "New Techniques in the Service of Mankind", and sociological theme, "Contemporary Sociological Problems of Women Engineers and Scientists"

The patron of the conference was the Minister of Science, Higher Education, and Technology, Sylwester Kaliski. The Organizers were the Central Technical Organization and the National Council of Polish Women. The arrangements were made by the Center for Progress in Technology—NOT Warsaw, NOT Cracow Division.

Izabela Zachwatowicz was the Chair of this fourth conference (see Appendix 1). A unique aspect was the availability of simultaneous translation in four languages for the plenary sessions: English, French, Russian, and Polish; for sessions, translation was provided from French, Russian, Spanish, and Polish into English.

Executive Committee of the Conference

- Stefan Farjaszewski, president of the Executive Committee/Polish Union of Civil Engineers
- Izabela Zachwatowicz, president of the Co-ordination Team/Association of Polish Architects
- Anna Sawicz, president of the Organization Team/Polish Association of Chemical Engineers
- Jacek Wawrynkiewicz, president of the Host Committee in Cracow/Polish Association of Metallurgical Industry Engineers

M. Frize et al., *Women's Contribution to Science and Technology through ICWES Conferences*, Women in Engineering and Science,
https://doi.org/10.1007/978-3-031-33800-7_5

- Wojciech Podwapinski, secretary of the Executive Committee/Polish Association of Mechanical Engineer/the Center of Progress in Technology

The opening keynote was given by Isabel Hardwich, president of the International Continuity Committee (ICC).

A welcome speech was given by Jan Kaczmarek (scientific secretary of the Polish Academy of Sciences and president of the Chief Council of the Central Technical Organization); Maria Milczarek (president of the National Council of Polish Women); Jerzy Pekala (president of the town of Cracow); Roman Ney (vice minister of Science, Higher Education, and Technology); and a lecture by Jan Szczepański (vice president of the Polish Academy of Sciences), on the Humanization of Technology Sciences.

The opening session was followed by a Polish fashion show!

Participants

The conference attracted 618 participants (excluding accompanying persons), of which 470 were from Poland and 148 from 29 other countries. The out-of-country proportion was 24% (see Appendix 3). Some regular attendees were as follows: Mahin Rahmani (Iran), Anna E. Amour, and Dorothy Mizoguchi (Japan); they had attended all first three conferences; Dormer Ellis (Canada) attended the last two and Nicole Bécarud (France) attended ICWES-II; she also attended ICWES-IV with eight other colleagues. Nicole was to propose France for the next conference at the Business Meeting. From the United Kingdom, Hilda Blount, Rose Winslade, Isabel Hardwich, Elizabeth Laverick, and Marjorie Bell were in Poland along with nine other women. From the United States, Carolyn Phillips, Isabel French, Aileen Cavanagh, and Beatrice Hicks are recurring names. Kathleen Harer, who became a SWE president in 1987, attended her first ICWES in Poland, and several more after this. There were 45 women from the USA attending ICWES-IV, the largest delegation from participating countries.

Another significant aspect of ICWES-IV was that it was held for the first time in a country considered "behind the Iron Curtain"; some delegates from Western nations, in particular Kathleen Harer and Carolyn Phillips, mentioned having met with the consular service in Warsaw to obtain information on what they could do and not do. A notable example was: Do not change money in the street. This appears in some of the testimonials from women who attended this conference (see section "Testimonials" below).

The two main themes of ICWES-IV were "New Techniques in the Service of Mankind" and "Contemporary Sociological Problems of Women Engineers and Scientists."

Elisabeth Laverick, in a presentation she gave later (ICWES-X, Complete Proceedings, p. 10), explained that the Opening Address was delivered by the Scientific Secretary of the Polish Academy of Sciences and President of the Council of the Polish Federation of Engineering Associations, Professor Jan Kaczmarek. He told us that in Poland 'women accounted for 46% of the labor force. They constituted 39% of manpower with University education, 54% with secondary school vocational education, and 70% with general education; 50% of medical doctors were women, 91% dentists were women, and 83% pharmacists were women. They account for 31% of the University teaching staff. They are active in light industry, food and agricultural industry, as well as in printing, electro-technical, and chemical industries. They play a more and more important role in the scientific life of the country. They are awarded about 25% of all university degrees'. Not surprisingly, 50 of the 106 papers were given by Poles (Laverick, in ICWES-X, Complete Proceedings, p. 10).

The editors of the ICWES-IV Proceedings were Izabela Zachwatowicz, Maria Grabszewska, Irena Kotela, and Maria Baczak.

ICWES-IV, S&T Papers

In the introduction of the ICWES-IV Proceedings, one can read the motivation for the main scientific topic of the conference:

> "'New Technique in the Service of Mankind' refers to almost all branches of today's economy and industry, thus establishing common ground for all the contributions and discussions of the conference which should therefore concentrate on the relationship between the achievements of technology and the welfare of all nations." (ICWES-IV, Complete Proceedings, p. 3)

To provide some context, the world population was rapidly growing at the time of the conference. It was 4 billion in 1975, 10 years after the first ICWES, when the population of the world was 3.3 billion. Everyone seemed concerned. It was also a time of rapid economic growth, industrialization, and development of computer applications.

As Mary Ellen Russell put it in the introduction of her presentation called "Satellites and the Geodynamics Project in Global Land Use Planning" (ICWES-IV, Complete Proceedings, pp. 383–396):[1]

> "Global land use planning is basic for all nations […] The earth's resources are limited. How can we [feed], house and clothe the world's present population and yet provide for future generations? Planning is essential." (ICWES-IV, Complete Proceedings, p. 386)

[1] Mary Ellen Russel introduced in this paper early examples of the Landsat program. According to Wikipedia ("Landsat Program" n.d.), "The Landsat program is the longest-running enterprise for acquisition of satellite imagery of Earth. It is a joint NASA/USGS [the U.S. Geological Survey (USGS)] program. On 23 July 1972, the Earth Resources Technology Satellite was launched. This was eventually renamed to Landsat 1 in 1975. The most recent Landsat 9 was launched on 27 September 2021." The images that Landsat provided constitute important resources for global research and applications in agriculture, cartography, geology, forestry, regional planning, surveillance, and education.

The 82 speakers at ICWES-IV S&T sessions were all women, among which 51 were from Poland, 12 from the USA, one from Latin America (Uruguay), six from Western Europe, two from Eastern Europe (Hungary) besides the host country, one from the Middle East and North Africa (Iran), six from Central Asia, one from Southeast Asia, and two from the United Kingdom, Australia, and New Zealand. Three speakers had two papers each.

Much in continuity with previous conferences, which put an emphasis on impacts on the populations, the five S&T sessions[2] (sessions 2–6) of ICWES-IV address various physiological human needs, and encourage interdisciplinary and international collaborations. The organizers were convinced that "Science and technology should serve mankind in satisfying its needs as well as increasing better and happier environment" (ICWES-IV, Complete Proceedings, p. 3).

The organizers specified in the introduction of the proceedings (ICWES-IV, Complete Proceedings, p. 6) that the languages of the conference would be English, French, Russian, and Polish, with complete simultaneous translation for the plenary sessions and English translation for the papers in another language. Papers in English, French, or Russian were accepted. A total of 74 presentations were listed in the conference schedule for the technical part, but the ICWES-IV, Complete Proceedings, contains only 36 papers.

Strikingly, the authors of the technical papers at ICWES-IV had interesting demographic particularities. From the 22 biographies showing the year of graduation in a first diploma at the university level, 80% of the authors (17) obtained their degree between the end of WWII and 1960. Also, among the 36 papers, 28 were written in English and 8 in French. All of the French language papers (except one) were presented by Polish participants. This unusual number of French papers can be explained by the recent history of Poland. In the nineteenth century and until the Second World War, knowledge of the French language and culture was the indispensable element in the education of the social elite of noble intellectuals in Poland: This was an element that served them as a professional instrument or as a sign of social prestige. The French language enjoyed a fairly high reputation in Poland until the 1970s, by culture disseminated through songs, books, and films. This language gradually lost ground in Poland (Ziolkowski 2004/2003).

As for the previous conferences, the ICWES-IV technical sessions provide a very interesting glimpse of many S&T developments happening in research centers and industries worldwide, notably in the host country (see Table 5.1). Such developments include not only technical innovation, but also the expansion of applications to sustain human life and even changes in the employment structure.

Recurrent topics since ICWES-I were present in ICWES-IV sessions, such as shelter and urbanization (Section 2), nutrition (Section 3), clothing (Section 4), and energy (Section 6). Automation and computer science, already introduced in

[2] In fact, a separate session on transportation had been announced in the call for papers (ICWES-IV, Complete Proceedings, p. 4); it was part of session 2 in the final schedule. Also, in the Proceedings, the word "section" is sometimes used to designate a "session" of presentations related to a same topic.

Table 5.1 Distribution of S&T sessions and papers in ICWES-IV, Complete Proceedings[a]

ICWES-IV S&T sessions	Session titles	# of papers expected	# of papers in proceedings	# of abstracts in proceedings
2	Spatial Arrangement of Human Environment	25	17	0
3	Problems of Nutrition	13	6	0
4	Material Supply and Equipment Problems	11	3	0
5	Automation and Computer Science Problems	12	4	0
6	Raw Materials and Power Engineering	13	6	0
	Total	*74*	*36*	0

[a]An additional section, "Transport. New means of transport. Motorization and roads. Airways and waterways," appears as Section 5 in a list at the beginning of the proceedings (ICWES-IV, Complete Proceedings, p. 4) but is not found elsewhere

ICWES-III, returned in Section 5. In ICWES-IV, these topics were adapted to preoccupations for the 1970s and to the host country's situation. This was the case, for example, regarding the preoccupation to create computer-based information systems, due to the explosion of collected data.

A Glimpse of ICWES-IV's S&T Content

Session 2, *Spatial Arrangement of Human Environment*, was very large, with 25 presentations scheduled for 3 days (September 9–12, 1975). Among the 17 papers that are in the proceedings for this session, three major topics can be found: (1) safeguarding cultural, historical, and architectural assets (2 papers from Italian participants); (2) territory development with 6 papers from Polish participants (with subjects such as water resources compatible with economic development in the Carpathians, photo inventory and photogrammetry for road systems survey, planning of a highway network connecting Poland with the rest of Europe, and urban road system that would preserve the environment); (3) the impact of the industrialization on our environment, with 3 papers from US participants who spoke of controlling chemical contaminants in wastewater and in the air around factories.

One of the authors is Carolyn Phillips, who was a hygienist at Shell Oil Co. in New Orleans, Louisiana, and a consultant on occupational health problems to refineries and chemical plants in the southern region of the United States. In her paper called "Rationale for Safety and Health Standards," she wrote: "Each year in the USA about 250,000 new chemicals are added to the two million already registered and above 500 new chemicals are used commercially" (ICWES-IV, Complete Proceedings p. 413). She was particularly concerned by the need to intensify research and control programs focusing on the problems of pregnant workers (ICWES-IV, Complete Proceedings, p. 419).

Mahin Rahmani, a professor at the Tehran University, was a participant in all ICWES conferences from the very first one in New York to ICWES-XII in 2002 in Ottawa. She set the tone on the second day of Session 2 with a paper titled "The Future Environmental Problems Associated to Land Use and Urbanism Through Industrial Expansions in Developing Countries" (ICWES-IV, Complete Proceedings, pp. 334–348). She started her talk with very strong statements: "Population monster, as has been termed by Norman E. Borlaug, the 1970 Nobel Prize Winner, should be viewed from all its angles of attack on the environment and on the people, such as food, education, clothing, transportation, employment, sanitation, population congestion, hygiene and other concomitant economic and social implications duly integrated and scrutinized. [...] Rapid population growth is usually accompanied by an important urban development, accelerated industrialization, excess production and consumption and naturally serious environmental complexities" (ICWES-IV, Complete Proceedings, p. 337). She offered eight very strong and surprisingly contemporary recommendations in her conclusion.[3]

In tune with the times, many authors in Section 2 mentioned the creation of international bodies to reach international collaboration and to set up universal standards. Mahin Rahmani, for example, lauded the efforts "to establish the international organization to be called 'The World Heritage Trust' [...] which primarily would safeguard the aesthetic needs of mankind and preserve and protect the relics of the past not [only] from the environmental and climatic aggressions, but also to protect them from man's modernized activities" (ICWES-IV, Complete Proceedings, p. 346). This organization is called UNESCO World Heritage today.

Another example can be found in the paper "Rationale for Safety and Health Standards," by Carolyn F. Phillips (ICWES-IV, Complete Proceedings, pp. 410–421). Phillips informed the audience that the Occupational Safety and Health (OSH, USA) Act was signed into federal law in the USA, on December 29, 1970, "to

[3] "After reviewing the entire aspect of the problems treated in the above, the following conclusions could be derived:

1. Population growth should be decelerated.
2. The planet's quantum resources should be used with utmost care, thrift and scrutiny.
3. Particular emphasis has to be put on the use and consumption of fossil fuels, particularly coal and petroleum. The development and application of nuclear energy should be given due considerations and the probable hazards associated to its use has to be properly considered.
4. The use and application of the biocides, germicides, etc. have to be reconsidered.
5. Thermal pollution of the planet, whenever possible, has to be prevented.
6. Excessive use of communication [transportation?] facilities such as cars, planes, etc. has to be discontinued.
7. Man's aesthetic and recreational requirements [have] to be given due weight and consideration. This will imply perfect maintenance and preservation of historic monuments, natural parks, pasture lands, forests, etc.
8. The preservation of wildlife has to be given due thought and consideration." (ICWES-IV, Complete Proceedings, p. 347)

assure safe and healthful working conditions for the nation's working men and women and to preserve our human resources" (ICWES-IV, Complete Proceedings, p. 416). OSH got involved internationally, particularly with two important organizations in the field of occupational health and disease control: the International Labor Organization, the oldest remaining unit of the World War I League of Nations (1919), and the World Health Organization, an agency of the United Nations that still exists today.

Papers presented in Session 3, *Problems of Nutrition (Agriculture, Forestry, Horticulture, Food Industry)*, at ICWES-IV explored specific problems and solutions.

In a well-documented paper called "Hunger and Malnutrition: A Psychological, Physiological and Sociological Survey" (ICWES-IV, Complete Proceedings, pp. 285–294), Lillian G. Murad-Bablanian, the second president of the Society of Women Engineers (1952–1953), and currently science educator in the Oyster Bay East Norwich School District, Nassau County, NY, postulated that "nutrition is one of the most pervasive factors influencing growth, mental and physical development, health, and social stability and progress" (ICWES-IV, Complete Proceedings, p. 298). She presented the situation of food availability and the alarming possibility that the earth might be running out of food in the future. She examined in detail the effects of malnutrition on the human body, and the effects of starvation on entire populations due to famine.

Antonina Rumińska (Poland) had worked since 1939 in the field of research in biology and agrotechnics, and cultivation of both medical and condiment plants. She is coauthor of papers on a variety of medical plants, and author of many scientific papers, scripts, and manuals; for example, a book was published in 1974 under the title "7 'Medical Plants'—Bases of Biology and Agrotechnics." She presented a fascinating paper called "Plant Condiment in Nutrition" (ICWES-IV, Complete Proceedings, pp. 321–333), where she shared with the conference participants her vast experience on plants and their physiological role.

Among solutions to nutritional problems, the author Kamalaveni Chandrasekhar from India, in "A New, Effective and Simple Technique for Enhancing Production of Calorierich Protein Food—Fish," saw the implementation of local fisheries as a solution to overcome India's and Bangladesh's hunger crisis (ICWES-IV, Complete Proceedings, pp. 321–333). Chandrasekhar is a scholar who built up a Neuroendocrinological Research Unit at the Institute of Medical Sciences at Banaras Hindu University. She had several research scholars working under her guidance for a PhD degree. The Polish author Jadwiga Grabczewska looked into the feasibility to grow fruit rose hips at home to overcome vitamin C supply problems (ICWES-IV, Complete Proceedings, pp. 269–284). Judit Thorma Asbót, a technological development engineer who, among other accomplishments, designed the experimental tobacco factory of the Hungarian Tobacco Industry's Research Institute, saw in quick-frozen food and meals a paramount solution to save time for working women ("The Production of Quick-Frozen Foods in the Service of Households and Public Supply," ICWES-IV, Complete Proceedings, pp. 303–320).

Various specialized industrial applications were reported in Section 4 *Material Supply and Equipment Problems*. Among them, there was a paper titled "Application of Ultrasonic Energy to the Textile Processes," written by Wanda Wiśniewska, a Polish specialist in raw materials, resistance, and behavior of fibers and products, in both sonic and ultrasonic fields. Her paper deals with using ultrasonic energy for textile production or cleaning processes (ICWES-IV, Complete Proceedings, pp. 247–256). In the same section, Ofelia Vito Bulaong and Asuncion B. Kalalo presented a paper called "Rationalization Process for the Philippine Apparel Textile Industry" (ICWES-IV, Complete Proceedings, pp. 217–228). Also, the proceedings include a paper from an assistant in the Institute of Topography of Politecnico di Milano in Italy, Baj-Agnoletto, on using photogrammetry of X-ray images to get 3D representations in medical applications ("Medical Application of X-Ray Photogrammetry," ICWES-IV, Complete Proceedings, pp. 229–246). Indeed, the same technology is still used today.

The four papers available in Section 5, *Automation and Computer Science Problems,* provide an overview of interest in computer science and engineering in the 1970s. Rashmi Rastogi, who worked in the Computer Department of Bhabha Atomic Research Centre, Bombay, India, described elements chosen for the design of a 16-bit data processor ("Design of a 16-Bit Data Processor," ICWES-IV, Complete Proceedings, pp. 158–178). Some of the elements she reported are well known today, such as accumulators, memories, BUS, CPU, etc. Hanka Wanda Sobczak Chryssafopoulos, from Rio Grande do Sul, Brazil, described ways to transmit and store information with remote sensing aerial systems (aerial photographs, satellite imagery, and spectral data analyses). This civil engineer provided many examples of applications that would help humans, such as detection of underground or surface fresh water, maps of soils for agriculture, land use and urban planning, pollution detection, and so on ("Remote Sensing and its Application to Human Needs," ICWES-IV, Complete Proceedings, pp. 194–206).

With the proliferation of computers and applications in the 1970s, the amount of information to conserve and analyze was exploding. It was becoming essential to create computer-based information systems, a subject presented by two Indian scientific officers in the Bhabha Atomic Research Centre, Kamala Balaraman and Roshan J. Begum, in "Information Retrieval Systems for Science and Technology in India" (ICWES-IV, Complete Proceedings, pp. 207–216). They provided examples of international information systems, some still existing today such as the International Nuclear Information System (INIS), established in 1971, that publishes information on the peaceful uses of nuclear science and technology, or the World Information System for Science and Technology (UNISIST), a network created under the auspice of UNESCO in 1971.

In Section 6, *Raw Materials and Power Engineering*, the 13 papers appearing in the final schedule concerned natural resources and energy production. Four Polish papers on how to process coal made an interesting contribution, dealing with its solid or gas subproduct forms, and on how to decrease its environmental impact.

Of the six full papers in the proceedings for Session 6, two were about nuclear energy. The first one, "Nuclear Power in a Mixed Fuel Economy" (ICWES-IV,

Complete Proceedings, pp. 84–93), was from J. Cicely Thompson, past president of WES and chairperson of ICWES-II. Thompson spent all her working life on power station design and construction. Her paper promoted nuclear power as a viable alternative energy resource: "if it is desired to maintain or increase the present use of energy, there are, for the time being, only two options open: either to develop the more inaccessible coal, oil and gas resources or to turn to nuclear reactors and in particular fast reactors" (ICWES-IV, Complete Proceedings, p. 93). She introduced different types of reactors and dealt with the safety and environmental aspects of nuclear power. As a nice follow-up to this presentation, Betty J. Miller (Engineer in the Fuel Handling Engineering Group, Westinghouse Hanford Company, Richland, Washington, USA), in her paper "The Test Reactor—Basic to US Breeder Reactor Development" (ICWES-IV, Complete Proceedings, pp. 70–84), presented a description of a breeder reactor that would be reliable and safe.

Possibly the most exciting paper of this session was written by a senior engineer in the Boeing Aerospace Company, Eliza Imislawa Wojtaszek-Piatkowski, who had completed a Master of Science in Electrical Science Degree with a specialization in microwaves and bioengineering in 1974. Her paper "New Uses Safety Actors for Microwave Energy in the Service of Mankind" (ICWES-IV, Complete Proceedings, pp. 94–108) surveyed different microwave energy potential uses and stressed the need for cooperation between engineers and scientists throughout the world in ensuring its safe usage. She pointed out that the first developments of microwave technology were for radars during WWII. Thirty years later, in the period of ICWES-IV, she presented two recently developed applications in the medical field that are still used today: the therapeutic use of electromagnetic energy at high frequencies (microwave diathermy), to relieve arthritics and muscle trauma, and the diagnostic tools based on microwave radiometry that can detect pathological conditions such as cancer, tumors, and circulatory pathologies. She also spoke of power transmission by microwaves (propagation of electromagnetic waves in the microwave frequency range as the carrier of wireless power). Wojtaszek-Piatkowski foresaw, for example, the transmission to earth of solar energy collected in satellites; but the feasibility of this technology is still under evaluation today (Wang and Lu 2016). However, she could not predict, in 1975, that using microwave power transmission would still be very useful, and a popular research subject, 40 years later, for portable electronic devices. Was she dreaming of the intensive usage of microwave technologies in the fields of telecommunication, and radars, that we have today, such as Wi-Fi, Bluetooth, cellular phones, and satellite communications?

Women Scientists' and Engineers' Contributions to S&T

Zsuzsa Szentgyörgyi's paper presented in Section 5 (ICWES-IV, Complete Proceedings, pp. 179–193) stands out with a subject she titled "Computer Technique and Social Responsibilities." This Hungarian scientist from the Computer and Automation Institute of the Hungarian Academy of Sciences (see Appendix 1) wrote several scientific papers and some papers on the professional status and career of Hungarian women engineers.

She spoke about the emergence of computing technologies, including production of hardware and software, as a new branch of the industry. She underlined the benefits this may bring. However, most of her paper was concentrated on the contradictions and dangers that might happen with the widespread use of computers. Among other things, she was preoccupied on the creation of a large number of new jobs related to the development and production of hardware and software and the effect it might have—or not—on the situation of women in S&T. She insisted that this change in the employment structure due to computer progress was not preventing the reproduction of old patters for women:

> "If we examine the hierarchy of employees in the computing technique, practically the same distribution can be observed in function of sexes than in the other, traditional, 'white collars' branches of industry. On the summits (systems engineers, chiefs, senior programmers), the ratio of males is determinant. The lower we go in ranks and salaries, the higher the number of females [there] is; and finally, at the lowest levels (card-punchers, junior programmers) almost [only] women can be found... The restoration of the 'regular' structure of employment in a brand-new branch of economy is not 'natural' at all. As a proof against this phenomenon, I mention here the names of creative women scientists as Grace Hopper, Jane Sammet, or the Hungarian Rózsa, 'Rose' Péter who are among the first of the best in the computing field" (ICWES-IV, Complete Proceedings, p. 192).

ICWES-IV, Gender Papers

In the introduction of the Conference Proceedings (ICWES-IV, Complete Proceedings, pp. 3–22), the Organizing Committee took the opportunity to underline that this meeting would take place in the year which had been recently proclaimed by the United Nations as International Women's Year. The fourth ICWES meeting would therefore aim to present the achievements of women engineers in science and technology and to discuss their "social function position" in the world of science.

Both of the conference's themes called on participants to concentrate on the relationship between the achievements of technology and the welfare of all nations. Science and technology should indeed serve mankind in satisfying its needs and help develop better and happier environments. According to the Committee, the "complex problems of psychophysical human needs" compelled the world of science and technology to acknowledge the "full significance of interdisciplinary collaboration and correlation with biological sciences, the humanities, and others in determining conditions of human life." What is more, concluded the Committee, "[t]he contribution of women to the successful solution of these problems seemed to be of paramount importance" (ICWES-IV, Complete Proceedings, p. 3).

What is interesting about the committee's perspectives regarding women's role in science and technology is their close alignment with the three official aims of ICWES (see Section Business Meeting), which were put forward and adopted at the ICWES-IV delegates meeting.

Table 5.2 Distribution of social and gender sessions and papers in ICWES-IV, Complete Proceedings

Social and gender sessions	Session titles	# of papers expected	# of papers in proceedings	# of abstracts without papers in proceedings
1	Contemporary Sociological Problems of Women Engineers and Scientists. Role of Women Engineers and Scientists in the Development of Science and Technology	19	8	1

As Table 5.2 indicates, 19 papers were expected under Section 1 of the ICWES-IV conference, titled "Contemporary Sociological Problems of Women Engineers and Scientists. Role of Women Engineers and Scientists in the Development of Science and Technology." Eight full papers and one abstract are present in the conference proceedings. All speakers but two were female (18 vs. 1 man and 1 unknown). Not surprisingly, a large number (8) came from Poland, 4 came from the United States, 1 from Latin America (Mexico), 2 from Western Europe (France and Italy), 1 from Eastern Europe (Bulgaria), 3 from Central Asia (India), and 1 from the United Kingdom. From the nine papers and abstracts found in the conference proceedings, one was written in French (from Poland), and the rest in English.

Roshan J. Begum and Kamala Balaraman, "Contemporary Sociological Problems of Women Engineers and Scientists in India"

Two papers were presented by speakers from India, who had participated to ICWES-III 3 years earlier. One was particularly enriching as it provided important insights into the general history of women in India, before moving on to a more focused discussion on the various contexts and factors underscoring the presence and role of women scientists and engineers throughout that history. Titled "Contemporary Sociological Problems of Women Engineers and Scientists in India" (ICWES-IV, Complete Proceedings, pp. 594–603), the paper was cowritten by Roshan J. Begum and Kamala Balaraman, two scientific officers working at the Bhabha Atomic Research Centre, located in Bombay. Women's education, they first reminded us, was almost nonexistent in India until the early nineteenth century; in their view, the greatest impetus to women's education came from the National Freedom Movement, in response to the "clarion call of the Father of the Nation, Mahatma Gandhi" (ibid., p. 599). Thousands of women then came out of seclusion, and once awakened, they did not fall back into apathy. On the contrary, they were still on the march, occupying high positions, with hardly any fields escaping their interest. During the last two decades, state the authors, India has been engaged in the formidable task of transforming an ancient tradition-bound people into a modern nation, in which science and technology will play a major role. The authors proudly recognize that Indian women were active participants in these efforts, contributing to the research field and to the advancement of science.

Sadly, however, women could not currently take full advantage of the various opportunities offered to them because of the prevailing socioeconomic conditions, and because they did not yet enjoy equality of status with men. Indeed, sons were still considered more important than daughters, and not working was viewed by many women of means as a symbol of status.

To celebrate International Women's Year, the Indian Women's Scientists' Association undertook a survey to assess the working conditions of India's women scientists and engineers. A preliminary study of 182 cases produced some very interesting findings. For example, more than 80% of the women surveyed had studied science because of their keen interest in the subject and because of career ambition. However, the survey revealed that these women received little help or support from their employers with respect to the challenging task of combining their demanding careers with a satisfactory family life.

The authors also reported on the survey's important findings regarding job satisfaction, opportunities, and promotion prospects, which were obtained through a comparison with the women's male counterparts. What transpired was that job satisfaction and equality with men diminished with each increasing year. While young girls were likely to have a status comparable to that of boys their own age and qualification, the discrepancies increased dramatically as they grew older. Promotion prospects were extremely discouraging at all stages. In fact, reaching a senior position was very difficult for any woman engineer or scientist.

Overall, mothers responsible for childcare during their absence from work faced the biggest challenge. Nursery schools were, for example, either lacking or totally inadequate. For their part, the coauthors suggested a revival of the joint family system where elderly relatives took care of the children; interestingly, they also proposed a model based on the USSR's, where the state provided a wide range of facilities to allow mothers to work while raising a family.

In the end, Begum and Balaraman concluded their paper on this somber note:

> "The social expectation—conscious or unconscious—is that women should remain in a relatively subordinate status and women, in general, are not yet aggressive or ambitious enough to register their protests. Perhaps co-education from the very beginning may help to put women on an equal footing with men and uproot the theory of mental and social inferiority complex in women" (ICWES-IV, Complete Proceedings, p. 601).

Halina Mortimer-Szymczak, "Feminine Occupational Work in a Socialist State. Selected Problems"

Halina Mortimer-Szymczak, an economist, and at the time Extraordinary Professor at Łódź University and Head of the Division of Planning and Economic Policy, delivered a paper titled "Feminine Occupational Work in a Socialist State. Selected

Problems" (ICWES-IV, Complete Proceedings, pp. 581–593). Taking Poland as an example, and more specifically Łódź, she assessed the employment rate of women in relation to socioeconomic factors. For the author, drawing on substantial statistical data, a common trait in socialist states was the "continuous and systematic increase of women's occupational activity which is the result of both political and socio-economic transformations" (ibid., p. 586).

According to Mortimer-Szymczak, women's occupational work has "its ceiling which should not be overpassed" (ICWES-IV, Complete Proceedings, p. 587), since they also have important familial duties to fulfill (though these too should be capped at some point). The idea then, both in politics and in science, was to find the balance, for women, between professional work, education, and family life and to come up with a corresponding model that would be grounded in a couple's shared responsibilities in their everyday life.

Mortimer-Szymczak also wondered about the reasons behind the clear division between typically masculine and feminine work sectors; the wage gap; the search for work; and the professional skills specific to women and how to improve them.

Though she reported that, in 1970, more or less half of Polish women were employed (outside agriculture), representing 40% of the overall workforce, Mortimer-Szymczak also underlined the gaps between larger and smaller agglomerations, or between sectors of activity. Solutions could be: (1) allocating the workforce more equally around the country; (2) making a distinction between the branches of economy; (3) preventing industrial monocultures; (4) expanding educational, cultural, and social centers more widely; and (5) aligning education and formation more closely with employment demands.

An important consequence of an important feminine workforce in large agglomerations was their "feminization," meaning that there were more women than men in big cities (e.g., 105 to 100, in Poland; 115 to 100 in Łódź). One problematic aspect of this phenomenon was that it often took shape through the dominance of the textile industry, which massively employed young and inexperienced women and set a standard of low incomes and qualifications. If the problems of gender-typical work division and lower incomes for women could be solved to some extent through a wider interest and access to polytechnical programs, employers also had important responsibilities in terms of work atmosphere and conditions: "there is a lack of good will or of willingness to employ women at posts suitable for them" (ICWES-IV, Complete Proceedings, p. 592).

By way of conclusion, Mortimer-Szymczak endorsed the idea that a socialist state could play a determining role in the education of children, namely, "through the development of the network of schools-dispensaries" (ICWES-IV, Complete Proceedings, p. 592). She also advocated for the inclusion of a specific section on women's employment within the Council of Mutual Economic Assistance (CMEA).[4]

[4] Also known as Comecon, economic organization (1949–1991), led by the USSR, and comprised the Eastern Bloc and other socialist states around the world; created in response to the Marshall Plan and the OEEC (later OECD) ("Comecon" n.d.).

Danuta Mischke, "Women at the Higher Technical Schools in Cracow in the Thirty Years Period of Poland"

Danuta Mischke, working at the Cracow High Polytechnical School, Institute of Construction Mechanics in the Land Engineering Department, spoke about "Women at the Higher Technical Schools in Cracow in the Thirty Years Period of Poland" (ICWES-IV, Complete Proceedings, pp. 668–685). She first reviewed the history of three main technical schools in Cracow, then addressed the presence of women both as students and teachers.

Though there were already talks of opening a technical school in Cracow going as far back as the end of the seventeenth century, the Polytechnical Institute was founded in 1824, transformed into a Technical Academy in 1854, and closed in 1875. A short-lived Mining Academy was set up early in Kielce (1816–1826), but plans for a like-minded establishment only took shape a century later and then postponed until after the war. The Academy of Mining was founded in September 1919; 30 years later it had ten divisions. During Nazi occupation, professors of the Academy lectured at the Cracow Higher Technical Mining-Metallurgical-Surveying School, where clandestine education also took place. Moreover, though agriculture was an early concern in higher education, formal teaching was only set-up during the nineteenth century, first through a program of agriculture studies in the Philosophical Division of the Jagiellonian University and, in 1923, in a separate Agriculture Division. An independent Agriculture School was founded in 1953, with the creation of a Water Improvement Division 2 years later. The Cracow Higher Polytechnical School was established in 1945 (mostly at the instigation of academics from Lwów[5] who had survived the war. It was not connected to the Mining Academy and was composed of four Divisions (Architecture, Land Construction, Water Construction, and Mechanics), to which chemistry was added in 1966.

As Mischke showed, women only started teaching at higher technical schools after World War II. "It is understandable because of the fact that in the pre-war period the number of studying women was not a high one" (ICWES-IV, Complete Proceedings, p. 676). At the Mining-Metallurgical Academy, the first woman docent (more or less an associate professor) was appointed in 1950, the first full professor in 1953. In 1974, women made up just under 23% of the teaching staff; none worked in higher management. At the Academy of Agriculture, 40% of all teachers were women; several held higher administrative positions. In the case of the Cracow Higher Polytechnical School, the first woman to be appointed full professor took office in 1952. In 1974, women amounted to 23.3% of the teaching personnel.

Though higher studies resumed right after the city's liberation in early 1945, not many women reintegrated into schools since few of them had been in attendance before the war. Mischke argued that an increasing economic demand, first related to postwar reconstruction, encouraged many to enroll in technical schools. The first woman to enter the Mining Academy began her studies in 1921, as extramural. To

[5] Now Lviv, in Ukraine.

ensure the "prevention of feminine competition" (ICWES-IV, Complete Proceedings, p. 681), a *numerus clausus* was decreed, and a maximum of eight women in the first year was established (against 80 to 100 men). Women graduating from the Mining Academy amounted to less than 1% in 1946, then to 19% during the years 1972–1974 (though gaps were sizable between divisions). The first woman to be awarded a PhD graduated in 1949. Around 9% of new doctors were women in 1974; the same year women made up 38.5% of all habilitated doctors. The first female student in Agriculture finished her program in 1907; in 1952, 28.5% of graduates were women, going up to 42% in 1973–1974. The first PhD in agriculture bestowed to a woman was in 1912; the first doctorate delivered by the agriculture school to a woman was in 1948, and the first habilitation was in 1954. In 1974, women made up 36% of new doctors and 33% of habilitated doctors. At the Cracow Higher Polytechnical School, the first female doctor received her title in 1956.

Mischke also traced general trends in the technical education of women: greater interest for chemistry (viz., ceramics and nonferrous metals), botany, zoology, and construction. Least attractive to women were areas in mechanics. The author concluded that women, based on their qualifications and achievements, have gained recognition, reaching even positions of higher administration, "where they show much initiative and interest in the units under their respective management" (ICWES-IV, Complete Proceedings, p. 684).

Delegates Meeting

Minutes of this meeting are not featured in the conference proceedings but are cited later in Elizabeth Laverick's opening talk at ICWES-X in 1996a:

> "At the Delegates meeting, where 32 countries were represented, the Chairman put forward the aims of ICWES as: *Firstly*, to bring together professional women engineers and scientists worldwide. *Secondly* to encourage the use of technology for the betterment of life. *Thirdly* to consider how to increase the participation of women. These aims were adopted, and although slightly differently worded, are still the aims today. In discussing the host country for ICWES-V, it was recorded that: 'there was a general feeling that Japan would be an ideal venue'. (There were six Japanese delegates present at ICWES-IV). However, the Japanese representative regretted she was unable to commit the Society of Japanese Scientists. In the event no definite offer to host ICWES-V was forthcoming. After the conference an offer was received from France and accepted by postal ballot among the delegates" (Laverick, in ICWES-X, Complete Proceedings, p. 10).

Grants for WES Students to Attend ICWES-IV

> "The Caroline Haslett Memorial Trust has allocated £500 to provide a number of bursaries to assist in sending student and/or junior members of the Women's Engineering Society to attend the Fourth International Conference of Women Engineers and Scientists in Poland in September 1975."

"These bursaries were intended specifically for those students who held passes which entitled them to a reduction in air fares." ("ICWES-IV" 1974–1975, p. 18)

"A wonderful donation of £50.00 has been given anonymously to open a fund to assist WES members to attend the Conference, mentioning that further donations would be very welcome."

"The donor also suggested that WES members, who would much like to be in Cracow next September, but who are prevented from attending by security regulations or because of other activities associated with their work, might be particularly glad to contribute" ("ICWES-IV" 1975, p. 2).

"Questionnaire on financial assistance received by delegates—Interim report.

A total of 48 forms were returned from delegates from outside Poland; 32 delegates had received help from employers or trust funds, 10 for complete costs, 10 for registration fees, 10 for travel and/or accommodation costs, 4 for leave with pay; 16 delegates had paid all their expenses, some of them coming from as far afield as USA and Australia" (Hardwich 1976, p. 3).

Testimonials

Isabel H. Hardwich, from the "International Correspondent's Report to Council":

"The 16 delegates from UK included 15 WES members, who were accompanied by two friends. Six WES members from outside UK were also present. Apart from the Poles, about 30 of the foreigners were from Eastern Europe, so that we had a far better opportunity to meet women from behind the iron curtain than we had at previous ICWES."

"During the week over 100 papers were read in six parallel sessions and summaries of each of these sessions were given at the final plenary session. Three papers were presented by WES members from UK and two by WES members from outside UK."

"At the closure of the Conference, our President said some suitable words and presented a lovely bouquet of carnations to the Chairman and then a posy to each of the working members of the Conference Committee. They also gave a posy to me" (Hardwich 1975–1976, p. 6).

Lee Arnold, "Reflections on ICWES":

"ICWES-IV, in 1975 in Cracow, Poland, stands out for me as it was the first time I ever really tried to be understood in another language. High school German helped, as did a Berlitz Polish phrase book and tape. Several of us tried at least a few words. One day when we watched some Russian dignitaries parading to a factory we were supposed to visit but couldn't, it was SWE Past President Isabelle French who lured us off the street for 'lody' (ice cream), and another time I was able to obtain the desired cocktails for Bea Hicks and Georgine Poultney. Poland was behind the Iron Curtain at that time, so one had to wonder what it would be like. In reality, the most noticeable effect of their version of Communism was the high degree of efficiency and promptness in our tour organization. Equally impressive was the seeming lack of concealment and apparent similarity of ordinary lives to our own, though some attendees reported that they felt 'watched'" (Arnold 1996a, p. 9).

Kathleen Harer on Attending ICWES-IV in Cracow, Poland, 1984 and Others Over the Years:

"Poland was my first ICWES and my first trip overseas (living in Seattle, going to Canada as we often did, didn't count as foreign travel) and my first and only trip to a communist country. As the world has changed over the years (the fall of communism, the changes in Iran, the end of the war in Vietnam, etc.), a question is: how have they impacted our experiences in other countries, with women from other cultures? For example, I remember the Iranian women in Poland as very sophisticated, polished, stylish; seeing some of them again in Japan, I was impressed with the change in their dress, including wearing of headscarves, which one stopped wearing after a couple of days. I also remember my roommate in Poland walking out of one of the plenary sessions because the woman scientist from Vietnam was making some pretty harsh statements about the US. I was not so much offended by the statements as I was struck by the differences in perspectives among the attendees. Someone told the Vietnamese woman that she had apparently offended some people and I was told she was surprised about it. She was speaking about her reality. Interestingly, what impacted me the most from a geopolitical standpoint was the 'freedom' we had in Poland vis-a-vis what my security people had warned me about, such as expecting to be followed and possibly having our rooms searched while we weren't there. And seeing almost 500 Polish women engineers and scientists–I didn't know there could be that many in one country. Such naïveté on my part, but that was one positive aspect of attending these conferences and meeting women from throughout the world and from such varying cultures."

"Pre-conference and/or post-conference activities were sponsored by the conference coordinators in Poland, which is where I met some wonderful people that I kept in contact with between conferences. In Poland, a woman from India decided, after she got to Poland, to take the train to another communist country on the evening at the end of the conference but had a lot of difficulty getting a visa in Poland to go; she decided to go to the train station and try to get on regardless although I never heard if she was successful. I also remember some of the African women who attended Budapest were required to get a visa (US women were not) and to fly first through Belgium rather than directly to Budapest. The same happened for their trip home."

"After attending a number of ICWES conferences around the world, in addition to two in the US, ICWES-VII and ICWES-XVI) what I got from attending them was an appreciation for our similar life experiences as women engineers and scientists, in contrast with our varying cultural identities. We were able to find common ground and appreciate our differences. And each host committee presented us with experiences that highlighted their countries, leaving us with an appreciation and better understanding of other countries and the professional women in them" (email to Monique Frize, March 6, 2021).

Dormer Ellis:

"Although every ICWES had the same general format as earlier conferences, each one had some unique features. ICWES-IV was held in Poland, a country where the proportion of women in the engineering profession was much higher than in the countries where previous conferences had been held. I believe that about twenty per cent of Polish engineers were women, but no one seemed to know or care about this statistic. There was no association of women engineers in Poland and ICWES-IV was hosted by an engineering society to which both men and women belonged. Polish engineers could understand having an organization exclusively for chemical engineers or for electrical engineers, but *women* engineers? [emphasis original] The idea seemed as amusing to them as we would find an association of left-handed engineers or red-headed engineers or any restriction based on some irrelevant characteristic!"

"The fact that an ICWES was being held in a communist country affected registration. There were many attendees from the USSR and other communist countries but very few from the USA. I felt quite comfortable in a communist country because, many years before, I had cycled around Czechoslovakia without government officials ever doing me any harm. The few Americans who attended ICWES-IV were fearful in their belief that their hotel room was bugged and that they were being followed and watched. I knew them well and was constantly assuring them that no one had any special interest in them."

"At ICWES-IV, I brought greetings from Canada at the opening reception and I also chaired the sociological session. There were papers from Poland, Bulgaria, USA, etc. and lively discussions of their content. When I presented my paper, 'Widening the Horizons of Canadian Young Women', a Czech delegate took over the chairmanship. The sessions were held at the School of Mines where the facilities were excellent. One event that stands out in my mind was a well-organized visit made by a dozen selected delegates to a death camp where four million people died. The Polish woman who arranged this trip asked me if I was willing to participate in the visit and to read a few lines at a memorial service afterwards. Of course, I knew about the camps from newspaper accounts of their liberation and from radio broadcasts, but nothing could really prepare anyone for the reality of the site. It was an experience I will never forget. Afterwards, the ICWES group put flowers on a memorial and each in turn said a few sentences from a prepared statement in a different language—Polish, Czech, Hungarian, French, English, Spanish, Hindi, Russian, etc. It was a moving experience for all of us, but especially so for the delegate from Brazil as she was Jewish. The final evening of ICWES-IV was joyous as we sang and danced together and promised to meet again at the next ICWES" (letter to Monique Frize, March 2021).

Carolyn Phillips:

"Poland was a bit of an issue for visas, but not for US citizens. We had very careful instructions from our Consul General in Cracow about changing money in Poland due to the political situation. Communications between attendees outside the meeting rooms was often the best part of an ICWES, particularly in Cracow" (email to Monique Frize, March 2021).

Gwendolen Howard, "The Magic of Poland":

"For the eight delegates who went to Poland by a Magic of Poland Tour, the start was anything but Magical. Fog prevented us landing at Cracow, and we eventually came down on the closed airport at Warsaw to spend the next six hours in a succession of very bleak and uncomfortable airport lounges. However, we did get to Cracow for a late breakfast around 11.00 a.m. followed by sleep and a super lunch at 15.30 and the magic of Poland did come true."

"Our present to the Polish Women Engineers was a new book called 'Britain Observed' by Geoffrey Grigson, published by Phaidon. It had 149 illustrations including 31 in colour. 'This is a book of the Landscapes of England through the eyes of many artists. We bring it to you with greetings from the United Kingdom because we feel that pictures do not need translating. We hope that this book will encourage you to visit our Country to see it for yourselves…'"

"The Conference itself was rather ambitious in that delegates had a choice of six themes to follow. This meant that if you wanted to hear a lecture in another section, you needed to make a sharp dash over highly glazed and slippery floors."

"On looking back over the whole conference, the first question is: did it resolve anything? Did we get anywhere? The immediate answer must surely be: Yes, in that we had met behind the Iron Curtain in a country with a much higher percentage of women in Science and Engineering, to find that they had much the same problems as ourselves. The encouraging fact was that we had all talked together about technical problems without politics and prejudices, in our own language mainly. Incidentally how wonderful to go to another country and find English as the official conference tongue."

"The lovely old city of Cracow was a perfect place for the Conference, but I was glad that we were shown something of the marvelous way the Polish people have resurrected their country after the horrors of the war. It was right that in the midst of friendships old and new we should see the remains of Auschwitz; the leaders of delegations were all present when the Polish President of the Organization Team laid flowers on the Memorial Shrine just as the sun was setting; it was a very moving moment."

"I was glad that we were able to say 'Dziekuje, w uznaniu', thank you in appreciation... with flowers to every single member of the organizing committee at the close of the conference because they had shown us hospitality, friendliness and appreciation without stint."

"Thank you, Women Engineers of Poland, we enjoyed our visit" (Howard 1975–1976, p. 7).

Dorothy M. Cridland, "Fashion Show":

"A lighter note was introduced before lunch by a Fashion Show covering a display of suits and two-piece autumn outfits, mostly in wool or man-made fibres in plain colours or checked material, effectively displayed by six attractive young models. Poland is obviously becoming very fashion conscious, which was indicated by the inclusion of three young men showing very smart suits, sports and leisure wear. In all the exhibits one noticed good line and clear colour, but with the omission of extreme features which is typical of the general attitude, clothes like everything else have to be severely practical as they cannot readily be discarded but must last a long time and wear well" (Cridland 1975–1976a, p. 8).

Shirley A. Fairweather, "Reception at the Banquet Hall etc. 'TV Ardowska'":

"The reception was held in the banquet hall of the television station on top of a hill overlooking Cracow. On arrival we were amazed by the spread laid out for us, including chicken, ham, cheese, fish-in-jelly, bread, tomatoes, Polish biscuits and fruit. Initially there were glasses of wine, but when emptied they were replaced by glasses of vodka. The interpreters and guides attended the party and it was they who gave us our first lessons in the art of drinking vodka!"

"The leaders of each national delegation brought greetings from their countries, often including small gifts. Our President, Mrs G. Howard, spoke in Polish and presented the organizers with a book. The evening was a great success in that we met people informally before the serious business of the conference was started" (Fairweather 1975–1976a, p. 8).

Dorothy M. Cridland, "Performance of Folk Songs and Dances":

"A memorable evening was spent on Thursday when the whole of the Bagatela Theatre was given over to a performance of Polish dancing, music and singing provided by the resident repertory company who proved most highly skilled; although composed mainly of amateurs, it was obvious from their spontaneous vivacity that they enjoyed every minute of their varied and colourful program, an experience shared equally by an enthusiastic audience; we could have happily watched and listened for hours."

"The brilliantly coloured costumes and vigorous and intricate dance routines conveyed such a vivid life pattern, somewhat at variance with the rather sombre atmosphere we had noticed earlier in the week. We gained an insight into the warmth and friendliness awaiting visitors such as ourselves, willing to learn and share in everyday events and ambitions which left a lasting impression we shall all cherish" (Cridland 1975–1976b, p. 11).

Cridland Also Wrote About the Banquet:

"The Hotel Cracovia was the venue for our final Buffet Supper which was crowded, and to our delight a large number of the Polish delegates were able to attend, so much fraternizing took place. It is customary for candles to be used extensively when entertaining at night and elegant candelabra lent an air of enchantment to the scene. As usual many colourful and varied costumes were in evidence and with vodka literally flowing, conversation soon became lively and vivacious. Beautifully decorated ad loaded tables filled the centre of the banqueting hall and we were intrigued to find many native dishes quite strange to our rather more conservative palates. Music was provided by the hotel resident orchestra so when tired of eating we could take the floor in a series of somewhat spontaneous dance routines which always seemed to resolve into a Polish version of the Conga" (Cridland 1975–1976c, p. 20).

Shirley A. Fairweather, "Conclusions":

"It was marvelous meeting other women engineers from all over the world and to discuss the success and problems associated with our profession. The Polish people were wonderfully helpful in everything, including selecting gifts for our friends back home. The Conference organizers gave us a full program to enable us to see as much of the country as possible in so short a time. We carried out a small survey amongst the younger delegates on the problems that they had found in entering "a man's world". It seems that, although many have found some obstacles, provided that a girl is determined, she will succeed anywhere in the world to become an engineer" (Fairweather 1975–1976b, p. 20).

Margaret Dalski, "Conclusions":

"I think it is obvious that I thoroughly enjoyed my stay in Poland. The warmth and friendliness shown to us by everyone, but most especially our Polish hosts, did much to restore my faith in the kindness of people and, hopefully inspiring me to be similar. Perhaps the most pleasing consequence of the Conference was the renewal of my enthusiasm to do Engineering. This was brought about primarily through the example and encouragement of the women I met, who were shining examples of what engineers should be and of how to combine femininity with one's career" (Dalski 1975–1976, p. 20).

Chapter 6
ICWES-V, Rouen, France, September 4–8, 1978

Theme: "Technology and Freedom"

Rouen, the ancient capital of Normandy situated on the River Seine, was the venue for ICWES-V. This conference included a session on "Information of women on scientific and industrial careers and their training in these careers," as well as sessions titled "Industrialized production—Wherefore?" "From the Transfer of Technology to Industrial Cooperation," and "Responsibilities of the Research Scientist." At the closing of the event, "it was emphasized that engineers and scientists have not only a responsibility to their professions, but also to the public. This responsibility is of the utmost significance to women scientists and engineers, who are in a better position to convey their knowledge to a seldom well-informed public and to help all women to realize they should not be unconcerned with a world ruled by technology but have a part to play in it" (Laverick, in ICWES-X, Complete Proceedings, p. 11).

The conference Chair was Nicole Bécarud, *Présidente du Cercle des* Femmes Ingénieurs (see Appendix 1). The patrons were Alain Poher, President of the Senate, as well as five ministries, and two Secrétariats from the French Government. Bécarud's organizing committee consisted of 15 women with high positions in academia and industry.

There were 146 participants from 30 countries, including 43 from France, 20 from the United Kingdom, 34 from the United States, and 49 from other countries. Two countries, Vietnam and Uruguay, sent their best wishes but could not come (see Appendix 3).

The opening was at the Palais des Congrès, with welcome statements from Isabel Hardwich and Nicole Bécarud, and several political and governmental representatives. Bécarud's speech can be found (in French) in the Conference Proceedings (ICWES-V, Complete Proceedings, pp. 21–25). The keynote delivered on the conference's main theme was by Jacqueline Nonon, from the Brussels Commission

M. Frize et al., *Women's Contribution to Science and Technology through ICWES Conferences*, Women in Engineering and Science,
https://doi.org/10.1007/978-3-031-33800-7_6

(Head of Division, "*Emplois des femmes*," Commission of the European Communities); her presentation was on "Technology and Freedom." The full text can be found (in French) in the Conference Proceedings (ibid., pp. 26–37). Nonon spoke about the positive and negative impacts of technology and described the programs being put in place by the EU to ensure that women are treated with equality in workplaces (see section "Gender Papers" below for a more detailed presentation of this lecture).

The organizing committee was made up of 16 women from the same number of eminent organizations. The two main organizations were the *Cercle Français des Femmes Ingénieurs*, in close collaboration with the *Association Française des Femmes diplômées des Universités (AFFDU)*. Nicole Bécarud then proceeded to thank all government representatives, the organizing committee members, and the local industries that provided technical visits for delegates.

ICWES-V, S&T Papers

In 1978, the scientific and technical (S&T) revolution had reached a cruising speed. To name a few scientific events, the first Test Tube Baby, a girl named Louise Brown, was born in England on the 25th of July from in vitro fertilization, 11 days before the conference ("Louise Brown" n.d.); the Fax was beginning to be used; the first *LaserDisc* for storage support was commercialized in North America ("LaserDisc" n.d.); and Bell Labs launched a trial of the first commercial cellular network in Chicago using Advanced Mobile Phone System ("Timeline of the Telephone" n.d.).

Simultaneously, environmental and poverty concerns grew. It was in 1977 that the Kenyan Wangari Muta Maathai, who would receive the Nobel Peace Prize in 2004, founded the Green Belt Movement to plant trees across Kenya, in an effort to ease poverty, and put an end to conflicts. "Poor people will cut the last tree to cook the last meal," she once said. "The more you degrade the environment, the more you dig deeper into poverty." She inspired the United Nations to launch a campaign that has led to the planting of 11 billion trees worldwide (Ighobor 2011).

If the S&T part of the preceding conference, ICWES-IV, was enthusiastic about new technologies and innovations, ICWES-V reflects a more critical look at the implementation and impact of the scientific and technical revolution. It would be interesting to investigate if this point of view was motivated by the work of the international bodies for scientific cooperation introduced in ICWES-IV. With a theme such as Techniques and Freedom, the organizers were calling for discussions on the side effects of the S&T revolution and on scientists' responsibilities. Jacqueline Nonon, Head of the Women's Employment Bureau at the Commission of the European Community, set the tone of the conference in her opening speech:

> "[…] the scientific and technical revolution gives us fabulous knowledge and power, [but] there is always a flip side to the coin: it is the aspects of the century of technology, which are now revealed through the action of environmental movements, organizations for the

> defense of consumers, organizations for the fight against the arms race, through backward fashion and 'back to basics'. So there is a problem. […] Well, [the technical progress] has created a new responsibility which, I am convinced, will constitute one of the major axes of [the participants'] reflection. In fact, it seems difficult to deal with the theme 'Techniques and Freedoms' without introducing the 'responsibility' dimension." (ICWES-V, Complete Proceedings, p. 29; authors' translation)[1]

This call for discussion is clearly reflected in the four Commissions (sessions) topics related to S&T listed in Table 6.1: Commissions 1, 2, 4 (partial, the paper of Okwakol has been placed in the gender category), and 5. Commission 1 focused on the dangers related to industrialization; Commission 2 was on the international transfer of technologies; Commission 4 on responsibilities of the research scientist; and Commission 5 was on the issue of information technologies, as they are related to freedom of choice. One of the papers in Commission 4, and its discussion has been included in the gender contributions.

Today, one can only guess if the use of the name "Commissions" underlined an intention on the part of the organizers to have working sessions with clear outcomes that could be useful to participants once they returned home. It is probably to achieve this end that the Session Chairs provided summaries of the session discussions, and that Nicole Bécarud, president of the conference, summarized the work of all commissions in her closing speech (ICWES-V, Complete Proceedings, pp. 40–47, all in French). Bécarud was an active member of AFFDU at the time of

Table 6.1 Distribution of S&T sessions and papers in ICWES-V, Complete Proceedings

ICWES-V S&T commissions	Commission titles	# of papers expected	# of papers in proceedings	# of abstracts present in proceedings
C1	Industrialized production—Wherefore?	8	4	4
C2	From the transfer of technology to industrial cooperation	9	8	1
C4	Responsibilities of the research scientist	5 (6)	5	0
C5	Technical development of information systems	4	4	0
	Total	*26*	*21*	*5*

[1] "[…] la révolution scientifique et technique nous donne un savoir et un pouvoir fabuleux [mais] il y a toujours un revers à la médaille : ce sont les aspects négatifs du siècle de la technologie, qui se révèlent désormais à travers l'action des mouvements écologistes, des organisations de défense des consommateurs, de lutte contre la course aux armements, à travers la mode passéiste et de 'retour aux sources'. Il y a donc un problème. […] Eh bien, [le progrès technique] a créé une responsabilité nouvelle qui, j'en suis convaincue constituera l'un des axes majeurs de votre réflexion. De fait, il parait difficile de traiter le thème 'Techniques et Libertés' sans y introduire la dimension 'responsabilité'." (ICWES-V, Complete Proceedings, p. 29)

the conference, and she organized ICWES-V to fulfill part of that organization's objectives.[2] It would be interesting to investigate whether the outcomes of the conference discussions have had a real impact.

The number of presentations expected at ICWES-V for the S&T sessions and the number of papers or abstracts reported in the proceedings are presented in Table 6.1. The 27 authors who are listed in the proceedings for S&T presentations were all women; one woman was from the home country, 11 from the United States, two from Latin America (Brazil), two from Western Europe (Italy), one from Eastern Europe (Hungary), one from French-speaking Africa (Senegal), two from English-speaking Africa (Nigeria, Sierra Leone), four from Central Asia (1 from Iran, 3 from India), one from Southeast Asia (Thailand), and two from the United Kingdom. Five of the 26 papers and abstracts were written in French. The ICWES-V, Complete Proceedings, does not contain biographical information.

A Glimpse of ICWES-V's S&T Content

The session Chairs of Commission 1, *Industrialised Production, [Why?]*, France Thomas, and Jacqueline Gastinne, introduced it as follows: "Owing to the immense advances in science and technology during the last few decades, methods of production have undergone radical changes which can be observed in all spheres of human activity: agriculture, industry, town planning, transportation, etc. The improvement in productivity, so necessary for today's world, enables the satisfaction of varied needs of an increasing larger number of individuals. However, this changing world must retain its human values. Women engineers and scientists can help with this" (ICWES-V, Complete Proceedings, p. 49).

Anna E. Amour opened this S&T Commission with a paper called "*Lutte au gaspillage et récupération des déchets*"[3] (ICWES-V, Complete Proceedings, pp. 51–60) on how to limit various types of industrial or domestic wastes. She reviewed wastes that could be recycled and suggested developing usages for some by-products. After her talk, three of the presentations in Commission 1 were focused on technologies and health: Kathleen Harer (USA) in "Why Safety and Health are Important to Production" (ICWES-V, Complete Proceedings, p. 73; abstract only), Veronica Milligan (the United Kingdom, president of WES) in "The problems and benefits for the health service derived from the transfer or adaptation of technologies" (ICWES-V, Complete Proceedings, pp. 74–75), and Maria Watkins, also from WES, in "Enhanced Freedom for the Disabled through High Technology Electronics" (ICWES-V, Complete Proceedings, p. 76; abstract only). Three other presentations were on specific technological subjects: shale oil to replace petroleum in Thailand,

[2]Auzac (2021, pp. 73–77): this edition of *Femmes diplômées* devoted a full section on Nicole Bécarud (1933–2020). The organization of ICWES-V was part of the work of the AFFDU Commission called "Organisation and participation to conferences."

[3]"Waste Control and Waste Recovery" (authors' translation).

alternative energy (solar and wind) for developing countries, such as India, and protecting aircraft against lightning.

One paper stands out in Commission 1, *Industrialised Production*. Aileen Cavanagh from the United States presented a paper called "The Competitive Elements of a Dynamic Equilibrium Between Advancing Technologies and the World Demand for Individual Freedoms" (ICWES-V, Complete Proceedings, pp. 61–72). This paper fits the Commission 1 theme, but it offers a sociological point of view rather than an S&T one. It revolved around the juxtaposition of the two words of the conference topic, "Technology and Freedom." She analyzed the interaction between the concepts of technologies—an economic concept—in conjunction with the need for individual freedom—a social concept. She first stated a paradox: "[…] it appears that, as the practical range of collective choice grows through worldwide technological advances in all fields, the practical or perceived range of individual choice narrows to those options which remain available within the society within which the individual lives" (ICWES-V, Complete Proceedings, p. 61). Her paper identifies two boundary conditions to achieve a system in which technology and freedom would come to a peaceful dynamic equilibrium, and in which advances in technology would be followed by advances in freedom: on one hand are the human needs, expressed by market demand and political programming, and, on the other hand, are limitations of technology development through acceptance or rejection, and political controls. Presented mainly with tables and lists, part of the message in this paper is difficult to follow today, but it was a good attempt to set a framework for the conference theme.

Commission 2, *From the Transfer of Technology to Industrial Cooperation*, was chaired by Monique Brylinski and Geneviève Charlet. Cooperation was the main operative word. As reported by the Session Chairs, "The presence in this Commission of engineers from developed countries, with experience on technology transfer, and enlightened representatives of recipient countries, allowed a very open dialogue and a remarkable continuity in the discussion. Emphasis was placed on several conditions that must be met if a technology transfer operation is to be successful" (ICWES-V, Complete Proceedings, pp. 92–96; authors' translation).[4]

First, it is essential for the receiving countries to be capable of mastering the advanced technology, in terms of experience, costs of energy, and finance. Technical education and a fit to the country culture must also be respected. Otherwise problems can occur, such as creating a rural exodus by taking away women's traditional work, causing dietary imbalances, or causing a loss of benefits that flow outside of the country.

In this Commission, scientists from developing and developed countries were sharing their experience. For example, Lydia Pickup (United States, past president

[4]"La présence dans cette Commission d'ingénieurs de pays développés, ayant l'expérience de transferts, et de représentantes éclairées de pays receveurs, a permis un dialogue très ouvert et une remarquable continuité dans la discussion. L'accent a été mis sur plusieurs conditions qu'il est indispensable de satisfaire si l'on veut réussir une opération de transfert de technologie." (ICWES-V, Complete Proceedings, p. 94)

of SWE), suggested in "Configuration management during technology transfer" (ICWES-V, Complete Proceedings, pp. 116–119) that success would come by using an appropriate type of management she called *Configuration Management.* For other participants, to limit the risk of problems, developing countries could also create their own technologies, adapted to their needs and capabilities. Many examples were provided: for example, Marie-Thérèse Basse (Senegal), in a paper called "*La technologie au service de la conservation des denrées alimentaires*"[5] (ICWES-V, Complete Proceedings, pp. 103–105), reported that the activities of the Institut de Technologie Alimentaire (ITA) in Senegal succeeded to improve the conservation of foods; this was due to techniques such as the mechanical production of flour, dehydration using the sun's energy, as well as improving the storage of fish.

From a standpoint between vertical transfer of technology and local development, one of the most interesting papers in this Commission was presented by Yvette E. Stevens (Sierra Leone), "Appropriate Technology in Developing Countries—The Role of Women" (ICWES-V, Complete Proceedings, pp. 134–145). The concept of *Appropriate Technology* aims to ensure that women who live in rural areas have time to devote to community development projects, to participate in adult education programs, to take better care of their children, and thus to improve their living standard. She first "stressed [...] that *Appropriate Technology* [for developing countries] does not have to be a compromise between Indigenous and Western Technology, or some kind of second-class technology. Indeed in some cases it could in sophistication compared with technology from the developed world. The emphasis here should be on the use of [local] interest both within and outside developing countries. It is seen as a means of improving living conditions in the rural areas and thus narrowing the gap between the rich and poor families of developing countries" (ICWES-V, Complete Proceedings, pp. 134–135).

Antoinette David chaired Commission 5, *Technical Development of Information Systems.* In a similar manner to the other three S&T commissions of ICWES-V, Commission 5 intended to address questions such as: to what extent do the technical means of handling information, together with the techniques of information presentation, contribute to the freedom of choice between technologies? How can the engineer effectively influence them?

This commission consisted of a variety of topics. Rose Mary Thompson (USA) presented a very technical paper called "Mass Memories for Environmental Data" (ICWES-V, Complete Proceedings, pp. 328–341). Her paper covered the challenge of providing computer accessible storage for the massive amounts of sensor data that are generated by environmental satellites, for example, in the LANDSAT program already introduced at ICWES-IV. In "Trends in the Use of Computers in Manufacturing Industry" (ICWES-V, Complete Proceedings, pp. 351–361), Sheena Inglis from the United Kingdom proposed an original contribution, illustrated with cartoons, to help companies to improve their manufacturing management by using computers.

---

[5] "Technology in Service of Food Preservation" (authors' translation).

The information to be stored and preserved is a preoccupation that had been raised in ICWES-IV already. In 1978, when ICWES-V occurred, it was a reality librarians had to deal with. Kamala Balaraman (India) closed Commission 5 in presenting the consequences of the growth in information from the point of view of a Librarian in "Technical Growth of Means of Information. Its Consequences" (ICWES-V, Complete Proceedings, pp. 362–373). She noted that the world population is living not only in an age of technological revolution, but also in one of information explosion. She reminded the participants that tremendous advances have been made in the last two decades in the development of the production of information, vital to all other disciplines of technology and research. She spoke on the great advantages of online systems for storing and accessing the sources of information; she also mentioned the issue of the potential loss of the information prior to the 1960s, and the cost of the hardware which is a major problem for developing countries.

As can be seen, the initial questions on information vs. freedom and scientists' responsibility were only partly answered in this final S&T commission. The Conference Chair Nicole Bécarud, however, commented on Commission 5, emphasizing that the engineering and scientific community has a responsibility to make advances in technology accessible to users at different levels: from highly trained and technologically savvy people to people who have received only rudimentary training. She went further by providing a conclusion to the entire S&T commissions. She was convinced that women engineers and scientists, who meet at their own events such as the ICWES Conferences, are fewer, and that this creates a favorable opportunity to exchange views among specialists in different disciplines. She expressed the hope that these activities, of vital importance, will be publicized, and that special efforts will be done to draw attention in their countries and their professional environment to the results of the work of this Commission, and of the Conference in general.

Women Scientists' and Engineers' Contributions to S&T

Zsuzsa Szentgyörgyi (Hungary; see Appendix 1) also offered a follow-up in a paper she presented at ICWES-IV, in "Informatics at the Service of Mankind" (ICWES-V, Complete Proceedings, pp. 342–350). She stated that the distribution of women in the job hierarchy in the informatics sector was still at the low end and may even have gotten worse. She advocated doing something about this issue.

Yvette E. Stevens (Sierra Leone), in her paper already summarized above in Commission 2, argued that it is necessary to get women from developing countries to participate in cooperative projects. Specific efforts are needed to overcome the suspiciousness of women in developing countries at the introductory stages of cooperation, to involve the participation of female social workers, and to include women in the design stage. "Since [women] would be the ultimate users of the devices, it would be of immense benefit if design is done according to their specifications. [...]

The mistake here is that men are always in mind when technologies for the rural areas are designed" (ICWES-V, Complete Proceedings, pp. 135–136). According to her, the needs for Appropriate Technologies would focus on devices and activities that generate income such as weaving, soap-making, and food processing, and in devices to help women make the home environment healthier and more comfortable. She provided examples such as solar energy for cooking, biogas, maximized firewood stove, rain water catchments, hydraulic ram pumps, and micro hydraulic turbines.

ICWES-V, Gender Papers

Jacqueline Nonon's Opening Lecture: "Technology and Freedom"

As mentioned earlier, the opening lecture was given in French by Jacqueline Nonon, from the Brussels Commission (Head of Division, "*Emplois des femmes*," Commission of the European Communities), on the topic of "Technology and Freedom" (ICWES-V, Complete Proceedings, pp. 26–37). Nonon stressed how these four conference days would address a "highly topical concern" (ibid., p. 26).[6] Tracing the broad history of energy, from muscular to technological (with the rise of new sources like petroleum and electricity), she wondered about the opportunities made possible in this new era of "telematics" and how these advances have impacted on our material, professional, and social lives. Nonon then underlined the more negative consequences of such progress, particularly in the job market and often to the greater detriment of women, but also in relation to the ever-increasing social disparities and the destruction of natural resources. From these concerns emerged the question of social responsibility, whether individual or collective, national or international. Nonon also talked about liberties and how new kinds, acquired through technological progress, could jeopardize traditional ones, such as access and the collection of information, or privacy. Finally, she addressed the manners in which technology was directed and financed, and how women were almost entirely absent in decision-making at all levels, even though they were voters and consumers equal to men and were more deeply impacted by decisions dealing with everyday life (either to their own disadvantage or to that of children). She argued that women had a part in decision-making processes, and that their different backgrounds and perspectives benefitted all parties (they "act as catalyst and multiplier," ibid., p. 35).[7] Nonon closed her talk by encouraging women to take an active part in every sphere of activity, at every level—"become rightful partners"—and by

[6] "un thème d'actualité brûlante" (authors' translation).

[7] "un rôle de catalyste et de multiplicateur" (authors' translation).

stressing that, to succeed, this effort must be done "in a spirit of solidarity, not rivalry, with other women […]" (ibid., p. 36).[8]

Nicole Bécarud's Closing Lecture

A final talk by Nicole Bécarud rounded off the five commissions. In addition to a summary of each of the commissions, Bécarud also echoed Nonon's last remarks on solidarity, which seemed "so natural to [them] in this conference" (ICWES-V, Complete Proceedings, p. 46),[9] stating that it should radiate toward less privileged women who had not benefitted from the same education, girls in need of guidance, young women entering the job market, and others looking to resume their careers, as well as all others to come after them.

Sociological and Gender topics were spread across two commissions (C3 and C4), each divided into five sessions: two general discussions, both in French and in English, and one session of papers. For Commission 3, 14 papers were given by 16 different speakers (one speaker presented two different papers), coming from France (4), Brazil (3), Canada (1), Thailand (2), the United Kingdom (2), the United States (3), and the USSR (1). For Commission 4, one paper dealt with Gender issues (out of a total of six); its author came from Uganda. All 15 Gender Papers were delivered by women and can be found in full in the conference proceedings: for Commission 3, 5 in French, 9 in English; for Commission 4, 1 in English (Table 6.2).

Table 6.2 Distribution of Social and Gender Sessions and Papers in ICWES-V, Complete Proceedings

Social and gender commissions	Commission titles	# of papers expected	# of papers in proceedings	# of abstracts without papers in proceedings
C3	Future prospects in different countries	14	14	0
C4	Ultimate objective of the research: Are women ever in a position to decide the priorities of the research goals?	1 (6)	1	0
	Total	*15*	*15*	*0*

[8] "dans un esprit de solidarité et non de rivalité avec les autres femmes" (authors' translation).

[9] "Cette solidarité, qui nous apparaît si naturelle entre nous dans ce Congrès … […]" (the authors' translation).

Commission 3, "Information of Women on Scientific and Industrial Careers and Their Training in These Careers"

This Commission was organized by Gisèle Hugues, President of the Association Française des Femmes Diplômées des Universités (AFFDU), and Marguerite Cordier, former director of the École normale supérieure (ENS) in Fontenay-aux-Roses (1961–1974). Sessions were chaired by Dormer Ellis (Canada), Marian Owen (Iran), Clara P. Steinberg (Brazil), Ila Ghose (India), and Susan Turi (Hungary). This commission was interested in how "[t]he position of women in the industrial world and in the scientific world is as much conditioned by the information to which they have access and to the training which they receive as it is by the attitude of society to career women" (ICWES-V, Complete Proceedings, p. 156).

Discussions on the "Presentation of Information—Goal and Description of Careers" and the "Psychological Attitude of Society to Women Engineers and Scientists" were carried out both in French and in English. The former was interested in finding ways to foster the importance and interest in science in girls and in women more generally, and in exploring whether women really want access to posts of major responsibility. The latter delved into religious, familial, social, and cultural views on a woman's career in science and engineering.

A session of papers addressing "Future Prospects in Different Countries" closed Commission 3. Talking about the various positions in science and technology generally available to women, this session investigated the possibilities opened with new industries and the advancement of research and development, as well as the work conditions that would help women achieve careers in STEM.

Elizabeth Laverick, "The Education and Training of Professional Engineers in the United Kingdom and Its Influence on Women Engineers"

Elizabeth Laverick, a faithful and valued participant since the beginnings of ICWES (see, for example, the paper she presented during ICWES-IV: "The Education and Training of Professional Engineers in the United Kingdom and its Influence on Women Engineers," ICWES-IV, Complete Proceedings, pp. 641–649), also took part in the 1978 edition, with a paper that explored "Modern Trends in the UK & their Effects on Women Engineers" (ICWES-V, Complete Proceedings, pp. 263–269).

Laverick reminded us at first that both WWI and WWII fostered an increase in the number of women in engineering in the United Kingdom; she then noted that some current trends were producing the same results, including the rapidly growing contribution of women, especially married women, to the economy, as a result of the "pill revolution" and of the smaller families generated by access to birth control. However, while the overall proportion of women entering the professions was increasing in the United Kingdom, the proportion of those selecting engineering as a profession remained small.

How could we explain such low numbers? Laverick responded by proposing a list of barriers detrimental to women, first at the school level, then at the post-school level, during graduation and training, and finally, during employment. For example, many girls of school age tended to choose arts over science, a decision that could be related to a teacher's general lack of appreciation of the engineering field and of industry more generally. Furthermore, when the time came to choose a career, it was unlikely, as a matter of tradition, that engineering would be selected by parents, teachers, and career advisors.

At the post-school stage, the lack of support for industrial sponsorship or for the practical training of women within industry represented a key barrier for aspiring women engineers. Indeed, many in industry believed that recruiting a woman professional engineer was not a good investment, especially if she was married with family duties or wished to return to her career as an engineer.

Fortunately, reported Laverick, there were changes on the horizon, which could improve the situation and status of women in engineering in the United Kingdom. These were taking place mainly in the educational and legislative fields. There was also a growing awareness by the industry of the need to make better use of women's contribution to the work force. In 1977, for example, the Conservative Party set up a task force to examine the relationship between education and industry. The main recommendations included a 5-year degree for engineers, the creation of special engineering schools, improvements in career advice, and closer links between industry and universities.

The enactment of the Equal Pay Act in 1970 and of the Sex Discrimination Act in 1975 imposed certain legal obligations on employers related to the recruitment, training, promotion opportunities, pay, and terms and conditions of service of women and men. The Sex Discrimination Act also set up the Equal Opportunities Commission, whose duties were to work toward the elimination of discrimination and the promotion of equal opportunities. Legislative efforts were also conducted by the Government to end tax discrimination against married women, introduce effective Maternity Provisions, and ensure equal treatment of women and men in occupational pension schemes.

Laverick concluded her paper in a positive note overall; with respect to the challenges still facing aspiring women engineers in the United Kingdom, she wrote: "Outstanding then, will be how to attract them into engineering and to encourage them to make the most of the opportunities, and to persuade industry to meet the needs of the married woman wishing to continue her career, after the family gap. It is in these fields perhaps above all else, that more needs to be done" (ICWES-V, Complete Proceedings, p. 267).

Anna Saengbangpla and Sasithorn Boon-Long, "Education & Information of Women Engineers in Thailand"

In another presentation for Commission 3, Anna Saengbangpla and Sasithorn Boon-Long discussed the "Education & Information of Women Engineers in Thailand" (ICWES-V, Complete Proceedings, pp. 240–250). The coauthors of this ICWES-V paper were both teaching at the time at Chulalongkorn University, Thailand. In 1961 (Phulporn), Anna Saengbangpla graduated as a mechanical engineer from the same institution; she then received in 1968 an MSc degree from Manchester University (United Kingdom). She worked and researched for several years in France and England, before returning to Chulalongkorn as an Assistant Professor in the Department of Mechanical Engineering. A registered mechanical engineer, Anna Saengbangpla, was also a member of the Women Engineering Society. Sasithorn Boon-Long received in 1971 a BSc in Industrial Chemistry from the University of South Wales in Australia. In 1976, she was awarded a PhD in Chemical Engineering from the Institut National Polytechnique in Toulon, France. In 1978, she worked as a lecturer in the Department of Chemical Engineering at Chulalongkorn University.

The two presenters devoted the first part of their paper to a discussion on the development of Thailand's educational system over the years. They first pointed out that this system needed to specialize in the field of science and technology in order to meet the requirements of current society. As well, the education system was designed to allow citizens to exercise their rights and duties in a democratic country such as Thailand. However, the authors reminded their readers that in the past, the country's religion, culture, and tradition had a marked and negative effect on women's lives; indeed, because they were considered inferior to men, women were barred from the schools.

Luckily, women's education began to grow, which enabled them to demonstrate that they were able to work side by side with men to develop the country and Thai society. Over time, scholars and economists realized that, if women were given the opportunity to participate fully in the country's economic progress, the goal set would be reached even faster. Presently, you could find women working in a large variety of professions, such as medicine, engineering, the civil service, and employed in industries.

On the other hand, only men could hold certain administrative positions, such as an officer-in-charge of a district or a governor of a province. These sharp restrictions of a woman's role in administrative types of work brought back the country's culture and tradition which, according to the authors, "Is buried deep down in every Thai man's outlook on Thai women: that is, a woman should be gentle, humble and be a follower rather than a leader" (ICWES-V, Complete Proceedings, p. 241).

The two authors concluded their introductory observations with the following key statements:

> "The majority of Thai women themselves regard paid work as second important, they do not wish to work wholeheartedly nor to bear the responsibility of work, they work only to supplement the family's income. To them, family is more important than work. However,

> education in Thailand at present has been designed so that men and women have equal opportunity to study. They have the liberty to choose their own fields of specialisation. At the same time, women are trained to have responsibilities, to be able to make a definite decision and to be leaders in various social activities and have a role in developing the country. It can be concluded that there is no discrimination in education in Thailand" (ICWES-V, Complete Proceedings, p. 242).

The second part of the paper was devoted to a brief description of Thailand's education system. It was divided into four levels: kindergarten, primary, secondary, and higher education. Colleges and universities were located at the fourth level, and their main goal was to create manpower in advanced educational and professional knowledge to help develop the country. At this level, both women and men had the right to equal education and to choose whatever was suitable for them. The percentage of female-to-male students in universities was 50/50. However, most women students usually chose to study arts, education, languages, and medicine, with only a small number choosing technical, technological, and engineering studies.

The two authors then focused on their own university's statistics to help illustrate the situation of women in engineering. While Chulalongkorn University had established its Faculty of Engineering 65 years ago, women's engineering education began only in 1947. At first, only one or two women students applied each year, and in certain years, there were none who did. Things improved after 1961, with women enrolling every year and their number increasing gradually. During the last 30 years, about 150 women students had enrolled in engineering, with 100 having already received their degrees. Other data from the Faculty of Engineering revealed that women students could be found in almost every branch of engineering, results which, according to the authors, indicated that women were free to choose their own specialization. The most popular department, with respect to the number of students enrolled, was the Department of Electrical Engineering. The reason given for this was that electrical engineering was a vast field where graduates could easily find a job.

After 1967, reported the authors, five smaller-sized universities and institutes were created in Thailand, each one establishing its own Faculty of Engineering. So far, these schools had produced only about 40 women graduates, their numbers remaining small.

The paper's last section was devoted to an interesting discussion on the employment of women engineering graduates in Thailand. We learn that most of this group worked in three main sectors: government offices, private firms, and state enterprises. The authors then argued that women were attracted to the first sector for the following reasons: they were submitted to little pressure because of a lack of competition and discrimination with respect to wages and status; furthermore, women holding these jobs were not expected to engage in serious decision-making. Many women engineers also chose the civil service because they preferred to follow other people's ideas rather than their own. Nevertheless, a large number had secured high-level positions as leaders and had been directly engaged in jobs requiring problem-solving and decision-making.

As for those women engineers working in state enterprises, they also played a role in developing the country, especially in the field of public works. There, they advanced in status, escaped discrimination (except, it was noted, for a restriction barring woman from late night shifts, as these were considered unsafe), and received the same wages as their male counterparts.

The final section of the paper addressed some of the major concerns shared by women in engineering. As expected, a major challenge was to attract more young women into engineering studies and then to encourage more female graduates to enter the profession.

The coauthors responded for their part by submitting these final thoughts:

> "However, another important problem arises, that is, how can a woman engineer put herself wholeheartedly into her work without relating to the culture and tradition of the country because women have a duty of caring for the family and bringing up her children, at present, that men have contributed some of their time to family activities and this should increase in the future. Both should share the family's responsibilities which would give [...] women [the] opportunity to work to the best of their abilities and not having to worry about trivial affairs as it is at the present" (ICWES-V, Complete Proceedings, p. 251).

Commission 4, "Responsibilities of the Research Scientist"

This Commission was organized by Marguerite Ribeyrolles, appointed director of ADUNI,[10] and Erna Hamburger, professor at the *École Polytechnique* Fédérale *de Lausanne* (EPFL).[11] The two discussions, in French and in English, dealt with "[t]he duty to keep the public informed about the research" and asked its participants: "Under what conditions does a women scientist consider the research to be a contribution to progress?" In the final session, papers were presented on the topic "Ultimate objective of the research. Are women ever in a position to decide the priorities of the research goal?"

Mary Jossy Nakhanda Okwakol, "Women's Role in Scientific and Technological Research in Uganda"

Mary Jossy Nakhanda Okwakol, at the time lecturer in the Department of Zoology at Makerere University,[12] spoke about "Women's Role in Scientific and Technological Research in Uganda" (ICWES-V, Complete Proceedings, pp. 293–300). Okwakol

---

[10] Information about this association has not been found.

[11] Erna Hamburger (1911–1988) was the first woman to be appointed full professor ("professeur ordinaire") in a Swiss polytechnic school ("Qui sommes-nous?" under "La fondatrice Erna Hamburger" n.d.).

[12] Okwakol had obtained an MSc in Zoology at the same university 2 years prior; she would become Senior Lecturer in 1988. After being awarded a doctorate from Makerere in 1992, she was appointed full professor ("Mary Okwakol" n.d.).

first looked back on a statement expressed during the first ICWES in 1964 about a substantial lack of engineers and scientists with proper qualifications in Africa: "To a great extent this observation still applies today" (ibid., p. 295). She then presented the number of Ugandan women conducting scientific and technological research, whether in universities or state institutes. Her survey was established through 55 completed questionnaires.

Broadly speaking, Okwakol found that there are more women employed at Government Research Institutes (22%) than in universities (7.5%), since the latter had much higher requirements. Women were also more likely to choose biology or related fields, while generally avoiding physical sciences. Okwakol explained this discrepancy through the relatively recent tradition of physical sciences in developing countries and the type of work it demands. She also underlined the complete absence of women in the Faculty of Engineering at Makerere University and related the consistently small percentage of women researchers to the equally low output of women graduates. Perceptions on the efficiency of women researchers varied; around half of the respondents noted similar efficiency between men and women, though a quarter of them thought women were less efficient due to the time they had to set aside for their familial duties. Women were also viewed as faring better in laboratories than in fieldwork.

Respondents generally felt that there was social prejudice against women scientists, especially those working in the field or industrial sectors. Okwakol argued that these positions emerged from cultural and religious views, namely, how a woman's "life-long happiness is expected to come from marriage and the family other than from a career" (ICWES-V, Complete Proceedings, p. 297). As such, marital status had important consequences on women's capacity to devote themselves to research: an increase in domestic tasks, long absences due to pregnancy and child-rearing, personal satisfaction in home and stability. In schools, career guidance was found to be severely lacking, which led women to pursue higher studies in what was held to be typically feminine fields, like education and humanities.

Okwakol concluded that there was an urgent need to raise women's awareness of "the value of science and technology, opportunities for research scientists and the role of the research in National development" (ICWES-V, Complete Proceedings, p. 299). According to the author, this was a role, namely, for women with experience in more advanced positions in science and technology. Education should also be adapted to train more women, already at the undergraduate level. Finally, positive social biases toward women researchers in science and technology should be widely encouraged.

In the end, Commission 4 established that all researchers could make concrete changes in order to take responsibility toward the wider public: encourage practical work in schools; ensure better use of appliances in homes; expand scientific and technological knowledge in mass media. Moreover, in the face of a deeply ingrained lack of confidence in their own abilities and a traditional division of familial duties to the detriment of their careers, women needed to "convince themselves to initiate change," whether in their own lives, in society or in politics (ICWES-V, Complete Proceedings, p. 273).[13]

[13] "Sans minimiser l'évolution à accomplir au sein du couple, la femme doit se persuader que c'est à elle à prendre l'initiative du changement" (the authors' translation).

Business Meeting

The Business Meeting of delegates was on September 5, 1978; the minutes are in French, so an English translation is provided herewith. The goal of the meeting was to study the best effective manner, to ensure the continuity of the conferences, and to transmit the former organizers' experience to the organizers of the next event.

Laverick said that "it was at this conference that the meeting of national representatives confirmed the aims of ICWES and laid down Guidelines which included the composition of the Delegates meeting and its purpose (mainly to decide the host for the next conference). They also regularized the ICWES Continuity Committee (ICC) set up at ICWES-II. Their responsibilities are to encourage organizations to put forward proposals to host the next ICWES, and, once the host is agreed, to assist in the dissemination of its details. These guidelines can only be amended at the Delegates meeting. Some minor amendments have been made in ensuing years." (Laverick, in ICWES-X, Complete Proceedings, pp. 10–11).

From the notes at the Business Meeting of ICWES-V (ICWES-V, Complete Proceedings, p. 39), the following decisions were made by the majority of the delegates:

1. At each conference, an International Committee will be created to insure a liaison between the current conference and the next one.
2. The composition of the committee is as follows: the chair of the current conference becomes Chair of this committee; she will be helped by the Chairs and organizers of previous conferences and include a representative from large regions, such as Africa, North America, South America, Asia, Middle East, Far East, and Europe. The members of the committee will be selected by the regional delegates present at this meeting.
3. The committee is charged with encouraging bids to hold the next conference and to study the potential candidates.

Three proposals were received for hosting a conference in 1981.

The following decisions were adopted by a majority of delegates:

- The proposal made by the delegate from India was accepted under the condition that it be confirmed by November 1, 1978.
- In case the proposal from India did not get confirmed in time, two other proposals were retained: one from the Philippines and one from the United States. The assembly ranked the United States as first in line and the Philippines as second.

According to the decisions made on September 5, 1978, the proposal from the delegate from India was accepted. ICWES-VI will thus be in Bombay, on September 7–14, 1981.

Resolution Passed by Participants: Reported by Gillian Skinner

> "Under the guidance of Mme David, a resolution was drafted along the lines that all participants at the 5th ICWES should be concerned with the development and implications of information processing. It is significant that an intergovernmental conference on Strategies and Policies for Informatics (SPIN) was held recently in Spain. A statement by the United Kingdom government says, 'Information processing in all its aspects is the critical technology for advanced industrial countries'. It would be fitting if the conclusions of ICWES-V could be considered in future strategies and policies" (Skinner 1978–1979, p. 11).

At the closing ceremony, Nicole Bécarud made the following remarks:

"My French colleagues from the organizing committee and I are extremely touched by the kind words expressed by Mahin Rahmani in the name of all of you and from all my heart I thank you.

I must now speak of the future.

An important outcome is the establishment of a committee of nine members representing all regions of the world to ensure the continuity of our international meetings and to help the country charged with organizing the next conference.

It is a pleasure to confirm that our friends from India have a firm hand on ICWES-VI, but we shall keep an eye on the invitation from women engineers from the Philippines and from the Society of Women Engineers in the USA.

I thank our colleagues who are taking up the flag and accept the task to help organize the sixth conference.

Now, dear friends, I wish you a happy return to your country, much success in your endeavors, and I truly hope we will see each other in three years" (ICWES-V, Complete Proceedings, p. 47; authors' translation).

Bursaries:

> "The WES is to provide three bursaries for members to attend the Fifth International Conference of Women Engineers and Scientists to be held in Rouen, France, from 4 to 8 September next year. The bursaries, drawn from the International Fund, consist of £60 each. They are open to the whole membership of the Society, but it is hoped that one will be drawn from each of the following age groups: over 60, 30–60, and under 30. Final selection will however be within the discretion of the selection committee. Council has also allocated a sum of £20 out of the International Fund for gifts to the host country" ("VICWES Bursaries" 1976–1977, p. 2).

Testimonials

From WES Journal:

> "The fifth International Conference of Women Engineers and Scientists was held from 4–8 September 1978 at Rouen, France. Held under the auspices of the Cercle Des Femmes Ingénieurs, the conference was attended by over 200 delegates representing 35 countries.

The Women's Engineering Society was well represented with over a dozen members attending and some giving papers. The following pages contain a report of the conference which, although somewhat disorganised in certain aspects, most agreed was an enjoyable and useful event. The following members contributed reports: Hettie Bussell, Isabel Hardwich, Josephine Parker, Ira Rischowski, Gillian Skinner, Alison Spencer and Maria Watkins" ("Fifth International Conference of Women Engineers and Scientists" 1978–1979, p. 1).

Maria Watkins, "Inaugural Session":

"Isabel Hardwich declared the congress open (in one sentence very sensibly). Nicole Bécarud, very elegant and diplomatic, welcomed us. Jean Lecannet, the Mayor of Rouen, talked at length about the old and beautiful town and its traffic problems (the one-way system is very difficult to follow. It was practically impossible to get from our hotel to the railway station). Ywan Mirochikoff, president of the European Federation of Engineering Associations, talked about the history of development of engineering institutions in France. There are now 21 regional ones; 18 countries including UK, got together to form an international organization of engineers. Its purpose is to define the role and definition of engineers. There are no special problems for women engineers. Hugette François, Director General of TOURCO, represented the National Council of Patronat Français. She gave a lot of statistics including a statement that 4.4% of engineers in France are women. Claude Rerrault, secretary general of the Seine-Maritime district gave some statistics. Jacqueline Nonon, from the Brussels Commission, talked about last century legislation on liberty and technology: The changes which the advent of electricity and running water made in everyday life. The problems of tomorrow and the need for new sources of energy. The consequences of increase in material wealth, social revolution, new production methods, transport and medicine etc. The decisive effect on equalisation of people due to contraceptives, new methods of working, increased leisure and also of the negative aspects and compromises necessary. How machines replaced men and men replaced women in the textile industry. The effect of excessive organization. That the rich get richer and the poor get poorer. The effect technological progress has on social economics and on individual and collective liberties" (Watkins 1977–1978, p. 5).

Isabel Hardwich, "Meeting of National Representatives" (for the Continuity of the Conferences):

"This meeting had been arranged for Tuesday evening. However, a new group led by Beatrice Hicks and styling itself 'the chairmen emeritus and their personally-named deputies' had issued some 'guidelines for the conferences' and had proposed that these should be adopted by written endorsement and celebrated by a 'toast to the future' at the Tuesday lunch. Because of this, conference chairman Nicole Becarud called a meeting for the national representatives and the new group during Tuesday morning for an open discussion of the guidelines and here it was decided to put some amended guidelines to the national representatives at their programmed evening meeting."

"The evening discussions followed an absolutely superb dinner and, whether it was due to the effects of this meal and/or to a certain amount of bewilderment among the representatives as to the reasons for the guidelines, the proceedings dragged on and on. Nicole Bécarud was the chairman, Erna Hamburger acted as translator and Monique Brylinski was secretary. Eventually, at nearly one o' clock on Wednesday morning, a form of guidelines for written endorsement by the conference delegates was approved. It included the setting-up of a committee to ensure the continuity of conferences under the chairmanship of the host conference chairman and with eight members, each representing a world-area."

"A quick glance was then given to the three would-be bidders for the sixth conference. On a show of hands, it was agreed to accept the invitation to India from the Indian Women Scientists Association provided they could reassure Nicole Bécarud by 1 November 1978 that all was well. After this, the voting was to accept the offer of the Society of Women Engineers for a conference on the west coast of USA. The third bidder was Women Engineers of the Philippines. Unfortunately, the WEP (Women Engineers of Philippines) representative was not able to be present at the meeting so queries about the situation in her country were not answered. The probable date for the sixth ICWES was to be 1981."

"Following this meeting, the revised guidelines were displayed in the conference reception hall for endorsement and national representatives were asked to choose the members for the committee to ensure continuity. However, the results of these exercises were never announced and no 'toast to the future' was ever proposed. When closing the conference, Chairman Nicole Becarud did not refer to these matters at all. She did state that it was expected that the next conference would be in India and, failing that, in the Philippines or in USA" (Hardwich 1977–1978, pp. 11–12).

Lee Arnold, "Reflections on ICWES":

"At ICWES-V in Rouen, France, in '78, the technology of communications and information and the transfer of information across national boundaries was the subject of the sessions I attended. Yet it took the personal to make it work, as it was largely thanks to the bilingual abilities of Jacqueline Juillard of Switzerland that we were able to synthesize the discussions into the final set of resolutions to add to the proceedings. By this time, it was long apparent to me that my career would never include the practice of engineering. But, by the same token, the connections which had been formed were too strong to give up simply because I was not involved in technology" (Arnold 1996a, p. 9).

Dormer Ellis:

"ICWES-V was held in Rouen, France. It was a delight to see many friends from previous conferences. The paper that I contributed was a report of a longitudinal research project I was directing in which the research subjects were all women engineering students at Ontario universities. It was titled, 'Women Engineering Students—Who are They and What Happens to Them?'. As well as representing Canada at the opening ceremony, I had several other responsibilities during the conference, for which I either volunteered or was conscripted. As had been the case at earlier conferences, there was simultaneous translation at the main events but not at all the other sessions. Papers were supposed to be submitted in either French or English with a summary in both languages. There must have been a slight hitch somewhere as I was called upon to quickly produce English summaries of a couple of French papers."

"It was interesting being the 'shepherd' for a small group of delegates who knew no French and for whom English was not the mother tongue. Two of my 'sheep' were Japanese. One spoke only Japanese but the other could communicate in English. There was also an engineering student from Ghana whose mother tongue was Ga but whose education had been in English. My group also included a lady from Bombay and a Hungarian. I remember looking after these delegates at a reception, taking them on a shopping trip, etc. and explaining slowly in simple English what was going on in French around them. Another 'shepherd' was a multilingual Swiss engineer who helped German and Italian delegates who knew no French. While at ICWES-V, I chaired a session in which my Jewish friend from Brazil gave a paper in French—her third language" (Letter to Monique Frize, March 2021).

Chapter 7
ICWES-VI, Bombay, India, September 8–12, 1981

Theme: "Science, Technology and Society"

The conference was opened by Indira Gandhi, the prime minister. Elizabeth Laverick recalls, 15 years later: "It was a great thrill to meet one of the few women Prime Ministers in the world." Laverick reports that "there was a large delegation from India, mainly from academia and research institutes and many of the papers they presented were highly specialized and concerned with very specific research problems" (Laverick, in ICWES-X, Complete Proceedings, p. 11).

The Organizing Committee was composed of chairperson: Sumati Bhide (see Appendix 1); Deputy-Chair-Person: Ila Ghose; Co-Chair-Persons: Kamala Rudran, Rajani Ray; Members: Sarada Gopalkrishnan, Lalit Narurkar, Malati Baichwal, Sindhu Joshi, Shanta Gandhi, Bharati Bhatt, Shantoo Gurnani, Sudha Gangal, Sunila Mathur, Kamal Ranadive, Sudha Padhye, Padma Joshi, Usha Joshi, Chandrakala Baliga.

> "It was the first [ICWES] to be held in Asia and in one of the developing countries. About 300 delegates attended the conference, mostly Indian, although there were fifty-five foreign delegates from 17 countries. Most were from France (hosts of the previous ICWES), the UK, and the USA. Although there were representatives: from Iran, the Netherlands, Belgium, Canada, Brazil, Italy, Sweden, Poland, Malaysia, Switzerland, Nigeria and Japan. There were 14 delegates from the UK" ("6th International Conference of Women Engineers and Scientists" 1982, p. 7).

Laverick reported in her 1996 talk that there were 394 participants at ICWES-VI (339 from India and 55 from 17 other countries). She stated that "[o]nly 18 papers were offered by delegates from overseas." She added that "[m]any of overseas delegates took the opportunity to extend their stay after the Conference, and to learn more about this fascinating continent" (Laverick, in ICWES-X, Complete Proceedings, p. 11; see Appendix 4).

M. Frize et al., *Women's Contribution to Science and Technology through ICWES Conferences*, Women in Engineering and Science,
https://doi.org/10.1007/978-3-031-33800-7_7

"Indian Women Scientists Association" (IWSA)

"The host association, the IWSA, had been in existence for 8 years, but was widely recognised in India and had a large membership among the scientific community of women. Unlike many of the other organizations collaborating with ICWES, the IWSA was formed out of a feeling of social responsibility for the problems faced when integrating values from rural 'tribal' to the atomic, which co-exist in India. They felt the best way scientifically educated women could help in narrowing this gulf was to try to develop a scientific temper among the people, by sharing knowledge."

"The IWSA encouraged its members to take science to the people, with radio broadcasts on such subjects as nutrition, hygiene, and applications of science, participating in panel discussions and publication of articles, etc. They formed links with schools, arranging meetings and lectures, and have formed a library with sub-units based in Bombay and elsewhere."

"IWSA accepted that part of its role was to promote professional accomplishments in the scientific field, and so try to inspire its members to make such efforts through contact with stimulating personalities and ideas. It aims to help members 'widen their horizons and deepen their understanding' by contact with others in different scientific disciplines, through a program of visits and lectures."

"It has also tried to be a representative body of women scientists in giving expression to their opinions on significant problems, and at national and international forums" ("6th International Conference of Women Engineers and Scientists" 1982, p. 7).

"Inaugural Ceremony"

"A particular triumph for the IWSA was the presence of Mrs Indira Gandhi at the Inaugural Ceremony. Her visit lent much prestige to the conference and the organizing society. It also stimulated wide interest among nationwide newspapers and radio."

"In her speech, Mrs Gandhi urged women scientists to shed any feelings of inferiority in relation to their male counterparts, as they had ably proved themselves through their work, and some of them had even reached the highest echelons of science. She regretted that science had raised as many problems as it had solved, by pointing out that scientific innovations hastened the path to degradation of the environment and contributed towards pollution and the irrational exhaustion of natural resources."

"While 'science has led man nearer to total annihilation', she said she never ceased to marvel at new scientific discovery, and was saddened that many of today's youth took such 'wonders' for granted and were bored by them."

"She referred to the disparity between the developed and the developing nations with people in some parts of the world concerned only for their survival, while 'over-abundance and over-indulgence' elsewhere had eroded 'spiritual, moral and aesthetic springs'. Eighty per cent of the world's scarce resources were being cornered by only 20 per cent of the global population. She deplored the 'brain drain' of talent expensively trained by a country which could ill afford it, to developed countries offering inducements in terms of affluence and modernity. Much work awaited such trained people in India. She said that one should look to science, not for deriving means of luxury, but to bring about equality of opportunity, to solve the problems of deprivation and to create greater social and economic justice."

"She stressed that the task of Indian science at present is to relate new knowledge to old experience. There was a tendency to reject as irrelevant and supercilious all that was old in the enthusiasm for modernity. She added 'people create new superstitions as they go along'. She also spoke of the role women scientists had to play in India."

"Her final message to the delegates at the conference was to apply their minds more determinedly to their tasks, as 'determination and effort are keys to success'" ("6th International Conference of Women Engineers and Scientists" 1982, p. 7).

ICWES-VI, S&T Papers

In 1981, the scientific and technical (S&T) revolution had reached most economic sectors. Space technology was a popular scientific topic. April 1981 marked the first flight of the American Space shuttle. Other countries were involved in space technology, namely, India, the host country. That year, the IBM Personal Computer Commodore 64 was made available with a 4.7 Mhz processor, 16 K of memory, and optional floppy disk readers for 5.25 in. disk which could hold 160 K of data ("The Way We Were in 1981: Technology" 2010). The arrival of personal computers would make the computer accessible to everyone, and it would change the way to do science, business, learning, communications, and play.

The ICWES-VI Conference Proceedings are complete in the sense that the number of presentations expected in the schedule is the same as the number of papers present in the Proceedings. Table 7.1 summarizes the papers in the S&T sessions and the session titles. A total of 84 papers were presented. Note, however, that six papers were treated as related to the Gender section of the present book, since their main subjects were related to women, education, or sociology (the authors of these papers are Joshi and Purandare in session III-A, Diskalkar in session IV-A, Meenakumari and Shirohi in session V-A, and Ellis in session IV-B).

A total of 126 authors are listed in the proceedings for the S&T: 63 women, 11 men, and 52 of unknown gender. A vast majority of 115 persons among the authors were from India, the conference home country. Eleven participants were from other countries: two were from the USA, three from Western Europe, one from Eastern Europe, one from English-speaking Africa, one from the Middle East and North Africa, and three from the United Kingdom, Australia, and New Zealand region.

The proceedings marvelously reflect the main preoccupations of the women scientists and researchers in the Indian subcontinent, their self-confidence, and dedication at the time of the conference.

The conference was organized in two blocks, A and B. The first seven sessions up to VII-A were related in one way or another to human health. They addressed many different topics such as alternative medicine, rural problems and tribal issues, food accessibility and quality, agriculture, medical drugs, pollution, and education. Some recurrent topics in ICWES conferences were present, such as food supply in session 1-A, pollution in session I-B, and dairy and sea farming in VI-A. It is interesting to note that the scientific content presented by the authors in session VII-A,

Table 7.1 Distribution of S&T sessions and papers in ICWES-VI, Complete Proceedings

ICWES-VI S&T sessions	Session titles	# of papers expected	# of papers in the proceedings
I-A	Nutrition in Health and Disease	8	8
II-A	Pollution	7	7
III-A	Population Problems	2 (4)	2
IV-A	Recent Advances in Global Medicine: Allopathic, Homeopathic, Ayurvedic, etc.	7 (8)	7
V-A	Rural Problems – Multi-aspect Studies	6 (8)	6
VI-A	Expanding Frontiers of Agriculture Dairy and Sea Farming	6	6
VII-A	Research Papers (Molecular Biology & Biochemistry)	8	7
III-B	Communication Media for Adult and Child Education and for the Physically Handicapped	6	6
IV-B	Technological Models for Betterment of Mankind	7	7
V-B	Technological Models for Betterment of Mankind II	6	6
VI-B	Technological Models for Betterment of Mankind III	5	5
VII-B	Energy	11	11
	Total	*78*	*78*

entitled "Research Papers (Molecular Biology & Biochemistry)," was so specific that no one except specialists in these fields could understand them.

The last blocks of the conference, sessions III-B to VII-B, were focused on various technological analyses or solutions proposed for diverse issues. Session III-B was entirely focused on communication media for adult and child education and for the physically handicapped. The possibility to create different spin-off technologies derived from the space industry recent developments inspired many authors. Papers on energy production, another recurrent topic in ICWES conferences, were presented in the last session VII-B.

A Glimpse of ICWES-VI's S&T Content

It is an impossible task to convey here all the richness of ICWES-VI's scientific and technological content; choices of subjects were made to write this section.

India, in the early 1980s, was engaged in a quest to join the Western scientific revolution, struggling to reach a developed country status. Doing so, nevertheless, the Indian scientists were taking great care to develop S&T to fulfill India's own needs. High-level universities and research centers fully participated in this quest, such as the Atomic Research Centre located in Bombay which contributed many of the ICWES-VI conference papers. Indeed, the scientific revolution created great hopes in developing countries such as India; this is clearly demonstrated in most of the ICWES-VI S&T papers.

In a typical comment, Sudha G. Gangal from the Cancer Research Institute in Bombay wrote in a paper called "Perspectives of Immunology in Health and Disease in Developing Countries" (ICWES-VI, Complete Proceedings, pp. 106–109): "In the developing countries, it is preferable to utilize the scientific resources – brain power as well as money, towards projects of direct applicability in Biomedical Research. From this stand point, the diseases prevalent in India and the problems faced by us such as population control need concentrated efforts towards their [eradication]. In this report, four such areas and the efforts Indian Scientists are making in combating them are outlined. These are Immunology of leprosy, Immunology of malaria and filariasis, immunology of reproduction and Immunology of Cancer" (ICWES-VI, Complete Proceedings, p. 106).

However, it is obvious as well that consciousness of side effects brought by some technologies was increasing at the time of ICWES-VI, such as the consequence of the pollution and the dangerous products entering human food.

Session 1-A, *Nutrition in Health and Disease*, is particularly rich. Judith A. Zwartz (the Netherlands), in a paper called "Naturally Occurring Toxicants in Foods" (ICWES-VI, Complete Proceedings, pp. 34–37), wrote: "Even though it may have been known that several natural foods contain toxic substances, by far the greatest concern and attention since the early 50s has been directed toward the possible toxicological hazards of chemicals, primarily the pesticides and food additives, that enter the food supply as a result of their use by man in food production and processing. […] The practices of drying, fermenting, pickling, salting and smoking food, introduced over a period of several centuries, further increased the number and the type of compounds to which humans are exposed through food. Although as yet poorly understood, genetic manipulation of food plants to improve their characteristics […] have no doubt resulted in further additions" (ICWES-VI, Complete Proceedings, p. 34).

Obtaining precise measurements or statistics to quantify the impacts of S&T developments on human health was a major preoccupation. Additionally, obtaining regional or nationwide data was a major concern in many sectors to get information that could help develop mitigation activities. The scientists were testing, measuring, analyzing, and offering clues.

A paper presented by Sarada Gopalakrishnan, from the Air Monitoring section of the Bhabha Atomic Research Centre (B.A.R.C.), in Session II-A, is representative. Her paper, "Trace Element Concentrations in the Urban Atmosphere in India and the Pollution Status" (ICWES-VI, Complete Proceedings, pp. 42–44), details different specific levels of stable trace elements observed in the atmosphere. She introduced her paper with: "The first important step in pollution control is the actual measurement of the level of pollution. For this, a countrywide network of monitoring stations in different regions of this country has been set up and is in operation. The suspended atmospheric particulates collected from these stations are sent to the

B.A.R.C. laboratories for analysis. The non-destructive multielement analysis of these samples are carried out by the method of Instrumental Neutron Activation Analysis with the help of high-resolution gamma ray spectrometer" (ICWES-VI, Complete Proceedings, p. 42).

Nalini Raja & J. P. S. Chatha from the same institution, in "Chemical Reactions in Urban Atmospheres and the Problem of Air Pollution" (ICWES-VI, Complete Proceedings, pp. 57–61), went further: "The problem of air pollution has been of growing concern in the recent years. The physical and chemical mechanisms for the growth and decay of major and minor pollutants must be well understood to develop control programmes for air pollution. In this paper some of the important reactions of primary pollutants which lead to the formation of secondary pollutants and radicals are discussed. The role of singlet oxygen as a reactive species with many of the atmospheric pollutants is explained. The importance of monitoring many of the atmospheric [materials] is discussed in connection with health hazards [...] Today much attention is focused on the effects of pollutants on human health, crops, buildings, etc. so that the levels of pollution at which various adverse effects start occurring can be quantified" (ICWES-VI, Complete Proceedings, p. 57). In this paper, reactions of primary pollutants like NO, S02, hydrocarbons, and CO were discussed using full reaction equations.

As mentioned above, a large place was devoted to social justice for rural and tribal districts of the Indian country in ICWES-VI. Jyotshnabala Kanungo, in a paper called "Acupuncture – The Olden-New Step in Therapeutic Methods," presented in session IV-A, justified this preoccupation in the following way: "According to David Morle (1976), although three-fourth of the population in the developing countries live in rural areas, three-fourth of the budget of the health organisation is spent in urban areas where three-fourth of the doctors live. Furthermore, three-fourth of the health budget is spent lavishly in the treatment of aristocrats of the society whereas the cause of three-fourth of diseases could be prevented by a much less finance [= with less money]" (ICWES-VI, Complete Proceedings, p. 94).

Papers in session V-A were specifically looking into the rural problems. For example, a very interesting paper by Gurmeet Dhillon, named "Health and Nutrition as a Way of Life," echoes session 1A preoccupations in discussing the results of a nutrition survey, and solutions that included scientific knowledge dissemination in rural areas (ICWES-VI, Complete Proceedings, pp. 158–160). Session VI-A explored *Expanding Frontiers of Agriculture Dairy and Sea Farming* with some agriculture novelties, as well as the negative sides of the pollution from pesticides and industrial effluents in irrigation waters.

The second block of the ICWES-VI conference explored more technical issues. Sessions V-B and VI-B were devoted to *Technological Models for Betterment of Mankind.* Some topics had already been seen in previous conferences: electronic instrumentation in farming, transportation, water supply, and energy. Proposed solutions were sometimes surprising but adapted to India. An example, in "Recent Progress in Modes of Transport as well as the Problems – Pertaining to the Same" (ICWES-VI, Complete Proceedings, pp. 315–317), the SMT. S. S. PAWAR, from the Health Physics Division of Bhabha Atomic Research Centre, summarized the evolution of transportation means, up to the vehicles that use petrol engines. Considering these vehicles as sources of pollution, she concluded her paper in the following way: "The conflict between increasing energy use for transportation and the consequent environmental deterioration will have to be solved in near future. In that case men may have to go back to bullock-cart era once again or else" (ICWES-VI, Complete Proceedings, p. 317).

A strong preoccupation developed at this conference was related to technologies derived from the well-advanced space science and technology in India. The country had just launched, in May 1981, its first successful satellite Rohini (RS-1).[1] Papers in session VI-B expressed hopes for the future brought by the Indian satellite program to develop spin-off technologies useful for everyday life. For example, Girija Rajaram, in her paper titled "Space-Flights – Their Contribution to the Quality of Human Life" (ICWES-VI, Complete Proceedings, pp. 323–327), outlined the history of man's venture into space and pointed out the benefits that were acquired from space flights: "Satellites are often regarded as instruments [...] of destruction to Mankind. On the contrary they have probably averted several world wars through rival nations keeping constant vigil on each others' defense activities, and remaining equally balanced. The beneficial uses of spaceflights and satellites far outweigh their sinister possibilities [...] India has been actively making use of satellite technology for communications. Remote Sensing and Weather Forecasting, to help solve the colossal problems of a large land and a large population. She has started using some of the spin-offs in industry, and for rural employment" (ICWES-VI, Complete Proceedings, p. 326).

Of course, technologies necessitate energy. Considering the crisis related to nonrenewable energy sources at the time of the conference, such as coal and petrol, the issue of renewable energy sources was again considered as a solution. At the turn of the 1970s, many research efforts were already allocated worldwide to solar and eolian energy. Elizabeth Laverick (UK) introduced this subject in her paper "The Energy Situation in the UK – A Review" (ICWES-VI, Complete Proceedings, pp. 346–356), citing the conclusion of the 1978 Conservation Commission of the World Energy, arguing that a large contribution from nonfossil sources of energy would be needed by the year 2000. "This is due to the increasing demand of the developing regions with their rapidly growing population combined with an increasing consumption per capita" (ICWES-VI, Complete Proceedings, p. 346). The final ICWES-VI session VII-B concerned developing and implementing such new technologies in the energy field. Although more descriptive than technical, many solutions were proposed in this session. For example, Bharti V. Dave, from the L. D. College of Engineering in Ahmedabad, in a paper called "Non-Conventional Energy Resources Prospects and Retrospect" (ICWES-VI, Complete Proceedings, pp. 364–370), described unconventional alternative energy sources such as biomass,

[1] "Rohini is a series of satellites launched by the Indian Space Research Organization (ISRO). The Rohini series consisted of four satellites, each of which was launched by the Satellite Launch Vehicle (SLV), and three of which made it successfully to orbit. The series were mostly experimental satellites" ("Rohini (satellite)" n.d.).

geothermal energy, satellites power stations, Ocean Thermal Energy Conversion System,[2] and magnetohydrodynamic (MHD) generators.[3] Besides biomass and geothermal energy, none of the other technologies mentioned are implemented on a large scale today.

S&T and Women in ICWES-VI

Preoccupations at the frontier between women's condition and S&T were rare in ICWES-VI. One paper called "Gainful Employment for Coastal Fisher Women of India" (ICWES-VI, Complete Proceedings, pp. 125–127) by B. Meenakumari, from the Central Institute of Fisheries Technology, advocated that more fisherwomen would be useful for the fish industry. The already mentioned author Gurmeet Dhillon, in "Health and Nutrition as a Way of Life" (ICWES-VI, Complete Proceedings, pp. 158–162), suggested for rural areas to counteract the desire for male children with old age pensions and other measures of security.

Kamal J. Ranadive, Anjali Bhide, Smita Deshpande, Vibhavari Shah, and Satish Ranadive, all from the Life Sciences & Technology Centre, in the paper "Multiaspect Pilot Study on Tribals of W. Maharashtra" (ICWES-VI, Complete Proceedings, pp. 140–148) reviewed an exploratory pilot study that was undertaken on tribal general health problems, food and other habits, usages, environmental factors, and lifestyle. Many of these issues concerned specifically women and children.

ICWES-VI, Gender Papers

For the sixth conference, only one session, titled "Open Forum," was entirely devoted to Gender topics. There were three speakers, all women, of which one came from the host country and two from France. Another six papers were scattered throughout the rest of the program, which was mostly composed of technological and scientific sessions. These papers were given by six women, five from India and one from Canada. For the present purposes, we have divided certain sessions between S&T and Gender topics: in Table 7.2, the first number indicates the

[2] Utilization of the Solar Energy stored in the warm surface of oceans. This technique was first proposed in the early 1880s by the French engineer Jacques-Arsène d'Arsonval. The OTEC system has a great potential but only a few demonstration plants have been installed. The best known are the Natural Energy Laboratory of Hawaii Authority (250 kW), and Saga University in Japan. Larger OTEC systems are in development or planned in several countries, mostly to supply electricity and desalinated water for island communities ("Hydropower Explained: Ocean Thermal Energy Conversion" 2022).

[3] MHD generators have not been employed for large-scale mass energy conversion [so far] because other techniques with comparable efficiency have a lower lifecycle investment cost ("Magnetohydrodynamic Generator" n.d.).

Table 7.2 Distribution of social and gender sessions and papers in ICWES-VI, Complete Proceedings

Social and gender sessions	Session titles	# of papers expected	# of papers in proceedings
III-A	Population Problems	2 (4)	2
IV-A	Recent Advances in Global Medicine: Allopathic, Homeopathic, Ayurvedic, etc.	1 (8)	1
V-A	Rural Problems – Multi-aspect Studies	2 (8)	2
IV-B	Technological Models for Betterment of Mankind	1 (8)	1
VIII-B	Open Forum	3	3
	Total	*9*	*9*

Gender-themed papers, while the number in brackets gives the total of papers for a given session.

Dormer Ellis, "The Entry of Women into Canada's Engineering Profession"

As an active member of ICWES, Dormer Ellis had so far consistently presented papers at all meetings, except the inaugural one. Ellis became the first woman professor in Electrical Engineering at the Ryerson Institute of Technology in Toronto, Ontario, Canada (Ricard 2018). Trained both as a professional engineer and an educational researcher, Ellis was concerned above all with the very low rate of women in the engineering profession in Canada and was seeking to find the various reasons for this problem. In the paper she delivered at the previous conference in Rouen in 1978, Ellis announced some positive changes on this front, as well as the launch of a long-term research project aimed at following the academic and professional careers of members of a complete cohort of women engineering students in Ontario, Canada's most populous province. In 1981, in a paper titled "The Entry of Women into Canada's Engineering Profession" (ICWES-VI, Complete Proceedings, pp. 260–265) presented in the session "Technological Models for Betterment of Mankind," Ellis provided some preliminary findings of the first 5 years of this longitudinal project, throwing light on the continuous challenges faced by young Canadian women interested in nontraditional occupations.

So far, Ellis's research had not identified any deliberate attempt to keep Canadian girls away from professional engineering. Instead, the basic roots of the problem seemed to be the persistence of old social customs and of traditional beliefs about what an appropriate occupation for a woman was. However, the dramatic increase in the number of women engineers in the United States, especially as a result of the federal government's affirmative action program, led Ellis to believe that the situation in Canada need not be a permanent one. Indeed, she was pleased to report in her

Bombay presentation that the number of women engineers in Canada had greatly increased in the past 10 years, although they did remain a rarity in that country.

The paper's second half focused on longitudinal studies, the methodological approach Ellis had adopted and developed over the years as an educational researcher. Two of its main features were formal survey questionnaires and individual, structured interviews with her selected research subjects; but Ellis also advocated the gathering of informal data through social interactions and general discussions with young women engineering students at all stages of their programs, an approach that revealed much on the latter's attitudes, impressions, and feelings.

The paper highlighted some of the interesting findings linked to Ellis's ongoing research projects. One dealt with the degree of support granted at home and in high school to young women wishing to study engineering in the early 1970s. Ellis reported that they received very little support from their mothers, grandmothers, and older sisters. Things were somewhat better with their male relatives, especially their fathers, who were substantially more favorable to their daughter's interest in engineering. But Ellis was especially concerned with the situation of female high school students considering engineering; as her research had revealed, they received very little or no encouragement at all from such key groups as teachers, guidance counselors, and principals of either sex, all of whom strongly promoted nursing and home economics as suitable careers for women.

The paper closed nevertheless on a positive note, with Ellis drawing on her abundant survey results to illustrate the significant progress registered by women in engineering since the early 1970s with respect to such critical issues as enrollment and training of female students, job opportunities, salaries, overt discrimination, and sexual harassment.

Ellis also proudly reported on the first Canadian Convention of Women Engineers, an event that took place in Toronto in May 1981 and which attracted delegates from across Canada. A vibrant French-Canadian engineer, Claudette Mackay-Lassonde, and her committee of women engineers working at the Ontario Hydro Commission were the main organizers of this landmark meeting, which drew considerable attention from the press. Ellis, who attended, was pleased to interact there with some of her research subjects; as she reports, when she asked them if they had any regrets about their decision to become an engineer, their typical response was: "Are you kidding? I've got it made!" (ICWES-V, Complete Proceedings, p. 265).

Somdatta Sinha, "Some Thoughts on the Present Science Education System and Its Relevance to Our Society"

Somdatta Sinha, at the time a PhD student in Theoretical Biology at Jawaharlal Nehru University, New Delhi ("Somdatta Sinha" n.d.), presented a paper titled "Some Thoughts on the Present Science Education System and its Relevance to our Society" (ICWES-VI, Complete Proceedings, pp. 430–431) in the Open Forum session. Sinha first posed several questions about the socioeconomic situation in India: (a) the discrepancy between India's high ranking in science and technology and the general quality of life of its inhabitants; (b) the antithetic growth between scientific progress and superstition; (c) the persistent poverty and illiteracy of the population, even after decades of an independent government promoting formal education; (d) the rural exodus of educated citizens. These are all interdependent issues.

After briefly stating that "[a]n integrated rural development programme should be the first priority" (ICWES-VI, Complete Proceedings, p. 430) in India's planning, Sinha turned to the problem of education, specifically science education. Criticizing its hermeticism, she argued that the current system, based on the superficial memorization of facts and figures, could never encourage any scientific interest in students. According to Sinha, a better approach would be to bring science and society back together, otherwise there will remain "a dangerous trend to accept science as a neutral process" (ICWES-VI, Complete Proceedings, p. 430). The author lists several examples of this questionable tendency: (a) easy acceptance of using the country as a dumping ground for toxic products banned elsewhere; (b) supremacy of multinationals over local companies with the same expertise in turnkey projects; (c) substantial work on defense projects for other countries; (d) unconsented testing for new drugs on third-world populations; and (e) deliberate avoidance of environmental or ethical laws of foreign countries by conducting research in the Third World.

Sinha also denounced the propensity of scientists to "treat science and technology as a sort of magic which will eradicate poverty and other problems by themselves while they remain in their ivory towers" (ICWES-VI, Complete Proceedings, p. 431). According to the author, there is a willful ignorance among the scientific community as to why there is such a discrepancy between technological progress and its betterment (or lack thereof) of daily life. What she called "our unscientific irrelevant education system" seemed to be the main cause of all these issues, since it failed to induce a scientific interest in the general population, while alienating rural students from their environment. Sinha advocated for a nonformal scientific education geared toward adult women in rural areas, which would be based on the practicality and concrete application tailored to their specific needs. Though there were already many initiatives to that effect, most were conducted in urban centers; the author argued that this type of work needed to be decentralized. This would be a good start in taking their "social responsibilities towards building a prosperous nation without international interference" (ICWES-VI, Complete Proceedings, p. 431).

Nicole Bécarud, "Successful Business Women Today in France"

The last paper of the Open Forum session was given by Nicole Bécarud, past president of the Cercle des Femmes Ingénieurs, on the topic of "Successful Business Women Today in France" (ICWES-VI, Complete Proceedings, pp. 432–437).

Bécarud first argued that there was, in Western societies, "a fundamental distrust towards women's abilities to assume major responsibilities" (ibid., p. 432). For instance, when interviewed about their careers, women needed to emphasize how they overcame adversity and how they managed to combine work and family life.

Citing the Middle Ages, Bécarud found that women were widely more trusted in higher offices than at the time of the conference, though this authority started to decrease from the fifteenth century onward. It reached its lowest point 400 years later, when Napoleon I passed laws to invalidate a married woman's (legal, economic, social) autonomy. With these measures, women could only aspire to "a household and ornamental upbringing" (ICWES-VI, Complete Proceedings, p. 433). This had a profound and lasting effect on mentalities.

First, high school education was closed to girls until the 1920s, whereas women only gained global access to engineering schools in the 1970s. Though there was at the time a more or less equal proportion of girls and boys graduating from high school, notably fewer girls pursued higher studies and even fewer in scientific areas. According to Bécarud, with the great increase of working women in the last decade, it became indispensable to secure gender equality (viz., in terms of pay, working conditions and insurance) through Europe-wide legislation.

To illustrate the current situation of "successful business women," Bécarud gave the example of women "cadres" (roughly translated as "executives") in France, whom she defined by their earning of salary, a specific competence (generally a university degree), a high office, and a considerable pay. Official numbers in 1978 put women at 16.5% of all management positions. In engineering and science areas, the numbers were much lower: only 7% of women graduates (themselves only 5.3% of all graduates in engineering and scientific fields) occupied positions in higher functions. Representation of women among engineers and senior executives in industry is barely registered. Salaries of women executives were also persistently much lower than men's; Bécarud noted that "staff widows" would collect "an average of 41% more than a working woman" who would have worked her whole life (ICWES-VI, Complete Proceedings, p. 435).

Based on a sociological model of a dialectic difference between men and women, in particular in the context of women acceding to higher offices, the author detailed three types of functions: traditional, new, and future. Traditional functions, representing 85% of all women executives, relied on typical feminine traits of motherly care and concerned administration, health, or welfare. Though they required a certain professional ability, these jobs tended to give women only a little space in decision-making. New functions, based on specialization, could include heads of HR departments, teachers, or assistant directors. Bécarud remarked that women's influence remained informal, despite their status, since they "may officially participate in decisions but they do not appear in the official decisions" (ICWES-VI, Complete Proceedings, p. 436). As for future functions, these would imply a perfect symmetry of titles and authority. Until very recently, however, the very few women who had reached these types of positions had to follow a "masculine model," grounded in combativeness and the sacrifice of a personal life.

For Bécarud, the solution resided not in quotas, but in policies that would ensure the same opportunities for men and women. The author concluded that "[t]he conditions of better efficiency, in and out the firm, is to create a new partition of responsibilities in professional and domestic life" (ICWES-VI, Complete Proceedings, p. 437).

Business Meeting

No minutes have been found for this meeting in the Proceedings. However, it is known that the USA was selected to hold ICWES-VII in Washington, DC, in June 17–24, 1984, jointly with a SWE Annual Conference. Notes on the business meeting were found in the WES Journal and as follows:

Notes from Isabel Hardwich, "Business Meeting"

"22 delegates from 14 of the 16 countries represented at the Conference met together in the afternoon of 11 September 1981 under the joint Chairmanship of Sumati Bhide (VI ICWES) and Nicole Bécarud (V ICWES). The UK was represented by Maria Watkins and Gillian Skinner. Other WES members present were Evelyn Murray-Lenthall (USA), Yeside Soyinka (Nigeria) and Isabel Hardwich (contact person for all the ICWES Conferences). Evelyn Murray-Lenthall presented the Society of Women Engineers' bid for VII ICWES in Washington DC, to be held between June 24 and July 1, 1984, with the theme 'Technological Advances Impacting Society'. There were no other bids and the offer was unanimously welcomed and accepted, although it was agreed that the proposed title was difficult to say and even more difficult to translate."

"Both Yeside Soyinka from Nigeria and Marthe Achy Brou from Ivory Coast expressed a desire to host ICWES-VIII; this suggestion of an ICWES in Africa was warmly received. To ensure that ICWES-VII does indeed occur, and to assist organizers in all ways possible, the following International Committee was set up:

Convener-Chairperson VI ICWES (Sumati Bhide)
North America – Dormer Ellis
Africa – Yeside Soyinka
Europe – Nicole Bécarud and Isabel Hardwich
Asia – Mahin Rahmani (Middle East), Sumati Bhide (Central Asia) and Katsuko Saruhashi (Far East)" (Hardwich 1982, p. 10).

Testimonials

Dormer Ellis

"ICWES-VI, held in Bombay in September of 1981, combined ICWES traditions with Indian customs. At the opening reception, attractive ice sculptures kept the room at a comfortable temperature and the colourful saris worn by our hostesses were admired by all the foreign attendees. At all conference meals, there was a vegetarian option available. (Many Indians are vegetarian, often for religious reasons.) At buffets, one could choose the vegetarian line or the non-vegetarian. Although unfamiliar to many delegates, the food was attractive in appearance and delicious."

"Just as was the case at earlier conferences, I brought greetings from Canada the first evening and gave a paper about Canadian women engineers at one of the sociological sessions. There were several optional, industrial tours and I visited an electrical machine production plant that was very similar to the one where I had worked in England thirty years earlier. At the English facility, generators and large motors had their conductors installed by men, but small motors were wired by women. The women were paid about two thirds of what the men earned. In Bombay, *all* the generators and motors were wired by men and *those working on the smaller machines were paid the most.* I was told that this was because the task required extra skill, care and finger dexterity. Just in front of the Bombay plant, a large expanse of old concrete was to be replaced. Men were breaking up the old concrete with pneumatic drills and the debris was being carried to a dump a half-mile away by *young women.* They supported trays loaded with hunks of concrete on their heads as they walked to the dump, some with a young child holding onto mother's skirts. When I expressed surprise at seeing women carrying such heavy loads, I was told that women can *naturally* carry heavy items on their heads and was reminded that, in rural areas, it is the young women who carry the household's supply of water from the community well on their heads!"

"What was most memorable for me about ICWES-VI was a very informal afternoon tea party with Indira Ghandi, to which about twenty women, each from a different country were invited. We were shown into a beautiful room where tea and finger foods were laid out, told to help ourselves and that Mrs Ghandi would join us shortly. There were uniformed guards standing against the walls, but they did not participate in any way. Wearing a plain white sari, Indira Ghandi entered the room with two Indian ladies from the conference organizing committee whose saris were lovely. They introduced the Swiss and French delegates who were nearby and left Mrs Ghandi chatting in French with them. (Indira Ghandi had excellent French, having attended a Swiss boarding school as a girl.) After a few minutes, I was introduced and we continued chatting. When I motioned Lydia Pickup (USA) to join us, Mrs Ghandi switched to English. We were all impressed by her friendly manner and interest in how we had become engineers and how we managed to combine our professional and family lives. One after another, we introduced each other to this gracious lady who chatted with us and posed for a group picture" (letter to Monique Frize, March 2021).

Bette A. "BK" Krenzer, "An International Conference: Impressions"

"A welcome by the prime minister and a ride on an elephant! Those were a highlight and a sidelight of ICWES-VI, the Sixth International Conference of Women Engineers and Scientists, which convened in India September 7. Many of the 60 foreign delegates (10 from the USA) arrived in India directly from home with no stopovers en route to offset the cultural shock. We were ill prepared to deal firmly (or economically) with the early morning (3 a.m.) hassle at the Bombay airport – to the profit of baggage-handlers, taxi-drivers, and others. But we all survived and, after some good sound sleep, even revived. What followed was a most interesting, colorful and sometimes even exotic conference."

"Registration was marked by receipt of printed and bound volumes of abstracts, lists of attendees, name-tags, and some lovely handcrafted gifts. Slight errors in spelling had crept into the data on attendees so some of the name-tags needed a few alterations. The tags also produced a few chuckles over the *real* first names of SWE members Adele Coyle and Lee Arnold. Some of the Indian women couldn' find their own names on the printed list until they remembered to look for the misspelled versions on their name tags."

"In an inspiring keynote address, the prime minister, Smt. (Indian title equivalent to Ms or Madame) Indira Gandhi recognized the superior potential of women technologists to solve the many problems of developing nations. She noted the efforts now being put forth by the women represented at the conference to tackle those problems. She also lamented that the women from India and similar countries contribute to the serious 'brain drain' their countries are suffering when they extend their student status in other countries to permanent stays."

"The conference itself included four excellent lectures by leading Indian scientists on Tuesday through Thursday and Saturday. There were also some 125 papers presented by delegates – two of them from the USA (Lydia Pickup and Mary Ellen Russell). Friday was set aside for tours of local industry and / or research institutes, all of them interesting and instructive. The visits also included some fascinating insights into Indian working life. In the factories (assuming that Crompton-Greaves Ltd was typical) men are more likely than women to be employed at the detailed handwork of production; all production work reflects the ready availability of workers, so the lines tend to be labor intensive vs. automated. Excellent lunches, interesting technical or professional discussions and some new ideas to take back were all byproducts of the visits. Seen in an Indian industrial plant:

"'There is only one magic to remove poverty. HARD WORK / IRON WILL / CLEAR VISION / STRICTEST DISCIPLINE – is the incantation to use.'

"To supplement the intellectual exchanges, opportunities to glimpse Indian culture were made an integral part of the conference. At the delegates' reception on Monday evening we were treated to a fiesta of women's folklore from the four Indian states of Maharashtra, Kerala, Bengal, and Gujarat. This program was presented by the Indian delegates, attired in colorful regional costumes. On Wednesday evening the program was dance by a 'renowned exponent of the Odissi style'. The words of the program are hardly adequate to describe the performer of this delightful event. 'She beautifully blends the tenderness and languor of the Odissi style with her graceful body and emotive excellence. Her recitals are not only a visual and aesthetic pleasure, but also an educative journey into the myriad beauties of the dance'. The Odissi style is the most lyrical of the classical dances of India and in this case, perfectly fitted the description of 'sculpture in movement'. The third event – Thursday evening – was dancing in the Kathak style of 'rhythm in motion' where the dancer and drummer mingle in a harmonious duet with rhythmic alteration. Our program was presented by one of its most talented exponents who 'infused this exquisite / dance form with intelligence, imagination and a remarkable amount of feminine grace' to make it a truly memorable experience. Her performance elevated her art to 'its original dignity and spiritual grace'."

"SWE's First Vice-President Evelyn Murray-Lenthall, the Society's official delegate at the conference, successfully presented the United States bid for the 1984 conference. Others from the USA were Nan Steel, Mary Munger, Emma Barth, BK Krenzer, and Marian Owens. Seven of us joined the 20-member group of foreign delegates for a post-conference tour which visited many of the famous points of interest, including the Taj Mahal by full moonlight. Nearby, in the abandoned city of Fatepur Sikri, encountering our Italian associate Anna Amour with another tour group was a delightful surprise."

"It would not be possible to get the full flavor of this conference without mentioning the food. It was plentiful, flavorful and (probably) fattening but... the caution of 'don't drink the water!' definitely needed a supplement of 'be prepared for hot seasoning!' Though some American stomachs succumbed, it was only a temporary problem which careful food selection and generous portions of the delicious desserts corrected. The Sixth International

Conference of Women Engineers and Scientists, with its theme of 'Science, Technology and Society' can be pronounced a complete success. Now it's time to begin looking forward to 1984 when the Seventh International Conference will deliberate in Washington, DC, on how technology interacts with humanity to provide 'Dignity for All' in contrast to the Orwellian prediction for 1984 of a technology taking over society" (Krenzer 1981, pp. 2–3).

Gillian Skinner (WES Member), "Impressions of ICWES"

"The organization was better than Rouen, mainly because there were 3 main sessions, instead of 5, of which 2 were run in parallel. Publication of the program and the abstracts was excellent, and with the Prime Minister able to be present, the opening ceremony was very impressive."

"The conference lectures were interesting, of a high standard and, with one exception, well presented. The exception was marred by the chairperson's unnecessarily long introduction; one or two other chairpersons were guilty of this and it should be noted for future conferences. I attended all the technological sessions and on the whole found them interesting, varied and reasonably well presented. I have a great admiration for people who deliver papers which are not in their native language, but even so, some are better at it than others!"

"The main criticism is that there were too many papers. Omitting the Open Forum, 125 papers were accepted, of which 9 were to be read by title. This gave an average of about 11 mins per paper, which is cutting things very fine. Also the number of papers per period varied between 6 and 11, which was unfair on the speakers; 80 papers would have been ample."

"However to criticize is much easier than to organize, so some praise. The IWSA did an excellent job of the conference technical visits, the social events and in ferrying us to and from our hotel and 1 for one enjoyed it immensely. Members of IWSA impressed me with their friendliness and hospitality and with their enthusiasm and dedication to their professions. This was most evident in the technical visit to the Haffuing Institute where research is carried out in conditions which, by western standards, can only be described as primitive."

"In fact, primitive has to be used to describe many aspects of India, and 1 returned with a greater awareness of the enormous problems the country has, of the many efforts being made to solve them, and a greater appreciation of what we enjoy in the UK" (Skinner 1982, p. 9).

Maria Watkins (WES Member), "Impressions of India"

"The 1981 conference in Bombay, India, has as one of its objectives to enhance the standing of Indian women scientists and engineers, and it undoubtedly achieved this objective. Enormous publicity accompanied Mrs Ghandi's opening of the conference and extensive support was given by Indian industry. If this will result in a change in the status if professional women in India is an open question."

"We the overseas delegates received a warm and kindly welcome and I was impressed by Indian hospitality as well as their ability to organize such a large event. A printed timetable and abstracts of papers greeted us on arrival. The only hiccup I remember is to inform us in time when social events took place in places other than the Zoig Mahet."

"The level of the papers varies tremendously from hardly disguised copies of popular papers to interesting reviews of present research. This is an issue which has worried me about women's conferences for some time. I am sure that if we are to obtain respectability, we must have the same standards as conferences organized by scientific societies."

"My travel through India encompassed Udaipur, Bombay, Jaipur, Agra, Delhi and Kashmir. I was appalled by the appalling poverty on one hand and commercial pushing on the other. While I found Indian architecture and scenery interesting, it was Nepal and its people that really captured my heart. It is to Nepal that I wish to return sometime" (Watkins 1982, p. 10).

Emma C. Barth (SWE Pittsburgh Section), "Women Engineers of Other Nations: Their Career Access, Professional Concerns"

"*(At the request of the editor, Emma Barth talked with women engineers and scientists attending ICWES-VI about their professional concerns and the access of women to engineering as a career. Here's what she learned.)*"

"**Ragnhild Wallin** of **Sweden**, recently retired, is a consulting engineer, specializing in search and survey of chemical literature and in translating documents. According to her, more and more girls are entering engineering every year. In 1940, one per cent of engineering students were women; by 1980 they represented 50 per cent of the enrollment. Swedish girls are not being specifically urged, however, to go into engineering."

"Theoretically, but not in practice, the opportunities for men and women are equal. There is no affirmative action program, and very little discussion about the situation. Engineering openings are scarce now, and they are more likely to go to men. Wallin felt that more recognition for women in engineering must come gradually, by discussion rather than by government edict."

"She explained that in Sweden, women workers do not have to choose between working and having a family. A woman's salary continues when she stays at home to have a baby. When she is ready to work again, she comes back on the job. Men also are given the privilege of staying off the job to help with the children. There has not been much experience yet with women engineers taking time out to have babies. It is not known how engineering firms will react to the arrangement. As for the engineering challenges of the country, Wallin said that air pollution is the biggest problem. The pollution is killing birch trees and poisoning the lakes. Air pollution is partly due to exhaust fumes but is also augmented by smoke coming from Poland and Germany."

"**Yeside Soyinka**, the representative from **Nigeria**, is a structural engineer and owner of a construction firm. At times she employs up to 200 engineers and technicians, none of whom is a woman. She got her engineering degree in 1971 and set up her own business in 1975. When I asked her whether there are any other women in Nigeria heading construction firms, she said she didn't know of any. She said girls are now showing an interest in studying engineering. Electrical engineering is the most popular, but some are choosing chemical or civil engineering. There are many engineers in Nigeria, joined in an Institute of Engineering, but there is no organization of women engineers."

"When I asked her what the greatest concern of engineers was, she said it is resentment toward the attitude of the national government. For big engineering projects, such as the construction of a dam, foreign engineers are brought in. The native engineers are employed only on smaller projects. This is greatly resented by the Nigerian engineers, who are not able to develop real expertise in dealing with the nation's projects. Almost as an afterthought, since I had noticed that she was not wearing a wedding ring, I asked her whether she is married. She said, yes, she is married to a lawyer, and that they have four children. Asked about her husband's attitude to her work, she replied, 'Who knows what a husband thinks!'"

"**Paula Talpaert**, delegate from **Belgium**, is a chemical engineer who has just retired from her teaching position. She had been teaching industrial chemistry to students of industrial engineering. She herself had decided to study engineering because it sounded interesting when she first heard about it. She worked as a chemical engineer in industry before becoming a teacher."

"For a long time, the number of Belgian women in engineering was one per cent. Recently the enrollment of engineering classes has become 10 to 15 per cent female. In her last two years of teaching, 25 per cent of her students were girls. According to Talpaert, girls seem to prefer chemical or electrical engineering. She has known only one girl who studied mechanical engineering, and that was because her father owned a factory."

"In Belgium, getting a husband apparently rates more highly than getting a job, even for a professional woman. Furthermore, there are few openings for engineers in the country, and these jobs are more likely to go to men. Laboratories seem to hire more of the women. With the scarcity of positions to be had, many young engineers are going to other countries to look for employment. When I asked Talpaert what the greatest concern among engineers was, her answer was the same as that of the women from other middle European countries: 'acid rain'."

"The largest delegation at the conference came from **France**, totaling 14 women (plus three husbands). According to Mme. **Josette Cachelou**, president of Cercle des femmes ingénieurs and head of the delegation, 5.6% of the engineers in France are women. Graduating classes now have 10 per cent women. Whereas many engineering schools used to be closed to women, at the present time all 154 of them except for the four or five connected with the Army admit women students. Engineers are so badly needed that every graduate is sure of being offered a position. Starting pay is equal for men and women, but after that it is hard to get pay raises, especially for women."

"Mme. Cachelou felt that girls have great illusions about their future in engineering. They think the associations in their working careers will be pleasant as they were in school. But firms do not favor women. There is no provision for part-time work if the women decide to have children."

"Women workers are asked to retire at the age of 58. Women engineers face the same problems as any typical French family: Who will take care of the children while both of their parents are at work? Each couple must find its own solution. Firms which hire more than 200 employees must make an annual financial statement and a social statement to the government. Companies do not like these reports because they show how the women are being treated."

"**Judith Zwartz** was the delegate from **The Netherlands**. She is a senior lecturer in human nutrition at the Agricultural University of Wagenlingen. There is no organization of women engineers in the country, but the National Association of University Women is encouraging girls to study science and engineering. By now a few multi-national firms have employed one or two women engineers each. These examples have been publicized. The government is urging other companies to do likewise but has not set any quotas. During the International Year for Women, the Business and Professional Women's organization, as well as the National Council for Women joined the Association of University Women in urging that more women enter the field of engineering. Brochures were directed at school counselors and students. All universities were urged not to discriminate against women when hiring employees in any capacity."

"The return of mothers to the field of employment has received much attention. The government made an extensive study, with many interviews. Brush-up courses subsidized by the government have been set up. The National Association of University Women has organized courses in management. On the whole, the situation of women engineers in the Netherlands has improved considerably in the last few years."

"The delegate from **Poland** was **Maria Grabczewska**, a recently retired architect. She was sent by the Association of Building Engineers and Technicians, which is a division of the Polish Federation of Engineering Associations. It so happens that the president of the Federation is a woman, but there is no separate organization of women engineers. She was not able to give me any statistics on women engineers. But she revealed an unpublicized problem which is giving engineers and architects grave concern. As in cities around the world, many new apartments have been built since WWII, to ease housing shortages. In Poland, fly ash was disposed of by being mixed in the concrete for the buildings. Recently it has been determined that this fly ash was radioactive, and that people are being subjected to much more radiation in their homes than if they were in the streets. Grabczewska had no idea how this problem is to be solved."

"Without interviewing any particular individual, I talked with some of the Indian women about girls studying engineering. The conference was held by the Women Scientists' Association, since there is no organization of women engineers. I did not see any figures about what percentage of the membership is in engineering. There are good facilities for training engineers and scientists in India, and girls are not barred from the schools. But in this country the pressure for a woman to have a family above all is exceptionally strong. A woman physician told me that after her marriage, much against the urging of her mother and mother-in-law, she postponed her first baby for four years, so that she could finish her medical education."

"**India** is strongly committed to science and technology, to help the country advance and overcome its problems. This point was emphasized by Prime Minister Indira Gandhi, who gave the opening address of the conference. She also mentioned the 'brain drain' in which 'the poor are giving to the rich', since so many of the Indian engineers and scientists find employment in other countries. Private opinion holds that the government loses these trained people because they are not paid enough. Since the conference was widely publicized by Indian newspapers, it will be interesting to see whether it will affect an improvement for women scientists in the future."

"There were no delegates from **Greece** at the conference, but since I had the opportunity of speaking to a nuclear engineer while I was visiting Athens, I asked him a little about women in engineering. He felt that there is 'no problem' for girls who want to study engineering. He himself is a professor at a university of engineering. He said that about 5 per cent of engineering students and 50 per cent of the architects are women. There are some women in chemical engineering, but very few in mechanical. Last year he had one girl in his nuclear engineering classes."

"I gained additional insight into the fly ash problem. Grecian engineers know that the ash is radioactive because the lignite which is burned in the steam generators of the power plants was radioactive when it was dug up. The radioactivity escaping into the atmosphere is not a major problem yet, but engineers are pointing out how much cleaner nuclear reactors would be, with no smoke or pollution."

"It would seem that the history of women engineers is running parallel all around the world: no women engineers at all 100 years ago, a breakthrough 50 or 30 years ago, a big rush of girls into the engineering field 10 years ago, and now an increasing number of young women trying to establish themselves as engineers. But the problems that face all women workers are still slowing them down."

"I would be inclined to be pessimistic if I had not seen that the passage of time does make a difference. Thirty years ago the guidance brochure of the National Society of Professional Engineers stated flatly, 'Engineering is not a suitable field for women'. Who could have guessed that in 1981 the NSPE would have its first woman regional vice-president!" (Barth 1981, pp. 3–5).

Chapter 8
ICWES-VII, Washington, DC, USA, June 17–24, 1984

Theme: "Technology – An International Bridge"

Washington, DC, was the next location for an ICWES conference, offering a multiple choice of museums, galleries, historical monuments, and, of course, the White House. The conference was combined with the Society of Women Engineers' National and Student Conferences, so out of a total of 1204, the majority of participants were from the USA (1150), with 54 from 12 other countries, which is a proportion of 4.5% (see Appendix 3). The conference included a job fair and a student paper competition for the US attendees.

Organizing Committee

Chair: Naomi McAfee, Suzanne Jenniches, Pat Patterson (see Appendix 1);
Secretary: Gwendolyn Hays;
Treasurer: Rosemary Lafrance;
Student Co-chairs: Marilyn Berman, Kathy Garber;
Protocol: Sue Lyon;
Section President, Baltimore/Washington SWE: Mary Jones.

Laverick reports that "the aim of the conference was to reduce the gaps in understanding and knowledge among the peoples of the world … [The] extensive program of talks was arranged in three parallel sessions covering 15 topics, ranging from Defense, through Technology and Society, to Transportation, and Our Planet Earth. There were also sessions on Undergraduate Education, and Women in Engineering and Science, and for papers which did not fit into any of 14 topics, there was a 'Pot pourri' session. Industry was very much to the fore at the Conference,

M. Frize et al., *Women's Contribution to Science and Technology through ICWES Conferences*, Women in Engineering and Science,
https://doi.org/10.1007/978-3-031-33800-7_8

running recruitment evenings and participating in an exhibition. Professional development workshops were available most afternoons, proving a great attraction to the younger delegates. Technical and other tours abounded, usually at least six to choose from every afternoon! We were spoiled for choice …" (Laverick, in ICWES-X, Complete Proceedings, p. 11).

The opening keynote speaker was Ruth M. Davis, who spoke about "Super Computing as an International Technology." Ruth M. Davis was president and founder of the Pymatuning Group, Inc., which specialized in industrial modernization strategies and technology development. Technological areas of concentration include microelectronics, computers, information, automation, and robotics. The company also assists synthetic fuel commercialization projects.

The Banquet keynote speaker, Estelle R. Ramey, was a professor of physiology and biophysics at Georgetown University Medical School. She had been vice president of the Washington Heart Association and was currently a member of the board of directors.

Geraldine V. Cox was the Student Banquet keynote speaker. Cox was vice president and technical director of Chemical Manufacturers Association. CMA was a nonprofit trade association whose company members represent more than 90% of the productive capacity of basic industrial chemicals in this country. Dr. Cox directed technical activities for the chemical industry association, which included occupational health and safety, toxic substance regulations, environmental programs, energy, transportation, engineering standards, and biomedical and special environmental research programs. She also directed the management of CHEMTREC—the industry-funded 24-hour-a-day emergency response network for hazardous chemical spills.

ICWES-VII, S&T Papers

The history of science records at least two important events for women in S&T in the period of ICWES-VII ("Timeline of Women in Science" n.d.). In 1983, the American researcher Barbara McClintock (1902–1992) received the Nobel Prize in Physiology or Medicine for discovering that the genome is a dynamic entity, genes being able to jump from place to place. Also, Dian Fossey (1932–1985), a well-known American primatologist and the world's leading authority on mountain gorillas, published her autobiography *Gorillas in the Mist*.[1] In the aerospace sector, in 1984, the space shuttle Challenger lifted off on February 3. Ronald Reagan relaunched the space station project, the cost of which turned out to be so high that it could only be realized with the cooperation of the Russians, Europeans, and Japanese, under the aegis of NASA.

[1] Dian Fossey would be killed 2 years later in Rwanda; her murder remains unresolved to this day.

As for the S&T background, the progress in computer technology was very rapid. The year 1984 saw the beginning of the development of the X Window System (X11) of the Massachusetts Institute of Technology (MIT) and the exploitation system GNU ("1984 en informatique" n.d.). The Domain Name System (DNS) was introduced on the Internet. Apple launched the Macintosh personal computer with a television commercial during the 1984 Super Bowl, which played on the theme of totalitarianism in George Orwell's book *1984* (published in 1949). Motorola announced its 32-bit processor 68,020. The first HP LaserJet printers reached the market ("1984 in Science" n.d.).

Software development also became a major activity in this period.

With its theme "Technology – An International Bridge," the ICWES-VII conference in Washington was timely. More than any earlier ICWES conference, ICWES-VII reflected the achievements of several American women in S&T. The joint SWE-ICWES event brought more than 1200 women working or studying in many industries, research centers, and universities to ICWES-VII. The conference provided an opportunity for women engineers, researchers, and students to present substantial and advanced scientific content. Indeed, there were a very large number of papers related to computer technologies in the proceedings (ICWES-VII, Complete Proceedings). More than any ICWES Conference so far, the 17th edition was a celebration of women's participation in S&T. It must have been a very exciting event for the young women engineers who could meet so many role models, not only at the ICWES conference, but also at the industrial booths and career development workshops organized by SWE.

The ICWES-VII proceedings offered a schedule of 13 regular sessions with 60 papers (Table 8.1) proposed by 62 authors (57 women and 5 of unspecified sex). This represented 49 authors from the host country, 2 from Canada, 1 from Eastern

Table 8.1 Distribution of S&T Sessions and Papers in ICWES-VII, Complete Proceedings

S&T session titles	# of papers expected	# of papers in proceedings
Medicine	4	1
Defense	5	5
Computer Applications	5	5
Energy	5	4
Potpourri	3 (5)	2
Nuclear Engineering	5	4
Effect of Technology & Society	5	3
Software Engineering	5	4
Our Planet Earth	5	4
Applied Mathematics	5	4
Manufacturing	5	4
Transportation	5	4
Education & Training	3 (5)	2
Total	*60*	*46*

Europe (Yugoslavia), 3 from the Middle East and North Africa (2 from Iran, 1 from Egypt), 6 from Central Asia (India), and 1 from the United Kingdom. All the abstracts for the scheduled papers are in the proceedings (ICWES-VII, Complete Proceedings), and 14 of the full papers are missing.

However, 13 supplemental papers, not present in the schedule of the conference, were added to the proceedings (see Table 8.2), adding 13 more authors (10 women, 3 of unknown sex). Among them, 11 authors were from the host country, two from Central Asia (India), and one from Eastern Europe (Montenegro).

A student engineering competition was organized for the first time at an ICWES conference. Nine papers were recorded in the proceedings (Table 8.3), written by women who identified themselves from US universities, except one (unknown).

Therefore, adding all papers, supplemental papers, student papers, and abstracts (even when the paper is absent), 83 authors from the USA (92%) and 7 (8%) from other countries presented a scientific contribution at ICWES-VII.

A Glimpse of ICWES-VII's S&T Content

Tables 8.1 and 8.2 included some topics of interest for ICWES attendees, such as medicine, transportation, energy, and pollution, topics that were already present at previous conferences. Strikingly, many papers' contributions had a very high

Table 8.2 Supplemental S&T Papers in ICWES-VII, Complete Proceedings

Paper title	Origin of speakers
Basic Control Systems Course in Multivariable Style for Next Generation	India
The Effect of Telecommunication and Data Processing Advances on Information System Design	USA
Microprocessor Applications	India
Energy – The Essence of Power	USA
Waste to Energy Recovery Systems	USA
Training Systems for Defense Preparedness	USA
Application of Participatory Work Schemes in Research & Engineering	USA
Transfer of Technology in the World of Computer	USA
Software Team Structures	USA
Software Quality Assurance – An Emerging Field	USA
Elaiographical Investigations of Autochthonous Cultivars of Olives in Subregion of Boka Kotorska	Montenegro
Effects of the Atlas Peak Fire on Surface Water Quality in Upper Cilliken Creek Watershed Napa County California	USA
Total Quality: The Hewlett-Packard Way	USA
Low-volume Production of Desktop Computers Using Jit[a]	USA

[a]Jit: just-in-time compilation

Table 8.3 Student competition at ICWES-VII

Paper title	Speaker	Affiliation
Formation of ZBLA Glass via Reactive Atmosphere Processing	Guiton, Theresa A.	Pennsylvania State University
A Velcro Projectile Sharp-Shooter	Gamble, Lorie	Memphis State University
Redesign of a Die-Cast Hinge	Brill, Naomi	University of Minnesota
An Engineering Analysis of the Drynoch Landslide[a]	Irmen, Michelle Anne	Michigan Technological University
Determination of the Effectiveness of a Physical Therapy Ankle Rehabilitation Program on Reducing Time out of Balance in Normal Subjects	Wenstadt, Patricia J.	University of Minnesota
Investigation of the Hydraulic Characteristics of the Scouring Film Reactor	Borland, Sharon	University of Missouri-Columbia
The Cost of the Three-Mile Island Accident to the Callaway Nuclear Power Plant	Nelson, Christine	University of Missouri-Rolla
A Robot End-Effector for Resistance Welding Circuit Board Connections in Electronic Components	Pepin, Ellen W.	Oregon State University
Obstacles and Trade-Offs of Duplicating a Wideband Amplifier for less than One Tenth the Cost	Seto, Susan L.	UCLA

[a]The Drynoch Landslide is located in British Columbia, Canada

technical perspective. More than ever, engineering and industrial applications were numerous.

Center to the theme of the conference "*Technology – An International Bridge*," some contributions at ICWES-VII presented an international perspective. It was the case, for example, of a paper called "The Evolution of the Hercules – The International Airplane" (ICWES-VII, Complete Proceedings, pp. 116–120), presented by Kay C. Cornelius, a Senior Aircraft Structures Engineer at Lockheed-Georgia Company. It focused on the improvement over time of this airplane designed in the 1950s as a hauling cargo for the US Air Force. Due to "ongoing efforts of Lockheed's engineers to incorporate structural changes and improvements to meet international customers' varied needs and keeping the aircraft at the forefront of air lifters design technology, the many versions of the Hercule were "utilized by over 55 countries as cargo troop carriers, flying hospital, aerial photo mapper, search and rescue, weather observer, and forest fire fighters" (ICWES-VII, Complete Proceedings, p. 116).

Another great international adventure was described in the paper called "Space Shuttle International Cooperation in Orbit" (ICWES-VII, Complete Proceedings, pp. 359–362), written by Anita E. Gale from Rockwell International Downey, California: "No less amazing than the technical achievements required to develop the Space Shuttle is the philosophy, adopted very early in its development, that it is not exclusively an American vehicle, but an international resource" (ICWES-VII, Complete Proceedings, p. 359). The author explained, for example, how important the joint venture was regarding the creation of the Canadian Remote Manipulator System (RMS),[2] essential for satellite retrieval and repair missions, and the launching of the European Space Agency (ESA) SPACELAB.[3] Together, these elements favored international scientific experiments in space and the launching of many communications satellites from the members' nations of ESA.[4]

It is very interesting also to see the importance given to software engineering in ICWES-VII. Many authors mentioned that the global effort had switched recently from hardware to software. For example, in "ADA – Past, Present and Future,"[5] Elisabeth A. Strong, a Computer Applications Engineer at Honeywell, Inc. Minneapolis, Minnesota, wrote: "As the computer industry passed the point from being hardware intensive to becoming software intensive, the software crisis was recognized. Software costs were increasing rapidly, while hardware costs were decreasing rapidly. The physical size of hardware was being reduced and hardware operating speeds were increasing. Meanwhile, software was still in the Dark Ages, comparatively speaking. Not only was software costing more than hardware, it was routinely being delivered late, not working, and not maintainable. Fixing one bug frequently introduced two or three other bugs. As hardware costs decreased, more applications were found for computers, thus increasing the demand for computer software. Software became a bottleneck in the scheme of things" (ICWES-VII, Complete Proceedings, p. 260).

[2] "Many of the Space Shuttle's unique capabilities are made possible only through use of the RMS [...]. The RMS and all of the mechanisms, displays, and controls required for its operation were designed and built in Canada. The device is used to remove experiments and payloads from the Space Shuttle Orbiter payload bay, and can also accurately return them to their initial locations for reinstallation." (ICWES-VII, Complete Proceedings, p. 359)

[3] "SPACELAB was developed, designed, and built in Germany by the European Space Agency (ESA). The eleven ESA member nations are Belgium, Denmark, France, Ireland, Italy, Netherlands, Spain, Sweden, Switzerland, United Kingdom, and West Germany." (ICWES-VII, Complete Proceedings, p. 360)

[4] The International space station is still in use today. Wikipedia mentions that the International Space Station (ISS) program evolved from the 1984 American proposal to construct a permanently crewed Earth-orbiting station, and the contemporaneous Soviet/Russian Mir-2 proposal from 1976 with similar aims. It is a multinational collaborative project involving five participating space agencies: NASA (United States), Roscosmos (Russia), JAXA (Japan), ESA (Europe), and CSA (Canada). The first ISS component was launched in 1998, and the first long-term residents arrived on November 2, 2000, after being launched from the Baikonur Cosmodrome on October 31, 2000. The station has since been continuously occupied ("International Space Station" n.d.).

[5] Ada is a registered trademark of the United States Government (Ada Joint Program Office). This language was named in honor of Augusta Ada, Countess of Lovelace. She is considered the first programmer, because of her work with Charles Babbage on his Analytical Engine in the 1800s.

Taken all together, these ICWES-VII papers provide an almost pedagogical overview of the role performed by different software engineers and how their work was progressively developed. The following are good examples.

In her paper "Software Engineering" (ICWES-VII, Complete Proceedings, pp. 250–254), Diane B. Revell, from the Boeing Aerospace Company, explained that "software engineering is a new field. The term 'software engineer' has been in use for about 15 years, but has only recently become recognized. However, many people who recognize the term software engineer still do not have a clear understanding of its meaning" (ICWES-VII, Complete Proceedings, p. 250). The purpose of her paper was to provide a detailed understanding of the need for software engineers, their education and certification, and the type of work they do.

Shivran Siddhu from Applicon Incorporated,[6] in a paper called "Computer Systems Analysis: An Exercise for Sherlock Holmes" (ICWES-VII, Complete Proceedings, pp. 264–268), explained the role of computer system analysts. When a computer system has become so complex that nobody understands it anymore or is able to maintain it, it becomes essential to use a method to analyze and document this system and to make comprehensive reports for it. The author provides details on such a method.

With time, the work of software engineers complexified, and it was necessary to organize work teams. Wendy Barta, a Senior Systems Programmer at the Johns Hopkins Hospital, in her paper called "Software Team Structures" (ICWES-VII, Complete Proceedings, pp. 273–278), described and compared two team structures with respect to seven different issues: difficulty, size, duration, modularity, time, reliability, and sociability. The Chief Programmer Team proposed by Baker and Mills was a hierarchical, controlled, centralized structure where the Chief Programmer controls all aspects of the design and implementation. By contrast, Weinberg's Egoless Team was a democratic, decentralized structure, where all team members participated equally in the design and implementation. Both have their advantages and disadvantages.

The author Karen Dillon, from Hewlett-Packard, presented the paper "Software Quality Assurance – An Emerging Field" (ICWES-VII, Complete Proceedings, pp. 279–283), about the emergence of software quality assurance (SWQA) in the computer industry, and the role of the SWQA engineer in this process. In this paper, a software product was explained in detail, along with the activities of the R&D group and the SWQA group. "There are four major phases in a general soft lifecycle starting when the product is first conceived and continuing until the product is finished. Different companies may subdivide these phases, or add additional phases, but the concepts are basically similar. The phases are requirements, design, implementation and testing. What make these stages unique to SWQA are the quality assurance activities during [each of] these phases, and the documents produced" (ICWES-VII, Complete Proceedings, p. 280).

According to Rona J. Prufer, in her paper "Total Quality: The Hewlett-Packard Way" (ICWES-VII, Complete Proceedings, pp. 349–353), the Hewlett-Packard Co.

---

[6] Applicon Incorporated was one of the first manufacturers of Computer Aided Design and Manufacturing (CAD/CAM) systems. It was cofounded in 1969 in Bedford, Massachusetts, by four founders working at the MIT Lincoln Lab. It eventually became a division of Siemens AG ("Applicon" n.d.).

had 49 manufacturing divisions throughout the world at the time of the conference, each of these divisions considered a separate entity with its own departments of Research & Development, Finance, Manufacturing, Marketing, Product Assurance, and Personnel. Over time, the Product Assurance group that had the responsibility for the quality of the division's products became very expensive. The Total Quality Control concept was implemented to solve that issue. In this strategy, each function (Manufacturing, R&D, Marketing, Finance, Product Assurance, and Personnel) was responsible for the quality of its own service. Prufer developed the contribution to Total Quality Control of these six groups from a product viewpoint.

S&T and Women in ICWES-VII

In addition to the papers belonging to the Gender group in ICWES-VII, it was rare to see a mention of specific concerns for women brought on by the S&T progress; they were generally presented in a positive way.

A. Kulshrestha was the principal at Government Polytechnic for Girls, and Managing Director of the Gujarat Women Economic Development Corporation in Ahmedabad at the time of the conference. In "The Role of Women for Optimum Utilisation of Solar Energy for Developing Countries" (ICWES-VII, Complete Proceedings, pp. 165–167), Kulshrestha explained that in India, 45% of the energy requirement comes from the domestic sector. This situation stressed how useful it would be for women in rural areas to have access to solar energy to replace wood, which is increasingly scarce.

Another concern for women was raised in the already-mentioned paper "Technology – Bridge to the Future [Or Is It a Barrier?]" written by the Consultant BK Krenzer from Kansas City: "Concerning the sex barriers posed [or perceived as due to] the omnipresent computers, the literature abounds with psychologists' and sociologists' reports of and comments on the problem. [...] Sexual stereotyping of the past which identified the congenital inability of the female to cope with mathematics is now reportedly projected to a machine which is the mechanization [as parallel to the personification] of mathematics. However, [...] these concerns may be exaggerated" (ICWES-VII, Complete Proceedings, p. 241).

As can be seen from this joint SWE-ICWES conference, it would be a correct assertion that the women who participated did not feel disadvantaged in practising computer science.

ICWES-VII, Gender Papers

In addition to the paper sessions, workshops were also held during this conference. In total, there were 38 different workshops (1 was held twice), given by 41 different speakers (5 speakers offered 2 separate workshops), 35 women and 6 men. All of

Table 8.4 Distribution of social and gender sessions and papers in ICWES-VII, Complete Proceedings

Social and gender session titles	# of papers expected	# of papers in proceedings	# of abstracts without papers in proceedings
Women in Engineering Science	4	3	1
Undergraduate Education	4	2	2
Education & Training	2 (5)	2	0
Potpourri	2 (5)	1	1
Total	*12*	*8*	*4*

the workshop moderators came from the USA. Subjects were varied, ranging from career guidance to stress management and skincare.

For ICWES-VII, 12 social and gender papers were spread across four sessions. Two sessions were entirely devoted to these topics, while another two also included scientific and technological papers. In Table 8.4 below, for Sessions 3 and 4, the first number under # of papers expected indicates the gender-themed papers, while the number in brackets refers to the total of papers delivered within a given session. There were 11 speakers, all women: 5 came from the host country, while the rest came from Canada (1), Switzerland (1), India (2), the Philippines (1), and the UK (1).

There were four additional papers in the proceedings relevant to this section, but they do not appear in the conference program:

- "Woman Architect in the History of Mankind" (Sena Sekulic, Yugoslavia)
- "Radiation Risk and Women" (Mangala P. Dhairyawan, India)
- "The Problem of Getting Girls Interested in Physical Science and Technology" (Daphne F. Jackson, USA)
- "Limitations of Contraceptive Techniques and a New Approach to Population Control" (Kamala Chandrasekhar, India)

Jacqueline Juillard, "Looking Back Twenty Years Ago with a Foreign Eye"

Jacqueline Juillard, a consultant in industrial and scientific literature for the École Polytechnique Fédérale de Lausanne (EPFL) and the Swiss Society of Engineers and Architects (SIA),[7] presented a paper titled "Looking Back Twenty Years Ago

[7]Trained in chemical engineering, Juillard graduated from the EPFL and qualified in 1987 for the title of "European engineer." In 1970, she founded and directed the Public Relations and Scientific Information Service at the Battelle Institute in Geneva (where she had started working 9 years prior). Juillard opened her own engineering consulting firm in 1976, while continuing as a free-lance science journalist (Juillard 1988, p. 25).

with a Foreign Eye" (ICWES-VII, Complete Proceedings, pp. 68–71) in the session "Women in Engineering Science." Juillard, in her capacity as a SWE Member-at-Large since 1960, recalled events, stories, and people from the very first International Conference of Women Engineers and Scientists that took place in 1964 in New York, around the theme "Focus for the Future: Developing Engineering and Scientific Talent" (see Chap. 2).

The author first listed a few milestones in STEM fields at the time, touching on a variety of advances like LED displays in communication hardware, heart valvular in medicine, or telecommunication via satellite and human spaceflight. Juillard also cited a few political events: For instance, Nikita Kruschev would be replaced by Leonid Brezhnev, China was conducting its first nuclear weapons test, the Vietnam War was still ongoing, Beirut was booming, Martin Luther King won the Nobel Peace prize, etc.

Juillard recalled how the "Women Engineers and Scientists Day" marked the opening of ICWES-I, as well as a fancy dinner at Patricia Murphy's Candlelight Restaurant in Yonkers (one of her best memories of the whole event), a memorable speech by Lillian Moller Gilbreth (on the value of women in engineering fields), and the presentation of the SWE Achievement Award to Grace Murray Hopper for her excellent work in computer hardware (UNIVAC) and software (COBOL). Looking at the 1984 technical program at ICWES-VII, Juillard underlined how "the fields covered by these four sessions were already in the foreground 20 years ago" (ICWES-VII, Complete Proceedings, p. 69).

Naming some of the organizers of the first conference, Juillard then paid tribute to "prestigious SWE ladies which became very dear friends" (ICWES-VII, Complete Proceedings, p. 69): Ruth Shafer, Elsie Eaves, and Winnifred "Winnie" White.

Brooklyn-born Ruth Shafer was a design engineer for Gibbs & Cox, Inc., working on submarine air-conditioning systems (see Appendix 1). Juillard had met her 5 years prior to ICWES-I, at the groundbreaking ceremony of the United Engineering Center. Citing Shafer's "dry and spicy humour," Juillard recalled her retort when she was told that a male colleague doing exactly the same work had a raise, and she did not, because he had five children: "I didn't know you were paid for production at home instead of at the plant" (ICWES-VII, Complete Proceedings, p. 70). Juillard cited the impromptu auction held on the last day of the ICWES-I conference as a perfect example of Shafer's drive within SWE. She met Shafer for the last time in September 1971, a year before Shafer passed away from breast cancer, when she was awarded SWE's very first Certificate of Recognition.

Juillard also talked about Beatrice Hicks, conference director for ICWES-I (see Appendix 1). Hicks had stayed with her family several times, before her death in 1979. According to Juillard, Hicks was "a soft-spoken person, attractively diminutive, wearing flowers, earrings and polka-dot linen pumps" (ICWES-VII, Complete Proceedings, p. 70). Juillard commented that, in the eyes of men, you would never suspect that Hicks, as a professional engineer, had invented and manufactured several sensing devices and scanners, useful in fields like environment and space.

As noted by Juillard, Elsie Eaves started working in the mining industry as a summer job and finally worked as a manager for the Business News Department of

McGraw-Hill for which she launched several publications. This "modest but very knowledgeable woman with a rather round face and a wonderful smile" continued to surprise people with her expertise on construction matters (ICWES-VII, Complete Proceedings, p. 70). In 1957, Eaves was the first woman to be admitted to the American Society of Civil Engineers as a full member. Juillard was a regular guest at Eaves' residence in Port Washington, Long Island, designed and constructed by Eaves herself; the last time she heard from Eaves was in a letter dating back a couple years prior to the 1984 conference.

When SWE was moving to the new Engineering Center, its first Executive Secretary, Winnie White, decorated the Society's first two-room office with discarded furniture from the other societies and 50 dollars. On ICWES-I's opening day, SWE's office "was filled with delegates from all over the world and she received compliments from wide-eyed male engineers from the other Societies of the [Engineering] Center" (ICWES-VII, Complete Proceedings, p. 70). White's husband, Bud White, launched the Men's Auxiliary SWE (MASWE) later in 1976. White was a very efficient administrator for SWE, and her mailing list in 1961 came up to 500, peaking at 2200 during her tenure. Juillard had a standing luncheon invitation from White whenever she would be in New York.

To conclude this "walk down memory lane," Juillard recalled and cited the resolutions agreed to at the first conference. She then spoke directly to the younger SWE members. As her generation had paved the way for them, namely, through seven ICWES conferences up to that day, she told them: "So now it's your turn to further pave the way for the next generation of your daughters, and you seem to be on the good run" (ICWES-V, Complete Proceedings, p. 71).

Susan M. Hobart, "Breaking Into Engineering"

"How does a young adult choose a career?" asks Susan Hobart in the opening paragraph of her article, "Breaking Into Engineering" (ICWES-VII, Complete Proceedings, pp. 93–95). Herself a mechanical engineer working at EG&G Idaho, Inc., Hobart argued that a career-based decision requires much thought and the development of the young adult, as it can alter the course of a lifetime.

In the first part of the paper, Hobart dwells on the series of factors that determine the young adult's capability to make such a decision. She began with "outside influences" and the "complete environment" of the young adult, which includes "locale and exposure to the area's dominant industry, guidance, family life, and all the other forces that shape the person" (ICWES-VII, Complete Proceedings, p. 93). Hobart considered guidance as an especially important element in the development of the young adult. It is provided by various groups, such as parents and family, friends, teachers, and sometimes strangers. High school counselors can also provide important guidance in their capacity as advisors who can direct young adults as students toward a profession that will satisfy their need to achieve success.

What happens once a female student decides to "break into engineering" was one of Hobart's major areas of interest. She first reminded participants that engineering is a profession that will always be needed, although some fields may attract more students than others at a particular time. Hobart then proceeded with a definition of "engineering," borrowed from the Random House Dictionary: "the art or science of making practical application of the knowledge of pure sciences as physics, as chemistry, etc., in the construction of engines, bridges, buildings, mines, chemistry plants and the like."

In her view, the decision to pursue a career in the profession must come from inside the individual person, based on her own interests and skills in math, science, research, and, building, with the benefits of the guidance and support received up to this point in her life.

Choosing which school to attend represents a key decision; Hobart is, without a doubt, a strong advocate of specialized colleges, where the emphasis is placed on engineering education. She also lauds the system of co-op programs; established by colleges and several companies, it enables the student to earn credits while working and studying engineering as a trade. At graduation time, "The student will have confidence in her engineering abilities and will be able to interface better with those already in the profession" (ICWES-VII, Complete Proceedings, p. 94).

Hobart contended that communication skills and the ability to deal with people become in time very important aspects of engineering. A person may attend the best college and/or graduate with top grades, but she may not be ready to be an engineer if she lacks the ability to communicate. Hobart insisted throughout the article: Engineers must be able to work with other people, to share and develop ideas together, and to compromise when necessary.

Hobart then moved on with some interesting insights into the status of women engineers and the main challenges they currently faced in the profession. She felt that while "discrimination and chauvinism" are "dirty words" in today's society, in some professions such as engineering, they are in fact a "reality." Parts of the male population, she went on, "still resisted the idea of women engineers." As a result, "The young female engineer may have to work twice as hard as her male counterparts to earn the respect of her working peers, but there is nothing that says she can't do it." Hobart adds: "Besides, a smiling female face has an advantage over a smiling male face, because it is a refreshing change for the men to deal with women occasionally, among all the men" (ICWES-VII, Complete Proceedings, p. 95).

Hobart also reflected on the impact of government measures aimed at bringing more women in the profession, starting with the push for Equal Employment Opportunities. In several companies, she contended, this push created many more job openings for young women, which provoked a "reverse discrimination" reaction. According to Hobart, these young female engineers, now labeled "token engineers," would have to overcome two major obstacles: "The first will be dealing with her injured pride after being put into an unflattering category, and the second will be proving that she is qualified for the job, that she was not selected just because she is a woman" (ICWES-VII, Complete Proceedings, p. 95).

Ila Ghose, "Women Engineers in India and Impact of Women's Polytechnics"

As ICWES presenters duly reminded their audience over the years, women from ancient India who belonged to families of kings and of Acharyas (i.e., preceptors in religious matters) were awarded a "status of honour" for a quite a long time; while they had no access to formal education, they could engage in the study of philosophy, religion, and literature, in addition to taking part in religious seminars and politics. However, with the passage of time and for various social and political reasons, women's status and role in Indian society underwent a deep transformation, which led to their continuing domination by men.

Things began to improve during the twentieth century, thanks to the spread of education and the influence of Western culture. The post-independence era, which began in 1947, was especially supportive of measures and policies aimed at improving the status of women in society. Ila Ghose, a mechanical engineer, professor at the Women's Polytechnic in Calcutta,[8] explained in her paper (ICWES-VII, Complete Proceedings, pp. 190–192) that the planners of the nation realized that women were indispensable for economic development and social progress and that the large population of women (nearly 50% of the total Indian population at the time, living mostly in rural areas) could not be kept in a state of degradation and depravation.

Ghose provided an account of the various strategies, measures, and policies that were examined and adopted by the Indian government to improve the condition of women. One of the most innovative recommendations put forward at the time was to train them technically so as to improve their employment rate in small and medium businesses. Facilities dedicated to the education of women were established throughout India. These measures were supported by an International Labour Organization (ILO) study, carried out in 1958, for seven countries in Asia. A national committee set up the same year (and a number of others created afterwards) was charged with assessing women's needs in technical education, training, and employment, depending on their various backgrounds. Such requirements were later reiterated and emphasized by the National Education Commission (or "Kothari Commission") (1964–1968), looking to improve and expand women's education nationwide, both in specialized and general programs. Institutes in New Delhi, Bombay, and Bangalore were founded "for planning and development of a network of specialised training institutes" (ICWES-VII, Complete Proceedings, p. 190).

[8] Ila Ghose (1930–2019) was the first alumna of the Bengali Engineering College (Howrah) when she graduated in 1951 and became India's first woman mechanical engineer. After her studies, Ghose worked at Barr and Stroud, an optical engineering firm based in Glasgow, and in an ordnance factory in Dehradun. She held lecturing positions at Delhi Polytechnic and at the Institute of Jute Technology in Calcutta, was principal of Calcutta's Women's Polytechnic, and was appointed CAO in the UNESCO-sponsored creation of a Women's Polytechnic in Dhaka, Bangladesh. Ghose was a regular participant of ICWES since the second edition (Cambridge, 1967) ("Ila Ghose" n.d.).

Prior to Independence, most Indian women, especially those living in rural areas, had limited access to general education, even less to technical training. Grasp of technology might well have been restricted to basic carpentry. As such, "the government's decision to induct women at all levels in the technological field in spite of their awareness of the tremendous difficulties has indeed been a momentous one" (ICWES-VII, Complete Proceedings, p. 190). Changes in society and economy have, more generally, led women to gain awareness and fight against systemic injustices and prejudice.

Due to the firm belief that the technological ability of a nation has great significance for the economic prosperity of the country, India established a broad system of technical education at every level. Women gained access to engineering schools after Independence, though only a handful were admitted. "These few women either came from elite class or were exceptionally motivated" (ICWES-VII, Complete Proceedings, p. 191). At the time of the conference, Ghose stated that women amounted to 10–15% of all undergraduates admitted to technical colleges and universities. Certain fields were still regarded as typically male, like Metallurgy, Mining Engineering, Naval Engineering, Production Engineering, Chemical Engineering, and Agricultural Engineering. Despite the low admittance of women in such schools, Ghose argued that there seemed to be an increasing desire in women to explore engineering and technical fields.

At the diploma level, women's presence came much later, since they were generally discouraged, or reluctant to enroll in coeducational polytechnics. This changed with the establishment of women's polytechnics. Although women used to pursue an education, generally liberal, before they would get married, an ever-increasing number of them were enrolling in secondary education, and more and more were admitted in higher technical programs.

According to Ghose, "The establishment of women's polytechnics paved the way for the technical education of women" (ICWES-VII, Complete Proceedings, p. 191). With the establishment of women's polytechnics, women at the technician level increased from 0.05% to 20% of all technicians. Compared to the early days where women had to be convinced to attend, now they joined freely and in large numbers since they realized that technical training could secure better employment than liberal education. An initial proposition was to set up one women's polytechnic per state, but this was largely surpassed: For instance, at the time of the conference, there were ten in Tamil Nadu, five in Kerala, five in Karnataka, three in Odissa, while cities like Delhi, Bombay, and Madras counted three each. Moreover, due to a large demand from students, courses in fields like electronics, radio & TV, architecture, machine design drafting, textile processing and printing, civil engineering, or instrument technology were added to the more traditional offer.

Ghose argued: "There is urgent need to upgrade the skills of these women for fulfilling job requirements created by modern technology" (ICWES-VII, Complete Proceedings, p. 192). To that effect, she suggested to (1) increase the number of women's polytechnics in all states; (2) offer a wide spectrum of courses to appeal to more candidates; (3) set up modular and part-time courses; (4) encourage women to enter general polytechnics in every field, like Mining or Aeronautics.

Business Meeting

Minutes of this meeting are not found in the conference proceeding but are presented in a report published in *SWE Magazine* by the Meeting Chair, Naomi McAfee:

"The international delegates to the VIIth International Conference of Women Engineers and Scientists met at the Shoreham Hotel, Washington, DC, on Friday, June 22, 1984. The meeting was called to order by the Chair, Naomi McAfee at 1 p.m.

The delegates present and the countries which they represented were:

Canada, Dormer Ellis;
Egypt, Sohier Sakr;
France, Véronique Raimbault;
India, Ila Ghose;
Iran, Mahin Rahmani;
Italy, Anna E. Amour;
Ivory Coast, Marthe Achy Brou;
Nigeria, Joanna Maduka;
Phillipines, Lydia Tansinsin;
Switzerland, Jacqueline Juillard;
United Kingdom, Daphne Jackson;
United States of America, Barbara Wollmershauser;
Chair Emeritus, J. Cicely Thompson (UK).

[...] There were two items of business: 1) the location of the next ICWES and 2) the approval of conference guidelines.

The Ivory Coast delegation proposed to hold ICWES-VIIIth in Abidjan, Ivory Coast, the latter half of July in 1987. This offer was accepted contingent upon the Ivory Coast delegation submitting a written proposal describing the exact date, description of facilities, accommodations, comparative current costs, plans, government/associations support, etc. to the VIIth ICWES Chair no later than December 1, 1984. If this was not done, the United Kingdom delegate offered to host the ICWES-VIII in 1988. However, the United Kingdom would prefer to host ICWES-IX in 1990.

It was stressed that any host country which does not have English as its major language must provide translation facilities, as most of the papers will be presented in English.

The second order of business was the approval of 'Guidelines for the International Conference of Women Engineers and Scientists (ICWES)'. These were a consolidation of the Rouen Guidelines, the Bombay Minutes and some minor current updating. These were to be implemented immediately upon approval. A significant feature was the selection of the ICWES Continuity Committee (ICC). The ICC consists of a Convenor (someone other than the current ICWES Chair, who would run the next Business Meeting) and 11 world area representatives. The areas and representatives selected were:

North America (USA and Canada): Dormer Ellis;
Latin America (Mexico, Central America and South America): Clara Steinberg, Brazil;
United Kingdom (also Australia and New Zealand): Daphne Jackson, UK;
Western Europe: Véronique Raimbault, France;
Eastern Europe: TBD
Africa (French Speaking): Martha Achy Brou, Ivory Coast
Africa (English Speaking): Joanna Maduka; Nigeria;
Africa (Arabic Speaking): Sohier Sakr, Egypt;
Indian Subcontinent: Ila Ghose, India
South East Asia: Lydia Tansinsin, Phillipines
Middle East: Mahin Rahmani, Iran.

Lydia Pickup, USA, was elected by the delegation to serve as Convenor for the next ICWES." (McAfee 1984, p. 11)

Testimonials

Dormer Ellis:

"ICWES-VII was hosted by the (US) Society of Women Engineers and held in Washington. Three Canadians gave papers and there were four other Canadian attendees. At the Bringing of Greetings Reception, I spoke for Canada and was thrilled to hear the greetings from Belgium, India, Nigeria, Britain, France, and Puerto Rico, as the speakers were all people I knew from earlier ICWES conferences."

"The 1984 ICWES was different from earlier ones both in size and composition. Held concurrent with SWE's annual convention, there were over a thousand participants, but only about fifty were not Americans. They came from fifteen countries. SWE and ICWES delegates mingled at many events, but reserved for ICWES registrants only were a visit to the Baltimore aquarium by bus and a delightful cruise on the river the final evening."

"At the meeting of the Continuity Committee, I agreed to continue representing North America, at least for one more ICWES. The delegates from Ivory Coast presented their case for holding the next ICWES in Abidjan and showed documents to illustrate the support they had from professional organizations and government. Their offer was accepted. ICWES delegates looked forward to meeting in Africa. The women from Ivory Coast were very young. It was their first experience of being abroad. As they had no English, it was not easy for them to explore Washington. I took them out one afternoon for a ride on the subway as they were excited by the idea of a railway underground. They were especially impressed by the escalators. They had seen a small one in a department store in Abidjan but the long escalators going down to the subway fascinated the young women and they rode up and down, up and down, while calling out to me to take their pictures." (letter to Monique Frize, March 2021)

Hilda Blount (WES Member), "Personal Impressions of the Conference":

"Arriving at Dulles International Airport and asking at the Information Desk where to catch a bus to the Shoreham Hotel, I was given the number and told it was just outside a few steps away. Then I was informed that another lady, a few minutes earlier, had made the same request, and by the description of her, I knew it was Paula Talpaert from Brussels. This was my first experience of American hospitality and friendliness as I had not visited the States before."

"The Shoreham Hotel lived up to its reputation of splendour and I was duly located on the Computer, given my room number and key and escorted by a bell-boy with my luggage to the room I was to share with my friend Lee Arnold. Two king-sized beds and other furniture in the bedroom, a bath room, a writing room, a small balcony and a wardrobe-dressing room were real luxury, but the long walk traversing the corridors to reach them was tiring every time."

"Voyages of discovery to find the rooms allocated for various activities were necessary and most enquiries at the Reception Desk resulted in 'Sorry, not at the moment, come back later when the right person will be present.' (All that glitters is not gold!) The SWE Registration Office was a bit chaotic not surprising when there were so many people wanting their papers etc., and those 'manning' the desks were SWE Members and therefore inexperienced in this kind of work. But all got sorted out eventually and the real business of the Conference began."

"The first snag was finding out that the pre-conference tour to the Luray Caverns was cancelled as only 4 people had booked to go. Fortunately, my room-mate said she would take us there in her car (we had the money back for the booking! and off we went). This was a worth-while visit which we enjoyed."

"The third morning, after breakfast, the restaurant manager informed my room-mate that she couldn't pay for her breakfast on her account as she had no money in her name. When we enquired at the Reception Desk, it was found that her deposit had been credited to me!! I had no idea I was so rich for those three days. But that was quickly rectified and all was well. (Not such a marvellous hotel after all.) There were three SWE Conferences proceeding at the Shoreham: (1) the ICWES, (2) the Student Conference, and (3) the SWE Annual Conference. Rather a lot to fit into one week."

"Most of the Technical Sessions began at 8.15 a.m. which meant a very early breakfast for those attending and only the dedicated ones turned up for the start."

"In the evenings there were many Receptions and Banquets for which one had invitation cards and these were very enjoyable, especially the Students Awards Banquet and the final Banquet on the last evening."

"During the week several tours had been arranged and we were able to visit Mount Vernon, President Washington's home, followed by Arlington Cemetery, where we saw the Changing of the Guard, and a wreath laying ceremony (only about 5 of these occur each year). A tour of Washington to sec the famous buildings and monuments included a visit to the Pentagon. A day-tour to Baltimore to see the reconstructed harbour and a huge shopping complex, also the Planetarium and the Aquarium. (Not many opted for the last two: we preferred the shopping precinct.) On the final afternoon a large number of us joined the Potomac River Cruise on a replica of a Mississippi Riverboat—with paddlewheel and the New Orleans Luncheon on board was superb, also the weather." (Blount 1984, pp. 4–5)

R.R. (WES Member), "1984 SWE Student Awards Banquet, 20 June 1984":[9] The article talks about some of the messages of its Keynote speaker, Dr. Cox. The author also reflected on the large number of students present and compared this to the situation in the UK.

"Dr. Cox spoke of the alarming ignorance of technology, particularly in the United States Congress, and of the great need for more technologists, which must be an incentive to encourage women into the profession. She then advised students not to aim for wealth and security in their jobs but to go for the work they enjoyed doing and to be willing to take risks. Wealth and security are by-products, she claimed."

"The number of students and the extent of the support given to SWE by industrial companies in the States seems extraordinary to women engineers from the UK and other Western countries. In Britain the relative number of professional women engineers is less than one percent. Why is this?"

"Ten years ago, the percentage of women engineers in the States was similar to that in this country (less than 1%) but since 1975 the number of women engineering graduates has risen exponentially and today the proportion is 12%. In the UK the proportion is 6% in

[9] R.R. could not be identified, but maybe Ruth Raymond or Rosalind Redwood.

universities, less than 2% in Polytechnics (1982 figures). In 1982, Dr. Sue Bullivant of Loughborough University went to the States on a Churchill Scholarship to analyze the reasons for the increase in the States and her report was summarized in the New Scientist of 30 September 1982."

"Dr. Bullivant pointed to three major factors: the Equal Pay and the Affirmative Action Laws (which required Federally-assisted educational establishments and large companies to develop plans to employ women at all levels in proportions reflecting the availability of qualified women), the Vietnam war, and its cessation, both of which had significant effects on the demand and supply of engineers, and on the high status of engineers in the States, where they command starting salaries higher than doctors and accountants. In effect, the supply of women engineers has increased because there is a demand for more engineers, because women are given strong and active encouragement, and because the incentive is high."

"The need for engineers grows with the development of new technology and the WES offers a means of giving the support and encouragement needed by young women in this country. The success of SWE in the States, as shown by the enthusiasm of the students at the conference, has been a heartening experience for those who attended." (R.R. 1984, p. 3)

"Another short comment on ICWES-VII was published in the WES Journal; there are some inconsistencies in the various reports about the number of participants and the countries they were from. The comment below lists 13 countries from which delegates came from: Fifty-three overseas delegates from 13 countries attended, namely, Belgium, Canada, Egypt, France, India, Iran, Italy, Ivory Coast, Nigeria, Philippines, Switzerland, United Kingdom and Yugoslavia. Nine members of the WES represented the United Kingdom."

"At a WES Council meeting in 1987, members mentioned that ICWES-VIII would be in September at Abidjan in the Ivory Coast, and because of the high cost and the unfamiliarity of the venue, it seemed likely that there would be few British delegates. The tentative proposal made at Washington in 1984 that the UK should host ICWES-IX was considered to be impossible of fulfilment in 1990, but there was support for the suggestion that the Society should consider offering to run the Conference in 1991." ("International News – VII ICWES" 1984, p. 6)

Chapter 9
ICWES-VIII, Abidjan, Ivory Coast, September 20–27, 1987

Theme: "Science, Technology and Development"

It was opened by the Head of State, preceded in procession by a group of Kings and Chieftains, resplendent in beautiful robes and carrying golden carved staffs and maces. The theme was explored at plenary sessions and working parties met to discuss 'Science and Women', 'Communication and Transfer of Technology', 'Science and Development' and 'Sciences and Techniques'. Recommendations from the working parties emerged at the final plenary session. Visits were made to the Port Authority, palm nut plantations and a power station. The Conference ended with a buffet banquet, with African dancers and fireworks, and an enormous ICWES iced cake!"

The Organization Responsible for ICWES-VIII: CIFISATS

The Cercle Ivoirien des Femmes Ingénieurs, Scientifiques, Architectes et Techniciennes Supérieures (CIFISATS) was created in 1985. Its president, at the time of the conference, was Marthe Achy Brou (see Appendix 1). The goal was to mobilize all women in these fields and encourage women to choose these careers. The organization holds information campaigns to promote women and recognize their qualifications and their careers. It also encourages collaborating with other organizations within Ivory Coast and internationally. The conference was supported by the Association des ingénieurs, techniciens, et agronomes de la Côte d'Ivoire (AITACI).

M. Frize et al., *Women's Contribution to Science and Technology through ICWES Conferences*, Women in Engineering and Science,
https://doi.org/10.1007/978-3-031-33800-7_9

The Opening Ceremony

First, the mayor of Abidjan welcomed the participants. This was followed by a talk by the president of the conference, Marthe Achy Brou. Her speech was in French and what follows is a summary in English of its highlights (ICWES-VIII, Proceedings, pp. 49–53).

Achy Brou stated that this conference was an exceptional occasion for Ivory Coast women to show their competence and skills in organizing such a major event. Through the main theme, "Science, Technology, and Development," the conference planned to display the important role that well-designed research and an appropriate technology can play in the fields of Agriculture, Energy, Electronic Equipment, Construction, and Public Works in building a country. The goals are to lighten the hardship of field work, improve the yield of cultures, and make essential infrastructures accessible to the population.

After thanking the President of Ivory Coast, Félix Houphouët-Boigny, for his vision on the importance of education, science, and technology, Achy Brou informed participants that she and her colleagues were very proud that the ICWES Continuing Committee had chosen the African continent for the first time. She informed the assembly that women came from 31 countries, 5 continents, and that the national organization of women in STEM (CIFISATS) had more than 350 women as members working in various fields such as chemistry, architecture, engineering, medicine, agronomy, informatics, and others. She ended her presentation by wishing, in the name of all African women, "akwaba" ("Bienvenue"/"Welcome") to all delegates from around the world. Achy Brou thanked her organizing committee who had worked very hard to make the event a great success. The vice chair was Issié Yvonne Gueye, and 23 other women were members of the committee.

The president of AITACI, Konan Lambert, said a few words of welcome. He deplored the fact that African countries were mostly exporting raw materials and imported manufactured products. He explained that, due to the poor economic gain of this situation, it was important for African countries to manufacture products themselves. He added that, when Ivory Coast became independent politically in 1943, the country had eight engineers, whereas now (in 1987) there were more than 3000, including several hundred women; he stated that CIFISATS is a full member of AITACI (ICWES-VIII, Proceedings, pp. 21–27).

The next speaker was the ICC Chair, Lydia Pickup (USA) (ICWES-VIII, Proceedings, pp. 13–15). She outlined how ICWES operates and explained that any organization with women scientists or engineers as full members can submit an invitation to hold an ICWES conference. There is a meeting of delegates (one person per organization) to review the invitations and select where the next conference will be held; the chair of the current conference and her deputy take the minutes of the business meeting, which includes past chairs of conferences and their deputy. Among participants who do not represent an organization, one person is chosen per country as their delegate for the meeting. One more task is to select the convenor of the next event and to appoint the members of the International Continuing Committee (ICC) that is composed of one delegate from each of the regions, which she defined as: North America, South America, United Kingdom with Australia and New Zealand, Western Europe, Eastern Europe, English-speaking Africa, French-speaking Africa, Arabic-speaking Africa, Indian subcontinent, Far East, and other Asian countries. Pickup mentioned the responsibilities of the ICC: to encourage organizations to propose holding a conference and to assist in the development of themes for the next event. She invited all delegates to attend the business meeting; all participants are welcome to attend, but voting is reserved for the official delegates. The next speaker was Mathieu Ekra, Minister of State and Chair of the Conference (ICWES-VIII, Proceedings, pp. 28–45). Then the conference was formally opened by the president of the Ivory Coast Republic.

Participants

Although Achy Brou mentioned in her opening remarks that participants came from 31 countries, the existing list identifies 59 participants from 23 other countries. Out of 221 participants, the majority came from Ivory Coast (166). The out-of-country presence at ICWES-VIII was 25% (see Appendix 3).

The Program

The conference had four main topics that were organized in working groups. Each group had a chair, a recorder, and two reporters. The four topics were as follows:

- Commission 1: "*Scientifiques au féminin*" (= Female Scientists)
- Commission 2: "*Communication et transfert de technologie*" (= Communication and Technology Transfer)
- Commission 3: "*De la science et du développement*" (= On Science and Development)
- Commission 4: "*Sciences et techniques*" (= Science and Technology)

Technical Program

As mentioned in Chap. 1, 16 tapes were recovered and subsequently digitized. From these 16 tapes:

- 10 tapes: recordings of 10 papers (mostly full texts; a few are recorded more than once)
- 2 tapes: opening ceremony, addresses, and lectures
- 2 tapes: the same discussion (without the paper); of particularly bad quality
- 1 tape: welcome session, practical information on commission work, list of papers
- 1 tape: less than a minute

A program has been partially reconstructed on the basis of these recordings, in combination with other sources, like the Conference Booklet and articles from WES and SWE. Information drawn from the recordings are twofold: either a paper was listed by the program chair (Safoura Fadiga) at the beginning of the conference, or the paper itself (full or partial) has been recorded. In particular:

- Commission 1 (5 papers): 5 titles/speakers taken from conference booklet, p. 4 (none recorded)
- Commission 2 (4 or 5 papers): 4 titles/speakers cited in Thompson and Maple (1987, p. 19), plus an additional title/speaker drawn from the chair's list (possibly moved elsewhere or not presented at all) (none recorded)
- Commission 3 (6 papers): 5 titles/speakers drawn from the chair's list (**3** are recorded); **1** additional paper is recorded (but is not cited in the chair's list)
- Commission 4 (13 papers): 13 titles/speakers drawn from the chair's list (**4** are recorded); 1 paper may have been moved to Commission 3

An additional three papers are recorded (mostly) in full but were not listed by the program chair, and it is impossible to determine in which session they were presented.

A total of 10 full papers were recovered from the recording, as shown in Table 9.1.

It should be noted that the list given by Safoura Fadiga was preliminary: speakers were invited to register their paper titles and names in the welcoming session. As such, it is impossible to determine how fragmentary it is with regard to the actual number of papers given at the conference. Similarly, the titles listed in this session were given in French, but the actual language in which the paper was delivered remains uncertain, unless this paper has been recorded. Information (name, sex, country) on speakers cited by the program chair or recorded has been supplemented with data found in the list of participants.

ICWES-VIII, S&T Papers

For ICWES-VIII, 26 S&T papers were spread across four sessions. In Table 9.2, we were able to recover 11 papers from the recorded tapes. A total of 27 speakers were expected (26 women, 1 man). Five came from the host country (5), while the rest came from Canada (2), the United States (2), Switzerland (1), the Netherlands (1), Bulgaria (1), West Africa (3), Tunisia (3) India (6), the Philippines (2), and the United Kingdom (1).

Table 9.1 Papers recovered in full from ICWES-VIII, recording

Paper	Speaker	Commission
Rôle des femmes dans l'autosuffisance alimentaire de l'Afrique	Marie-Laure Lohoury-Guigui	3
The Role of Women Engineers and Scientists in Combatting the Hazards of Modern Scientific and Technological Development	Kamala Chandrasekhar	3
Nutrition and Agriculture: A LAK Marriage	Judith Swartz	3
Science and the Law. A Comment with Reference to the Environment	Rosalind Cairncross	4
Cultural Conservation Considerations for Environmental Development	Laverne Wells-Bowie	4
Facteurs d'environnement à considérer pour la conception et la mise en opération des plateformes de pétrole immobiles en mer[a]	Joyce Table	4
Utilisation de la culture d'embryons zygotiques in vitro dans la collecte de germplasm de cocotiers. Méthodes de prélèvement et de conservation et raisons d'abandonner une culture in vitro	Béatrice Assi	4
[festival panafricain des arts et cultures]	Aminata Sahko	?
[géobéton]	Jacques Ando	?
L'utilisation de la palynologie comme outil de recherche pétrolière	Juliette Téa	?

[a]Paper delivered in English, but full title only found in French

Table 9.2 Distribution of social and gender sessions and papers in ICWES-VIII, recording

General theme: Science, technology, and development	# of papers expected	# of papers in proceedings
Communication and transfer of technology	4	0
Science and development	6	3
Science and techniques	13	5
Others	3	3
Total	**26**	**11**

ICWES-VIII, Gender Papers

For ICWES-VIII, five social and gender papers were expected for Commission 1, but none were recovered from the audio tapes (Table 9.3). A total of five speakers were expected (four women, one unknown) distributed as following: none from the host country, 1 from Canada, 1 from Poland, 1 from India, 1 from the Philippines, and 1 unknown.

Table 9.3 Distribution of social and gender sessions and papers in ICWES-VIII, recording

General theme: Science, technology, and development	# of papers expected	# of papers in recording
Science and women	5	0
Total	**5**	**0**

Meeting of Delegates

No minutes have been found about the meeting. However, there was a resolution passed by the delegates:

> "After having studied a UNESCO report which showed that in spite of recent progress, everywhere in the world fewer are entering engineering and scientific fields than in the classical studies.
>
> This present congress considered ways of improving the situation. Work was based on the experiences of the different participating countries.
>
> Resolutions on information which aims at encouraging young women to take up and to succeed in the engineering and scientific studies are as follows:
>
> 1. Invite governments to encourage girls to choose scientific studies by offering special scholarships, grants or other incentives to young women as it is actually done, for example, by the French Government. (The incentives offered would follow general national policy in this matter.)
> 2. Invite governments to pay special attention to the school programs in order to make sure that both girls and boys receive the same quality of education, especially in scientific studies.
> 3. Encourage all women engineers and scientists to be members of national associations. One of the main objectives of these associations would be to orient young women to take up scientific studies, through role models, information leaflets and through all, sensitize the young as well as their families and those who provide guidance to them. The message should also inform them on future scientific opportunities.
> 4. Suggest that associations use the most appropriate means to ensure a regular interchange of information.
> 5. Women scientists and engineers should strive to assist women, especially rural women, to ensure that they master the technical innovations offered to them."
>
> "The congress notes that, women engineers and scientists, in most countries, are still few in member and are yet exposed to prejudices and stereotypes which stem from the traditional image of women.
>
> However, all women scientists and engineers, whether members or not of associations, are involved in a slow but and continuous move towards equal rights in professional fields.
>
> Therefore, the conference recommends:
>
> 1. That we each strive to convince decision makers that qualified women can be as good a choice as men for technical and scientific career openings.
> 2. That careful attention be paid to the implementation of laws to assure family protection in order that those laws not cause discrimination in women's employment.
> 3. That women engineers and scientists be encouraged to become role models in order to publicize their ability to combine successfully both professional and family life.

4. That women be encouraged in their desire to have full professional career.
5. That contacts between national associations be developed in order to improve communication between the groups and to welcome and guide visiting members.
6. That associations be encouraged to participate in ICWES meetings.
7. That in each country, a day be established to recognize the accomplishment of women in science and technology.
8. That associations be encouraged to find awards which be given in recognition of outstanding work by a woman scientist and engineer.
9. That the national associations should have close contacts with other professional women.
10. And that the participants at this conference should present these recommendations to their respective relevant government ministries i.e. industry, science and technology, women's affairs, etc., and to representatives of national and international organizations and corporations."

"Women engineers and scientists recognize a responsibility for the growing need to ensure exchanges between developed and developing countries.

Therefore, the conference recommends:

1. That all women scientists and engineers should be encouraged to have an active role in any organization that provides training for women to become skilled workers.
2. That whenever a new technology is to be introduced, especially in agriculture, women should first be consulted to ensure that the project is appropriate and effective and improves the status of women.
3. That open exchange and collaboration between agronomists and nutritionists be encouraged in order to reduce malnutrition and the food problem in developing countries, particularly on the African continent.
4. That in the field of construction there be more reliance on indigenous technology and greater use of raw materials available in the developing countries" (ICWES-VIII, "Final Resolutions")."

On the Location of the Next Conference

"A meeting of the ICWES delegates resulted in the selection of Great Britain to host ICWES-IX in 1991. The fall-back selection was Lausanne, Switzerland in 1990. These were offered by C. Thompson and J. Julliard, respectively. Both France and Zaire also presented bids but were asked to rebid next time. The delegates voted to revise the Guidelines to add committee members – a separate one (from Dormer Ellis, reelected the North American representative) for the USA was proposed by ICWES regular and SWE member, Lee Arnold. SWE past president Aileen Cavanagh was appointed to that spot with BK Krenzer as alternate. Others added were a second Latin (Central and South) American one so that both Spanish and Portugese speaking countries would be in the communications net and a Portugese one for Africa was added as well, since Arabic, French and English are already covered there. US representatives, particularly A. Cavanagh, promised to assist the new ICC Convener, M. Achi Brou, in identifying contacts in those areas of the world which were not represented. Latin America, Japan, China, etc. still needed representatives. Other proposed revisions of the guidelines are to be submitted to the committee between meetings so that delegates can be prepared for faster action when it is time to vote on the changes" (Krenzer 1988, p. 7).).

Testimonials

BK Krenzer, "ICWES VIII – A West African Experience"

"When you find yourself on a US airline domestic flight waiting for a translation of the pilot's announcement, you know that you were in French West Africa too long. Even those for whom French was a familiar second language had trouble understanding it, so for barely-able-to-read-a-little-French attendees at the 8th International Conference of Women Engineers and Scientists (ICWES) it was a continuing case of' 'waiting for the translation.' But there was no difficulty for anyone in appreciating the visual and intellectual effects of the Conference: Even before the sensational opening on Monday, September 21st, the color, rhythm and excitement of the Sunday evening session had set the mood. The color was provided by the beautiful native dress of the participants, many of whom were spotlighted as they brought greetings from their respective countries."

"[...] That formal event [= the Bringing of Greetings] was followed by a reception where the color and music again prevailed. Refreshments were contained in hollowed-out pineapples each topped with a bright red hibiscus blossom and the music was from an African 16 string 'mandolin' made from half of a large leathercovered gourd."

"Even more impressive was the opening session with a whole contingent of robed local chieftains with their staffs of office, a full orchestra and a group of women leaders in costumes of various areas/periods. Participants who stayed at the Hotel Ivoire which adjoins the Convention Center were fortunate to be involved in the preparations of this group – the drums started just after daylight. This was all preliminary to the arrival of the President of Cote d'Ivoire who, with a number of his ministers, attended and officially opened the 8th ICWES. He and SWE's Lydia Pickup led the procession of dignitaries into the hall – very impressive. The work of the conference was divided into four 'Commissions'. One on technology was under the direction of a representative from Zaire, and the other three on various professional and social aspects of careers for women in engineering and science were led respectively by Nicole Becarud of France, by Winifred Karagwa Byanyima of Uganda, and by Bette K. Krenzer of the US. The latter three had a goal of developing resolutions for adoption by the Conference with carefully paralleled versions in both French and English. These were subsequently approved and are included (in English) at the end of this article."

"This report would not be complete if it did not also include information on the many cultural, entertaining and educational aspects of the meeting. Food was, of course, a continuing highlight. Lots of area specialties – fresh fruits, unfamiliar but tasty vegetables, wonderful French breads and rolls and meat cooked in spiced sauces – characterized the many meals which were part of the conference. Fish, chicken, and mutton were the main 'meats' but it was the papayas, mangos, bananas, pineapple and even the fried plantain that could become addictive for the US attendees. Those who stopped in Senegal also had a chance to try the very spicey national dish, fish and rice, in several forms and in all cases, there seemed to be no lack of appetite among the US contingent. Lunches were a regular part of the meeting and at least three gala dinners were highlights. In addition to fruit and cheeses, there were wonderful pastries and mousses to finish off the meals (and the diners)."

"Tours of the city (Abidjan), the port, and the Blohorn factory (which produces margarine, soap, etc. from palm oil) were taken and reported by Renee Stone, Betsy Osborn, and Helen Marie Chapman during the early part of the week while others were tied up in sessions. The port was especially notable for its use of long-range planning being now well into its third master plan after completing implementation of the first two. By mid-week all participants took time out for a tour of a fabric printing plant. The Uniwax factory is one of the largest producers of batik prints in the world. We had an excellent tour with behind-the-scenes views of all the technology and artistry involved in the production of fabric. We found, unfortunately, that women (the major users of the product) were not employed or even consulted in the plant anywhere from design to, production. A couple of secretaries and the cleaning women struck us as an inadequate use of females, denying some valuable inputs to the business."

"Our next venture into industry was a tour of a palm oil factory and plantation. In this case we studied the business from planting/growing the trees through harvesting the nuts to actually producing the palm oil and palm fiber byproducts. This event was capped by an outdoor lunch hosted by the plant manager. A subsequent discovery was of the use of palm leaves (we'd call them branches) for fences. The leaves are placed with their spines in parallel about as far apart as the length of a frond (approx. 8″) and the adjoining fronds are interwoven. It makes an attractive, sturdy, ribbed fencing material. But much more unique and colorful were the fences, chiefly in Senegal, which used the leftover sheets of 1″ foam after the soles of rubber thong sandals had been stamped out. The resulting grills in shades of white, blue, red or green struck the USA contingent as making best use of by-products. Some of us had always thought those sandals were all made in the Far East."

"SWE's Director of Yankee Region, Sally Osborn, proved to be an invaluable asset to the US contingent since she was not only adept at French but was uninhibited by familiarity with a Parisian accent, which seemed to hamper the Europeans for whom French was a second language. Her skill was particularly useful in the local markets where bargaining is de rigueur and language differences can get you into real financial trouble. She and her daughter, Betsy, bought some of the official conference fabric and had outfits made – very smart looking. They weren't ready in time for the two major conference events outside the city of Abidjan."

"An all-day tour to the 'new capitol' of Yamoussoukro was the first of these. The trip began with a visit to a modern hydro power plant fed by a lake formed by damming the River Bandama. Then we were on to Yamossoukro, hometown of President Houphouët-Boigny. Especially designed from the bare land up to include lovely well-lighted wide avenues, public buildings, schools and residences, the 'Brazilia-like' city was impressive and beautiful to behold. Highlights were a President's Park on land which belongs to the President's family, a university which has completely modern and esthetically pleasing facilities, a beautiful reception hall where we were welcomed by the mayor and a banquet hall where we were entertained during lunch by a very talented group of local dancers and musicians. Many of the conferees joined in the dancing."

"The second, a real treat, was a post-conference picnic at Grand Bassam hosted by the Mayor of the city, Madame Achi-Brou, our conference chair. The event had everything – sightseeing, greetings on behalf of the mayor at the 'town hall', a ceremonial welcome at the Mayor's home by local chiefs in their formal regalia, a fabulous picnic lunch and entertainment at the mayor's pavillion, a walk on an ocean beach outside her compound, and a tropical rainstorm to round out the day."

"[…] Four US women scientists and architects who were not SWE members found out about and attended the Conference. They have agreed to be contacts to help SWE get the word out better for the next ICWES. A chance encounter with a Los Angeles MD, Dr. Williams, who traveled from Abidjan to Dakar with the SWE contingent, demonstrated the need for better publicity in the USA. She did not hear of the Conference in advance and would have participated while in Abidjan had she known of it."

"Finally, L. Pickup and D. Ellis traveled to Cote d'Ivoire through Paris together and departed first for stopoffs in Europe. L. Arnold and H.M. Chapman traveled to Africa with the main contingent but stayed on in Abidjan for a couple of days post-conference sightseeing before returning directly to the US. A. Cavanagh arrived a day after the Conference opening and accompanied the main group to Dakar but left a day early for the US. Family weddings were the draw for her."

"The rest, BK Krenzer, S. & B. Osborn and R. Stone, enjoyed a full two weeks in French West Africa climaxed by an invitation to lunch in Dakar from a fellow Conferee, Aminata (Ami) Sakho, who made that the second highlight of our visit to Senegal. The other was our first evening meal at the hotel in Dakar. Four French residents had reserved the terrace for a special anniversary meal so we were consigned to a round table in the (all-glass walled) corner of the dining room. We had, it turned out, ring-side seats for a full evening of classic music, dancing and singing arranged for the special party. It was even better than if we'd hired them ourselves because the entertainment was both excellent – and free."

"The concensus was that the VIIIth ICWES was colorful, informative, successful and a smash hit with the US contingent. Attendees each received a brass-plate-topped white marble paper weight bearing the conference logo, an attractive memento of a very unique experience" (Krenzer 1988, pp. 6–8).

Cicely Thompson and May Maple (UK), "Eighth International Conference of Women Engineers and Scientists Abidjan, Ivory Coast 20–27 September 1987: A Report"

"The Conference was organized and hosted by the '*Cercle Ivoirien des Femmes Ingénieurs, Scientifiques, Architectes et Techniciens Supérieures*' (CIFSATS), who, under the Chairmanship of Mme Marthe Achy Brou, Vice-President of the National Assembly, gave all the delegates the warmest possible welcome and arranged interesting technical sessions and visits interspersed with splendid social events. Some 200 delegates from about 30 countries attended. After the 'Bringing of Greetings' on Sunday evening from all the countries represented, the Conference was opened on Monday morning by the Head of State, President Félix Houphouët-Boigny, in the presence of other Ministers, representatives of Ivory Coast women's organizations, local dignitaries and the diplomatic Corp as well as the Conference delegates. The afternoon was spent setting up the technical sessions on the overall Conference theme 'Science, Technology and Development'."

"Apart from formal lectures at plenary Sessions on three days, the technical sessions took the form of working parties (Commissions) with the following subjects:

Commission 1: 'Science and Women';
Commission 2: 'Communication and transfer of Technology';
Commission 3: 'Science and Development';
Commission 4: 'Sciences and Techniques'.

Appropriate papers were presented and discussed at these commissions and, arising from the discussions, each was asked to submit recommended resolutions for adoption at the final plenary session."

"In Commission 2 we heard interesting papers from Paola Pigrucci (UNIDO and Italy) on the database which is now being compiled by the UN Industrial Development Organisation (UNIDO) in Vienna, comprising statistics and a bibliography on the employment of women in industry; from Chebaâne Hassiba (Tunisia) about appropriate technology transfer and education programs by USA and France in Tunisia; from Jacqueline Juillard (Switzerland) on the role of technical journalism; and from Marie Anne N'Dendé (Gabon) about her experiences as an architect working in Europe on African projects and in Gabon working on European projects."

"Each paper provoked lively discussions out of which common themes emerged; the exploitation of women; the need for technology transfer programs to be appropriate to the local environment and local industries; the need to develop and support local ideas for use and improvement of technology in local industries; and the need for educated women to find ways of helping their less fortunate sister. It was reported that in Africa 90% of agricultural work is done by the women yet their experience is ignored, and the aid programs frequently failed to consult their consumers."

"The resolutions adopted at the final plenary session centred mainly on the need to open the educational system fully to women on equal terms with men and on ways of encouraging and enabling girls to enter careers at all levels, in science and technology. Since these sentiments are familiar to WES members these resolutions are not quoted here in detail but the final three which dealt with exchanges between developed and underdeveloped countries, are given below:

Le Congrès recommande que:

1. *Les femmes ingénieurs et scientifiques soient actives dans toutes les associations ayant pour objectif de faire acquérir par les femmes non qualifiées des qualifications valorisantes.*
2. *Chaque fois qu'une nouvelle technologie est introduite spécialement en agriculture, s'assurer que les femmes sont effectivement consultées au préalable et que le projet est adapté, en particulier, veiller à qu'il améliore le statut de la femme.*
3. *Encourager une coopération active entre agronomes et nutritionistes en vue de réduire la malnutrition et d'améliorer la situation alimentaire dans les pays en voie de développement, particulièrement sur le continent africain."*

"The last item on the Conference Agenda was the announcement that, subject to confirmation, ICWES-IX would be hosted by WES in England in 1991" (Thompson and Maple 1987, p. 19).

May Maple, "Social Aspects – Random Jottings"

"Cicely Thompson and I [May Maple] arrived in Abidjan on Saturday 19 September at about 8 pm. It was very hot and humid but the air-conditioning in the excellent 5-star Hotel Ivoire was such that, at times, we nearly froze to death!"

"We had travelled from Paris, via Geneva, with the French and Belgian delegates and joined them on Sunday for an outing to Grand Bassam, a small beach resort several miles east of Abidjan. The sea was too rough to swim so we lazed in the shade, lunched in an open-sided, thatched roof café on the beach whilst being entertained by some splendid male dancers and musicians. En route, we were able to visit a busy craft centre and watch skilled artisans making artifacts in bronze and brass, wood carvings, hand painted batik, furniture and the like. We also found a delightful costume museum where, unfortunately, we were not allowed to take photographs."

"That evening, after the 'Bringing of Greetings' presided over by the 'Minister for the Promotion of Women', Madame Aka Angui, there was a 'Welcome' Cocktail Party in the hotel. Not only was the hospitality terrific but we were struck by the considerable number of African women engineers and scientists, the beauty and elegance of the African women generally and their superb carriage – they moved like queens. I must confess we felt rather drab in comparison."

"On the Monday morning, Conference opening day, local people gathered under the huge portico outside the Congress Hall and several groups of dancers, singers and musicians performed for the general benefit whilst awaiting the arrival of the President. He was preceded by a group of kings and chieftains resplendent in beautiful robes, wearing crowns or other headgear denoting rank and carrying golden carved staffs or maces. The President, in civilian suit, was escorted into the building by a bodyguard reminiscent of our Household Cavalry in uniforms of red, green and gold. They looked superb."

"Tuesday was the first real workday but, after an excellent buffet lunch, Cicely and I decided to join the Port Authority visit. This turned out to be the right decision because after we had been told about the land-based facilities we were ushered on board one of the Authority's launches and taken on an exhilarating trip round the lagoons as far as but not into the open sea. In the evening we were treated to a typical African meal in a local restaurant which was interesting and enjoyable – spicy meats and lots of strange fish. Incidentally, we found that most drinks were fresh fruit juices although there was a very good locally brewed beer called 'Flag' and mainly French wines were served with meals."

"Wednesday found us working again in the morning and on a town tour in the afternoon when we visited the big local market where practically everything and anything was on sale. You did not, however, pay the initial price asked. Bargaining was the essence and added to the fun. There were practically no shops as we know them but everywhere stalls and markets large and small. There were many fine buildings in the business part of the town including a rather lovely modern cathedral of unusual shape and design, seating 4000 and with beautiful stained-glass windows. It was floodlit at night and, sited on a hill, looked as though it was floating. On the other hand, the outskirts of the town were rather sleazy but the people looked clean, well-fed and happy."

"The town museum was well worth a visit. There were wonderful carved wooden masks and legendary figures (gods), native costumes, furniture, utensils and weapons through the ages but I must tell you of one thing in particular which stands out in my memory. We came to a small bridge over a stretch of water (I don't know whether it was a river or part of the many lagoons), and were amazed to see 60–80 men busily doing the washing on slabs of stone and up to their knees in the water. This was the local laundry and the washing was carried in and out in large bundles on the heads of the men concerned. We were told that all paid work was done by men. For instance, all the staff at the hotel, including chambermaids, were men, whilst the women did the hard unpaid work in the fields, etc.!"

"Thursday, all delegates visited a palm nut plantation. Row upon row of tall palms bearing large clusters of orange-coloured nuts about the size of a Brazil nut. These clusters were cut off by men each wielding a scythe-like knife attached to a very long bamboo pole. The clusters were then taken to the factory, roasted, crushed and made into oil, detergent, soap, etc. We learned that Lux soap was made there. An excellent lunch was provided in an alfresco setting with thatched roofing to shield diners from the very hot sun. The evening was passed at the Golf Hotel, across the lagoon from the Ivoire, on the invitation of the Mayor of Abidjan. Many ambassadors were present, but alas, not the UK; cocktails were served in the open air on the terrace followed by a sit-down buffet supper in the dining room."

"Next morning, Friday, we all embarked on coaches to visit the new administrative capital of Yamoussoukro some four-hour drive inland through lush green plantations of palms, bananas, pineapples, mangoes. Everything seems to grow in great profusion and, further north even rice and cotton. En route we visited the Barrage de Taabo and were shown over the power station."

"Our arrival, somewhat late, at the Yamoussoukro town hall was heralded by a white-suited brass band lined up on the steps and as soon as our African friends left the coaches and heard the music they started to dance up the steps and we proceeded in this fashion up the inside stairs and into the very elegant council chamber. The Mayor greeted us and when the Welcoming speeches were over a very good buffet lunch was served in a large dining hall with a stage at one end on which appeared a series of native groups who danced, sang and made music for our entertainment."

"At the end of the proceedings, the mayor, a large jolly man and an advocate by profession, rose from his table and insisted that everyone dance, himself included. It was hilarious and a most fitting end to a thoroughly enjoyable episode. Before making the long drive back to Abidjan we toured this new city and found it bright and spacious with beautiful public buildings amongst which was the new House of Assembly (Parliament), a two-tier circular building which shone in the sun and was surrounded by green well-manicured lawns, trees and flowers."

"Saturday, of course, saw the closure of the working part of the Conference and was followed in the evening by a magnificent buffet-banquet in the Hotel Ivoire. African dancers entertained us, an enormous ICWES iced cake was wheeled in and fireworks set off around it before the President of Congress, Marthe Achy Brou, cut the cake and distributed slices to each person present; a wonderful end to a memorable week."

"The next day, Sunday, Cicely and I, with Paula Talpaert from Brussels, flew up north to Korhogo for a few days escorted by a friendly African courier named Patrice. He had been born in the area and his family and friends still lived in the villages. The red earth and lush fertility of the region were striking. Villages of round thatched roofed mud houses lay at the end of narrow dirt tracks, scarred with gullies made by the rain in season. The main road was merely a wider dirt track which went straight on to the horizon, a red gash in the greenery. A sense of space was tangible with nothing man-made to distract the eye."

"Each village seemed to specialise in certain crafts as well as tending the fields, to which families walked several miles every day with equipment and food on their heads – even the youngsters carried their share – returning in the afternoon to the village where the older people left behind had prepared a huge cauldron of food for the evening meal in the open air. We visited villages which made cotton goods of all descriptions from the raw cotton they had grown through spinning and weaving; others who hand painted stylised birds, animals and fish, or strange masked god-like creatures on hand woven cloth and pinned them on the outside mud walls to dry; yet others who made coco-butter from the nuts, bark and leaves of the karitier trees. Wooden sculptures and intricate masks were carved in a side street of Korhogo, the raw materials piled outside the house as branches taken freely from local forests and, here again, stalls and markets in the main street selling everything under the sun. The hotel in Korhogo was a ranch style house with a swimming pool and four of the largest tortoises I have ever seen in the garden. One was 100 years old, quite 2 ft high and more than 2 ft long!"

"We returned to Abidjan for the flight home excited and thrilled with all we had seen. Looking back now, one remembers a beautiful green country, red earth, wonderful vistas as far as the eye could see and charming kindly people who did their utmost to make our stay a happy one – an experience I am glad not to have missed" (Maple 1987, pp. 19–20).

Chapter 10
ICWES-IX, Warwick University, Coventry, UK, July 14–20, 1991

Theme: "Communication"

ICWES-IX was hosted by the Women's Engineering Society (UK) at the University of Warwick in Coventry, UK, in 1991. The theme was "Communication," and topics included Transport, Satellites and Telecommunications, Basic Sciences, Information Transfer, Technology Transfer, Education, Career Development, and Gender issues and Demographics, including the education, training, employment, and status of women engineers and scientists worldwide. It attracted 466 delegates, including 305 from the United Kingdom and 141 from 41 other countries, a proportion of 30.3% and the highest number of countries represented to date outside of the host country (see Appendix 3).

Jacqui Brookes Provides an Update on the Organizing of ICWES-IX

> "The ICWES-IX executive committee welcomed its new chairman in December, Betty Laverick, of the Women's Engineering Society and the Institute of Physics. Thanks are expressed to Hilda Blount for her term of office and for setting up the registered company, ICWES-IX."
>
> "The committee has been further expanded to include Margaret Farago, representing the Royal Society of Chemistry and women chemists in the UK and abroad."
>
> "The organization of a conference which seeks to attract up to 500 people from around the world is a mammoth task, and therefore a professional conference organizer has been appointed – Conference Services Ltd, who will report to the committee and provide the detailed organization needed."

M. Frize et al., *Women's Contribution to Science and Technology through ICWES Conferences*, Women in Engineering and Science,
https://doi.org/10.1007/978-3-031-33800-7_10

"Thanks are due to the generosity of WES members who have become guarantors and who have provided donations. We are seeking further funds in the form of a sponsorship package for company participation."

"The conference theme, 'Communication,' will cover all aspects of science, engineering, education and social issues. If you would like to suggest a topic or present a paper, please gather your ideas now ready for submission to the technical committee in summer 1989."

"Many WES members have friends and contacts abroad; do let them know of ICWES-IX and invite them to participate."

"The ICWES-IX committee welcomes any offers of help from WES member for fundraising, publicity" (Brookes 1989a, p. 17).

Jacqui Brookes Provided a Further Update on the Conference Organization

"We are pleased to announce the patronage of eight eminent and respected scientists, engineers, and industrialists in connection with the 1991 Conference. They are: Dame Anne Warburton of Cambridge University. Baroness Gardiner of Parkes, Baroness Platt of Writtle, Eric Ash, Rector of Imperial College, Robert Malpas of BP and Chairman designate of Little G., Cicely Thompson, an inspiring force within the ICWES movement, Basil Butler, Managing Director BP and Sir William Barlow BICC."

"The major effort at this point in the Conference build-up is fundraising. To attract as wide participation as possible for a realistic conference fee, substantial funds and sponsorship are required. ICWES-IX is seeking to appoint a fund-raiser to head up the campaign. The aim is to raise at least £50,000. In association with the Conference there will be an exhibition with industry and education promoting careers in science and engineering and illustrating the Conference theme, Communications. Beverley Stamford of Courtaulds is coordinating the exhibition" (Brookes 1989b, p. 17).

Elizabeth Laverick, "ICWES One to Nine"

"On the first evening of the conference, a 'Bringing of Greetings' ceremony was held followed by a party in the form of an Old English Fayre, which included stalls providing such Old English delicacies as Game pie, Syllabub and Mead (a honey-based drink), and entertainment from Morris Dancers and a Jazz band. We were honored by the 'visit' of the famous Lady Godiva who in the 11th Century rode naked through the marketplace of Coventry in order to persuade her husband to reduce the taxes he had imposed on the town."

"Needless to say a variety of tours, technical and social, were available; learning from the Washington experience, a career guidance and training seminar was held for the younger delegates. With a view to increasing the number of girls interested in engineering, and to tell delegates more about the country, the opportunity was taken to hold an exhibition in parallel with the conference with stands from industry, government and educational and professional organizations. A series of educational events was also held during the week to which local schools were invited and in which delegates were also able to participate. These included a WISE (Women into Science and Engineering) bus, (a travelling exhibition offering hands-on experience), part of a project organized by the Engineering Council and the Equal Opportunities Commission to tell schoolgirls about Science and Engineering. A lecture for schoolchildren organized by the Institution of Mechanical Engineers was given by the Post Office. This was attended by some 1,000 pupils who also visited the exhibition" (Laverick, in ICWES-X, Complete Proceedings, p. 11).

Having reviewed all previous nine conferences, Laverick poses the question: "So what can we learn from the ICWES series so far in terms of its stated aims?".

She responds: "Firstly, the figures of attendance and of countries represented show that our first aim, that women active in the fields of engineering and science shall know each other worldwide, is being achieve. [Authors' note: Laverick includes a table with attendance ICWES-I to IX]. More needs to be done to promote the conference worldwide, and to help us all to keep in touch. [...]

"Secondly, the themes chosen have been wide ranging and, in line with our second aim, have enabled us to consider and discuss areas where technology can be used for the betterment of life. Not only has this encouraged a large number of papers, and attracted large audiences, but also the papers have provided an impressive record of the growing technical contributions made by women in these particular areas over the last 27 years.

"This impinges on our *third aim* – that the participation of women in engineering and science shall be increased. In addition, many of the papers in the sociological sessions discussed the different ways in which countries are tackling the problem of the shortage of women engineers and scientists, and the waste of talent that this implies" (Laverick, in ICWES-X, Complete Proceedings, p. 12).

Opening Ceremony

A message was given by HRH The Duke of Kent:

"The Engineering Council is determined to recruit more women into science and engineering. Although the number of women entering these professions is rising gradually, the total is still lamentably small: only around five per cent. For this reason, I welcome the Ninth International Conference of Women Engineers and Scientists to be hosted by our Society at Warwick University in July, which should do much to raise the profile of women in these fields. By meeting together to discuss the theme 'Communication' your delegates will be demonstrating the significant contribution made by women to this important subject."

"Communication implies informing people, whether of good news or bad, the mundane or the dramatic. Like it or not, much of the information we receive today concerns afflictions of one kind or another; famine, ill-health, deprivation or natural disaster. Some of these afflictions are inescapable, but many can be either prevented or ameliorated through the application of technology. This is a prospect which I believe ought to have a strong appeal to women, especially to women who are qualified as scientists or engineers."

"I know that your delegates are gathering from many countries and in the hope that your discussion will highlight the contribution that women can make to solving some of these complex problems, I wish you a highly successful and fruitful conference" (ICWES-IX, Conference Report, p. 2).

The Welcome from the President of WES: Dorothy Hatfield

"ICWES-IX was, without doubt, the most significant and impressive experience of my Presidency of the Women's Engineering Society. It was certainly a week to remember and the presence of over 400 delegates from 42 countries must make it the most cosmopolitan event I can ever hope to attend."

"Although I did not get to as many of the working sessions as I would have liked, I did see a few excellent snippets and thoroughly enjoyed the social events. But for me, and I suspect for many others, the best thing about the week was just having the opportunity to meet and talk with so many women who shared a common interest in science and engineering which bridged the differences between our various and differing cultural backgrounds."

"The conference certainly came as a wonderful ray of light in the middle of a year which was not a happy one for many people throughout the world. Many countries were in the depth of recession and many places were torn by political troubles and by disaster both natural and man-made."

"Of course, all this fantastic happening only came about because a lot of people put in a very great deal of hard work over a long period. All gratitude and credit are due to Elizabeth Laverick and the members of the organizing committee, and also to numerous others, both in the UK and from other countries, who gave time and effort to make it all such a success. Three years ago we knew it would be a tremendous undertaking – we were certainly proved right when the recession and the Gulf war mitigated so strongly against us."

"It is ever more apparent that the only way for a voluntary organisation to make things happen in this day and age is to maximise the use of all the will, skills and experience available to the project in hand. This certainly proved true for ICWES-IX and must increasingly be used as the means of making the world an acceptable place to live."

"In these days when there is almost universal concern for our environment, it sometimes seems that many people do not realise that we cannot hope to achieve the goals and improvements we seek without scientists and engineers. It is only they who can develop the facilities and systems which will give us the essentials of food, comfortable shelter and the ability to keep ourselves healthy and active in ways that will be sustainable and acceptable in the future. Of course, they must also be the source of providing the means of our enjoying the non-essentials the world of the 21st century will except, entertainment and relaxation as well as the means of survival."

"To do this we must learn how to make the best possible use of all the resources available to us, food, fuels and above all our most precious resource – people. This means that the half of humanity who are female, must be encouraged to make a full contribution in all areas and at all levels. Their talents cannot be allowed to be channeled into relatively few specialised areas; they are needed everywhere."

"The breadth and depth of the papers presented at ICWES-IX certainly demonstrated the capabilities of women who are […] involved in scientific and engineering pursuits. The continuing need is to extend that involvement to many other women and to show the world at large that these disciplines are not only essential but exciting and interesting as well."

"So, COMMUNICATION proved an excellent choice for the theme of the conference and must continue as a core objective for us all. We must communicate with each other and not lose the valuable links forged at ICWES-IX. We must communicate with the rest of the world to show how we, as women, scientists and engineers, can help towards a secure and comfortable future" (ICWES-IX, Conference Report, pp. 17–18).

Elizabeth Laverick, "Organising ICWES-IX"

Dr. Laverick explained the process followed for organizing ICWES-IX, which can serve as an excellent model for anyone undertaking such a huge responsibility. It is likely the only ICWES to date to set up a company to protect the funds of the Society and of the Organizing Committee Members.

"Final Thoughts"

> "So, we did not go bankrupt! There were no obvious hitches apart from the lack of translation facilities, and everyone seemed to enjoy themselves. But how successful was it and what can we learn for the future? We, the Executive Committee, examined our own navels and came up with quite a few thoughts which we have passed on to Nigeria, the potential hosts for ICWES-X. We also circulated a questionnaire to delegates with a view to learning from them, the receptors. 173 delegates responded out of a total of some 400" (ICWES-IX, Conference Report, pp. 17–18).

The process followed by the ICWES-IX organizing committee is an important message for anyone tasked with organizing a conference, especially an ICWES (see Appendix 6 for the complete text from Elizabeth Laverick on this subject).

ICWES-IX, S&T Papers

The beginning of the 1990s was marked by the twentieth anniversary of Earth Day[1] and the dissolution of the Soviet Union in 1991. Regarding the women in STEM issues, it was also a turning point in many respects. Women scientists broke some glass ceilings: on May 18, Helen Patricia Sharman became the first British astronaut as well as the first woman to visit the Mir space station. Doris Malkin Curtis became the first woman president of the Geological Society of America, and the Indian geologist Sudipta Sengupta became the first woman scientist to receive the Shanti Swaroop Bhatnagar Award in the Earth Sciences category ("Timeline of Women in Science" n.d.).

Innovations in S&T were buzzing; history retained that the first call using the Global System for Mobile communications (GSM), was made. On February 26, the English computer scientist Tim Berners-Lee (Professorial Fellow of Computer Science at the University of Oxford and professor at the Massachusetts Institute of Technology) introduced the World Wide Web, the first web Internet browser. The

[1] "EARTHDAY.ORG's mission is to diversify, educate and activate the environmental movement worldwide. Growing out of the first Earth Day in 1970, EARTHDAY.ORG is the world's largest recruiter to the environmental movement, working with more than 150,000 partners in over 192 countries to drive positive action for our planet." ("About Us – Earth Day" n.d.)

Table 10.1 Distribution of S&T sessions and papers in ICWES-IX, Complete Proceedings

ICWES-IX S&T sessions	Session titles	# of papers expected	# of papers in Proceedings	# of abstracts only
ST	Satellites and Communications	14	9	5
TR	Transport	12	10	2
BS	Basic Sciences	12	11	1
IT	Information Transfer	12	9	1
TT	Technology Transfer	10 (12)	8	3
	Total	*60*	*47*	*12*

first website was built at the CERN labs and published online on August 6. While Microsoft was working on MS-DOS and Windows, the Finnish student Linus Torvald announced, on August 25, the creation of the free and open-source operating system Linux kernel.

The ICWES-IX proceedings (ICWES-IX, Complete Proceedings) present a schedule of five sessions (Table 10.1), with 47 submitted by 70 authors (63 women, 1 man, and 6 of unspecified sex). Forty-two authors were from the host country, five from the USA, two from Canada, one from Latin America (Uruguay), three from Western Europe (France, Ireland, and Germany), one from Eastern Europe (Poland), two from English-speaking Africa (Kenya, Zimbabwe), nine from Central Asia (eight from India and one from Pakistan), three from Far East Asia (China), one from Australia, and one from New Zealand.

Also, 18 posters (with 15 papers and 3 abstracts) were added to the proceedings (ICWES-IX, Complete Proceedings). Table 10.2 lists them, adding 28 more authors (14 women, 6 men, and 8 of unknown sex). Among them, 16 authors were from the host country, one from Western Europe (Belgium), two from Eastern Europe (USSR, Yugoslavia), one from English-speaking Africa (Kenya), three from Middle East and North Africa (Egypt), four from Central Asia (India), and one from Far East Asia (China).

A Glimpse of ICWES-IX's S&T Content

The conference's main theme in Warwick was "Communication", a very broad subject since communications are everywhere. As the author Monica Dawson (UK) from Leicester Polytechnic in the UK expressed this in her paper called "Scientific Visualisation – The Graphical Presentation of Computed Data" (ICWES-IX, Complete Proceedings, pp. 389–392):

> "The theme of this conference is 'Communication' and this can mean different things to different people. In the world of physics and electronic engineering, it would be interpreted as telecommunication. In the advertising world, it would be considered to be the means of conveying ideas, images and, product information" (ICWES-IX, Complete Proceedings, p. 389).

Table 10.2 Distribution of S&T posters in ICWES-IX, Complete Proceedings

S&T poster sessions	Poster session titles	Poster titles	Country
ST	Satellites and Communications	The Work of the Telecommunications Vocational Standards Council	UK
TR	Transport	Dimensioning Method for Elastic-Plastic Road Pavement	UK
BS	Basic Sciences	The Design and Development of Molecular – Sensors based on selective macro cyclic systems	UK
		The Synthesis and Characterisation of Liquid-Crystalline Crowns	UK (x3)
		Biomacromolecule as an Acoustic Emission source	USSR
		Simultaneous Optical and Acoustic Measurement of Bubble Cloud Evolution in Tap and Saline Water	UK
		Minimization of the Transverse Effect in Electro dynamic Vibration Measurement	Egypt (x3)
		Optical Fibre Sensor for Liquid Density	China
		Synthesis, Properties and Applications of Metal (O) Polymer Composites	UK (x3)
		Study of the Mechanical Properties and Microstructure on Aircraft Material AI-Li Alloys	Belgium
		Continuous Optical Monitoring of Electrical Breakdown in Synthetic Resins	UK
IT	Information Transfer	Communicating about Design: Architectural Features of Expert Systems which cooperate in Design Activity	UK
		Quality Design	UK (x2)
		Recursive Adaptive Lattice Algorithms in Estimation and Control	Yugoslavia
TT	Technology Transfer	Energy Conservation in Leicester City	UK (x2)
		Developing Countries – Communication Priorities, Bottlenecks and Appropriate Strategies	India
		Technology Transfer and Its Impact on Human Development	Kenya
		Relevance of Technology Transfer as an Impetus to Technology Innovation in Developing Countries	India (x3)

Therefore, the subjects are going from popularization to development into very specialized scientific fields. These last were often very technical and not accessible to someone outside the field of the author. It is only possible to provide a short review of the rich S&T content proposed in ICWES-IX.

The first session was on *Satellites and Communications* (ST). Among the 15 papers in this session, nine were written by researchers. They cover many aspects of different subjects, such as satellite and deep space communication or multispectral imagery for geological mapping.

Focusing on communication technologies, an example is a paper from Gill Waters (University of Essex, UK), called "Integrated Multi-Service Communication Networks" (ICWES-IX, Complete Proceedings, pp. 366–369), which states: "Recent advances in computing and communications enable voice, video and data services to be carried on a single network [...] Information of many kinds can be encoded in digital form and transmitted over telecommunication links." Her paper summarized the development of integrated service networks, examined dome-user applications, and discussed the communication protocols that enable these applications to be designed.

More specifically, the paper "The Metropolitan Area Network, A Powerful New Telecommunications Vehicle," written by Sandra Armentrout, from A T&T USA (ICWES-IX, Complete Proceedings, pp. 375–378), explored the implementation of MAN networks. A MAN is a computer network that interconnects users with computer resources in a geographic region.

The author wrote: "The MAN is the next evolutionary step beyond the LAN [Local Area Network]. It provides a voice and data communications. It operates at speeds and across distances greater than a LAN and with more versatility than today's public switched WAN [Wide Area Network]" (ICWES-IX, Complete Proceedings, p. 375).

Armenhout foresaw that MAN would be widely used one day, which is the case: "For the information revolution to take its next step in a way that most businesses can use, a wideband, high-speed, public network service that can connect business premises within a single metropolitan area and that can provide WAN connectivity via the inter-exchange carriers extending beyond the metropolitan area is required. As such, Metropolitan Area Networks may one day revolutionize the information industry" (ICWES-IX, Complete Proceedings, p. 378).

Following a recurrent theme at ICWES conferences, the second session on *Transport* (TR) contained papers written half by women in the industry and half by women in research centers. Among these papers, seven concerned roads and traffic, while the others presented technologies for cars and planes, issues in ports and harbors, and noise reduction for railways. Subjects on transport already seen at previous ICWES conferences included the paper of Rosemary Wanjiku Kung'u (Sapamo Consultants, Nairobi, Kenya) describing the building of a road system in Kenya "The Rural Access and Minor Roads Programmes in Kenya" (ICWES-IX, Complete Proceedings, pp. 433–436), and the paper of the Indian scientists Vaishali Nashikkar and A. S. Sal (National Environmental Engineering Research Institute (NEERI), Nagpur, India), which concerned the environmental impact "Environmental Considerations in Road Transportation" (ICWES-IX, Complete Proceedings, pp. 442–446).

The UK women mainly wrote on the emerging traffic issues. Among them, the paper of Christiane Bielefeld (The MVA consultancy, Waking, UK), called "Applying Road Transport Informatics in an Integrated Road Transport Environment" (ICWES-IX, Complete Proceedings, p. 431), clearly linked transport and communications. She compared different programs written to solve increasing traffic issues based on information technology, telecommunications, and broadcasting in America, Japan, and Europe, all based on the development of integrated systems. According to the author, the most important project in Europe was PROMETHEUS (Programme for European Traffic with Highest Efficiency and Unprecedented Safety). It was a very ambitious and visionary program,[2] with a number of industrial and basic research subprograms.[3] Today, the idea behind some of the subprograms has been implemented, such as PRO-CAR: in-vehicle systems, based on sensors, to keep cars in lane and to avoid collision with other vehicles or any obstacles on the road.

Authors from different fields were represented in the third session, *Basic Science* (BS): biology, acoustics, physics, chemistry, electricity, electronics, mathematics, statistics, etc. Many of the papers in this session were fundamental or very narrow in scope.

On the technical side, new sensors for all sorts of applications were presented. A good example was the design of a molecular sensor at the frontier between chemistry, biology, electronics, and signal processing: "Molecular Sensors" (ICWES-IX, Complete Proceedings, pp. 11–14) by Gillian M. Greenway (University of Hull, North Humberside, the UK).

It was interesting that some of the papers in this third session aimed at training the participants. For example, the paper "Optical Neural Networks for Pattern Recognition" (ibid., pp. 15–18), from Karen P. Russ (Pilkington Group Research, Ormskirk, Lanes, the UK), provided a nice introduction to neural networks and optical neuron networks. Moyra McDill, who would be one of the leaders of the founding meeting of the International Network of Women Engineers and Scientists, and her colleague Alan Oddy (both from Carleton University, Ottawa, Canada), presented the basic concepts of the automatic finite element analysis (ibid., pp. 40–43). Betty P. Preece's (Science Education Consultant, Indialantic, FL, USA) presentation must have been very popular, with her tutorial for engineers on elementary particle physics. According to her abstract, she led participants:

> "[…] through cosmology (the history of the universe), the "Standard Model" which depicts relationships between elementary particles (quark and lepton families), the four forces

[2] The PROMETHEUS research program took place from 1987 to 1995 (Stuart and Sanders n.d.).

[3] "The Eureka PROMETHEUS Project (*PROgraMme for a European Traffic of Highest Efficiency and Unprecedented Safety*, 1987–1995) was the largest R&D project ever in the field of driverless cars. It received €749,000,000 in funding from the EUREKA member states, and defined the state of the art of autonomous vehicles. Numerous universities and car manufacturers participated in this Pan-European project." ("Eureka Prometheus Project" n.d.)

(E/M, gravity, weak and strong), and their carriers; parts of operations of proton/anti-proton and electron/anti-electron particle accelerators, especially the ones at Fermi National Laboratory and Stanford Accelerator Center; and some basic information about the suggested superconducting supercollider" (*ibid.*, p. 35).

The papers in the fourth session were on different subjects related to *Information Transfer* (IT). Worth mentioning, two papers discussed the changing world of education in S&T due to technology, such as the "Role of the British Library in Supporting Science and Engineering" (ibid., pp. 303–306), proposed by Stella Pilling Undine (British Library Document Supply Centre, Boston Spa, West Yorkshire, UK), and "Multimedia Development for Engineering Education and Training DA" (ibid., pp. 313–314), by Kate Ward (Institute of Irrigation Studies, University of Southampton, UK). This last paper was concerned about the availability of technology for end users: "Multimedia technology has much potential for enhancing the educational process. However, more attention has been given to the acceptability of multimedia to the end user, the student, rather than to the instructor." (ibid., p. 313).

A technical preoccupation for authors in this session was about linking electronic devices for different purposes concerning communication, at home or within industries, as it can be seen by paper titles such as "Microcomputer Networking and Local Area Network" (ibid., pp. 315–318), by Rosemary West (Computer Centre, University of Technology, Loughborough, UK), "Communication Within the Home" (ibid., pp. 319–322), by K. Y. Crowhurst (GEC-Marconi Limited, Hirst Research Centre, Middlesex, UK), "A Computer Based Process for the Exchange of Information Between CAD Systems and Manufacturing Systems" (ibid., pp. 327–330), by R. Singh (Brunel University, Uxbridge, UK) "Communicating about Design: Architectural Features of Expert Systems which cooperate in Design Activity" (ibid., pp. 341–345), by Janet McDonnell (Brunel University, Uxbridge, UK).

The fifth and last session was on *technology transfer* (TT). To introduce what is meant by the technology transfer, Elizabeth Robinson (British Technology Group, London, UK) provided a description in the introduction of her paper "Technology Transfer. A User's Guide" (ibid., pp. 461–464):

> "It is important to understand Technology Transfer and the mechanisms of the many different forms it can take. In what follows I shall concentrate on transfer for profit although there are many important non-profit transfers, for example, whenever you have a paper published in a scientific journal" (*ibid.*, p. 361).

Many examples of technology transfer were proposed in the session: international collaboration in the nuclear fuel manufacture sector (Sue lon, ibid., pp. 465–468), feasibility, impact, or political issues for developing countries (respectively Sue Birchmore, UK, ibid., pp. 469–472; Gurmeet Bambrah, Kenya, ibid., pp. 328–331; Durdana Habib, Pakistan, ibid., pp. 473–479). As in previous conferences, environmental issues captured a lot of interest at ICWES-IX, especially on how to raise awareness through communications: among engineers (Martha E. Munzer, the USA, ibid., pp. 482–485), in the industry (Jan Brooke, Julia Everard, and Helen Paipai, UK, ibid., p. 353), in rural areas (Anuradha Gadkari, India, ibid., pp. 491–494), within governments (Ana Cazzadori de Ferencz, Uruguay, ibid., p. 390), and others.

Another aspect developed in this session was linked to the human factor. In this category, a striking paper on communication problems was introduced by Josephine Parker (Thames Water, Reading, UK) (ibid., pp. 486–489):

> "After 10 years of war, Afghanistan's infrastructure needs substantial inputs of aid. However, communication problems can affect the success of projects there. These problems include overcoming both the problems of interpersonal communications and the logistical and long distance problems. In addition, media communication is necessary to ensure support from donors" (*ibid.*, p. 486).

At least two papers in this session addressed the need to improve public understanding of science: "Improving People's Understanding of Our Planet – Involving the Public in Scientific Research Expeditions – by Corinne A. Locke (Geology Department, Auckland University, New Zealand) (*ibid.*, pp. 495–498) and "Energy Conservation in Leicester City, UK," by Tina Thorley and Paul Fleming (ibid., pp. 499–502).

These last authors (City Engineer's Department, Leicester City Council, New Walk Centre, UK), wrote:

> "Energy conservation is the responsibility of every citizen. Many channels are available to raise public awareness of the technical, financial, and environmental implications of energy saving. The lack of technical understanding in many households is a barrier to information transfer" (*ibid.*, p. 499).

S&T and Women in ICWES-IX

As in previous conferences, technology transfer toward rural women was present in some of the papers, this time from the perspective of efficient communication. Anuradha Gadkari (NEERI, India) reported, in "Rural Women and Environmental Awareness: Effective Means of Communication" (ibid., p. 491–494), on a project for generating environmental awareness among rural women, launched in nine villages around Nagpur:

> "The project involves usage of various communication means. The effectiveness of these means is evidenced by the active interest shown by the rural women in the environment development activities being planned" (*ibid.*, p. 491).

She commented on the effectiveness of the communication means put in place:

> "The communication means adopted for the environment education seem to be yielding the desired effects, viz. generation of environmental awareness in the selected group. These

women are also interested in communicating what they have learned to others in their village and want to motivate the community to participate in environmental projects" (*ibid.*, p. 492).

Going beyond this transfer at the end of the communication chain, a preoccupation to include women at all levels of S&T projects was presented in an abstract called "Women in Projects or Projects for Women?" Rosemary Hilhorst (The British Council Project Development Department UK), ibid., p. 460). The author questioned some projects funded by international and bilateral aid agencies that developed policies regarding the participation of women:

> "[Do they fund] in reality for the greater involvement of women in all projects and the opportunities these provide, or no more than the implementation of a few women-centred projects for women – such as those based on mother and child health care and small scale-operative/income-generating activities? The latter type of project is essential and to be welcomed but should not be a substitute for encouraging the equal involvement of women, particularly as scientists, engineers and technicians, in mainstream development projects." (*ibid.*, p. 460).

Hilhorst thinks that the potential has not been tapped, and she suggests actions, particularly in the first stages of the cycle (ibid., p. 460).

ICWES-IX, Gender Papers

For the ninth conference, three whole sessions dealt with social and gender topics. Another "Technology Transfer," featured S&T and gender papers. In Table 10.3, the first number indicates the gender-themed papers, while the number in brackets under # of papers expected gives the total of papers for a given session. In total, there were 32 papers discussing gender-related questions across four sessions. Of the 45 speakers, 40 women and 5 men, nearly half (19) came from the host country, while 13 came from North America (11 USA and 2 Canada), 2 from Western Europe (1 Greece and 1 Sweden), 7 from English-speaking Africa (3 Ghana, 2 Kenya, 1 Nigeria, 1 Zimbabwe), 3 from India, and 1 from Australia.

Table 10.3 Distribution of social and gender sessions and papers in ICWES-IX, Complete Proceedings

Social and gender sessions	Session titles	# of papers expected	# of papers in proceedings	# of abstracts without papers in proceedings
E	Education	24	17	7
HC	Career Development and gender Issues	5	3	2
D	Demographics	1	1	0
TT	Technology Transfer	2 (12)	1	1
	Total	*32*	*22*	*10*

Table 10.4 Distribution of social and gender posters in ICWES-IX, Complete Proceedings

Social and gender poster sessions	Poster session titles	# of posters expected	# of posters with papers in proceedings	# of abstract-only posters in proceedings	# of posters missing in proceedings
E	Education	3 (5)	3	0	0
HC	Career Development and Gender Issues	8 (9)	6	2	0
D	Demographics	14	9	4	1
	Total	*25*	*18*	*6*	*1*

In addition to these papers, posters were also presented during the sessions. A certain number of posters were accompanied by a paper. In Table 10.4, for Sessions 1 and 2, the first number indicates the gender-themed posters, while the number in brackets under # of posters expected gives the total number of posters for a given session. There were 25 posters in total, presented by 40 participants (33 women, 6 men, 1 unknown): 16 came from the host country, 2 from Canada, 6 from Mexico, 7 from Western Europe (2 Denmark, 1 France, 2 Italy, 2 Spain), 2 from Eastern Europe (1 Poland and 1 USSR), 2 from English-speaking Africa (1 Nigeria and 1 Zimbabwe), 1 from Pakistan, 4 from Far East Asia (2 China, 1 Hong Kong, 1 Japan).

There were three keynote sessions featuring papers dealing with the conference's main themes, which served to open the concurrent working sessions. For the sessions of interest in this section, these were the keynote lectures (they are included in Table 10.3, as an abstract):

- Education (Session 2): "Women, Science and Technology: Shifting the Paradigm from Role Modelling to Mentorship" (Eileen M. Byrne, Australia)
- Technology Transfer (Session 3): "Women in Projects or Projects for Women?" (Rosemary Hillhorst, UK)

There was also a plenary session on the topic of "Demographics" chaired by May Maple (UK). This session consisted of short oral presentations and a general discussion. These were the presenters:

- Olivia Wellesley-Cole (Australia)
- Paula Talpaert (Belgium)
- Dormer Ellis (Canada)
- Élisabeth Thouret-Lemaitre (France)
- Kishwar Jafri & Durdana Habib (Pakistan)
- Jacqueline Juillard-Feyler (Switzerland)
- Dalene McShane and Alison C. Clarke (UK)
- Evelyn Murray (USA)

Monique Frize, "Women in Engineering in Canada"

The year 1991 was a significant year for Canadian engineer Monique Frize. An expert in the field of biomedical engineering, Frize had worked for some 20 years as a clinical engineer in the provinces of Quebec and New Brunswick. In 1991, she attended her first ICWES conference.

That year also marked the second year of Monique Frize's appointment as the first holder of the Northern Telecom/NSERC Chair in Women in Engineering, located at the University of New Brunswick in Fredericton. As Frize explained in her paper, the Chair led to traveling across the country, speaking in schools, universities, and industry on how to increase the participation of women in this profession and how to retain them once they had selected this career. She also worked on abolishing stereotyped gender roles and subtle discrimination. Along the way, Frize built friendships that granted"the strength and determination to keep pushing the envelope of progress for future generations of women." They also provided the motivation to make a proposal later on to organize an ICWES (ICWES-XII, Complete Proceedings, p. 5).

At ICWES-IX, Frize presented a paper titled "Women in Engineering in Canada" (ICWES-IX, Complete Proceedings, pp. 125–128).[4]The article contains a wealth of information on the work performed at the time by the Canadian Committee on Women in Engineering (CCWE), established in 1990 with Monique Frize as Chair. Frize first reminded her readers that, in that same year, fewer than 4% of professional engineers in Canada were women. She contended that the main underlying reason for the underrepresentation of women in this field "was the lack of awareness on the suitability of engineering careers for women and on the particular contributions they can make." Despite an increase in the number of women in non-traditional careers in the workplace, "societal attitudes and pressures," Frize argued, still channel women into traditional roles (ICWES-IX, Complete Proceedings, p. 125). For those women who did choose engineering, many still faced systemic barriers that compromised their professional progression.

The second part of the paper described the various steps and approaches endorsed by the CCWE to collect the information and data it needed, first to examine the current situation of women in engineering and, secondly, to recommend ways to improve their status (on the CCWE, see also Chap. 18).

The CCWE thus held regional public forums, where briefs were presented. Private briefs were also received, with many describing the experiences of women who studied and worked as engineers. The CCWE also commissioned research into effective practices to recruit, retain, and promote the careers of women in engineering across the country.

The CCWE then reviewed the briefs, refining and prioritizing the submitted recommendations. The next step was the development and implementation of action

[4] See further Frize (1992), and Frize and Inch (1992).

plans; four main sectors of intervention were targeted to this end: pre-university, engineering education, workplaces, and professional associations.

The long list of draft recommendations included in Frize's article dealt with education issues only. The examples provided in the paragraphs below illustrate the content of some that were submitted by the CCWE. General recommendations are reproduced integrally, while specific actions have been excerpted:

1. "Educators in the pre-university sector should prepare girls for engineering careers by the following actions:
   - Enhance the mathematics and science curriculum by including references to women's scientific achievements [...].
   - Ensure girls interact with more female role models [...].
2. To enhance the recruitment of female undergraduate and graduate students, engineering faculties should:
   - Initiate specific efforts to recruit and retain women in engineering programs until at least 35% of the new students at all Canadian engineering schools are women.
   - Recruit and encourage qualified female students to pursue graduate studies by offering challenging and well-paid research projects, instituting cooperative work/studies programs, and instituting part-time graduate programs.
   - Encourage mature students, including those with non-traditional backgrounds, to enter engineering programs [...].
3. To retain students, faculties should:
   - Create an environment in the engineering faculty that is free of sexism, racism, and homophobia by adopting and enforcing a written code of behaviour [...].
   - [...] [Recognizing] male and female role models who behave in a non-sexist fashion and use gender inclusive language and examples in class.
   - Ensure the engineering curriculum is relevant to current social realities and needs.
   - Creating innovative ways in which engineering and arts students and faculty can interact so that engineering students develop an understanding of the arts and arts students are more comfortable with technology.
4. Recruitment and retention of female faculty:
   - Ensure all faculty understand the logic of employment equity measures and the benefits women faculty bring to the teaching of engineering; instituting gender harassment/sexual harassment orientation programs for all teaching and administrative staff.
   - Adjust tenure and promotion criteria by recognizing the impact of family responsibilities; [...] injecting flexibility into the tenure and promotion system and their time frames" (ICWES-IX, Complete Proceedings, p. 128).

In her paper's final remarks, Monique Frize strongly expressed her own views regarding Canadian engineering and the role and participation of women in this profession. She firmly believed that the engineering profession and its image would be "enhanced" by the greater participation of women. In her opinion, it was imperative that the engineering profession change its image to one that "is more dynamic, human, and visible." This change was the responsibility of all stakeholders, "as individuals and as representatives of the profession." Women, she concluded, "must have an equal opportunity to pursue careers in engineering and must be respected in the university and the working environments as well as by all professionals in the field" (ICWES-IX, Complete Proceedings, p. 128).

Margaret-Ann Armour, "WISEST—An Initiative at the University of Alberta"

In early 1982, Dr. Gordin Kaplan, who was at the time the University of Alberta's Vice President of Research, noted that, of the 150 people attending a seminar on microprocessors, only one participant was a woman. Already worried by the low participation of women in science and engineering careers, Kaplan took immediate action. With the assistance of academics, educators, and engineers, WISEST (Women in Scholarship, Engineering, Science and Technology) was established that year. Its founding committee welcomed about 20 members, including academic staff, students, professional engineers, educators, and representatives from the Alberta Women's Secretariat. Appointed as Founding Vice Chair was Margaret-Ann Armour, who had joined the University of Alberta's Department of Chemistry in 1979 after completing her PhD in Organic Chemistry at the same university.

WISEST's first mandate was to take action to increase the percentage of women in decision-making roles in all fields of scholarship. Since women were at the time significantly underrepresented in science and engineering, a large proportion of the studies and projects it conducted focused on these fields in its early years of operation.

WISEST was especially interested in explaining why young women stayed away from careers in science and engineering. While some contended that women were not as good at science as men, others argued that there was something intrinsic to the nature of science as it was practised since the late twentieth century, which excluded women from these fields. Finally, others cited the various social, cultural, educational, and occupational barriers that confronted women entering science and engineering.

WISEST believed that women had a great deal to offer in these areas, which led it to initiate several programs and activities that could help build bridges across science and engineering, wherever the expertise and opportunity existed. The following paragraphs describe two very well-known and successful WISEST programs.

1. **Summer Research Program for High School Students**

"One of the barriers which has been identified as keeping girls from considering the sciences and particularly engineering as possible careers, is their lack of knowledge of what a scientist or an engineer does in their everyday work. One of the ongoing programs of WISEST was intended to break this barrier. In the program, students who have completed their second to last year in high school spend six weeks during July and August working in research groups at the University of Alberta. The girls work in the Faculties of Science, Engineering and Agriculture, and the boys in the Faculties of Nursing and Home Economics" (ICWES-IX, Complete Proceedings, p. 214). Launched in 1984, the program enjoyed considerable success in its attempt to attain a balance of gender diversity in all fields of scholarship. It was also unique in that students joined research groups and worked on ongoing projects rather than taking specially developed courses.

Over the years, the program has received strong evaluations from the supervisors in charge of an increasing number of enrolled students. Margaret Ann-Armour observed in her article (ICWES-IX, Complete Proceedings, pp. 213–216) that "[p]erhaps the most rewarding aspect for the supervisors and coordinators of the program has been the surprise and delight expressed by the female students when they realise that 'This is engineering (or science); it is important; I can do it!' (ICWES-IX, Complete Proceedings, p. 214). This led Armour to observe: "Our experience with the program leads us to believe that it is successful both in encouraging young women into the sciences and engineering and also in increasing the likelihood of their being successful at university and continuing on to post-graduate studies" (ICWES-IX, p. 214).

2. **University of Alberta Women in Science and Engineering (UAYs)**

An acronym for the University of Alberta Women in Science and Engineering (UAYs), the UAYs formed a support group for all women in these fields at the university, including undergraduates, graduates, postdoctoral fellows, and academic and technical staff.

Its aim was to encourage young women who had successfully completed a degree in science and engineering to attend graduate school rather than move into a more traditional career. The group met about once a month during the academic year and produced a newsletter. The meetings included panel discussions on a variety of topics, such as "Managing a Career and a Family" and "Gender Role Development." There were also talks by leading visiting women scientists and engineers, including Rose Sheinin, a molecular biologist and vice-rector academic at Montreal's Concordia University; Ursula Franklin, Professor of Engineering at the University of Toronto; and Marie Morisawa, Professor of Geology, State University of New York at Binghamton. According to Armour, the UAYs had the potential to provide role models and informal networking that could help convince some young women of the viability of pursuing their studies and of envisaging an academic career.

3. **Annual Conference**

In 1991, the University of Alberta inaugurated day-long meetings for high school and undergraduate students that were held on campus to discuss various matters related to the lives of women, with considerable interest in the issue of "work-life balance."

> "The meeting provides the students with both the opportunity to meet and talk informally with women in a variety of non-traditional careers, and to have hands-on experiences in the sciences and engineering. Titles such as 'Science, Engineering and Technology: They are for Women Too', and 'Choosing Your Career the WISEST Way' have been used.
>
> The format of the conference was a plenary session with a keynote speaker or speakers, followed by small group sessions in which the students could question three or four women in non-traditional careers about the requirements of their work, the qualifications they needed, how they managed their career and their family, and why they chose their career. The small group discussions are facilitated by a teacher. Discussion continues over lunch, and then there are a variety of hands-on activities for groups of about ten students. Each

student participates in two of these activities. They have included working in the water resources laboratory of the civil engineering department, using a computer program to test the viability of a new product and how many need to be sold to make a profit, doing one of several experiments in the chemistry laboratories, making models of strands of DNA, exploring superconductivity, learning about the Hubble telescope, and examining plant cells" (ICWES-IX, Complete Proceedings, pp. 214–215).

Armour proudly reported on the results of this first venture, which had attracted about 100 students and 40 women scientists and engineers participating in each meeting. This kind of success had indeed led to the planning of such a meeting on an annual basis.

In her concluding remarks, Armour insisted on the importance of encouraging young women to consider careers in the sciences and engineering and of investing considerable time and effort to attain this goal. There were several reasons to support her call for action. First, WISEST believed in equal choice and opportunity for all members of society. Secondly, a shortage of trained personnel in science and engineering was projected for the next 5–10 years. This situation would trigger an increase in the number of jobs, many of which would replace traditional women's jobs, such as switchboard operators and stenographers. Thirdly, Armour predicted that pressure would be lifted from men, who had traditionally been expected to provide a good standard of living for their families, once it became more wholly accepted that women could have careers that allowed them to share this role. Finally, WISEST expected that increasing the number of women in the sciences, engineering, and technology would bring to these disciplines "a broader spectrum of human values" (ICWES-IX, Complete Proceedings, p. 216).

Lotte Algreen Reimer, "Women in Engineering in Zimbabwe"

The paper given by Lotte Algreen Reimer, member of the Zimbabwe Institution of Engineers (ZIE),[5] was titled "Women in Engineering in Zimbabwe" (ICWES-IX, Complete Proceedings, pp. 235–238). Reimer noted that in the early 1990s there was an important shortage of experienced engineers in Zimbabwe and that foreign nationals were not only making up for it but were also central to training local engineers. At the time, almost all technical staff in the engineering sector were men; according to Reimer, there were "no more than 25 women engineers in Zimbabwe, including expatriates" (ibid., p. 235). Looking at ZIE membership, women amounted

[5] Reimer (MSc Eng) graduated from Aalborg University, Denmark in 1979 in structural engineering. She worked for Ove Arup & Partners (a professional services firm) in London and, from 1982, in Zimbabwe. She would become president of the ZIE in 1995–1996; her tenure was pivotal in setting the Institution's Environmental Committee. Reimer would later go back to the UK, where she would undertake a Ph.D. ("European Graduates – Aalborg Universitet, Ålborg Area, Denmark" n.d.; Mason 2009).

to 1.5% of all engineers and 0.6% of all technicians. Reimer acknowledged that these numbers might not be true to life since certain women engineers and technicians, working a second shift at home, were reluctant to take part in the institution in addition to everything else. These numbers were more or less reflected in university and college admissions: women made up 1.1% in 1985 and 1.4% in 1989.

Reimer posed the matter of a substantial decrease in women enrolled at the Engineering Faculty of the University of Zimbabwe during the last few years: 35.5% in 1979, down to 24.2% in 1986. Although girls have been attending high school in higher numbers since the recognition of the country's independence in 1980, this decline in women engineers needed to be addressed. For Reimer, this could be explained by (a) a 3:1 ratio of boys and girls; (b) an ingrained tendency for boys to take mathematics and sciences and a disinterest of girls in these subjects; and (c) university requirements.

Another negative consequence of having significantly expanded the country's education system since 1980 was, in the face of a great increase in general enrollment, a critical lack of teaching staff and equipment. These effects were particularly visible in rural areas, especially in girls' education. Since fees for attendance and examinations at secondary level were already quite high in comparison with monthly wages, a "family would most certainly favour the son" (ICWES-IX, Complete Proceedings, p. 236). In addition, men from rural areas tended to go work in the cities, leaving women with all domestic work, childcare, and farm work. Women would rely on their children, especially girls, to help them. A similar situation could be seen in urban families with lower income: girls in high school "have to undertake domestic chores which the boys would never be asked to do" (ibid., p. 236). Conversely, wealthier families could afford domestic help as well as equal education opportunities for boys and girls.

Reimer also pointed to the urgent need of educating the wider population in what engineering actually was. Women role models as engineers were so few that engineering was persistently viewed as a typically masculine field. This meant that women, even at university level, were subjected to higher standards since men were felt as "natural engineers" (ICWES-IX, Complete Proceedings, p. 237). In addition to general misinformation and lack of representation, inadequate course and career guidance in high schools and fewer teaching staff in science and engineering also contributed to this problem.

Funding was a further impediment to women's training and higher education in engineering. Since they were expected to marry, have children, and leave school to take care of them, women were "often seen as a bad investment" (ICWES-IX, Complete Proceedings, p. 237). By contrast, men would easily obtain bursaries, even with lower marks. At home, women also faced expectations of domestic work and childcare, no matter the studies they undertook or the professional position they held. This pressure to maintain traditional roles was deeply ingrained in Zimbabwean society, making even supportive men very reluctant to take part in household chores "for fear of ridicule—and presumably also because of the inconvenience" (ibid.).

Work trips, essential to getting promoted, were yet another obstacle to women's careers since their husbands tended to forbid them from going on the basis that this was, for an unaccompanied woman among a group of men, "a morally compromising situation" (ibid.).

Reimer then turned to future action, noting, for instance, the importance of meaningful representation in career guidance aids, both in images and in text. Women should also take part in advisory teams. The Zimbabwean government pledged to hire women in a third of the management positions in public services. This goal would be achieved through further training and education for women and positive discrimination towards women in the case of equally qualified candidates. Though Reimer acknowledged that this aim might not be attainable in engineering sectors, she nevertheless argued that these measures would also benefit these fields, not least in transforming social conventions and public opinion on the roles of women and men. Finally, higher living standards would inevitably serve to bridge several gaps in the education and opportunities available to girls and boys.

Business Meeting

The delegates meeting, held on July 18, 1991, appointed the new ICWES Continuity Committee. They agreed that the subject of their scrutiny of detailed proposals, the Nigerian offer to host ICWES-X in 1994, should be accepted. However, because of political instability in Nigeria, the Continuity Committee was not able to confirm the acceptance of the Nigerian offer. Fortunately, Hungary stepped into the breach and offered to host ICWES-X in 1996.

Minutes are taken from the Conference Report:

> "At each ICWES a meeting is held of the delegates who represent recognised associations or societies of women engineers and scientists in the different countries. It is chaired by the Chairman of the current conference and includes the Chairmen Emeriti of past conferences with their personally named deputies, and one delegate representing the unaffiliated registrants at the conference (i.e., those registrants from countries that have no official delegate).
>
> The main purpose of this meeting is:
>
> - to consider offers to host the next ICWES;
> - to appoint the ICWES Continuity Committee whose responsibilities are to assist the organizers of the next ICWES by disseminating the details in those areas of the world they represent;
> - to encourage organizations to submit offers to host the next ICWES;
> - to discuss matters relating to the ICWES Guidelines and the organization of the Conference.
>
> At ICWES-IX, the Delegates Meeting was held on Thursday 18 July, chaired by the Conference Chairman, Dr. Elizabeth Laverick. Fifteen official delegates attended from Belgium, Canada, Denmark, France, Germany, Japan, Mali, The Netherlands, Nigeria, Switzerland, Uganda, United Kingdom, USA, Zaire and Zimbabwe. Professor Mahin

Rahmani from Iran represented the unaffiliated countries. Three past chairmen (or their deputies) from ICWES-II, VI, and VIII attended; there were also 21 observers from 11 countries.

As no bids for ICWES-X had been received in writing prior to the meeting, it was agreed to accept informal bids at the meeting and decide an order of preference by ballot. Detailed proposals would then be prepared to be submitted within six months and a final decision would be made by the Continuity Committee.

Presentations were made by Nigeria, Uganda, India, and Scandinavia; the latter two were offered as a possibility, depending on decisions after they returned home. Both the Ugandan and Nigerian organizations were confident of the support of their Governments and, in the case of Uganda, official approval and international support had already been secured. However, the Ugandan delegate made it clear that, should Nigeria submit a written proposal, the Ugandan bid would be withdrawn and reissued for a subsequent ICWES. The order of preference resulting from the ballot was first Nigeria, then Uganda, India, and Scandinavia.

Finally, steps were taken to appoint the Continuity Committee, which was to be composed of:

Dr. Elizabeth Laverick, Convenor;
Professor Dormer Ellis, North America (Canada and the USA);
Angeline Lopez, Latin America;
Mahin Rahmani, Middle East;
Ila Ghose, India;
Chebaane Hassiba, Arabic-speaking Africa;
Joanna Maduka, English-speaking Africa;
Loteta Dimandja, French-speaking Africa;
Marjo de Kryger, Western Europe.

The United Kingdom, Australia and New Zealand would be represented by the President of the Women's Engineering Society. Subsequent to the meeting, Professor Mitsuko Kazuno agreed to represent the Far East, but no representative was found for Eastern Europe" (ICWES-IX, Conference Report, pp. 37–38).

Testimonials

Lee Arnold, "Reflections on ICWES"

"At ICWES-VII in Washington, DC, we celebrated twenty years of ICWES, in conjunction with SWE's 1984 Convention. There I renewed many acquaintances; afterward I hosted Paula Talpaert of Belgium, Hettie Bussell of UK, and Jacqui Juillard of Switzerland at my home in New Jersey. At ICWES-IX, near Coventry, UK, in 1991, I saw Mahin Rahmani of Iran again after some years, and met many WES members, some of whom I had not seen since 1967. There is no doubt that these personal connections have practical value. It has been a long time since I used a hotel in the UK, and after ICWES-X in Hungary this fall, I hope to visit Jacqui Juillard in Geneva. To visit a friend's home and area obviously is an improvement over hotels and package tours. There is also great mutual value for those who are active professionally to be able to exchange information and do business with people who may be known personally and whose country you know at least a little about. To me, aside from rewarding personal friendships, the value lies in something which is hard to describe. Physical distance seems less a barrier when you have seen the other side of the sea. Cultural differences seem understandable and reconcilable when you have met and talked with individuals. Countries are less something you read about in the political news and are more real as collections of people like yourself and your own community. It makes a difference when you read about happenings in another country to think of someone you know and wonder how your friends are coping. I think these travels have influenced my thinking about the politics of my own country – about foreign aid and trade, for example, or urbanization and terrorism. In turn, this influences my voting and my participation as a citizen. I would like to think that ICWES makes each of us a better citizen of her profession, her country, and the world community" (Arnold 1996a, pp. 9–10).

"Conference Impressions" from Three Delegates (ICWES-IX, Conference Report, pp. 45–48)

Jacqueline Juillard, Chambesy, Switzerland

"We were nine Swiss delegates, comprising eight engineers and one architect. Eight grants were given to eight of these people in order to attend the conference. I managed to have them granted by prestigious Swiss institutions such as the Swiss Academy of Technical Sciences, the two Zurich and Lausanne Federal Schools of Technology, and the Swiss Society of Engineers and Architects. Five of them were from the French-speaking part of Switzerland, two from the German-speaking one and one from Tessin, the Italian-speaking one."

"Jacqueline Perrotet, an agricultural engineer, found – and it was confirmed during ICWES – that women talk a lot more than men before making a sound decision, which should therefore be a better one. Christina Zanini, a civil engineer, was interested to meet a lot of women in the profession, when usually one is surrounded by men. Veronique-Jost Gara, a mathematician, was surprised by the great number of highly qualified delegates from far-away countries, especially from Africa."

"The general impression, which was also mine, from the eight ICWES's that I have attended, is that the presentation of the subjects was a funny mixture: either highly technical or very much popularized. But the main interest of the conference is in fact the contacts established professionally, friendly and especially with colleagues from far-away continents. A sort of beginning of a professional women engineers' and scientists' networking the world-over."

Mitsuko Kazuno, Miyama, Japan, "From Japan with Love"

"It was a bit of a surprise to me to find the spectacular scene of so many lively and colorfully dressed women from all over the world gathered on the campus of Warwick University between 14 and 19 June 1991. Of course, the conference was for women engineers and women scientists. People from African countries are particularly colorfully dressed and create an air of gaiety."

"I myself participated for the first time in this 9th International Conference from Japan. For many years, Japan has not sent a delegate to the ICWES, although there is an organization called 'The Society of Japanese Women Scientists,' to which I belong. For me, having only attended international conferences on high energy physics or nuclear physics, the atmosphere was completely different from what I was used to."

"All participants were cheerful and friendly. I soon became friendly with a number of women from different countries. A young Russian woman who is a physicist showed me some photographs of her lovely children although she was worried about her country's serious economic condition. We talked about how she could manage both childcare and her professional work. Such a conversation would never be brought up in other conferences. A young woman from Denmark was worried about her career because she is laid off from her professional work while her children grow up. I realized that all women there at the conference had completely different problems and yet common problems too."

"The conference was well organized. The white denim shoulder bag bearing the conference symbol and an umbrella were handed out together with the conference proceedings and related documents. The sessions were so organized that one could attend most of the major presentations of the program without overlapping. The subject of talks was related to the vast fields of Research and Development and Education. From their presentations, I felt many women are indeed supporting today's technology world. They talk about canal construction, traffic management and satellite communications. The Gender issues were unique in this conference with lectures on 'women returners fellowship scheme' or 'career guidance,' 'girl's technology training' and so on."

"We enjoyed visiting historical places and saw the beautiful scenery around Warwick between the lectures. I enjoyed visiting Shakespeare's village. We felt the long traditions and deep culture of the country. I also enjoyed talking to a few members of the organizing committee, who were so busy during the conference to make sure that everything should go well without any trouble." "Nevertheless, they extended a warm hospitality to the participants who came from the far ends of the world. The Conference Chairman, Dr. Laverick, suggested to me that Japan could host a future ICWES. I wondered why Japan had not done so already. There are a number of excuses I suppose, such as the language problem, the high cost of living and the distance from most countries etc. However, Japan today is considered as one of the major technological countries."

"Why shouldn't Japanese women engineers and scientists come forward to host a future ICWES? I know there are a number of difficulties on our side. Having returned from the conference, I reported the features of this successful ICWES-IX to The Society of Japanese Women Scientists, a very academically oriented society. We seriously thought of this task. At the moment we have decided to form the 'Japanese Women Engineers Forum' and we will discuss the possibility of hosting ICWES sometime in the future."

"The Women's Engineering Society, established in 1919 in the UK, has about 700 members and was the host of this conference. I had an opportunity to talk with a past-President of WES, Dr. Cicely Thompson, who kindly provided me with some information on past ICWES conferences at my request. She has also written a book about the history of the Society. In the last but not the least, I thank again all the people who conducted and supported this successful ICWES-IX."

Olivia Wellesley-Cole, Australia

"The ICWES-IX Conference was an event I had spent four years looking forward to – and I was not disappointed. I now look back on it as a week spent with old friends, reminiscing about previous meetings, and making new friends in a benign environment."

"I was in England in 1987 at the time of ICWES-VIII, but was unable to attend that conference (in the Ivory Coast) as I was in the process of moving to Australia and couldn't really be in two places at the same time. As Chairperson of the London Branch of the Women's Engineering Society I was very keen that the next ICWES be held in the UK, and it was an opportunity for me to return to England to attend the conference."

"What particularly impressed me about ICWES-IX was that so many delegates attended, from so many countries. There was a marvelous feeling of international camaraderie. In particular, a number of the African delegates got together and held meetings relating to the progress of science and technology in those countries; this was an aspect of the conference that had not been planned, but with so many African states represented (some ten or more) it seemed too good an opportunity to miss. So, even with the mix of languages (we had to use French as well as English; one of the other Australian delegates acted as interpreter!) we were able to consider setting up efficient avenues for networking between the scientific and engineering organizations in the various African countries, which was very exciting."

"Another plus point for me was to be able to catch up with the activities of my colleagues in England, some of whom I had not seen for four years. A number of them too had moved overseas as part of their jobs (to Pakistan, Hong Kong, Europe and even another to Australia!); for those I was able to catch up with."

"I found that it was a time of consolidation in our career plans. Some had already changed jobs, or career direction; some (like myself) were considering change, and debating the best way to go about it. For me, the strategy of staying with the same employer and looking for an internal move (from design engineering to product marketing) has recently paid off; I wonder how others, contemplating similar moves, have fared? While there were not the twenty or thirty delegates from Australia that I had hoped for, there were three of us, a very creditable number considering the distance we had to travel to reach the UK. I was able to meet the other two delegates, and have been in touch with them since returning here."

"So all in all, I enjoyed ICWES-IX, and it was well worth attending. I look forward to the next one, and am already saving for the plane fare!"

ICWES-IX Organizer, Elizabeth Laverick, on "The ICWES Phenomenon"

"What is it that brings together women engineers & scientists from more than 40 countries every three years? The desire to meet, to confer, to learn and to demonstrate their technical achievements? The International Conference of Women Engineers & Scientists which started in the USA in 1964 has just completed its ninth event. In the interim, the conference has visited the UK (67), Italy (71), Poland (75), France (78), India (81), the USA (84) and

The Ivory Coast (87). Each conference has been sponsored by the host country and is organised on a voluntary basis with no permanent secretariat, but with an interchange maintained between the various international correspondents providing some form of communication channel in between."

"**Origins.** For most of us, the first ICWES, organised by SWE in New York in 1964, was our first international conference. At that time, only the USSR had more than a token number of women engineers, though possibly a few more scientists and even fewer had reached the status which leads to attendance at international conferences. Instead of being the odd one out, suddenly to find oneself in a peer group of 500-plus from all over the world created an immediate bond between delegates and with the American Genius for welcoming strangers produced a wonderful atmosphere of confidence and happiness" (Laverick 1991, p. 8).

Chapter 11
ICWES-X, Budapest, Hungary, October 8–10, 1996

Theme: "Towards the Third Millenium: The Role of the Intellectual Potential in a New World"

The Opening Ceremony

It was quite appropriate for one of the founders of ICWES, Elizabeth Laverick, to describe its past events. Her talk appears as the first article in the proceedings of the conference. Excerpts were quoted in several previous conferences as testimonials from Laverick (Laverick, in ICWES-X, Complete Proceedings, pp. 8–12).

Zsuzsa Szentgyörgyi (see Appendix 1) followed with her talk, "Why Hungary?" After describing some of the aspects that Hungary does not have, Szentgyörgyi listed many positive aspects that arise from the culture and the people, all the talent, counting many eminent scientists and mathematicians. She concluded that the delegates could expect very warm hospitality from the organizers and the people of Hungary (ICWES-X, Complete Proceedings, pp. 14–15; see section "ICWES-X, Gender Papers" below for a more detailed presentation of this talk).

Baroness Platt of Writtle gave a plenary lecture titled "Women Into Science and Engineering" (ICWES-X, Complete Proceedings, pp. 16–22).

Organizing Committee:
Conference Chair: Zsuzsa Szentgyörgyi
NOC Chair: Zsuzsa Pinter
Editor: Csilla Bany
Members: R. Bars, M. Fodor, Györgi Ronto, Anna Tompa, M. Wellek

Sponsors:
Hungarian Academy of Sciences
Foundation for Industry

M. Frize et al., *Women's Contribution to Science and Technology through ICWES Conferences*, Women in Engineering and Science,
https://doi.org/10.1007/978-3-031-33800-7_11

National Committee for Technological Developments
Soros Foundation

Participants
There were 82 participants, including 17 from Hungary and 65 from 20 other countries, namely 14 from the USA, 3 from Canada, 11 from Western Europe, 5 from Eastern Europe, 2 from French Africa, 3 from English Africa, 1 from Middle East, 2 from Central Asia, 9 from Far east Asia, and 15 from the UK. Out-of-country participation was 79.3% (see Appendix 3).

ICWES-X, S&T Papers

To provide S&T background for this conference, researchers in numerical analysis were both shaken and excited by a new awareness. In 1996, Ilya Prigogine, a physical chemist, and Isabelle Stengers,[1] a philosopher of science, had published the book *La Fin des certitudes* (The End of Certainty). Prigogine was already an authority; he had won a Nobel Prize in 1977 for his work on the thermodynamics of non-equilibrium systems. In the introduction, Prigogine argues that irreversibility can no longer be identified with a mere appearance that would disappear if we had perfect knowledge. Instead, it leads to coherence. Hence, this book rejected determinism, which, until then, was considered a rule in all physical processes. One can extrapolate that the capacity of traditional sciences to predict complex issues linked to the environment or meteorology, for example becomes less certain over time. Nevertheless, numerical analysis was becoming a main tool of academic and industrial R&D.

Another type of awareness was also emerging at the time of the conference through international efforts to mitigate environmental and global warming issues. In 1972, the first United Nations Conference on the Human Environment, attended by 113 delegates and two heads of state (Sven Olof Palme of Sweden and Indira Gandhi of India), was held in Stockholm, Sweden. Twenty years later, that is, 4 years before ICWES-X (June 1992), the United Nations held in Rio de Janeiro, Brazil, the Conference on Environment and Development. This conference focused on the state of the global environment and the relationship between economics, science, and the environment in a political context. It ended with the Earth Summit, which brought together leaders from 105 countries to demonstrate their commitment to sustainable development (Meakin 1992). Following this last event, the first Conference of the Parties (COP) to the United Nations Framework Convention on Climate Change was held from March 28 to April 7, 1995, in Berlin. It became obvious then that the economy and politics influence science and technology.

Nevertheless, women scientists and engineers enthusiastically presented their work at ICWES X. S&T papers in ICWES-X proceedings (ICWES-X, Complete Proceedings) were organized in two sections: "New Trend in Natural Sciences and

[1] Isabelle Strengers has not been credited as co-author of this book, but Ilya Prigogine considers her as such at least for the original French version ("La Fin des certitudes" n.d.).

Table 11.1 Distribution of S&T sessions and papers in ICWES-X, Complete Proceedings

S&T session titles	# of papers expected	# of papers in proceedings
New Trend in Natural Sciences and Technology	18	18
Transfer of Knowledge	6 (8)	6
Total	*24*	*24*

Technology" and "Transfer of Knowledge." As can be seen in Table 11.1, this represented 24 papers altogether.

There were a total of 36 authors (27 women, six men, and three of unspecified sex). Among them, eight authors were from the host country: four from the USA, (three from Canada (4 papers), six from Western Europe (five from Greece and three from Switzerland), two from Eastern Europe (Romania and Russia), one from English-speaking Africa (Botswana), three from Central Asia (India), three from Far East Asia (two from Japan and one from Hong Kong), one from Australia, and four from New Zealand.

A Glimpse of ICWES-X's S&T Content

New Trend in Natural Sciences and Technology

Respecting a main topic so broad as "Towards the Third Millennium," the papers in ICWES-X S&T were often prospective. The numerical trend, typical of the 1990s, was clearly visible. Many authors wrote papers on numerical simulations where systems of linear or non-linear equations to be solved were explicitly written. This is the case for the paper in physics "Formation of a P-N Junction in TiO2 upon Voltage Application," written by Demetrios Korfiatis and his colleagues from the Department of Physics of the University of Patras, Greece (ICWES-X, Complete Proceedings, pp. 73–75). They applied a non-linear finite difference model to study Fe Doped TiO2, a material defined as a mixed conductor at elevated temperatures. Another paper, "Modelling and Simulation for a Fly-By-Wire Aircraft," by Padma Madhuranath, a scientist at the Flight Mechanics and Control Division, National Aerospace Laboratories in Bangalore, India (ibid., pp. 131–138), presented a generic non-linear flight dynamic model as well as a linear model for flight simulators.

On the interdisciplinary side, two research studies using computer science to support medical personnel were proposed. Agnes Drahos and Zsolt Oreg (Hungary) wrote a paper called "Meteorology in Medicine and Biology" (ibid., pp. 77–81). They explained that a meteorological parameter becomes a biological parameter when an adaptative mechanism in the living organism begins. This paper presented a survey of the influence of the atmosphere as an environmental factor that affects humans and other organisms (season, climate, weather, etc.), a methodology to analyze hourly information over 6 years in Budapest, and results in terms of the circadian and seasonal rhythm of the meteorological conditions. It is well known that births occur more frequently in autumn and winter, and mortality due to upper respiratory infections and to the cardiovascular system is maximal in winter and minimal in summer. However, it is less obvious that the effect of a warm front can start 4–6 hours before arriving and that cardiac infarction happens mostly at 9–10 a.m., coinciding with the maximum downslide motions of warm and cold fronts.

Monique Frize, professor at the University of New Brunswick, Canada, and her students summarized their research in a paper called "Computer-Assisted Decision Support Systems for Medical Applications" (ibid., pp. 97–99). They used case-based reasoning and neural network techniques on existing medical databases to provide information to aid physicians in making decisions about the management of their new patients. Frize et al. showed that their system has resulted in improved accuracy for an intensive care environment in predicting mortality, length of stay at the hospital, and duration of artificial ventilation.

Two recurrent themes came back at ICWES-X. One of these themes was related to new energy production. In "Energy – Now and for the Next Century," Wilma C.M. Labelle (Senior SWE member in Wisconsin, USA) (ibid., pp. 101–104), provided a review of different new energy technologies; in "Direct Conversion of Solar Energy into Electric Power," Meera Ramrakhiani (Rani Durgavati University, Jabalpur, India) (ibid., pp. 105–101), presented a more technical paper.

The other recurrent theme was traffic, either regarding safety ("Development in Traffic in Botswana with Special Regard to Safety," by Vera Vokolkova, Civil Engineering, University of Botswana, ibid., pp. 117–120), or to assess the impact on the environment ("Impact of Congested Traffic on Emissions from Roads Vehicles," by Judith Ann Cox, UK, ibid., pp. 127–130), or including both topics as in the paper ("Ice Prediction in Roadway Covering: Both Traffic and Safety in Environment Problems," by Tatjana V. Samodurova, Voronezhh State Academy of Construction and Agriculture, Russia, ibid., pp. 121–126).

Transfer of Knowledge

The second part of the main topic, *"The Role of the Intellectual Potential in a New World"* was reflected in the six papers in this *Transfer of Knowledge* section.

One of the aspects developed was transferring technologies from governmental research laboratories to industry. In "Research Industry Interaction for Effective Technology Transfer" (ibid., pp. 237–238), Vaishali J. Nashikkar (NEERI, India) introduced a new challenge encountered by the Board of Scientific and Industrial Research (CSIR), a scientific body created in 1940, responsible for R&D activities in India. The origin of the challenge was due to new technologies flowing into the country following the liberalization of import and export modalities. She wrote: "In India, with a long tradition of mathematics and astronomy, the technology is not science based in the way that in the industrialized west contemporary research continuously informs and transfers technologies." The CSIR identified two ways to support the Indian industry: to utilize its own expertise in evaluating the incoming technologies and transferring this information to the industry, as well as to transfer the indigenous know-how and technologies developed at the CSIR.

Kathleen F. Harer, from the U.S. National Aeronautics and Space Administration, in "An Innovative Approach to Technology Transfer Business Incubators" (ibid., pp. 239–241), echoed Nashikkar's effort to present the point of view of a developed country. She surveyed a number of initiatives the US government laboratories had developed to transfer technology and uses into the commercial arena, such as providing low-cost rental space, reasonably priced business services, and advice to help overcome the obstacles experienced by new companies.

The other aspect in this section concerned four initiatives undertaken by different countries or organisms to enhance science and technology education through knowledge transfer. For example, Japan wanted to have more foreign students to improve their level of research and education and to develop their skilled manpower (Akiko Tsugawa, International Liaison Office of Science at the University of Tokyo, ibid., pp. 243–245). Another initiative came from Betty Preece, who explained that NASA opened its resources on space topics to international educators (ibid., p. 247).

ICWES-X, Gender Papers

A total of 18 papers dealing with gender-related themes were given during the tenth conference. Most of them are included in the conference proceedings (15); two have only an abstract, and another is missing entirely. These 18 papers were given by 33 different speakers, of whom 1 gave two separate papers (Lai Yin Ng). The great majority were women (31 vs. 2). Only 1 speaker came from the host country, 9 from the US, 1 from Canada, 7 from Western Europe (3 France, 2 Switzerland, 2 Spain), 2 Eastern Europe (1 Poland, 1 Georgia), 10 from East Asia (4 Hong Kong, 5 Japan, 1 Mongolia), and 3 from the UK (Table 11.2).

Table 11.2 Distribution of social and gender sessions and papers in ICWES-X, Complete Proceedings

Social and gender session titles	# of papers expected	# of papers in proceedings	# of abstracts without papers in proceedings	# of papers missing in proceedings
Women in Sciences	11	10	0	1
Chances of Women for a Career in the Different Countries	5	4	1	0
Technology Transfer	2 (8)	1	1	0
Total	*18*	*15*	*2*	*1*

Two opening lectures launched the tenth conference. First, Elizabeth Laverick, a leader and regular participant in ICWES, presented a retrospective on the previous nine conferences. In a paper titled "ICWES One to Nine" (ICWES-X, Complete Proceedings, pp. 8–12; see Appendix 4 for full text), Laverick reviewed each ICWES between 1964 and 1991. Noting themes and invited speakers, Laverick also featured each conference's resolutions and traced back how ICWES had come to be through trial and error and the gradual structuring of its organization. In her closing remarks, she declared that the conference's first aim, "to bring together professional women engineers and scientists worldwide" (ibid., p. 10),[2] was successful, while advising that more should be done in order to enhance the conference's visibility at an international level and in order to help women keep in touch with each other. Through a wide variety of themes addressed at each event, the second aim of the conference, "to encourage the use of technology for the betterment of life" (ibid.), had also been achieved: as Laverick sketched out, all the papers "provided an impressive record of the growing technical contributions made by women in these particular areas over the last 27 years" (ibid., p. 12). The success of ICWES's third aim, "to consider how to increase the participation of women" in this technological era (ibid., p. 10), was also made clear by the vast array of sociological papers that exposed "The different ways in which countries are tackling the problem of the shortage of women engineers and scientists, and the waste of talent that this implies" (ibid., p. 12).

Zsuzsa Szentgyörgi talked about the interest in holding ICWES in Hungary for the conference's tenth iteration (ICWES-X, Complete Proceedings, pp. 14–15). Starting with "a sorrow list of 'having-nots'" (e.g., the GDP per capita at the time was 4.5 thousand dollars, little ore and coal, depleted resources in oil and gas, an enduring unstable political situation), Szentgyörgyi countered with a series of "treasures": fruitful agriculture, an abundance of thermal sources and springs, an international harbor on the Danube, a strategic position at the junction of roads and states. However, for Szentgyörgyi, the true wealth of Hungary resided in its place as a hub for "ideas, knowledge, and culture" (ibid., p. 14); for instance, the country gave birth to 11 Nobel prize winners, though most had then left Hungary due to political reasons. According to Szentgyörgyi, the country's "windy" character, both literal and figurative, led the country's population to become "extraordinarily adaptive, docile, quick to understand, and enterprising" (ibid., p. 14). The 1996 Conference was a good opportunity to visit and feature Hungary as "one of the newly independent, postcommunist countries, perhaps the most open among them" (ibid., p. 14). Hungary was also, at the time of choosing the next location for ICWES-X 5 years prior, just announced as the venue for the 1996 World Exposition. Since Hungary could not, in the end, host the World Exposition, ICWES-X was postponed from May to October. So, "why Hungary?" asked Szentgyörgyi again. Because the country, though small, is "beautiful, not rich but developing country,

[2] These were formalized and adopted during the Delegates meeting at ICWES-IV (Cracow, Poland, 1974).

with charming landscapes, nice cultural memories, and hospitable people" (ibid., p. 14).

This tenth conference had four plenaries:

- "Women into Science and Engineering" (Beryl Platt, the UK; ICWES-X, Complete Proceedings, pp. 16–22)
- "The Influence of Science and Hi-Tech in Modern Literature" (Miri Kubovy, USA; ibid., pp. 23–29)
- "Socio-economic Situation, Depression and Morbidity in the Hungarian Female Population" (Maria S. Kopp, Arpad Skrabski, Sandor Szedmak & J. Lake, Hungary; ibid., p. 31, abstract only)
- "Is the Solar Radiation Beneficial or Harmful?" (Györgi Ronto, Hungary; ibid., pp. 33–35)

In addition, there were two workshops headed by Betty P. Preece (USA): "Low Budget Activities to Interest Girls in Science and Engineering" and "Successful Networking between Professional Engineering, Scientific and Educational Organizations" (ICWES-X, Complete Proceedings, p. 37 and p. 39).

Finally, a panel discussion about "The Role of the Professional Women in the New Order of Hungary" also took place. The panel was divided into five presentations, all given by Hungarian participants:

- "The Internal Conflict of Women Today – Balancing Careers and Family Life," Csilla Csoboth, Maria S. Kopp & Sandor Szemak (ICWES-X, Complete Proceedings, p. 41)
- "Life Expectancy and Working Conditions among Hungarian Women Employees," Anna Tompa (ibid., p. 43)
- "Feminism-Related Dysfunction in Socialization," Miklos Hernadi (ibid., p. 45)
- "Quantitative and Qualitative Characteristics of Way of Life among Hospital Female Doctors," Balint Boga (ibid., p. 47)
- "Gender Differences in Cognitive Functions of the Brain," József Hamori (ibid., p. 49)

"Women in Georgia," Elizaveta Buadze

Elizaveta Buadze[3] presented a paper on the historical roles and activities of women in Georgia, with a focus on women's sacrifice (ICWES-X, Complete Proceedings, pp. 219–223). Budaze first opposed women's historically difficult position with their occasional divine or royal status. A prevalent representation of women is that of the mother, whether "Mother Sun" or "Mother Georgia." Another famous image

[3] Buadze was by 2006 part of the administration for the Faculty of Technological Engineering at the Akaki Tsereteli State University, where she had a position as full professor. She would later be professor emerita ("History" n.d.).

of Georgian women is the Greek mythological figure Medea: beautiful, wise, and intelligent, the princess Medea was the determining factor in the Argonauts' quest for the Golden Fleece, but she was also jealous and, ultimately, beastly in her own quest for revenge when Jason scorned her. Buadze also talked about Christian legends, in which a young woman called Nino brought Christianity to Georgia in the first half of the fourth century. Saint Nino is frequently invoked in association with "truth, kindness, and miraculous power" (ICWES-X, Complete Proceedings, p. 220). Nino is tied to another famous figure: Nana, who was Queen Consort of the Kingdom of Kartli (or Iberia) and was converted to Christianity by Nino. Still, women were denied power by all religions and were relegated to subservient roles. Buadze then gave the example of Shushanik (d. 475), who was tortured until her death by her husband for staying true to her Christian faith.

However, according to Buadze, "attitude to a woman is much more elevated in Georgia than in the west and the east" (ICWES-X, Complete Proceedings, p. 220). In support of this argument, she listed a number of queens who marked the history of Georgia. For instance, the rule of Tamar (thirteenth century), a descendant of David the Builder, was a high point in Georgian politics, economy, and human rights: in addition to expanding the territory, encouraging trade with foreign states, and solidifying political alliances, she also outlawed execution and whipping. The Georgian flag was named in her honor: "Tamariani." Ketevan (d. 1624) is another example of might and beauty; she was tortured to death by the Shah of Persia for refusing him. Ketevan, as Buadze pointed out, defied established ideas of women being controlled by their emotions since her choice was based on rational thought, not passion.

Though Georgian history is rife with military and political struggles, "Georgian woman's purity, gentleness and beauty are known all over the world" (ICWES-X, Complete Proceedings, p. 221). Women are also the guardians of the mother language, giving great importance to reading and writing in Georgian. Buadze asked why Georgian women had "such a great part in cultural process"? (ibid.). Despite many similarities with women from other cultures, women in Georgia had the responsibility of child rearing and farming, through which they could ensure the preservation of Christianity, tradition, and nationality.

Buadze then talked about women's status in more recent times. A study from 1987 showed that Georgian women were more educated than other women living in the Soviet Union, save for the Baltic states. Internationally renowned chess players, like Nona Gaprindashvili and Maia Chiburdanidze, were a great source of pride. There was apparently a "Georgian phenomenon" around these high achievements in chess (ICWES-X, Complete Proceedings, p. 222).

For Buadze, the main attribute of Georgian women was a "great love towards the homeland" (ICWES-X, Complete Proceedings, p. 222). Sacrifice in the name of the country was an enduring trait in Georgian women. In addition to early martyrs, one recent example of such selflessness is the participation of women from all sectors in the events leading to the Sukhumi riots in 1989 ("Abkhazia" n.d.). Using Georgian traditions, which viewed the mantilla (a lace veil) as sacred, women stepped onto the battlefield to drop their veils to stop the fighting.

In the aftermath of these wars, where millions of women abandoned the capital Tbilisi, a "new type of woman appeared [, the] woman who looks with envy at everything. It's very difficult to make her believe" (ICWES-X, Complete Proceedings, p. 222). Neverteless, Buadze cited the particular situation of Margarita Mchedlidze, a scientist and doctor in agriculture who managed, even as a refugee at 80 years old, to pursue her work and publish a new book.

Several associations were created with the aim of helping refugees return peacefully to their homeland. Another organization, "Georgian Women for Peace and Life," which was led by Nana Tsagareishvili-Schevardnadze, First Lady of Georgia from 1995 to 2003, sought to study "the historical and ethnic causes of psychological and social conflicts" in order to find solutions and ways to avoid them altogether (ICWES-X, Complete Proceedings, p. 222).

Buadze concluded her paper with a bittersweet statement: Georgian women have become so used to struggling that they have learned to play the roles of both men and women. For Buadze, needing to be mothers and fathers at the same time "is our fate and as a heavy cross we must carry it to the end" (ICWES-X, Complete Proceedings, p. 223).

Anita M. M. Liu and Lai Yin Ng, "Sex Culture of Female Students in Hong Kong"

Anita M. M. Liu worked at the time at the University of Hong Kong in the Department of Surveying. After obtaining a bachelor's degree from the University of Reading, she worked for some time as a quantity surveying consultant. She then pursued an MSc and a PhD at the University of Hong Kong, where she later obtained a professorship at the Department of Real Estate and Construction.[4] Lai Yin Ng worked at the Department of Building and Construction at the City University of Hong Kong. Their paper, titled "Sex Culture of Female Students in Hong Kong" (ICWES-X, Complete Proceedings, pp. 197–201), talked about the sexual awareness and gender perception among women studying in post-secondary institutions in relation to the wider social discourse on sex and contraception.

Although women's social and economic conditions had improved over the last decades, traditional views on sex and gender were enduring and upheld both by men and women alike. Citing three past surveys from 1981, 1986, and 1991, Liu and Ng noted that despite an increasing freedom in sexual matters among young people, there was a marked drop in knowledge and awareness. According to the authors, this discrepancy could be explained by the "gap between main culture and sub-culture, i.e., the traditional societal moral standards towards sex and the youngsters' conceptual permissiveness" (ICWES-X, Complete Proceedings, p. 198).

[4] Anita Liu was also Chair Professor of Commercial Management and Quantity Surveying at Loughborough University (Fellows and Liu 2022, p. ix).

To "investigate how these social factors influence the sexual awareness amongst female students and to examine how females perceive their role in gender equality" (ICWES-X, Complete Proceedings, p. 198), a survey was conducted at the City University of Hong Kong by Liu and Ng. Out of 100 questionnaires sent, there was a response rate of 65%, of which two-thirds were 18–20 years old and the rest 21–23. On the topic of religion, 60% stated not belonging to any particular faith, while 40% identified as Christians or Buddhists. The analysis comprised three parts: sexual education background, sexual awareness, and perception of gender roles.

Looking first at sexual education background, 42% had had classes during high school, against 58% who either chose not to or did not have the possibility. Among students who did attend such classes, only half found them useful, while the other half deplored the generic content and teachers who were not open enough in their discussion of sexual matters. The controversy and taboo around sexual education being part of the regular curriculum was even reflected in how these classes were called, for instance, health education, or family education. However, all of the respondents were in agreement on the necessity of good sexual education classes as "a legitimate venue to obtain knowledge" (ICWES-X, Complete Proceedings, p. 198) on these matters. The authors were also of the opinion that a sexual education program should be mandatory in schools and argued that it should be built around the objective of "self-empowerment, where individuals, by understanding the consequence of their actions, can take responsibility for controlling their own lives and enhancing their self-esteem" (ibid.).

On the topic of sexual awareness, almost 70% of the respondents felt knowledgeable enough to navigate through their relationships. Most of them, perhaps surprisingly, did not attend sex education classes. Their main sources of information were mass media (60%), peer groups (22%), school (10%), and parents (8%). These numbers could be explained by the fact that schools were ill-equipped for sexual education programs, and parents would avoid discussing this topic with their children. Western and more liberal media from other parts of Asia became the main channel for young people to gain knowledge about sexual matters since this type of media was so easily accessible. Obtaining information through peers was also a consistent figure, since young people spent twice as much time with friends as with their parents.

Sixty-eight of the respondents did not agree that people should wait to be married to have sex, while the vast majority (82%) had some dating experience. An appropriate age for girls to start dating was 16–18 (51%), 19–21 (45%), and above 21 (4%). A 1991 survey from the Family Planning Association found that the average age of girls for a first experience in dating was 13.7 years old, explained by views like "dating is common" and "can't refuse." These reasons are also cited for dating in the research led by Liu and Ng, but come second and third after "dating to find long-life partners." This motivation was likened by the authors to some form of courtship, since most of the respondents felt like dating was more suitable for females older than the actual average at the time. The questionnaires also featured a series of true or false statements (with the option of saying "don't know") about pregnancy, birth control, and AIDS. Though general knowledge of the risks of getting AIDS was noted in all respondents, low success was noted in the question about the chance of pregnancy when sexual intercourse took place 2 weeks before menstruation.

Based on the premise that women, in traditional Chinese society, held an "instrumental and supportive role in managing home and supplying male heirs" (ICWES-X, Complete Proceedings, p. 199), the survey asked whether the students were expected to take up domestic chores (82% said 'yes'). Among children in a family, daughters took care of 20–30% of housework, while sons were at 5–10% of work distribution. Mothers had the majority of domestic work, with 70–100%, and fathers ranked at 5–15%. This division was consistent with traditional gender roles (the father as a provider and the mother as a homemaker), but a wide gap was also visible in families where mothers were working full time.

A large portion of the questionnaire featured various statements on gender roles and perceptions, such as "Achievement of a successful career is more important to males than females" and "Qualities such as gentleness and obedience are important characteristics of females" (ICWES-X, Complete Proceedings, p. 200). Respondents' views on gender equality were generally mixed; for instance, a little under half found that higher education was more important for their male counterparts. Liu and Ng saw this perception as the result of social and, more closely, parental pressure and expectations. Though traits like submissiveness, kindness, and care were generally seen as typically female, perceptions on whether men should be the sole "breadwinner" of the family tended to change quite substantially. For the authors, this transformation could be related to the strong presence of women in the country's workforce (46.5%) and to their increasing numbers in more specialized or technical fields (30%), as well as in administrative and managerial positions (20%). This, in turn, had positive repercussions for women's education.

When asked if being a woman in the 1990s was more advantageous than in previous decades, a little over a third agreed. This ambivalent perception might be linked to the previous statement of the survey, "A successful woman is one who is independent and devoted to her career" (40% agree), on the basis that there are ever-increasing opportunities for women, but there are now also "dual demands on women in today's modern society, where a woman is expected to perform the roles of homemaker and employed worker" (ICWES-X, Complete Proceedings, p. 200).

Even though men and women in Hong Kong were given equal opportunities in education and careers, there remained a strong prevalence of men in higher learning and better jobs. Gender roles were "socially and culturally constructed with parenting, families, school, the media, etc." (ICWES-X, Complete Proceedings, p. 200) and enforced a marked perception of typical attributes in men and women. Though Hong Kong was a "well-developed country with an educational and economic achievement that parallel any country in the west," traditional Chinese values persisted. For the authors, social progress and gender equality were only attainable through suitable sex education in schools, which is held to be essential in cultivating relationships and learning social responsibility.

Ágnes Fényi, "Changes in the Role of Women During the [Past] 50 Years"

This paper (ICWES-X, Complete Proceedings, pp. 203–204) by Ágnes Fényi, from the St Stepan Hospital ENT Department in Budapest, sheds light on the evolving social, political, economic, and cultural situation of women in Hungary during 1945–1980, with a focus on the conditions of the current, post-communist society. We are first reminded that after 1945, an increasing number of women were engaged in full-time employment, partly due to political reasons and partly owing to expansive economic growth. However, as a result of the underdevelopment of the infrastructure necessary for the smooth and untroubled lives of families, serious social tensions emerged. Divorces grew by leaps and bounds, and the organization of everyday life became almost impossible, even in the remaining families. Women, reports Fényi, felt the burden both in the primary and secondary economies, a situation that, she concludes, "resulted inevitably in their mental and physical deterioration" (ICWES-X, Complete Proceedings, p. 203).

But other critical factors were also at play, this time on the international front. These included the collapse of the communist party states of Central and Eastern Europe in the late 1990s and their move away from socialist planned economies towards the creation of modern state economies and parliamentary democracies. Hungary was one of these countries, undergoing deep social and political transition and coping with severe economic constraints. These barriers adversely affected the enforcement of economic, social, and cultural rights, the effective recognition of women's rights, and the development of an attitude conducive to the full acknowledgement of female equality.

Fényi reports that "[u]nder state-socialism, the equality of women existed for a long time as an unquestionable political objective. It was primarily based on the right to a gainful occupation as the most important criterion" (ICWES-X, Complete Proceedings, p. 203). Opened to debate was the question of how important "ideology, the vast labour-demand generated by extensive industrialization, and the political endeavor to maximize control over the individual and minimum wages through the 'two-earners-model'" (ibid.) were as underlying motives. "When and how much women needed an opportunity to work and when and to what extent it was inflicted upon them by the state was another important issue under discussion at the time. It is likely that there were innumerable factors, named and unnamed, that contributed to the final result. Elements of imposition, such as an increased workload on women and weakening domestic functions, undeniably had a legate impact. [This] continuing discrimination against women manifested itself in the structure of female employment and in wages lower than those of men. Their involvement in economic activities did help improve [...] the position of families, and in most cases, work outside the domestic setting meant an enrichment of women's lives as well as alleviating to a great extent female poverty" (ibid.). Finally, there were notable increases in the number of women enrolled in secondary and higher education.

Fényi pointed out that the countries in transition in Central and Eastern Europe "have been surprisingly quick to establish political structures typical of modern democracies fundamentals of a multi-party system, free elections, and the rule of law. With that, political and civil rights, which previously existed 'de jure,' have become a reality. Factors which primarily influence the status of women are related to ideology, the economy, and party politics. Economic decline in Eastern Europe has reached critical proportions. In the majority of these countries, the rate of output loss and unemployment growth is higher than during the Great Depression in 1929–1931. There is mounting inflationary pressure. By now Hungary has reached a point where the two earners model is no longer prevalent. In more than 50% of families, there are now single or no earners at all. As a result of economic decline, about two thirds of the population has been exposed to falling living standards. Those afflicted the worst include the unemployed, families with children on their own [= single parents]. The objective deterioration of the general situation is being accompanied by a heightened sense of insecurity" (ICWES-X, Complete Proceedings, pp. 203–204).

Fényi discusses how:

> "Hungary's first freely elected government restored the traditional family with the mother caring for hearth and home as the centrepiece of its values, as opposed to the ideological pressure in the 1950's for female employment. The already wide range of maternity benefits were further expanded while, with previous state prevalence gone, there was a substantial decrease in subsidies for childcare institutions. About half of the crèches were closed. Although there was a limited reduction in nursery capacity, families were charged more for meals, both here and in schools. The economic stabilization measures which the current socialist-liberal government is forced to take are not expected to yield improvement in any areas."

> "Of Hungary's population of 10,3 million, there are 5,4 million women and 4,9 million men. According to the 1990 census, 85.9% of families were full families [= families with two parents], 12% of couple-based families had the wife as active earner. In 80% of single-parent families it was the mother who lived with the child. In 1990, in 26.5% of all households, women acted as principal earners."

> "[Since] 1996, the female retirement age [has been set at] 62 years, [whereas] before 1996 it was 55 years. The level of women's general educational background reaches and, in young age groups, even exceeds that of men. Girls typically choose grammar school (two thirds of grammar school students are girls). There is gender-based market segregation by gender of various occupations, e.g., girls prevail in health care and in secondary commercial studies, whilst boys dominate heavy industry and construction occupations" (ICWES-X, Complete Proceedings, p. 204).

According to Fényi, "[t]his is not a consequence of institutionalized discrimination. Rather it is a matter of parental and children's preference. In the 1993/94 academic year, 52% of university and college students were females. The percentage of female students significantly varies by faculty. In medical schools, female students continue to outnumber males" (ICWES-X, Complete Proceedings, p. 204). The same can be said for faculties responsible for the training of a large majority of women in fields such as special education and primary and secondary school teacher training.

Fényi observed that "[a]mong women in managerial positions or coming from families of intellectuals, the percentage of those remaining in this segment of society has been steadily rising, which is not the case with men" (ICWES-X, Complete Proceedings, p. 204). In her view, "[t]he reason for [this] continued increase is that in 1992, 42% of the daughters of men in managerial positions or working as intellectuals became managers or intellectuals themselves, versus 28% in 1973. The small business sector provides livelihood for women to a growing extent. Recent years have seen a substantial decline in female employment in the case of both active-aged women and those outside the age of employability" (ibid.).

Fényi's ending remarks provide further information and raise further interest in the history of Hungary and in the situation and role of Hungarian women since 1945. For example, she contends that "the extensively developing economy [in Hungary] raised women up to work without considering what kind of social work would suffer [with respect to] bringing up children and taking care of the elderly" (ICWES-X, Complete Proceedings, p. 204). For a period of time, she explains, these services were "compensated by the lifestyle in which different generations lived together, but later on by the breaking up of this, these tasks become uncared [for]" (ibid.).

Fényi then discusses how "the increasing economical prosperity of the 1960s helped build up the nursery school system, schools and colleges and maternity benefit" (ICWES-X, Complete Proceedings, p. 204). However, in the late 1970s, Hungary had to face a serious political and economic crisis, and "with this the system of social allowances also faced a crisis" (ibid.). In fact, the crisis was still continuing, "manifesting itself in the disintegration of a free educational and health system, and in the reduction of allowances and grants" (ibid.). This situation had particularly hurt Hungarian families, because of a sharp increase in inflation. Fényi thus concluded her remarks on this somber note: "After the 90s the input of women's work into the economical sphere has increased the burden of our society by depriving us of rights and advantages. Most of all the burden of intellectual women affects their career badly. The prospective time to escape from it is hardly predictable. In the new economic sphere it would be very important to establish a real stability in the social and legal security of women" (ibid.).

Business Meeting

Minutes of the meeting taken from ICWES-X, Minutes:

> Meeting of the ICWES-X Delegates, held at the Society of Measurement, Automation and Informatics (MATE), Budapest, Thursday, October 10, 1996, at 1600 hrs.
>
> **Present:**
>
> Elizabeth Laverick, Chairman and ICWES-IX Chairman
> May Maple, Deputy ICWES-IX Chairman
> Zsuzsa Szentgyörgyi, ICWES-X Chairman

Anna Tompa, Deputy ICWES-X Chairman
Cicely Thompson, ICWES-II Chairman
Vera Vokolkova, Botswana
Dormer Ellis, Canada
Mitsuko Kazuno, Japan
Meera Ramrakhiani, India
Kante Assa Bore, Mali
Anna Szemik-Hojniak, Poland
Tatjana Samodurova, Russia
Abell Montse, Spain
Isabelle Kalt Scholl, Switzerland
Elizabeth Rhodes, UK
Peggy Layne, USA
Marie Theresa Lukusa Nonyabo, Zaire

Plus observers from six countries: Canada, Japan, Mali, the UK, the USA, and Zaire.

1. Dr. Laverick explained that she had been asked to conduct the meeting by and on behalf of the ICWES-X Chairman.
2. **Minutes of the last meeting held on July 18, 1991**: These were circulated to the meeting together with the guidelines.
3. **Confirmation of the mandates of the delegates**: The Chairman named those members present whom she understood were delegates representing their various countries, together with the ICWES Chairmen Emeriti and their personally named deputies. It was noted that all others present were observers.
4. **Amendments to the Guidelines**:

   Three amendments to the guidelines were put forward in advance of the ICWES-IX Delegates meeting (see Item 4.1 of the minutes). Following discussion, it was agreed at that meeting that these should be sent to all official ICWES-IX delegates within six months of ICWES-IX, inviting written comment. With no further comments received, the amendments were declared adopted.

   A further proposed amendment to the Guidelines was received from Ase Skov of Denmark for consideration at the ICWES-X Delegates meeting. This reads as follows:

   Section I: add "G. At least two official languages, English and French, should be used for documentation and presentations at the Conference"

   After discussion, it was agreed to modify the proposal as follows:

   Section 1: add "G. The official language used for documentation and presentations at the Conference should be English, but a second language could be used at the discretion of the host country."
5. **Consideration of bids for ICWES-XI**: The Chairman announced that only one bid had been received, from Japan, offering to host ICWES-XI in 1999. The Japanese proposal had been circulated to all conference participants. There being no other proposals, Professor Mitsuko Kazuno was asked to say a few words in support of the proposal. Following this, the delegates voted in favor of accepting the Japanese proposal.

   The American delegate then spoke of the intention of the USA to bid for ICWES-XII to be held in 2002 on the West Coast. This would precede the SWE Conference, and delegates would be invited to stay on and attend this also.
6. **Election of Convenor and Members of the ICWES Continuity Committee (ICC)**: The Chairman of ICWES-X had expressed her wish not to stand for election as Convenor on the ICC. The Chairman proposed that Minako Tanaka, who was attending her second ICWES, be elected Convenor. This was unanimously agreed upon.

   Area representatives for the ICC: The following arrangements were agreed.

North America (Canada and USA): Dormer Ellis
Latin America: Convenor to consult with Angeline Perez-Lopez
Middle East: Convenor to invite Mahin Rahmani to continue
Far East: Mitsuko Kazuno
India subcontinent: Convenor to consult with Ila Ghose
Africa: *Arabic speaking*-vacant (no news from Chebanne Hassiba since ICWES-VIII); *English Speaking*: Convenor to consult Joanna Maduka; *French Speaking*: Loteta Dimandja
Western Europe: Convenor to invite de Kryger to continue
Eastern Europe: Zsuzsa Szentgyörgyi
UK, Australia, and New Zealand: The president of WES or her personally named deputy

7. **Any Other Business**: A paper by Loteta Dimandja entitled "The Development of Africa, Myth or Reality" was circulated to attendees. There being no further business the meeting was closed."

Testimonials

Lee Arnold, Mid-Atlantic Region MAL, "Special Report: Reflections on ICWES-X"

"All that my friends heard from me for months was, 'I'm going to ICWES-X, in October, and it's in Budapest.' Except for the fourth International Conference of Women Engineers and Scientists in Poland, I'd never ventured that far east before."

"I felt about Hungary as I did in Poland in 1975, that the people are rooted in a national identity that does not depend solely on geographic space or political entity. Like Poland, they have been invaded, expanded or partitioned, added to empires, and regained independence so many times in their history that their strength as a people lies in their shared tribal history rather than in place and boundaries. Indeed, the country was celebrating its 100th anniversary while we were there."

"I was fortunate to be introduced to this before leaving by a Hungarian colleague who lent me maps and books and gave me all kinds of information. SWE members did not organize a travel group, but most of us seized the opportunity to plan personal sightseeing tours. Two others traveled with me, Helen Marie Chapman and Jo Webb. We eased our way into Europe by flying into Frankfurt and then travelling by rail to Vienna and by hydrofoil down the Danube to Budapest. What a fine way to see the countryside and the approaches before reaching the city itself. After the conference, we took a sleeper train to Venice for a stopover on our way to Geneva, where we spent two days visiting and touring with Jacqui Juillard before returning home. Jacqui, who frequently represented Switzerland at past conferences, was very interested in learning about ICWES-X, and she sends greetings to all her SWE friends."

"In any ICWES there are always two major levels of functioning. One is the practical management of its schedule and events; the other is the camaraderie, the spirit that flows as we all come together. This one was high in both. On the practical side, sessions started generally on time and were well moderated. Whatever the glitches may have been I have never been so little aware of them; they were simply handled without fuss. And the unscheduled was coped with magnificently. A canceled workshop left a free afternoon on the second day, but a bus tour was arranged. As a result, we had a wonderful trip to both touristy and historic sights outside the city. I don't know when it was planned, but it went as though expected all along. That was the way they did things!"

"The conference opened on Oct 8 with greetings on behalf of the Hungarian Academy of Sciences and the Scientific Society of Measurement, Automation and Informatics (MATE), the sponsoring organization, by Professor Laszlo Keviczky, Secretary General of the Academy and Chair of MATE. He was followed by Dr. Elizabeth Laverick (UK), Chair of the 1991 ICWES, who reviewed the past nine conferences. The final speaker was the ICWES 10 Chair, Zsuzsa Szentgyorgyi, who addressed 'Why Hungary?' She spoke of what Hungary does not have – coasts and harbors, natural resources (depleted in war and by Communism), but more hopefully of the rich heritage of scientists, artists and others that Hungary has given to the world. Hungary has its deeply rooted culture, yet they are very aware of being in transition today between political and economic worlds. 'To live in the crossroads is often dangerous,' she said, 'on the other hand, however, probably just this sharp and difficult situation made Hungarian people extraordinarily adaptive, docile, quick to understand, and enterprising. Maybe this was one of the reasons that the ICWES Council chose Hungary.' There could hardly be a better one!"

"An interesting variation was introduced into the opening plenary session. One of the two addresses was 'The Influence of Science and HiTech on Modem Literature' by Maya Bejerano, a leading contemporary Israeli poet. She sought to bring humanism into the engineering conference, feeling that technology is destroying intimacy, human contact and 'realness.' Regrettably, it consisted mostly of another poet's work, and failed, I thought, to illuminate the desired connection between the arts and sciences. The other address was given by Baroness Platt of Writtle (UK) on 'Women into Science and Engineering.' She is always an excellent and inspiring speaker."

"After the plenary session each day, we divided, as our interests took us, into the technical sessions on the theme 'New Trends in Natural Sciences and Technology' or the sociological sessions on 'Women in the Sciences.' One of the more interesting sidelights on the latter theme arose out of the technical sessions. An excellent presentation was given on 'Successful Implementation of Integrated Systems' by Erzsebet Erdelyi of Arthur Anderson Consulting in Budapest. Later she was interviewed by a journalist who involved two of us in the conversation. The journalist remarked that she heard the same sort of question so many times about the progress of women and the obstacles we face; was any progress being made? It showed again how necessary these conferences are, that change comes very slowly and at different levels in different countries and cultures."

"Ms. Erdelyi spoke in the interview of her family – parents, husband and child – of the arrangements she makes to enable her to work, of the support network she needs, of the pressures about managing family and career. All these are familiar issues to us, but Hungary and other countries are in a different place, dealing with issues at a level that countries like the USA and the UK dealt with almost a generation ago. We could envy the young women of Hungary, though, for having such a superb role model as Erzsebet."

"Another view on that same theme was presented on the third day. In the interests of 'fairness,' three men had been invited to submit papers. The first was titled 'Feminist-Related Dysfunction in Socializatio' by Miklos Hemadi, PhD, a Hungarian sociologist. He took what little truth there is in criticism of a small segment of radical feminism and generalized madly to attribute every possible societal ill, relevant or not, to anything called feminism. It wasn't intended to be funny, but the laughter from the otherwise polite audience began about halfway through. Fairly obviously he was talking from his own unhappy experience, and the grapevine confirmed later that he was just completing a messy divorce. Not one scrap of data was presented with this diatribe and, when challenged on that point, he replied, 'Evidence! didn't have time for evidence.' This to a group of engineers and scientists whom he obviously did not respect. One could admire his nerve, I suppose."

"The other two male speakers presented topics entitled 'Quantitative and Qualitative Characteristics of Way of Life Among Hospital Family Doctors' and 'Gender Differences in Cognitive Functions of the Brain,' They managed to be a bit more low key and to sound at least pseudoscientific, but their conclusions seemed a bit of a leap from the data."

"There was plenty of time for fun, too. The first day closed with a reception at which there was a pleasant welcoming speech and then lovely entertainment by an all-women choir, followed by supper. The afternoon of the second day gave us the guided tour to Szentendre (St. Andrew's) for a tour of a sculpture museum, for shopping, and then on to Estergom, site of the largest Catholic basilica in Hungary. The tour concluded with dinner at a restaurant on the resort of Szentendre Island, complete with gypsy band, dancers, and singers."

"The third day brought the closing of ICWES-X followed by a banquet at Hotel Gellert, a luxury hotel and home of the most famous of Hungary's many geothermal baths. We were gloriously fed and there was a pleasant minimum of formal speeches, but the affair was quickly allowed to become informal. A vote of thanks was proposed spontaneously by me. Baroness Platt delivered an extempore speech of thanks; two delegates from Zaire entertained us with a song; and there was much table-hopping, picture-taking, and endless goodwill mixed with the farewells."

"The final event was a half-day post-conference tour within the city, led by one of the guides who took us to Szentendre and Estergom. We visited Heroes Square, where we saw the changing of the guard at the Tomb of the Unknown Soldier, then St. Matthias. Church, and also a bright arcade housing a market with shops of all vanities where we could do some purchasing and were treated to small sandwiches of local ham and salamis and a sample of wonderful sweet Tokaji wine. Hungary is rich in history and is very lively in telling by its own people."

"SWE was officially represented at ICWES-X by Society President Peggy Layne. Three SWE Past Presidents and four MALs were among the attendees. A preliminary list of registrants, supplemented by business cards collected in Budapest, showed the following members at ICWES-X."

"Marjorie Aelion (Western Carolina), Lee Arnold (MAL-E), Susan Cavin (New York), Katluyn Cunningham (Chicago Regional), Dormer Ellis (MAL-X Canada), Kathleen Harer (Spacecoast), Lucy Hsu (Santa Clare Valley), Wilma Carol LaBelle (Wisconsin), Peggy Layne (Florida's First Coast), Evelyn Murray (Boston), Linda Schmidt (Baltimore-Washington), Siiri Sloat (MAL-X Sweden), J. Cicely Thompson (MAL-X United Kingdom), Josephine Webb (Pacific Northwest)."

"With perhaps fewer than 200 delegates, including the Hungarians, it was a small assembly compared to past ICWESes. However, I counted 21 countries from an incomplete list, and that makes it far from the smallest in representation. It was very good to see countries represented who have not often been – such as Greece, Georgia, Romania, Hong Kong, and Russia. The spirit of communication among countries and individuals was well fostered. We look forward to ICWES-XI, to be held in 1999, in Chiba (near Tokyo), Japan" (Arnold 1996b, pp. 5–6).

Peggy Layne, "Report From ICWES-X, October 8–10, 1996, Budapest, Hungary"

"I was very excited to be able to attend and represent SWE at the l0th ICWES meeting in Budapest, Hungary, this October. Women from around the world gathered to exchange experiences, ideas and hopes for the future. A networking workshop provided an opportunity for women from different countries to share information about the organizations supporting women scientists and engineers in their country, and a workshop on low budget activities to interest girls in science and engineering had many enthusiastic participants. What was most striking to me were the similarities in the challenges that face women in professional fields, regardless of where they come from."

"The first International Conference of Women Engineers and Scientists was convened in New York, NY, 1964, with nine subsequent conferences held about every three years since. The aims of the conference, explicitly defined at the fourth conference in Cracow, Poland, in 1971, are to bring together professional women engineers and scientists worldwide, to encourage the use of technology for the betterment of life, and to consider ways to increase the participation of women in engineering and science."

"Do these goals sound familiar? They should! They are very similar to the goals of the Society of Women Engineers."

"The program of ICWES-X included close to 50 paper presentations over the course of three days, addressing topics from the role of professional women in the new order of Hungary to technical presentations on energy use and environmental protection. The conference opened with plenary sessions giving some background on the history of the ICWES and an overview of women in science in engineering, and continued with workshops, a panel discussion and concurrent sessions addressing technical and social issues."

"Opportunities to network at the conference were many. Delegates were welcomed to the conference with music of the women's chorus of Hungary and a buffet of traditional Hungarian delicacies. The technical sessions were held at the international trade center, on the Pest bank of the Danube adjacent to the Hungarian parliament, with a wonderful view across the river to the castle district on the Buda side. On Wednesday, the second day of the conference, attendees were treated to a visit to the historical village of Szentendre, now an artists' colony, and dinner at a restaurant with folk music and dancing. The conference concluded with a banquet at the Gellert hotel, home of one of Budapest's famous thermal baths."

"While the presentations at the conference were very interesting, with many topics addressed and something for everyone, the greatest value of the conference was the opportunity to interact with women from four continents and a variety of cultures. Two groups of women from Africa were particularly interested in the low-cost activities to attract girls to science and engineering demonstrated by SWE members from the U.S. They asked when SWE members could come to Mali to help start career guidance programs there! It was also interesting to learn about the 'neighborhood engineer' program in the UK in which engineers work with local schools both to solve technical problems and to inform children and their parents about preparation for engineering careers."

"Japanese women representing the Society of Japanese Women Scientists, the Japanese Women Engineers Forum, the Women Researchers' Society of the Agency of Industrial Science and Technology along with the Ministry of Agriculture, Forestry and Fisheries presented a proposal to host the 11th ICWES meeting in Japan in 1999. The technical theme for the conference will be 'Science and Technology for Global Ecology.' The venue will be the Japan Convention Center in Chiba, a newly developed high-tech town between Tokyo and Narita airport."

"Based on my rewarding experience in Budapest, I'm looking forward to attending ICWES-XI in Japan and encourage you to begin planning now to do the same!" (Layne 1996, pp. 10–11).

Elizabeth Laverick Reports on ICWES-X

"Flying in, we were welcomed by the 'Summer of the Old Ladies' – an Indian Summer. Budapest looked magnificent! Our hotel was in the pedestrian zone: Vaci Utca is the Bond Street of Budapest with elegant shops and cafes, and only a stone's throw from the river. After a brief rest we set forth on foot along the banks of the Danube – it was a magnificent blue admiring the famous Chain Bridge en route. Opposite the Houses of Parliament, which are a symbol of the grandeur of the Austro-Hungarian Empire, we found the headquarters of the Scientific Society of Measurement Automation and Informatics, or MATES for short, the venue for ICWES-X. We registered, collected our Conference pack, paid our outstanding dues and chatted to old ICWES colleagues as they drifted in."

"On the way back we ventured on to the tram which proved a most efficient mode of transport. It was somewhat crowded in the rush hour, but no worse than the Underground. Back at the hotel we eagerly investigated our Conference pack. The largest item was the bound preprints of the Conference papers; these proved invaluable at the Conference sessions where sometimes the spoken English was difficult to comprehend. Next came the Final Programme. Plenty of interest here, both in the Technical sessions and the Workshops. The latter were designed to encourage delegates to exchange information on 'Low budget activities to interest girls in science and engineering' and also to encourage 'networking between professional engineering, scientific and educational organizations.' There were also tickets for the social events, the Welcome Reception on the Tuesday evening, an excursion to the captivating town of Szentendre on Wednesday, the Banquet at the famous Hotel Gellert on Thursday and a sightseeing tour of Buda and Pest on Friday."

"At the reception we were welcomed by Lajos Nyiri, Executive President of the National Committee for Technological Developments, and, suitably laced (?) with champagne we relaxed to the most harmonious voices of a female choir, Vox Hungaria, singing a delightful selection of Hungarian folk songs. This was followed by a typical Hungarian buffet supper. I particularly enjoyed the Tóltött Káposzta (cabbage leaves stuffed with rice and pork mince and served with a paprika sauce) washed down with plenty of Hungarian red wine. We took the opportunity to catch up with news of old friends and to introduce ourselves to new ones."

"The visit to Szentendre was again something different. It is known as the tourist capital of the Danube Bend, and is a charming C18th Baroque village painted all colours of the rainbow. With its interesting architecture, its museums, galleries, restaurants. Coffee houses and shops; it has a special atmosphere, frequented as it is by artists, writers and poets. On the way home we were given dinner with a typical Hungarian floor show. After the show, some of our party were invited to join in the dancing which they did, albeit with more vigour than expertise. It was all great fun!"

"So ICWES-X lived up to the ICWES ideals. The Hungarian Organising Committee are to be congratulated on the success of the Conference, particularly bearing in mind the difficulties they had to face, cancellation of EXPO 96, rampant inflation and all that entails in organising a Conference of this kind. The attendance, from 16 other countries, must have been particularly gratifying with about 100 participants, including 13 from the UK."

"The delegates' meeting at the end of the Conference unanimously agreed to accept Japan as the Hosts for ICWES-XI in 1999. So, give yourselves a treat. Start saving now for the next ICWES experience and another holiday of a lifetime' I recommend it!" (Laverick 1996, p. 3).

Elizabeth Rhodes, WES Member, "Destination Budapest—ICWES-X"

"After a lapse of five years the 10th International Conference of Women Engineers and Scientists was held in Budapest in October 1996. The Organising Committee chaired by Dr. Szuzsa Szentgyörgyi had compiled a programme of some fifty technological and sociological papers with plenary sessions intermingled with the more technical sessions and workshops. The meeting was held at the Scientific Society of Measurement, Automation and Informatics (MATE) on the Kossuth Lajos Terrace opposite the Hungarian Parliament Building and overlooking the Danube River."

"On registration the 90 participants from 16 countries were given Preprints of the Conference papers entitled "Towards the third millennium: the role of the intellectual potential in a new world." The conference began with a welcome from the director of MATE, Dr. Keviczky Laszlo. This was followed by Dr. Elizabeth Laverick, who spoke about the nine preceding ICWES meetings… *Attempts to form a worldwide network of Women Engineers and Scientists is still proving difficult.*" [our emphasis]

"Dr. Szuzsa Szentgyörgyi told us about the position of scientists in Hungary: a country which is relatively small but lies at the cross-roads of Europe and has a record of Nobel prizewinners."

"A main theme of ICWES-X was recruitment of women scientists and engineers and this was introduced by Baroness Platt who spoke about the WISE campaign and Equal Opportunities for girls and women in schools, industry, career training and career breaks. This was followed through by a workshop on interesting girls in science and engineering organised by SWE."

"Monique Frize (Canada) had evidence that humanizing the engineering curriculum, the culture and teaching style, could create a more women friendly environment. Presenters from France and Switzerland outlined their countries' efforts. Peggy Layne, President of SWE gave recent survey data from the USA where though young women in their twenties start out at parity with their male counterparts in salaries and status, older women report very different and discriminatory experiences."

"A series of papers (10) described the roles, social culture and attainments of women technologists in various sciences and industries and their management experiences from countries as diverse as Hong King and Japan, the UK and the USA, Poland and Hungary. The general conclusions seemed to be that girls tend not to be greatly interested in the sciences and engineering; often qualified women are not as ambitious as men and do not go for promotion, whilst inadequate childcare often prevents them from achieving their full potential."

"A section on education and transfer of knowledge had 8 papers. These were a mixture varying from the intelligent use of chemicals in the home to the use by international educators of the resources of NASA."

"There was a panel discussion on the role of professional women in the new Hungary, tackling such subjects as life expectancy among women employees, the conflict between careers and family life, feminism related dysfunction and its social effects."

"A curious paper by Miri Kubovy of Harvard University explored the influence of science and hi-tech on modern literature. The effects of consciousness and imagination trying to unite with reality's system and laws were exemplified in the poem 'To the Star of the Soul' by Baruch Spinoza in The Ethics. Here the language used in the poem leads to the idea that cosmic creation is being renewed rhythmically."

"There were 18 technical papers varying from the Integrated Communication and Information Systems to Computer Graphics. Modelling and Prediction. Traffic problems such as the factors affecting the formation of iced roadway surfaces, road safety and prediction of accidents and control of road vehicle emissions were discussed. Environmental safety and control in the nuclear industry, the availability of energy, the use and provision of solar cells in developing countries and small electronically commuted and controlled motordrives were considered."

"Finally, a meeting of ICWES Country Delegates reviewed the work of ICWES and enthusiastically accepted the invitation of Japan to hold ICWES-XI near Tokyo in July 1999."

"ICWES-X was not all work and we enjoyed the hospitality of the Hungarians at receptions and the Conference Dinner and a visit to the picturesque village of Szentendre [...]" (Rhodes 1997, p. 11).

Chapter 12
ICWES-XI, Chiba, Japan, July 24–27, 1999

Theme: "Science and Technology for Global Ecology"

The welcome remarks address the choice of the theme for the conference and the fact that it is being held for the first time in the Far East. Each of the opening talks is quite interesting, so we include them as delivered in July 1999. A few typos were corrected, but the texts are in their original form.

Her Imperial Highness Princess Takamado, "Message"

> "It is a source of great pleasure for me to be with you at the opening ceremony of the 11th International Conference of Women Engineers and Scientists. I am delighted that it is being held here in Makuhari, Chiba Prefecture, and I warmly welcome all the distinguished participants, especially those from abroad."
>
> "Today, as we stand on the threshold of the 21st Century, it is imperative that each of us give urgent thought to the Ocean Planet that gave us life and nurtured us. Therefore, I am delighted that the theme of the Conference is 'Science and Technology for Global Ecology' and that papers will be read and problems discussed by leading women engineers and scientists. It is to be hoped that this will result in the creation of a network between countries that will encourage us to exchange experiences and cooperate in ecological efforts. We should not be averse to voicing our thoughts on environmental issues nor to taking part in conservation efforts, for nothing could be more damaging for our future than collective inertia or a mass disinclination to take action."
>
> "While giving praise to the scientific endeavours of our forefathers and the sheer intelligence and enthusiasm that led them up along the path of scientific discovery, and while giving thanks for the many benefits and comforts we enjoy in our daily lives as a result, we must always give thought to the fact that, as a species, we have taken more than our share. Somewhere along the way, homo sapiens sapiens ceased to be ruled by wisdom (L-sapientia) in the true sense – that is to say, not high intelligence per se, but a compassionate, respectful

M. Frize et al., *Women's Contribution to Science and Technology through ICWES Conferences*, Women in Engineering and Science,
https://doi.org/10.1007/978-3-031-33800-7_12

and noble wisdom. Instead, we began to be ruled by greed. Perhaps now, as the species grows older, we will gain the wisdom to understand that life on earth is based on harmony and balance, not just among ourselves, but among all life forms on this planet."

"I take this opportunity to thank and congratulate the organizers of this Conference and take great pride in the fact that it is being held in Japan. It is extremely important that science and technology should contribute to the creation of an environment in which the human species is a harmonious part of nature and I wish all the participants success in their endeavours on behalf of global ecology" (ICWES-XI, Complete Proceedings, p. 4).

Michiko Ishii, Chairperson of ICWES-XI, "Welcome Address"

"Her Imperial Highness Princess Takamado, distinguished guests, ladies and gentlemen: Today it is the greatest honor and pleasure for the Organizing Committee of the 11th International Conference of Women Engineers and Scientists and its members to have Her Imperial Highness Princess Takamado at this opening ceremony of the Conference. On behalf of the Organizing Committee, I would like to express our heartfelt welcome to all members gathering from many countries of the world."

"The ICWES conference has been held every three years since the 1st Conference held in the USA in May 1964. The 6th ICWES was held in India in 1981. Since then, no ICWES has been held in Asia. It is a great pleasure and honor for our organizing committee to host this conference for the second time in Asia and for the first time in a Far-East Asia country. We have been organizing this conference in cooperation with the Society of Japanese Women Scientists and the Japanese Women Engineers Forum, and honorarily with the Science Council of Japan. Furthermore, we are supported by the Japan Women's Pharmacist Association, the Women Consulting Engineers Society of Japan, and many other organizations, such as the Ministry of Education, Science, Sports & Culture, the Ministry of Labor, the Science & Technology Agency, the Environmental Agency and many scientific societies. This conference provides an excellent opportunity for women engineers and scientists to share their knowledge with the international community, to help facilitate networking, and to promote cultural interchange among women in the areas of science and technology."

"We are proposing 'Science and Technology for Global Ecology' as the theme for ICWES-XI. We would like to extend our network to protect the Earth on a global scale via the exchange of experience and ideas. We will have three keynote lectures, two invited lectures, seven planetary lectures and more than 150 presentations. In addition, a symposium entitled 'Women's Role in Today's Environmental Issues,' a 'Junior Scientists Conference' for children, and various social events are planned."

"We, the Organizing Committee, hope that this conference will be a bridge to the 21st Century in research through provoking, active discussions among the delegates over these 4 days. Thank you very much" (ICWES-XI, Complete Proceedings, p. 5).

Hiroyuki Yoshikawa, President, Science Council of Japan, "Welcome Address"

"Your Imperial Highness Princess Takamado, honorable members of the International Conference of Women Engineers and Scientists, ladies and gentlemen."

"Today it is our greatest honor and privilege to have Her Imperial Highness Princess Takamado at this opening ceremony of the Conference. On behalf of the Science Council of Japan, it is also my great pleasure to express our sincere welcome all of you, who represent 39 nations from all over the world."

"The Science Council of Japan was established in 1949 as a governmental organization to represent the qualified Japanese scientists in all fields of cultural, social and natural sciences. The aim of the Council is to promote scientific research and to enhance its influence on social activities, industry and civil life."

"Since the foundation of the Council, we have been working to contribute to the progress of science in cooperation with academic organizations throughout the world, by sponsoring various international congresses here in Japan, and by sending Japanese delegates to international congresses held overseas. We do this because we believe that the promotion of international scientific exchange is one of our most important missions."

"We have opened today the 11th International Conference of Women Engineers and Scientists jointly with the Society of Japanese Women Scientists and Japanese Women Engineers Forum. It is my great pleasure to meet with distinguished engineers and scientists from all over the world and to have an opportunity to attend fascinating lectures and presentations."

"We regard it very delightful and meaningful that the 11th International Conference of Women Engineers and Scientists is held for the first time in Japan, in which approximately 400 participants have gathered from almost 39 nations."

"At the present Conference, it is expected that a number of research results will be reported across various research fields, and that intensive discussions will be held on the future direction of research, research themes of primary importance, international research collaboration, and so on. Particularly, we should think about the relationship between our research and human impact on the biological systems of the earth."

"In Japan, the number of women academic researchers in universities is not still great. The Science Council of Japan made an appeal for the improvement of women scientists' status in 1994 and established the special committee for the improvement of environment for women scientists during the 17th term. In the World Conference on Science held in Budapest (June 26–July 1), it was stressed that the improvement for participation of women scientists in scientific activities implies that the approach, and even the content, of scientific advances would change to focus more on the needs and aspirations of humankind. Especially, with the recent demographic change in Japan such as a decrease in the youth population, a shortage of young workers and the advent of a super aging society, the time has come for women scientists' power, which has so far not been fully recognized, to really contribute to the promotion of science and engineering in Japanese society."

"In conclusion, I sincerely wish the great success of the 11th International Conference of Women Engineers and Scientists, and for those participants from foreign countries, I hope your stay in Japan is enjoyable and you will form a friendship with Japanese engineers and scientists, making the present congress a very memorable one for you. Thank you for your kind attention" (ICWES-XI, Complete Proceedings, pp. 6–7).

Mitsuko Kazuno, President of the Society of Japanese Women Scientists, "Welcome Address"

"It is a great honor that Princess Hisako Takamado has graced us with her presence today for the opening of the 11th International Conference of Women Engineers and Scientists (ICWES). This is the first time in the history of the conference that we have met in the Far East. I am also very pleased to be the incumbent President of the Society of Japanese Women Scientists (SJWS), one of [the] host organizations."

"When I look back 8 years to the 9th ICWES conference held in England when I was asked by the chairperson if Japan would consider hosting the 11th conference, I knew that Japan had not hosted this conference in the past despite the fact that we had met in many other countries previously. I, therefore, felt that we should do it, but, for the 20 or so directors of the Society of Japanese Women Scientists, this task seemed to be too big."

"Since then, our great effort has been to publicize this conference to as many people as possible, setting up more organizations of women scientists and engineers, and establishing a network among them. In 1994, when we included about 700 members in our network, we held a trial international conference to test our capability in the following areas: how to organize an international conference, how to encourage attendance, and how to raise the necessary funds. We continued our expansion to the 2000-member mark, and finally in 1996, we took the plunge at the delegate meeting held in Budapest and offered to host the 11th conference."

"Today, I am so pleased to be able to again meet many old friends from various countries. On behalf of the ICWES-XI [I] sincerely hope that all participants here enjoy this 4-day conference, make new friends, and discuss the urgent problems of 'Science and Technology for Global Ecology.' Furthermore, we would like to encourage you all to exchange information about enhancing our working conditions."

"Thank you very much indeed for your attention" (ICWES-XI, Complete Proceedings, p. 8).

Susan Bird, Representative of ICWES and Former President, the Women's Engineers Society, "Message"

"I am deeply honoured to have been asked to make this address at ICWES-XI."

"The first International Conference of Women Engineers and Scientists was held in New York in 1964, and the second took place in Cambridge in England in 1967. Since then, these Conferences have taken place every 3 years or so. Other countries who have hosted

ICWES have been Poland, Italy, France, India, the Ivory Coast, Hungary, and now Japan. Six Japanese delegates attended ICWES-I, and 8 attended ICWES-II, and throughout the years Japan has had a strong interest."

"One of the high points in the past was the opening of the Indian Conference by Indira Ghandi, the Prime Minister. Now in Japan we have another landmark in the ICWES history by the presence here of H. I. H. Princess Takamado."

"My first visit to Japan was in 1994, when I spent 24 hours here as a stopover on the way to New Zealand. Through the WES and the ICWES network, two members of this ICWES organizing committee arrived to take me out and show me around. I was forever grateful for their friendship, as I had not foreseen how much of a culture shock it would be to land somewhere I did not speak the language, and could not ever read the writing. Plans for ICWES-XI had started even then."

"I have been involved with the Women's Engineering Society in the UK for a long time and I was involved with the organization of ICWES-IX. The good contacts that we have with Japan were made at that Conference. I know how difficult and what hard work it is organizing a Conference like this, but I also know what a fantastic experience it was, and I know that the organizers here will find the same after all the work is finished. I congratulate Japan on the' organization of what promises to be great ICWES, and I know that all of us here will enjoy it and be affected by it. I am so happy that the Conference has at last come to Japan" (ICWES-XI, Complete Proceedings, p. 9).

Organization

Honorary president: Her Imperial Highness Princess Takamado
Chairperson: Michiko Ishii (former Minister of State and director of General of Environment Agency)
Vice-chairpersons: Maki Iwakuma, Mitsuko Kazuno, Yoshiko Magoshi, Sachiko Okuda, Mieko Oshima, Makiko Tanaka, and Akiko Tsugawa
Members: Miyo Akishima, Sumiko Hamanaka, T. Iwakuma, M. Ishida, M. Ishikawa, Yukiko Konami, Yoko Nakamura, Y. Nishijima, M. Niwa, Hisako Ogura, I. Ohno, and Misao Ohi

Cooperating Organizations

Japan's Women Pharmacist Association
The Women Consulting Engineers Society of Japan
Ministry of Education, Science, Sports & Culture
Ministry of Health & Welfare
Ministry of Labor
Science & Technology Agency
Environmental Agency
Chiba Prefecture, Chiba City, Chiba Convention Bureau

Earth Science & Technology Organization
Meteorological Society of Japan
The Ecological Society of Japan
The Japan Science Society
The Japanese Society of Animal Reproduction
And another 36 Japanese Academic Societies

Participants

There were 349 delegates, including 203 from Japan and 146 from 39 other countries. Most delegates came from Far East Asia, like South Korea and China (26) and several from the USA (23); three came from Canada; 10 each from Western and Eastern Europe; and 10 from the UK. There were nine from Southeast Asia, eight from the Middle East, three from French Africa, four from English Africa, seven from Central Asia, five from South America, and two from Australia. Out-of-country attendance amounted to 41.8% (see Appendix 3).

ICWES-XI, S&T Papers

The main theme of ICWES-XI was "Science and **Technology** for Global Ecology." **Global Ecology**. Of all ICWES conferences to date, the eleventh was the one most focused on ecology and the environment. The scientific context is very interesting. Not only was this theme in harmony with the period of the conference, but it has also received much attention in Japan. The chairperson of ICWES-XI, Michiko Ishii, in her introductory speech, "Environmental Policy in Japan: Current and Future," explained to participants that "the Environmental Impact Assessment Law was passed [in Japan] in 1997 to establish regulations for preventing environmental damage, and measures were launched to promote co-existence with nature, such as the designation of world nature reserves and the development of strategies for preserving biological diversity" (ICWES-XI, Complete Proceedings, p. 26). The choice of the ICWES-XI main theme may also have been triggered by the work of the third Conference of the Parties to the United Nations Framework Convention on Climate Change (COP 3), which took place in Kyoto in December 1–11, 1997. The Kyoto Protocol that was adopted at COP 3 was the first major international treaty to slow global climate change through state commitments to reduce greenhouse gas emissions.

Related to the participation of women in science, just before ICWES-XI, from June 26 to July 1, 1999, the *World Conference on Science for the Twenty-First Century: A New Commitment*, was held in Budapest under the auspices of UNESCO and the International Council for Science (Cetto et al. 2000). This event led, in part, to the creation of the International Network of Women in Science and Engineering (INWES) in 2002 (see Chaps. 14 and 15). The *World Conference on Science* brought together 1800 delegates representing 155 countries, including 80 Ministers of Science and Technology, Research and Education, or their equivalent. Moreover, 28 intergovernmental organizations and more than 60 international scientific non-governmental organizations (NGOs) were represented ("Science for the Twenty-First Century" 2000). The main aim of this conference was "to help strengthen the commitment of UNESCO Member States and other major stakeholders towards science education, research and development, and to define a strategy that would ensure that science responds better to society's needs and aspirations in the twenty-first century" ("Science for the Twenty-First Century" 2000). Two non-binding but useful texts were adopted by consensus at this conference: The *Declaration on Science and the Use of Scientific Knowledge* and the *Agenda for Science – Framework for Action*. Among the topics covered, special attention was given to the need to promote more equitable access to science and its benefits, which included the need for greater participation of girls, women, and indigenous people.

S&T papers in the ICWES-XI proceedings (ICWES-XI, Complete Proceedings) were organized in five sessions, as can be seen in Table 12.1. Although 82 papers were expected, 78 papers and 3 abstracts are present in the proceedings; one paper is missing. Eight abstracts are related to three symposia on science and technology.

Table 12.1 Distribution of S&T sessions and papers in ICWES-XI, Complete Proceedings

ICWES-XI S&T sessions	Session titles	# of papers expected	# of papers in proceedings	# of abstracts in proceedings
A.1	Human impact	10		
A.2	Energy, Industry, Transportation, and Information	18		
A.3	Quality of Life	19		
A.4	Education and Ethics	12		
B.2	Women in the Medical Fields	23		
	Total papers	**82**	**78**	**3**
Symposium	Women's Roles in Today's Environmental Issues			3
Symposium	The Partnership between Ecological NGOs and Local Governments for Developing a Recycling Conscious Society in Japan			4
Symposium	International Partnership and Women's Roles in Establishing a Global Recycling Conscious Society			1
	Total Symposium			**8**

Table 12.2 Distribution of S&T posters in ICWES-XI, Complete Proceedings

S&T poster sessions	Poster session titles	# of posters expected[a]	# of posters with papers in proceedings	# of poster abstracts
A.1	Human Impact	6		
A.2	Energy, Industry, Transportation, and Information	12 (13)	12	
A.3	Quality of Life	11 (11)	11	
A.4	Education and Ethics	4 (7)	3	1
B.1	Women and Science	4 (14)	4	
B.2	Women in the Medical Fields	5 (11)	5	
	Total	**60**	**35**	**1**

[a]In parentheses: the total number of papers in a given session

There were a total of 174 authors (108 women, 45 men, and 21 of unspecified sex). Among them, 96 authors were from the host country, 14 from the USA, one from Canada, 11 from Latin America (8 from Cuba, 1 from Peru, and 2 from Mexico), 4 from Western Europe (3 from Austria and 1 from Germany), two from Eastern Europe (Russia and Ukraine), one from French-speaking Africa (Guinea), three from English-speaking Africa (1 from Uganda and 2 from Kenya), eight from Middle East and North Africa (3 from Israel, 2 from Iran, and 3 from Egypt), six from Central Asia (India), 15 from Southeast Asia (8 from Vietnam, 5 from Thailand, 1 from Malaysia, and 1 from the Phillipines), 8 from other countries than Japan from Far East Asia (4 from China, 3 from Taiwan, and 1 from Mongolia), and 5 from the UK, Australia, and New Zealand (2 from the UK and 3 from Australia and New Zealand).

Also, 38 posters on the same five themes were expected (Table 12.2), although only 31 have papers or abstracts in the proceedings. The content of posters and symposia is not summarized in the present section. There were a total of 61 authors (22 women, 1 man, and 38 of unspecified sex). Among them, 48 authors were from the host country: 5 from the USA, 3 from Canada, 2 from Eastern Europe (Russia), 1 from French-speaking Africa (Congo), 1 from Far East Asia (Mongolia), and 1 from thc UK.

A Glimpse of ICWES-XI's S&T Content

The importance of the participation of women in science was an issue discussed worldwide in 1999. Londa Schiebinger, a leading international expert on gender in Science and Technology, was most probably in the process of writing her famous book, *Has Feminism Changed Science?* This book posed the important question, "Do women do science differently?" One cannot doubt it after reviewing the scientific content of ICWES-XI.

As never before, ICWES conference content highlighted the substantial presence of women in the life sciences and in education, in Japan and elsewhere, compared to women scientists engaged in technical disciplines such as engineering and physics. Together, sessions A.1, A.3, A.4, and B.2 contain 60 papers in life sciences and education out of a total of 81 papers and abstracts (74%). Even in the more traditional topic *Energy, Industry, Transportation and Information* gathered in session A.2, papers are often related to genetic engineering, biochemistry, and chemistry, domains that attract women.

As mentioned previously, a major focus at ICWES-XI was related to *Environment and Ecology*. One can say that environment refers to the description of the physical, chemical, and biological components and their interactions, while ecology is the science of studying the interactions between living organisms and their milieu. The Chair, Michiko Ishii, in her keynote lecture (ICWES-XI, Complete Proceedings, pp. 26–27), listed three of the main issues contained in the environment theme: pollution associated with urban life and waste disposal, hazardous waste from industry, and global warming. She spoke of strategies to protect the environment, such as collection of scientific and technical knowledge, preventive measures, promotion of lifestyle changes, eco-businesses, and environmental policies. Gracheva Raissa (Russia), in "Strategy and Tactics of the Ecology Problems" (ICWES-XI, Complete Proceedings, pp. 64–65), defined the ecology issue as the natural balance infringement in cases of interaction between nature and humans. Setting aside the session called *Women in Medicine*, most papers within the S&T sessions were related to one of these categories.

Invited and Plenary Lectures

Ten authors from Japan, Korea, Laos, China, and the USA presented an invited or plenary lecture. Particularly worth mentioning is the plenary paper entitled "Epidemiological Study on Health Effects of Atomic Bomb Survivors" (ICWES-XI, Complete Proceedings, pp. 54–55), that was presented by Yukiko Shimizu.

At the time of the conference, Yukiko Shimizu worked as Assistant Chief in the Department of Epidemiology of the Radiation Effects Research Foundation, a cooperative Japan-United States research organization. This short paper discussed the effects on the health of survivors of the atomic bombs launched on Hiroshima and Nagasaki in 1945. The organization had access to a very rich information base for over 50 years on cancer incidence, morbidity, and mortality among 120,000 individuals, some of whom were residents of the two above-mentioned cities and others who were not exposed to the radiation.

The main finding of the study was an increase in cancer risk with an exposed radiation dose of 1.2.[1] The survey showed that leukemia, well known as a disease

[1] The paper did not explain this scale further.

that can be caused by radiation exposure, has a short latent period of one to seven years. The research also demonstrated that radiation-induced solid cancers, such as cancer of the lung, breast, stomach, and thyroid, began to appear after attaining the age at which the cancer is normally prone to develop and continued to be more frequent in comparison to the reference group. The author reported that survivors who were exposed in their first or second decade of life exhibited a higher relative risk in association with the radiation dose, but how it evolved over time was still unknown at the time of the conference. The fetuses exposed in utero at the time of the bombing showed a higher frequency of small head circumference and, later on in life, an increase in mental retardation corresponding to an increase in radiation dose; their cancer mortality increase was similar to that of those who were 9 years of age or younger at the time of exposure. The offspring of the survivors had not shown significant differences in frequency compared to the reference group by the time of the conference.

Shimizu concluded that the findings of this long-term study were used not only to improve the health of the survivors but also to estimate risks from occupational exposures to ionizing radiation within international or state organizations such as the United Nations Scientific Committee on the Effects of Atomic Radiation, the International Commission of Radiological Protection, the International Atomic Energy Agency (IAEA), and the Committee on Biological Effects of Ionizing Radiation.

ICWES-XI Papers on Renewable Energy

Renewable energy is among recurrent issues presented at ICWES conferences. Specific developments were discussed in the ICWES-XI A.2 session.

One example is the paper "High TC Super Conducting Filament" (ICWES-XI, Complete Proceedings, pp. 90–93), written by Tomoko Goto of the Nagoya Institute of Technology. The author argued in his abstract that "Superconducting technology is one of future revolutionary applications in the area of saving energy and of new energy resources for global ecology. The practical application of high TC superconductor for new energy resources requires the fabrication of brittle ceramic materials into tapes or wire with excellent superconducting properties" (Ibid, p. 90). For him, this application could be one of the most cost-competitive and environmentally friendly technologies that will help reach the goals of the Kyoto Protocol on global warming. Research on superconductivity at ambient temperatures is still happening today.

A more practical technique was presented in "The Applications of Nuclear Techniques to Geothermal Energy Development and Environmental Management" (ICWES-XI, Complete Proceedings, pp. 94–96). Jane Gerardo-Abaya of the IAEA informed the audience that geothermal energy was already an important source of energy for electrical and non-electrical uses in the Philippines, Mexico, Nicaragua, and El Salvador, where it contributed to about 25–40% of the total national energy requirement. At least 37 other countries were considering geothermal exploration. She explained that the IAEA provided nuclear techniques that help to understand the physico-chemical processes in hydrological systems. She explained in detail how isotopes 180 and 2H, together with chloride, can provide an indication of mixing between fluids of various origins, like meteoric water, reinjected wastewater, and geothermal fluids, which can be useful to formulate strategies for the development of reservoirs and for prioritizing areas for geothermal exploitation.

Providing a more social perspective, Maria Rodriguez Games et al. of the Centro de Investigations de Energja Solar (CIES) at Santiago de Cuba spoke of successful photovoltaic electrification to improve the life condition of people living in rural areas of Cuba in a paper called "Experiences in the Photovoltaic Rural Electrification in Cuba" (ICWES-XI, Complete Proceedings, pp. 122–125). The electricity produced was used for housing, schools, medical clinics, water pumping, buildings for cattle, tourist facilities, a forest keeper's house, and others. She witnessed that rural electrification has led to improving health services and employment and decreasing migration to urban zones and fuel expenses, all of this without affecting the environment.

ICWES-XI Papers on Women in the Medical Field

B-2, the last of the oral sessions at ICWES-XI, attracted 23 very technical papers in medical fields.

One of the invited lectures served as an introduction to this medical session. Prof. Fengsheng He's presentation was named: "Impacts of Pesticide Use on Human Health" (ICWES-XI, Complete Proceedings, pp. 30–33). She was the Honorary Director of the Chinese Academy of Preventive Medicine in China. She first reminded the audience of the definition of a pesticide: "A pesticide is defined by FAO as any substance or mixture of substances intended for preventing, destroying, or controlling any pest, including vectors of human or animal disease, unwanted species of plants or animals causing harm during or otherwise interfering with, the production, processing, storage, transport, or marketing of food, agricultural commodities" (ibid., p. 30).

These pesticides contain many compounds, such as insecticides, fungicides, herbicides, rodenticides, molluscacides, nematocides, plant growth regulators, and others. However, as she explained, insecticides represented the greatest proportion of pesticides used in developing countries at the time of the conference, whereas herbicide sales were greater than those of insecticides and fungicides in developed countries. The author explained that occupational exposure to pesticides can happen in manufacturing, storage, and the spraying of pesticides, both for pest control in agriculture and for vector-borne disease prevention programs, as well as through accidents. She also described the negative impact pesticides may have on human health. She concluded her presentation with a wish that is still very relevant today[2]: "In view of the major problem of acute pesticide [...], efforts are needed on the part of national authorities, non-governmental organizations and industry to control the health impacts of pesticide use."

Moreover, this same well-informed author provided more details in session B-2 on a specific type of insecticide, the organo-phosphorous. Organo-phosphorous pesticides form the majority of pesticides in agriculture, a class responsible for acute poisoning in many developing countries ("Clinical Features of Acute Organophoshate Pesticides Poisoning," ICWES-XI, Complete Proceedings, pp. 320–321). In this paper, she described the symptoms encountered with this type of poisoning, such as an acute cholinesterase crisis, delayed polyneuropathy, and intermediate syndrome.

S&T and Women in ICWES-XI Oral Sessions

It is striking that many S&T papers presented at ICWES-XI, contrary to previous conferences, spoke explicitly of impacts on women, their participation and role, and actions to improve their participation. Among the expected papers included in the oral sessions and lectures (papers, invited papers, plenary papers, symposium papers), 14 out of 96 (15%) had "woman," "women," or "gender" in their titles. They were present in all sessions. Therefore, the separation between gender and S&T papers in ICWES-XI is more blurred than in previous conferences.

Worth mentioning as an example of this intersection between gender and S&T issues, in the second paper in session A1, written by Sun Xiaoli (Center for Science and Society, Peking University, China), the author underlined the 35th anniversary of the death of Rachel Carson (1907–1964), an American marine biologist and author of *Silent Spring* (ICWES-XI, Complete Proceedings, pp. 60–63). Xiaoli wrote: "Her book, *Silent Spring*, published in 1962, was one of the most influential books of the last fifty years in the 20th century. The publication of this book [triggered] the beginning of the modern environmental movement in the world. Rachel Carson's influence reaches beyond the boundaries of her specific concerns in this book; she brought us an important idea that we must protect the natural environment for [human beings themselves, and for] the future of the earth" (ibid., p. 60).

A second example in session A.1 is a paper called "Depletion of Forests: Impact on Tribal Women and Role of Participatory Technology" (ICWES-XI, Complete Proceedings, pp. 74–77), by Swati Banerjee, who was a lecturer at the Research Centre for Women's Studies in Mumbai, India. This feminist researcher argued that the rapid depletion of natural resources in the forest already affected tribal populations, more specifically women of the Katkari tribe of Raigad District in Maharashtra. The consequences were impacting women in many aspects: their health, the

[2]One of the targets agreed upon at COP15 in Montreal on December 9, 2022, is to reduce risk from pesticides by at least 50% by 2030.

distance to fetch fuel, food, and water, an increased susceptibility to exploitation, and lower wages than their male counterparts. The author wrote: "In this changing context, the issue of devaluation of women's work and status in the tribal society, which was relatively egalitarian, is one that stands at the conjunction of depleting forests and changing economy" (ibid., p. 76). The study suggests that gender-focused technology, combining indigenous knowledge and advanced technology, would foster global ecology.

Another paper on women, in the A.2 session, is called "Awareness and Action of Japanese Women for Dioxin Problem – A Study on Environmental Opinion Poll Among Japanese Women" (ICWES-XI, Complete Proceedings, pp. 111–114), written by Ryoko Yanagibori and Kinko Koyano, from Aichi Gakusen University in Japan. As women are susceptible to have a significant impact on preventing the formation of dioxin from burning house waste, the study made use of a questionnaire to investigate the relationship between women's knowledge and their personal backgrounds. The author showed that many Japanese women were preoccupied by the presence of toxic dioxin as a health threat in their environment but lacked knowledge about the major source of dioxin exposure. In fact, the main source (95% of dioxin) came from incinerators burning chlorinated waste. More than 80% of the women considered that the dioxin causing air pollution and cancer originated from incinerators, but only 21% of them knew that chlorine is related to dioxin formation.

In session A.3, Sow Aissatou Billy (*Centre de recherche de Rogbane*, Guinea) presented a paper called "Women of the Coast, the Improved Technologies and the Agenda 21" (ICWES-XI, Complete Proceedings, pp. 177–180). Her subject focused on women's activities near the Guinean coast and their impact on the region's ecosystem. She showed the necessity to introduce sustainable technologies in accordance with Agenda 21.[3] This paper demonstrated well how scientists can use international plans of action in developing countries.

Finally, a text in session A.4 papers on education and ethics, "Women Proceed Environmental Science Education" (ICWES-XI, Complete Proceedings, pp. 219–220), from Kazue Tazaki (Department of Earth Sciences, Faculty of Science, Kanazawa University, Japan), drew attention because it directly targeted the training of women citizens in environmental science. The author argued that "[t]he air, water, soil and human problem facing the planet can only be tackled by teams of people from young to old expertise at local community. Women are sensitive for environmental hazards, because women bear children and raise future generations" (ibid., p. 219). Schoolchildren, undergraduate and graduate students studying environmental science at the Department of Earth Sciences at Kanazawa University, and the author have published three books presenting a wide variety of

[3] Agenda 21 is a comprehensive plan of action to be taken globally, nationally and locally by organizations of the United Nations System, Governments, and Major Groups in every area in which human impacts on the environment. Agenda 21, the Rio Declaration on Environment and Development, and the Statement of principles for the Sustainable Management of Forests were adopted by more than 178 Governments at the United Nations Conference on Environment and Development (UNCED) held in Rio de Janeiro, Brazil, June 3–14, 1992 ("Agenda 21 – Sustainable Development Knowledge Platform", n.d.).

topics on how to keep the earth clean. The author did not explain how the distribution of the books was made or if they were popular among women, but certainly, involving the communities through students was an inspiring approach to ensure the population is aware of environmental issues.

ICWES-XI, Gender Papers

Twenty papers addressed gender-related themes under the topic "Women and Science." For this eleventh conference, there were a total of 29 speakers discussing gender issues: 25 women, 2 men, and 2 that we could not identify. Of these 29 speakers, 6 came from the host country: 4 from the USA, 1 from Canada, 4 from Western Europe (2 from France and 2 from Turkey), 4 from Eastern Europe (1 from Poland, 1 from Lithuania, and 1 from Russia), 1 from Nigeria, 1 from Iran, 3 from Vietnam, 3 from Hong Kong, 1 from China, and 1 from the UK. The Women Consulting Engineers Society of Japan also delivered a paper, but the names of the presenting members were not indicated in the proceedings (Table 12.3).

A total of 18 posters on gender topics were presented in four different sessions by 50 different people. The majority of posters were featured in the "Women and Science" session. All the posters were accompanied by a paper. In the Table 12.4 below, the first number under # of posters expected indicates the gender-themed posters, while the number in brackets refers to the total number of posters delivered within a given session. Of the 50 presenters, 37 came from the host country: 5 from the USA, 3 from Canada, 2 from Russia, 1 from Congo, 1 from Mongolia, and 1 from the UK.

Table 12.3 Distribution of social and gender sessions and papers in ICWES-XI, Complete proceedings

Social and gender sessions	Session subject	# of papers expected	# of papers in proceedings	# of abstracts without papers in proceedings
B.1	Women and Science	20	17	3
	Total	**20**	**17**	**3**

Table 12.4 Distribution of social and gender posters in ICWES-XI, Complete Proceedings

Social and gender poster sessions	Poster session titles	# of posters expected	# of posters with papers in proceedings
A.2	Energy, Industry, Transportation and Information	1 (13)	1
A.4	Education and Ethics	3 (7)	3
B.1	Women and Science	10 (14)	10
B.2	Women in the Medical Fields	6 (9)	6
	Total	**20**	**20**

There were three keynote speeches:

- "Environmental Policy in Japan: Current and Future" (Michiko Ishii, Japan);
- "Women in Science and Engineering: Changes, Choices and Successes" (Yvonne T. Maddox, USA; abstract only);
- "From Space to Earth" (Toshibumi Sakata, Japan).

In an invited lecture entitled "Look more Lady than Heard" (ICWES-XI, Complete Proceedings, pp. 34–35), Sea-Wha Oh,[4] from South Korea, recalled how a company president in the early 1980s had told her upon meeting her that she looked more ladylike than he had expected. Oh deplored the popular opinion that competent women scientists could not be "proper ladies." Oh also reflected back on her own experience as a young girl thriving in the sciences and later as a woman looking to pursue her formal education and searching for a job that would meet her needs and her qualifications.

In one of the eight plenary sessions, Keobang A. Keola, Deputy Director of Cabinet, Science, Technology, and Environment Organization in Vientiane· talked about "Lao Women's Status in Society and Environment" (ICWES-XI, Complete Proceedings, pp. 50–53) and about the changing economic situation in Laos, namely how the increasing number of men moving to the cities impacted the lives of women, children, and elderly people in rural areas. Keola also noted that young rural women had started to migrate towards the cities in order to work in the textile industries. In 1986, there was a conscious effort in the Lao PDR to shift from a centrally planned economy to a market economy; these measures were known as the New Economic Mechanism and had generally bad consequences for women and children (labor exploitation and illegal employment, dismissal of women holding positions in the government, appeal of potentially dangerous jobs for rural women moving in urban areas, HIV/AIDS epidemic, etc.). This unequal economic development also had negative effects on the environment, a marked increase in monocropping practices, and the "feminization and graying of farming" (ibid., p. 50). In the face of these problems, Keola advocated for gender-sensitive social and economic policies, which would ultimately be beneficial to the entire country.

Meihe Zhou, "Working Circumstances of China's Women Workers"

In the session "Women and Science," Meihe Zhou, from the Institute of Policy & Management, Chinese Academy of Sciences, presented a paper titled "Working Circumstances of China's Women Workers" (ICWES-XI, Complete Proceedings,

[4] Oh holds a MA in chemistryChemistry (1967) and a PhD in organic Organic chemistry; she worked in the Dyestuff, Dyeing & Finishing Research Center, at the Korea Research Institute of Chemical Technology (ICWES-XI, Complete Proceedings, p. 34).

pp. 267–270). Her paper reflected on how the working conditions of women changed since the 1978 Chinese economic reform ("reform and opening-up"), a policy that made the country shift from a planned economy to a market economy. One important point was the emergence of private-owned and foreign-funded companies alongside the state-owned firms. This fast economic development gave women the opportunity to work and emancipate themselves, namely a good portion of younger and rural women. However, due to the fact that these new businesses were often privately owned, this rapid economic growth and the increasing employment of women also had negative consequences, like work injuries and insufficient protection measures for workers.

According to Zhou, "China's large-sized state-owned enterprises are better in providing good working circumstances for women workers" (ICWES-XI, Complete Proceedings, p. 267). A good example was the Woman Iron and Steel Company, which had held in recent years several workshops on health and labor safety and set up physical examinations on a regular basis. Factories and mines were equipped with high-performance air conditioning and dust removers. In certain more dangerous sectors, like underground oil depots, which had a higher concentration of women workers, this company modernized high-pressure pump houses in an effort to reduce noise and temperature. In new projects, it also took into account women's health and safety and often solicited workers' opinions.

Not all state-owned companies prioritized their workers' well-being over profit, and a certain number made use of a more aggressive profit-based business model. Zhou argued that in such companies, women workers were generally more affected by the employers' indifference to their safety, partly because the reduced income was "not enough to pay the expense of labour protection" (ICWES-XI, Complete Proceedings, p. 268). Women working in these enterprises were often subjected to harmful conditions like low or high temperatures, dust, toxic materials, noise, vibrations, high frequencies, and high magnetism. One important cause is obsolete equipment that is not replaced or modernized. Zhou cited a survey conducted in 1996 in which over a third of the respondents said they considered labor protection to be either at a standstill or had worsened over the years.[5]

In the case of the country's poorer rural companies, women workers' conditions were even worse. Many women even engaged in prohibited jobs; for instance, workers were in direct contact with toxic and dangerous materials and worked in noxious environments that caused diseases and arthritis. In a shoe factory that employed women to bind shoe parts with a benzene emulsion, there were 18 cases of chronic benzene poisoning, leading to one death. These poor working conditions not only harmed women but also had an impact on the health and development of future generations. In a red lead manufacturing plant, high levels of lead concentration caused several abortions and high infant mortality since women would bring their

[5]This survey covered 30 large state-owned companies across four cities, which employed a high percentage of women workers.

children to their workplace in order to work during the day, or workers would not be able to decontaminate properly before going home.

Another survey from 1996, conducted in private and foreign-funded firms in the textile, light, and electricity sectors across eight cities in Southeast China, showed that only two-thirds of the respondents had labor protection, even though they were in contact with dust, gas, noise, radiation, and vibration. Chronic benzene poisoning was also an important problem in these enterprises since the same benzene emulsion was in use. For Zhou, the shoe industry was "a disaster area of labour protection" (ICWES-XI, Complete Proceedings, p. 269). Benzene is a dangerous chemical that can quickly harm the skin, the respiratory tract, the stomach, and the intestines; it is also toxic for the blood circulation and the nervous system, and it can even cause leukemia. Tests conducted in 20 factories owned by 10 shoemaking companies in Quanzhou (Fujian Province) showed that benzene and methylbenzene concentrations were much higher than standard levels (respectively 23.3 and 13.9 times). Since benzene is a volatile substance, it not only affected workers but also impacted the surrounding residential areas.

In many of these companies, there were no physical examinations for workers at risk in order to hide the reality of these dangerous working conditions. Shoemaking companies, in particular, were neglectful of their workers' health and safety. Zhou cited the example of a shoe factory in Chengdu (Sichuan Province) that made employees work, eat, and live in a closed yard without proper ventilation or protection equipment.

For Zhou, a major cause of this disregard for workers was profit, since it was easy to cut expenses for labor protection. Regarding the shoemaking industry, an effort was made by the government in 1995 to replace the benzene emulsion with a non-toxic water-based adhesive. However, it seemed that many companies did not react to this measure; Zhou suggested that the government resort to coercive means. Another reason was the limited means of certain companies that could not afford to modernize their outdated equipment. A third factor was the difficulty to reconcile measures established during the planned economy period with the reality of the market economy. In the case of women workers, especially young women from rural areas, another problem was that they often entered the workforce as contract workers and, despite laws such as the "Regulations for Labour Protection of Women Workers," they "voluntarily engaged in harmful and extra long time [sic] jobs only to want earning more money" (ICWES-XI, Complete Proceedings, p. 270).

Labor health and protection in foreign-funded firms should be, according to Zhou, subjected to Chinese state laws. Another important measure was to raise the awareness of women workers about work safety. It would also be necessary to provide women workers with proper training; benzene, for instance, is a fragrant substance but highly toxic. She concluded: "It is by no means allowed to get foreign investments at the expense of safety and health of women workers" (ICWES-XI, Complete Proceedings, p. 270).

Susan Bullivant, "The Athena Project"

Susan Bullivant was Athena Project Director and worked for the Committee of Vice-Chancellors and Principals' (CVCP) in London, UK.

Paper taken from ICWES-XI, Complete Proceedings, pp. 315–318, with original references:

"Introduction"

"Whilst there has been an overall increase in the proportion of women entrants to higher education (HE) in the UK, this is not reflected in the number of entrants to science, engineering and technology (SET) subjects. This is compounded by the falling off in the number of women at postgraduate levels and the further high dropout rate for women throughout their academic careers. For example, 1994/95 statistics[6] show that women made up only 2.5% of the number of full-time (FT) full professors in biological, mathematical and physical science, engineering and technology. This compares with a figure of 7.3% across all subjects and with 19% in education. At the FT lecturer grade, women made up approximately 15% of the SET total but in engineering and technology alone the proportion was only 8%."

"A Government White Paper 'Realising Our Potential'[7] highlighted the importance of SET for the country's future economic growth and together with the findings of a subsequent report 'The Rising Tide'[8] concluded that 'women are the country's biggest single most under-valued and under-used human resource in SET' and that there is massive potential for attracting more women into SET. Consequently, the Government established a Promoting SET for Women Unit within the Department of Trade and Industry (DTI) to address the areas of attracting more girls into SET subjects at school; encouraging women into HE courses; identifying ways of improving women's chances of progression up the academic career ladder and encouraging women into careers in industry and at all levels of academic qualification."

"In 1995, the Scottish Higher Education Funding Council (SHEFC) established a 'Winning Women' initiative, which aimed to encourage the development, dissemination and adoption of good practice in relation to women in SET. One of its outcomes was the publication of three guides to good practice in improving access to, participation in and subsequent progression through careers in SET for women in higher education institutions (HEIs). The Athena Project will build on this experience and findings. This project focuses on the current loss to the HE sector of the skills and talents of women in the fields of SET." In particular:

- Those women who do not enter, or consider entering careers in HE.
- Those women who are not retained in the sector.

---

[6] Separate Tables, DfEE, UK, 1997, p. 39.

[7] Realising our Potential, A Strategy for SET (Cm2250), HMSO, London, 1993.

[8] The Rising Tide. A Report on Women in SET, HMSO, London, 1994.

"The project's aim is to increase the number of women in SET in HE at all levels beyond undergraduate recruitment and ultimately to achieve the representation of women in academic appointments which reflects the percentage of women in the undergraduate intake in each SET subject area, by:

- Reducing wastage and improving career retention and promotion prospects;
- Disseminating and encouraging good practice on participation in, and progression through, careers in HE;
- Raising the profile, visibility and contribution of women."

"The project is part of the Committee of Vice-Chancellors and Principals' (CVCP) Commission on University Career Opportunity's (CUCO) action agenda to remove the barriers to discrimination for women at all levels across HE. It is core funded by the Higher Education Funding Council for England (HEFCE) with support from the other UK HE Funding Councils, DTI and Research Councils. To achieve its aims the project is required to raise further funds through sponsorship."

"A sister project, launched at the same time, is the Women in HE Register (WHERE). The register's core business is the collection, analysis and dissemination of information on women in HE, collectively and individually. Whilst the Register will be open to all women in HE, it will initially concentrate on women in SET. The Register will therefore provide data and information analysis and IT support for the Athena Project, which in turn will respond to identified needs, and issues through the development and delivery of its programmes and through direct communication to the women via the WWW and email."

Issues for the Sector

"The HE sector in the UK has been the focus of many changes in recent years and, like many other sectors, has had increasingly to respond to its many stakeholders including Government and its diversifying student base. Whilst finance is an obvious important factor for HEIs with the need to broaden their income base, a critical driver is quality. Institutions are differentiated through quality and hence their ability to attract quality students and staff research, teaching and professional development are all indicators of quality with the weightings dependant on the individual HEIs desired strategic positioning in the marketplace. Both potential students and faculty are attracted by the perceived quality of education and 'does it seem friendly to me' i.e. the culture of the department and institution. Qualities sought in 'best staff' include those who are most active in research/teaching etc., those who motivate and inspire students who then stay and also promote their department and institution externally and those who offer continuity and commitment to the HEI. Modes of learning, widening participation, continuing and life long learning, new technology are all impacting on the sector with implications for flexible study and working modes. These are supported by the Government's agenda for raising the profile of family friendly policies and indeed promoting best human resource practice. Women undoubtedly have much experience, both technically and socially, to contribute to addressing these issues. The Athena Project and its aims are therefore directly congruent with the key HE sector drivers and hence the achievement of institutions' strategic goals."

Issues Pertinent to the Project

The following are some of the factors, which are familiar to many in this field, that have arisen through consideration of research documents, conference proceedings and indeed from women in the system as well as from initial interviews undertaken with vice Chancellors, representatives of professional institutions and bodies, Research Councils and Trusts, industry, Government Departments and other stakeholders.

"Culture, Attitudes and Behaviour"

"It is recognised that there is an accepted 'laddishness'/macho culture in many of the male dominated SET disciplines and departments. This often results in behaviour and indeed management styles that are not conducive to encouraging women to join or stay in the institution. This behaviour can be consciously or unconsciously appreciated by the men involved."

"Institutional Policies, Systems and Practices"

"Recruitment and development procedures and practices may inhibit women applying and progressing. Are job specifications too rigid? Are job shares and fractional appointments considered? Are interviewers conscious that women frequently undersell themselves and trained accordingly? What arrangements are in place to support the management of dual career/career break/family caring responsibilities? Are they flexible enough and capitalise on the potential for using new technology? Are women encouraged and developed for senior posts? Do management and other development programmes incorporate equal opportunity/gender issues and the benefits of consensus management behaviour and skills which are more typically demonstrated by women and often dismissed in strong macho cultures as weak? Are statistics collected and monitored to indicate areas for concern such as imbalances between women undergraduate and women staff populations in different disciplines, are women not applying for positions? Are women encouraged and assisted to apply for research grants? Are areas for concern followed up for more qualitative feedback? Is there gender bias in teaching and assessment?"

"Research Councils"

"A recent study of peer-review scores for postdoctoral fellowships at the Swedish MRC demonstrated that Women had to be 2.5 times more productive than their male colleagues to get the same peer-review rating for scientific competence. The Well come Trust in the UK have audited their decision making on grants and demonstrated that there is no evidence of sex-discrimination in the awarding of project grants, programme grants or Senior Research Fellowships in Basic Biomedical Science. However, there is evidence that women do not apply to the Trust for project or programme grants in the proportions that would be expected from the number of female academics working in UK universities. Together with the major

Research Councils in the UK they have commissioned a research programme to identify why this is the case and interim results will be available by the end of the year."

"Senior, well-qualified women are also increasingly being sought by programme evaluation committees for research and other public bodies. Identification of suitable candidates is an issue as well as encouraging organizations to actively seek women candidates."

"Personal and Career Issues for Women"

"The following, whilst not specific to every individual, are accepted as having general applicability: isolation issues particularly for women 'survivors'; loss/lack of self confidence; acquiring general career and personal development skills and knowledge; availability of networking, role models and mentoring schemes; development for higher professional recognition e.g. FRS, F.Eng; managing careers and family; addressing the macho culture; awareness and marketing of SET careers in academia to different target groups."

"Addressing many of the above, which are not exhaustive, would benefit men as well as women and lead to more competent people and institutions which are adopting best practice human resource and organization development approaches. It may also help to identity why women are not applying for SET research and academic positions as well as grants."

Athena Project's Role and Programme

"Neither Athena nor WHERE are highly resourced relatively in terms of people or finance and hence will need to work effectively with other organizations in the field and be focused initially on which and how to address some of the previous issues. The following, not in priority order, are roles and areas to be adopted and addressed:

1. Information Conduit

   Through its emerging register of women in SET in HE and developing its contacts with government, senior HE management, research councils and trusts, industry, and professional and learned bodies, Athena will be able to gather and channel information appropriately. This will assist in the identification of training and development needs as well as candidates for posts, grants, etc., joint partners for initiatives, and potential funders/ supporters.
2. Training and Development Program

   This will range from identification of needs, facilitation of delivery through others, or indeed the development and delivery of processes and programs for target groups, possibly men and women, e.g., personal assessment and counseling program for identified individuals, and possible awards for meeting the identified needs. Partnerships will be sought with the different stakeholders and potential funders to further this area, including providing mentors and expertise.
3. Wellcome Trust/Research Council's Initiative

   Monitoring the findings of the study and working with the organizations to develop programs to address the needs in addition to using the Athena information channels to publicize/progress the recommendations and other requests/information directly to SET women and vice versa. To sit on cross-council groups where appropriate to facilitate information transfer.
4. Research and Consultancy

Data exist in various sources and from different viewpoints with respect to women in SET. A collation of information would be beneficial to assist in benchmarking the project and individual institutions' progress. International data for comparison purposes and industrial good practice will also be gathered. The information will primarily be used to inform the development and implementation of initiatives. A consultancy to provide an independent audit' of qualitative issues at the institutional, faculty, or department level will be developed and offered.

5. Good Practice Development Project Grants

   One of the first initiatives will be to invite HEIs to bid for pump priming grants to support the development of good practice in this area in their institution. Evidence of matched resources will be required as well as the written commitment of senior management to the project.
6. Facilitating Initiatives on Cross-Sector Issues

   Mentoring programs and developments are a common theme, and this will be a first-year priority theme to be considered. A think tank meeting of key influencers in the Women in SET area as well as integrating EO issues and appropriate leadership/coaching/counseling management styles into management program at all levels in HEIs and other pertinent organizations will be promoted.
7. Marketing Issues and Special Projects

   Newsletters will support communication through well-developed websites. The Athena: 'How did you get there?' Interviews are planned to address the broader public's understanding of science and role model issues, which allow other women in SET and those contemplating a career in it to hear about personal and professional issues and decisions women at all levels in SET have made. The above gives a flavor of some of the initiatives planned, which the presentation will expand upon" (ICWES-XI, Complete Proceedings, pp. 315–318).

Components of and Findings from the Paper

Components:

1. The Athena Project was a research initiative launched in 1999 by the UK Resource Centre for Women in Science, Engineering, and Technology (UKRC) to address the underrepresentation of women in these fields. It aimed to identify and remove the barriers that prevented women from entering and progressing in STEM careers.
2. The project held several key components aimed at making a difference to the career progression of women in science in UK universities:

   (a) To conduct research and collect data to better comprehend the experiences and perceptions of women in STEM and the factors that compromised their participation and progression.
   (b) The development and implementation of best practices and strategies to support the recruitment, retention, and progression of women in STEM careers.
   (c) Create partnerships and develop collaborations with universities, industry, and government organizations to promote gender equality in STEM.

(d) Communication and dissemination of findings and best practices to large audiences, including policymakers, educators, employers, and the general public.

Findings:

1. Women face several barriers that prevent them from entering and progressing in STEM careers: unconscious bias, a lack of role models, and a lack of flexible working options.
2. Women in STEM need more flexible support through mentoring and networking and more effective communication and dissemination of information about career opportunities.
3. Gender equality in STEM requires greater collaboration between universities, industry, and government.

Concluding Remarks: The Athena Project promoted the development and implementation of practical solutions as an effective means to better understand and remove the barriers against gender equality.

Sumiko Sasagawa, "The Matthew Effect on Women Scientists: Case of Harriet Brooks"

Sumiko Sasagawa, PhD, worked at the Institute for Environmental Sciences in Rokkasho-mura, Aomori Prefecture, Japan.

Paper taken from ICWES-XI, Complete Proceedings, pp. 541–543 (it was accompanied by a poster):

"Introductive Description"

"The heroine of the biography, Harriet Brooks (1876–1933), born in Exeter, Ontario, Canada studied at McGill University, Montreal, in 1898, [where she obtained both] a BA in mathematics and natural philosophy. Then she began to conduct research with the young and rising Professor Ernest Rutherford (1871–1937) who had just arrived at McGill from the Cavendish Laboratory at Cambridge University. At the early time of radioactivity research, she was the first person to realize that one element can change into another one, namely radioactive decay, discovered radium A, B, C, D, and E, first attempted to determine atomic mass of radon, and observed recoil of radioactive atoms. She was a woman of rare experimental physicists and the first Canadian researcher in the field of radioactivity. Her research was done first with Rutherford at McGill, then with Joseph John Thomson (1856–1940) at Cavendish Laboratory at Cambridge University, UK and later with Marie Curie (1867–1934) at Curie Laboratory in Paris. Thus she did contribute to their Nobel prizes; Rutherford in chemistry 1908, Thomson in physics 1903, Marie Curie in physics 1903, and in chemistry 1911. Rutherford remarked that, next to Curie, Brooks was the most outstanding woman in the field of radioactivity. In 1907, she married Frank Pitcher, gave up her active participation in scientific work and raised three children. She turned her directions and served significant roles as President of the Women's Canadian Club, as a member of the Scholarship Committee of the Canadian Federation of University Women, and as a member of McGill Alumnae Society, and so on. She died of probably a malignant blood disease. The obituary in the McGill News described her later life more than the former one. The biography of Harriet Brooks provides us some aspects of the sociological science as well as ones of the circumstance of nuclear science at that time" (ICWES-XI, Complete Proceedings, p. 541).

The Brooks Biography points out Sumiko Sasagawa, has raised several topics and challenging questions for further discussion, especially those regarding women, marriage, and career, and women involved in science more specifically.

"Results and Discussion"

1. "It has been said for a long time that there is a crown, which might be worth it all – love, home, and children. A lot of women have suffered from this crown and faced with the alternatives of marriage or career. We know that is a sad choice, which still exists today.
2. Second, what spurred Brooks on to academic prowess? There are two factors proposed here. One is 'lack' of opportunity and parental support and encouragement. Another is self-motivation, a strong ambition. After World War II, many countries including Japan seem to have concentrated on the economy-for-economy principle, and consequently it is found that they are now 'full of economy,' especially in Japan. We must become seriously aware of 'to live poor but honest,' which we rarely meet today as well as 'lack'" (ICWES-XI, Complete Proceedings, pp. 541–542).

This brings us, thirdly, to the "Matthew effect," a crucial problem for women scientists, which Sasagawa considers as the "most emphatic theme of the biography" (ICWES-XI, Complete Proceedings, p. 542):

3. "The 'Matthew effect' was named after the saying in the New Testament, the Gospel according to Matthew 'For unto everyone that hath shall be given, and he shall have abundance; but from him that hath not shall be taken away even that which he hath.' The phenomenon of transferring credit to the famous person and away from the lesser-known person occurs irrespective of sex. This transference is the Matthew effect. Why has Harriet Brooks been overlooked? Science history has been looked at from the hero/heroine perspective, and in accounts of early nuclear science, the names of Rutherford, Thomson, and the Curies have always dominated the discussion. All the others, such as Brooks, were viewed as mere secondary players, or "invisible" scientists who performed much of the actual work, many of whom were women and there exists one additional factor "gender" operating in case of women. Although the Matthew effect occurs irrespective of sex, excellent women scientists have traditionally been ignored, while men scientists of the same calibre have been remembered for their particular contribution. However, the Matthew effect might have ignored Rutherford, who was very rare and particular to give credit for the exact authorship of any idea, thus supportive supervisor in encouraging both women researchers and young researchers at that time. Another important aspect is that, as Rutherford noted in his obituary for Brooks, the value of her work could not really be appreciated until the phenomenon of radioactivity was better understood."

4. "In addition the biography of Harriet Brooks suggests us that we should not forget that many old women created 'foundations' that we enjoy now. True [progress] could be brought [by] learning [about] the invisible past. [The] [f]irst step of true [progress] is to appreciate old women who created foundations and to reward them according [to] their merits. Typical cases of women behind [men nominated as] Nobel laureates, [like] a famous [philosopher], and a great [artist], will also be discussed because there exist Matthew effects in various fields" (ICWES-XI, Complete Proceedings, p. 542).

Sumiko Sasagawa sums it up with the following warning and words of advice:

> "We have experienced World War I and II, explosive development of sciences and technologies, and rapid extent of apparent abundance of life during the 20th century. However, many distortions are now occurring from those here and there in the world. The biography which visits a pioneer woman scientist who lived in the social circumstance just before one hundred years ago is anticipated to encourage not only women but also men, both who lived between the old and the new centuries" (ICWES-XI, Complete Proceedings, p. 542).

References (as Cited in ICWES-XI, Complete Proceedings, p. 543)

1. Rayner-Canham, M. F. and G. W. Rayner-Canham. Harriet Brooks: Pioneer Nuclear Scientist, McGill-Queen's University Press, Montreal & Kingston, 1992, Japanese translated by S. Sasagawa and M. Aratani, Maruzen, Tokyo (1998).
2. Merton R. K. "The Matthew effect in science – the reward and communication systems of science are considered." Science 159 (1968) 56–63.
3. Rayner-Canham M. F. and G. W. Rayner-Canham. "Pioneer women in Pioneer nuclear sciences." Am J Phys, 58 (1990) 1036–1043.
4. Nobel Foundation homes page through Internet.
5. The Rokkasho-mura Reading Circle. Messages from Rokkasho Women, 2 (1997).
6. The Rokkasho-mura Reading Circle. Messages from Rokkasho Women, 3 (1998).
7. Folsing, U. Nobel Frauen – Naturwissenschaftlerinnen im Partrat, C. H. Beck'sche Verlags buchhandling Munchen (1991), Japanese translated by M. Tazasa and T. Matsumoto, Gakkai Shuppan Center, Tokyo (1996).
8. Ettinger E. Hanna Arend//Martin Heidegger, Yale University Press, New Haven and London (1995), Japanese translated by K. Ohshima, Misuzu Shobo, Tokyo (1995).
9. Delbee, A. Une Femme, Presses de la Rainaissance (1982), Japanese translated by M. Watanabe, Bungei Shunju, Tokyo (1989).

Business Meeting

Although no minutes for this meeting have been found, the agenda was as follows:
Meeting of Delegates, July 27, 11:00–12:00

1. Minutes of the meeting held on October 10, 1996
2. Confirmation of the mandates of the delegates
3. Proposed amendments to the "Guidelines" (postponed to next conference because of lack of time at this meeting)
4. Consideration of bids for ICWES-XII
5. Election of the Convenor and members of ICC
6. Any other business

Delegates Present:

• Chair: Michiko Ishii
Vice-chair: Akiko Tsugawa

Australia, Vera Lukic; Austria, Annick Carnino; Canada and ICC, Dormer Ellis (and Monique Frize to present bid for ICWES-XII); China, Fengsheng He; Congo and ICC, Agnes Loteta, Diamndja; Cuba, Maria R. Gamez; Denmark, Dorte Lerche; Egypt, Alia A. Shakour; France, Marianne Rodot; India and ICC, Ila Ghose; Iran and ICC, Mahin Rahmani; Israel, Mira Baron; Ivory Coast, Issié Yvonne Gueye; Japan and ICC, Mitsuko Kazuno; Kenya, E.W. Omulokoli; Korea, Sea-Wha Oh; Laos, Keobang A. Keola; Lithuania, Daiva Dumciuviene; Malaysia, Khalijah Salleh; Mexico, Tatyana Pozniak; Mongolia, Butmunkh Lamzav; Nepal, Neena Ammatya; Nigeria and ICC, Joanna Maduka; Philippines, Fillipina Sotto; Poland, Anna Zsekik-Hojniak; Peru, Karla Solis-Garcia; Russia, Galina S. Netitailo; Switzerland, Anita Phaff; Thailand, Morakot Tanticharoen; Turkey, Sahinde Demirci; Uganda, Wilbrod Birabwa; Ukraine, Olena Shubravaka; UK and ICC, Petra Godwin; USA, Roberta B. Gleiter; Taiwan, Yu Wang; Vietnam, Phan This Kim.

Action item: Choice of location for the next conference. Monique Frize (Canada) presented a set of slides for the bid to hold ICWES-XII in Ottawa, Canada, in July 2002. The conference was to be organized by the Women in Science and Engineering (WISE) Ottawa Chapter, with support from the national organization: Canadian Coalition of Women in Engineering, Science, and Technology (CCWEST). The organizers proposed the Ottawa Congress Centre for the venue and an interesting program and tour.

The ICWES-XI Meeting of Delegates voted to hold ICWES-XII during July 2002 in Ottawa, Canada. SWE members have already volunteered to support Canada in hosting ICWES-XII in Ottawa. It was also suggested that the Chair of the next conference be the Convenor. However, Monique Frize eventually asked Dr. Moyra McDill to be the Convenor for ICWES-XII, and she accepted. Moyra was on the organizing committee and had been a co-host of the meeting in Merickville to discuss the creation of a transnational organization of women in STEM.

A declaration was approved by the delegates:

Declaration

1. ICWES-XI calls upon its members to expand and strengthen international networking to achieve a better global environment by incorporating women's perspectives.
2. The network will promote gender equality worldwide towards a better 21st century.
3. We call upon scientists and engineers to share their enthusiasm and knowledge of natural sciences and technology with the next generation.

Testimonials

"ICWES-XI: Lasting Impressions," 1999, pp. 17–22

"A delegation led by SWE Immediate Past President Roberta Banaszak Gleiter represented American women engineers and scientists at ICWES-XI held this past July in Chiba, Japan. The SWE delegation included three other past presidents: Kathleen Harer (FY88), Kathryn Cunningham (FY94), and Isabelle French (1964–66), at the time, the presidency was a two-year term), and members were: Mary McCarthy, Dormer Ellis, Betty Lou Bailey, Lana Rochelle Fountain, and Yoshiko Fukui. The following pages contain the images and text in which Roberta Banaszak Gleiter, Betty Preece and Kathleen Harer have captured their experiences" (p. 17).

Roberta B. Gleiter: "Over 400 women engineers and scientists participated in the International Conference of Women Engineers and Scientists recently held in Japan. Twenty-two U.S. women participated in the conference, including nine SWE members, two of whom brought guests."

"ICWES-XI Honorary President, Imperial Highness Princess Takamado, graciously accepted a memento from SWE. Other dignitaries at the conference also were presented SWE mementos."

"My lasting impressions of ICWES-XI range from humor and awe to a touch of sadness. From a humor standpoint, folks chuckled when they looked at my 'dead' SWE wristwatch, which had truly drowned (the temperature was 90 °F and the humidity was at least 90 percent) on the Tokyo subway during the 'human crush' of a morning rush hour. After this, the term 'a sardine in a can' takes on new meaning. The organized conference tours were wonderful and truly impressive; the people of Japan were always helpful and kind."

"One leaves ICWES with a sense of awe. Women engineers and scientists from forty countries gathered and, despite language barriers, held meaningful discussions on environmental concerns and issues specific to technical women in the workplace. Communications were clear and focused, and the involvement was inspirational. However, the fabric being woven throughout the conference clearly showed the long path that technical women still have to travel in all countries to reach even a modicum of parity with men. One can't help but feel a touch of sadness to recognize this. But we are up to the challenge. It is clear that governments, internationally, are awakening and starting to recognize that their countries will be left behind in the technology race if they don't utilize technical women."

"While chairing a conference session, I presented a paper, 'Trends: Women in Engineering' which showed that while the USA graduates one of the largest number of women engineers, the USA has not demonstrated the most progress in the ratio of male to female graduates. For example, in 1995 France had a 3.8/1.0 ratio in engineering, whereas Canada had 5.9/1.0, the USA 6.1/1.0, and Japan had 27.1/1.0."

"Both the USA and Japan are graduating into engineering only 2 percent of all of their degreed females each year, whereas the Czech Republic and Canada are graduating 14 percent and 12 percent, respectively. Our keynote speakers touched on the point that a correlation to the low prestige and low pay for engineers may exist in those countries with larger percentages of women engineering graduates. [...]" (pp. 17–18).

Betty Preece: "The official language of the conference was English, and we were humbled by the ability of so many women to express themselves so well in a language other than their native one. To capture the flavor of ICWES, I gathered the following comments from participants" (p. 18).

Joanne Maduka, Nigeria): "I'm an electrical engineer and I work with rural women in Nigeria, teaching them simple renewable energy systems and maintenance of their port-harvest machinery. I use solar and biomass energy. When our cooperatives go out, the women are very receptive, but we must always go through the men who are not always so receptive. There is a national education plan in Nigeria, but it is not detailed, so teachers and students do not know well how the system fits together" (p. 18).

Yoho Sendoka, Malaysia: "I'm working on a project to find ways to inculcate physics ideas in young children of kindergarten age. Three undergraduate students are studying with me" (p. 18).

Marianne Rodot, France: "I'm with AFFI (Association Française des Femmes Ingénieurs) which has about 50 members. AFFI has three major activities:

1. Collect statistics of women engineers
2. Has developed a presentation package for schools to use in presenting engineering to pupils, teachers, etc. The parts of the packets show engineering of common items like ball bearings, windmills, elevators. I would like to correspond with people interested in school competition and technical systems
3. Makes awards for technical projects of undergraduate girls" (p. 18).

Sea-Wha Oh, Korea: "Our Korean Association of Women Scientists and Engineers began in 1993 and today has 478 members. The title of my paper, 'Look More Lady Than (I) Heard,' (translated from the Korean) was taken from a comment made to me by a man after I took part very successfully in a technical meeting. Not many Korean women are in science activities, and fewer are in science professions. The attitudes of top management are critical for women. The objective of women in Korea is to promote science and engineering careers and research through networking. Most women are in chemical and biology fields, so it is difficult to find women role models ... Much of our effort goes into influencing mothers so they understand that science is easy and interesting for daughters, since the family environment is so important" (p. 18).

Kathleen Harer, USA: "The 11th ICWES in Japan was the fifth ICWES I've attended since going to my first conference in Poland in 1975. All of my ICWES experiences have been worthwhile, but this year's ranks among the most memorable. I traveled to Japan with Mary McCarthy of the Santa Clara Valley Section and my niece, Colleen Miller, who had just graduated from high school in Seattle. The trip was her graduation gift."

> "This was the first trip to Japan for all three of us. Once there, we experienced a very different culture, but one that has also become quite westernized over the years. We met people who thought nothing of going out of their way to help us when we needed it. We made new friends at ICWES and built up our networking contacts."
>
> "We left from Portland, on Sunday July 18, and arrived in Tokyo on Monday July 19, courtesy of crossing the International Date Line during our flight. We spent a few days in Tokyo before going on to Chiba, the location of the conference."
>
> "Tokyo has a wonderful subway and train system, which is needed to provide transportation for the large numbers of people who live and work in the city. However, the system's efficiencies still did not prevent me from leaving the station near our hotel by the wrong exit. Mary and Colleen had decided to do some shopping on the Ginza, and I had returned on my own. Since we had had no trouble getting to the station in the morning, I did not see how I could have trouble returning on my own."
>
> "At this point, I discovered for myself how helpful the Japanese people could be. I asked a young woman on the corner where we were waiting for the light to change, if I was going in the correct direction to reach my hotel. I showed her where the hotel was on my map of the city. Her English was not very good, but she understood my question. She took the map and motioned for me to start walking with her. Fifteen minutes later, we arrived at the hotel."
>
> "I thanked her in my best Japanese, which was definitely very poor, only a few simple phrases. I was particularly impressed that this woman went a half-hour out of her way to make sure that a lost stranger found her way" (pp. 18–19).

Petra Godwin, Immediate Past President WES, "My Personal Experience"

> "My personal experience of *ICWES-XI* was charged with emotion. I was...
>
> > ...**nervous** with anticipation both of travelling to (almost) the other side of the world – flying 12 hours – and of participating in a momentous event.

...**thrilled** (and grateful) to be 'bumped up' to Business Class (*I think they thought I was your Editor's dippy PA!*)
...**awed** – at approximately 400 delegates
...**surprised** by the familiar – Michiko Ishii's introduction: 'Environmental Policy in Japan: Current and Future' reminded me so much of my experience promoting Energy Efficiency through a government programme in the UK.
...**worried** by the perception of engineers and scientists and their status – a keynote address from Yvonne Maddox 'Women in Science and Engineering – Changes, Choices and Successes' suggested that we have made very little impact since *ICWES-IX* in 1991.
...**amazed** by the application of satellite technology in archaeology and personnel mine clearance – Toshibumi Sakata: 'From Space to Earth'.
...**punch drunk** from the workshops – according to my notebook I attended 22 in three days!
...**excited** by my first trip to Tokyo on the train to take Sunday tea with the British Consul for Science and Technology, who gave Sue Bullivant and myself a personal tour of the city, including a whisk around the British Embassy.
...**flabbergasted** to bump into one of my lecturers from 14 years ago, whose wife was giving a paper.
...**immensely honoured** to open the Banquet with some suitable words on how engineers are the people who make dreams come true.
...**hugely impressed** by the closing session of Dr. Sea-Wha Oh of Korea.
...**sad** to bid farewell to some new acquaintances (only to bump into a couple on Kyoto station!)
...**looking forward** to Ottawa. I've started writing my paper. See you there!" (Godwin 1999, p. 19).

Issié Yvonne Gueye, Ivory Coast

"This meeting of ICWES-XI, which coincides with the end of the 2nd millennium, should have been an occasion for hope for all the women engineers and scientists in the world. Sadly, although the much-wished integration of women into the realms of science and engineering has developed positively on other continents, it hasn't happened in Africa. Effectively, the problems which have always confronted African women have never been easy to overcome because they are of a cultural and structural nature, and touch the base of African society, which is fundamentally phallo-centric and strongly discriminatory towards women."

"Traditions and customs have always weighed heavily against women studying generally, but particularly against women engineers probably because they have had the audacity to challenge men on their own ground. And this is often seen as a challenge to society as a whole. They emerge marginalised and poorly regarded – even by their own families. The marginalisation is developed in a way that impacts negatively on their careers, often with and through a systematic blocking of any promotion. All the advances made over a period of years have been cancelled out and we are almost back to where we started. As a result, as we approach the start of the 3rd millennium, women engineers are not a role model for young people looking to plan their future. The danger for Africa is, in future, young girls will look towards more attractive professions and which pay better than technical careers."

"Today, with the crises that threaten most of Africa, coupled with the armed conflicts which destroy many of them, African women find themselves, despite their own wishes, at the bottom of the socio-economic plans. With their professional career blocked, hundreds of women engineers are abandoning their careers in order to take up jobs that will offer them a chance to compete more equally or just make ends meet at home. The sad thing is that, at the end of the day, not only are their dreams lost, but there is nothing to give hope to young women trying to follow them" (Gueye 1999, p. 20).

Joanna Maduka, Nigeria

"The opening ceremony was rather low-key, without the pomp that usually accompanies the opening of such international conferences, however, the special guest was HIH the Princess Takamodo."

"Mrs Michiko Ishii, Chair of ICWES-XI, said that the major advantage of a conference like ICWES-XI is the opportunity for women engineers and scientists to share knowledge and develop long-lasting networks."

"President of the Science Council of Japan, Dr. Hiroyuki Yoshikawa, pointed out that, with an ageing population and fewer young people, women must be encouraged into science and technology, and his comments were followed by goodwill messages from the Prime Minister of Japan and a host of other well-wishers."

"In a powerful keynote address, Mrs Ishii discussed the Japanese environmental policy and the issues involved in creating an acceptable environment, especially in the C21st, which included the re-emphasising of re-cycling, lifestyle changes and eco-business."

"In spite of language difficulties, I must congratulate the organisers for successfully hosting the conference. I hope I have established relationships which will be long-lasting in the areas of uplifting rural women in Nigeria and enriching our projects in renewable energy sources among them. I also intend to reproduce some of the conference papers for the Association of Professional Women Engineers in Nigeria."

"Finally, I wish to thank my sponsors, WES (UK), especially Dr. Elizabeth Laverick, for getting me to Japan in spite of all odds" (Maduka 1999, p. 20).

Ila Ghose, Calcutta, India

"My room on the 29th floor of the Makuhari Hotel afforded a breathtaking view – with Tokyo Bay on one side and a modern township, which glittered like a sea of diamonds at night, on the other. The architecture and the ambience of the conference venue at 'Makuhari Messe' was enthralling for one coming from a developing nation."

"Although the theme of the conference was 'Science and Technology for Global Ecology', an inescapable underlying theme was the discrimination women engineers continue to face the world over, more so in developing countries. Reflecting on our achievements, I feel we emphasized correctly on a changing global environment and gender discrimination. The challenge before us is to get our message across to the community, that women can help in evaluating and solving ecological programmes and in educating communities about environmental issues. In my opinion, topics discussed at conference are often not pursued further." Hence the implementation of the following holds interest for me:

1. The next generation of women engineers and scientists.
2. A global programme for environment and education.
   "The deliberations at the conference were hampered by a dearth of time and communication barriers imposed by different languages. However, the bonhomie and camaraderie amongst the women engineers transcended language and geographical barriers."

   "The social and technical tours were interesting, however they overlapped with the afternoon sessions. The meticulous arrangements made and the hospitality extended to us during our stay is commendable. A prompt medical service and the day-care service were thoughtful and praiseworthy gestures from the organisers."

   "Post-conference, I joined the tour to Kyoto, which offered an insight into Japan's rich cultural heritage and centuries-old civilisation. The rapid modernisation also on display – multi-storied Kyoto station, 'Bullet' trains, shopping malls and arcades – speak volumes about Japan's technological achievements. It was surprising to note that my colleagues from developed nations were equally enthralled. I would say my trip was extremely informative and interspersed with pleasant surprises" (Ghose 1999, p. 20).

Jacquie Carpenter, UK, "Renewable Resources and ICWES-XI"

"The theme for ICWES-XI was 'Science and Technology for Global Ecology' so I found it surprising that there was so little emphasis on moving towards the use of renewable resources. Much of the emphasis was on monitoring and evaluating problems. I do not often attend conferences where so much scientific work is presented and I was left with a deep impression that measuring the results of the assault of humankind on the planet is a political stalling tactic and a complete waste of money and effort."

"The Japanese in particular seem so fixed in the 'Mass produce, mass consume, mass waste' paradigm that they cannot begin to see a way out, although many are very worried indeed. Papers on renewable energy were few and those on other aspects of sustainable resource management, such as organic farming of not only food but of renewable raw materials, were even fewer."

"Women from the poorer countries, however, seemed very aware of their responsibility as engineers, mothers and educators in moving their countries towards a wise, sustainable future. Most were quite clear that they did not want to follow the richer countries towards a dead-end."

"Speaking to many women from many different countries, I felt–that we really were divided into two camps – those who felt that 'Global Ecology' is the biggest issue of all time, not only for women alive now but especially for their children, and those who were happy to be high-tech scientists, making a good living by studying the problems."

"There was a widespread feeling that women engineers and scientists can become agents of change and role models. The tension continued between the belief that women technologists can lead forward towards a better technological future and the belief that they need to help men to develop a simpler, more holistic, society. There was considerable mention of moving towards a high-tech, totally recycling society. But even though a minority clearly thought so, the Conference did not conclude that the development of renewable energy and of renewable raw materials was the most important future work of scientists and engineers" (Carpenter 1999, p. 21).

Karla Solis Garcia, Peru

"ICWES-XI was the most impressive conference I have ever attended. With more than 150 speakers, an audience of about 400 women participants and representing 36 countries, the result was an impressive four days of scientific work, technological thinking and sharing experiences."

"The conference was organised in two parts: Lectures and Sessions. The Oral and Poster sessions were very informative and it was amazing to realise and to understand that there is much to do in terms of research and transfer studies. As a Chemical Engineer, focused on the environmental problems caused by industrial sources and related with developing countries, I tried to attend most of the ones on Energy, Industry, Transportation and Information. It was interesting to find out about the application of geothermal energy as a source of electrical and non-electrical energy. I attended sessions on indoor air pollution in Egypt, outdoor air pollution in Japan, atmospheric transportation of air contaminants around Europe and noise pollution from transport sources in Teheran."

"The oral session on a small village in Uganda, where they are intensively polluting their own atmosphere through the use of kerosene lamps as a source of electrical energy, and where the cost of photovoltaic units is prohibitively high, was a good example that the ecological solutions are always possible in developing nations."

"I would like to take this opportunity to say that I am so grateful to [WES] for allowing me to attend this extraordinary conference. Without their generous sponsorship I would not have been able to experience this great occasion. ICWES-XI was extremely useful on two basic levels: the professional and the personal. I was able to make interesting contacts for my research and made me realise that I am, as a woman, in the same position as others around the world. It has given me the necessary incentive to continue my career in environmental issues and was a very rich source of inspiration for many of us. Many thanks for that experience [WES]!" (Solis Garcia 1999, p. 22).

Dormer Ellis, Canada and ICC Representative

"The ICWES conferences are as valuable for the information shared in the presentations as they are for meeting the people who attend them. The representative countries are in various stages of addressing the issues facing women in technological careers. We all shared experiences and successes and learned from each other. It was a growth experience for all the participants.

"Her Imperial Highness Princess Takamado opened the conference on Saturday morning. We were particularly impressed that, upon giving the opening address to the assembled engineers and scientists from more than 40 countries throughout the world, she stayed for the rest of the plenary session that morning. She participated in the opening luncheon reception and attended the meeting of science students from Japan that was being held concurrently in the afternoon."

"The papers presented ranged from highly technical topics to examinations on the status of women engineers and scientists in various parts of the world. There were poster sessions as well, and cultural activities such as demonstrations of the proper way to wear a kimono, and how to make an origami beating heart out of paper."

"The closing ceremony provided a chance to say goodbye to new and old friends. Following an excellent buffet dinner, we were entertained by an all-girl (except for the leader) group of drummers. They were such an inspiration that when several of us were enticed to come up on the stage to 'help' them drum, we beat the drums as hard as we could. There is something to be said about taking your frustrations out by beating on something -and the drums provided an acceptable outlet!"

"Following the final sessions, the next day, the Meeting of Delegates was held, with one person representing each country. The Delegates selected Ottawa, Canada, for the next ICWES. It will be held in July 2002" (Ellis 1999, p. 20).

"The ICWES held in Japan was the tenth I have attended. As each one was organised by a national engineering or scientific body they have all been somewhat different, although adhering to certain icwes customs. The cultural component of an ICWES adds to its value. At the reception following the opening ceremony in Chiba, delegates enjoyed some regional delicacies while watching traditional Japanese dances. The closing banquet included some Japanese dishes and an open-air performance by Japanese drummers."

"There were also several excursions from which to choose. One group went on a walking tour of a beautiful park noted for its Japanese-style gardens. Other groups went by coach to museums and galleries. I joined the group going to Boso No Mura, an interactive, outdoor museum – a village preserved as it was a century ago. Explaining some of the sights we saw during the trip, our guide said that Chiba has effective irrigation systems for its many rice fields."

"It being a hot day, Ila Ghose and I felt the need for a refreshing cup of tea, so we went off on our own to find a tea shop. Of course, we removed our shoes before entering. We sat on floor cushions at a very low table and enjoyed a popular jellied fruit dessert with our Japanese tea. We were the only non-Japanese clients in the shop. As we were leaving the coach after a delightful afternoon, we were each given a little souvenir – an example of origami, cleverly folded to resemble a kimono. Whatever our personal choices for these trips, we all appreciated the patience and kindness of the Japanese women engineers and scientists who included a cultural element in the programme" (Ellis 1999, p. 22).

Monique Frize, Canada, ICWES-XII Chair

"At ICWES-XI, the bid from Canada to host ICWES-XII in 2002 was well received, with the vote to support the choice of Ottawa unanimous (31 to 0). So ICWES-XII will be held in Ottawa, Canada, between 27 and 31 July 2002.

The main theme is "Women in a Knowledge-Based Society," with several sub-themes:

- Women in Research and Development
- Women and Technology
- Balancing Work and Family Life
- Achieving Harmony

"Technical papers and papers on gender issues will be welcome."

"Ottawa is a beautiful city, with two rivers, a canal, many parks, 29 museums, 60 festivals and live performances of every kind. The region is host to more than 900 high tech firms. It is bilingual (French and English) and thus displays many aspects of these two cultures and their charm. Social and industrial tours will be offered as well as several possible travel tours in Canada and the USA. In previous years, Ottawa was voted the best city in Canada and the 6th best in the world."

“The conference will also include a Youth component and a Student Paper Competition and several young people are expected to attend. A special reunion of participants at the first ICWES and at several other of these events will be organised. Links will be made with Women’s Studies to bridge research on women in science and engineering and women scientists and engineers themselves. It was heartening to see how excited people already were about coming to the conference in Ottawa, so plan your trip! We promise you a rich and rewarding experience at icwes12” (Frize 1999, p. 22).

Chapter 13
ICWES-XII: Ottawa, Canada, July 27–31, 2002

Theme: "Women in a Knowledge-Based Society" / *"Les femmes dans la société fondée sur le savoir"*

"Welcome" from the Conference Chair, Monique Frize

"It is a great honour to have been given the responsibility of organizing the 12th of a long series of successful ICWES conferences. It is with both anxiety and trepidation that I accepted the challenge. The most important goal that I set for this event is that each and every one of you find ICWES-XII to be a forum that provides great enrichment in knowledge and in friendship. This conference is an opportunity to share your personal experiences, your field of expertise, and knowledge on how to help women progress in your country and around the world. The information may also be useful for younger women to choose their future career path and to achieve all that they wish. For women to progress at all levels and everywhere, we must come together, network and share successes and failures. This makes us stronger in the face of adversity, discrimination, and sexism. It helps us to find resilience, strength to persist and go forward."

"In 1991, I attended my first ICWES (9th in the UK) and was hooked. I then attended the 10th (Hungary) and 11th (Japan). Since 1991, I have been thinking that Canada should host an ICWES conference. But we needed a strong network of women scientists and engineers, not only in Canada, but around the world. In 1991, I had just begun my role as national Chair for Women in Engineering, travelling across the country, speaking in schools, universities and industry on how to increase the participation of women in this profession and how to retain them once they have chosen this career. I worked on abolishing stereotyped gender roles and subtle discrimination. Relating success stories and sad ones, I was trying to make women more aware of the great satisfaction that awaited them in these fields and also how to deal with problems when or if they occur. I began to realize how my past life in engineering had contained so few contacts with women. Now women of all ages are becoming part of my life and these new friendships enrich my professional and personal life greatly. They provide me with the strength and determination to keep pushing the envelope of progress for future generations of women. It also provided the motivation to organize ICWES-XII."

M. Frize et al., *Women's Contribution to Science and Technology through ICWES Conferences*, Women in Engineering and Science,
https://doi.org/10.1007/978-3-031-33800-7_13

"ICWES-XII is what it is because of many special people who made it all happen. First, a committee of dedicated women and men volunteers shared the vision to make this a most memorable event for all participants and accompanying persons. Each committee member provided outstanding help for each project. An excellent Program Chair (and her assistants) put together the largest, most ambitious program to date and an effective peer-review process. Dedicated and competent staff, with Cheryl Cadrin as main coordinator, Cate Palmer and Diane Smyth as resources for the finance and website, and Moyra McDill as Convenor of the next conference, coordinator of the competition for developing countries, and the pioneer program (with Dormer Ellis, our Canadian pioneer!) Our two host universities (Ottawa and Carleton) provided help in several ways, amongst which were the negotiation of insurance coverage, provision of several volunteers and university services and the help of the Communications office at Ottawa University for the media relations portion of our program. Finally, this event would not have been possible without the great generosity of our sponsors!"

"I invite you to enjoy the program and our Canadian hospitality; visit our parks, lakes, mountains, and cities. Meet our organizing committee, our volunteers, our staff, and the other participants. I hope ICWES-XII is a most memorable event for all of you!" (ICWES-XII, Complete Proceedings, pp. 5–6; French version at pp. 3–4)."

"Welcome" from the Program Chair Claire Deschênes

"It is my great privilege to invite you to discover the program of the 12th International Conference of Women Engineers and Scientists. This program, designed for a global audience of women and men engineers and scientists, focuses primarily on developing a new vision of women's role and of their contributions in these fields. Never, in the history of ICWES, has internationalization been so vast. We bring together more than 500 participants from more than 45 countries."

"The program includes more than 200 papers and events, both technical and non-technical. There are two symposia, with presentations from internationally renowned speakers, on themes that are current concerns: Climate change and Ethics and Science. There will be 13 panels, 3 round tables, and 5 workshops, all on topics of current interest and significance. Examples are: Scientific policy, women in industry, women and information technology, and transmission of knowledge by Aboriginal women (keepers of the circle). Almost every aspect of women and science will be covered, from youth programs for attracting young women to these fields, career development, balance between life and work, status of women in various countries and sectors, to name a few. Three tracks have been created on the themes of Women's studies, led by Sue Rosser (US); another is on Youth, led by Submaranian Saraswathy (India); and Women and Health led by Monique Bégin and Miriam Stewart."

"I would like to warmly thank Monique Frize for having given me her trust and to have provided personal help at critical times in the development of the program. I also thank all those who helped in its creation, and particularly the authors who have responded positively to all our requests. I also wish to thank all the nearly 100 persons who competently reviewed the papers. Their enthusiasm has more than once impressed and motivated me! Several wonderful women have made contributions to the program and although I cannot name them all here, they have my eternal gratitude. I will mention Grazia Toma who, with the greatest patience, received the papers and responded to thousands of emails. I think she can recall the author's name and paper title by their number!"

"I wish that this conference will allow everyone to participate in the discussions, to assess where we are on this important question of the role of women in a knowledge-based society, and to leave with new ideas and strategies. Finally, I hope the conference will personally enrich your life as it has already done for me!" (ICWES-XII, Complete Proceedings, p. 8; French version at p. 7).

"Welcome" from CCWEST President Hiromi Matsui

"On behalf of CCWEST, the Canadian Coalition of Women in Engineering, Science and Technology, I am pleased to extend a warm Canadian welcome to all the delegates who have come to Ottawa to attend ICWES-XII. Welcome my friends! CCWEST is a national organization that represents several thousand women in science and engineering who belong to organizations across Canada, from British Columbia on the West Coast to Newfoundland and Labrador on the east coast. Our members have been pioneers and leaders in research, industry and volunteer organizations. We share a profound belief that diversity enriches and expands opportunities for everyone. We celebrate the contributions and achievements of Canadian women in science and engineering."

"Many CCWEST members will be presenting papers and actively participating in ICWES-XII. We want to share our experiences with you and to hear about your challenges and successes. We are fortunate to be holding our 9th CCWEST Annual General Meeting in conjunction with ICWES-XII."

"We would like to acknowledge the work of the ICWES-XII organizing committee and in particular, the leadership of Dr. Monique Frize who has been an outstanding champion for women in engineering. Monique has demonstrated the passion and commitment of Canadian women in science and engineering throughout her career, and for the breadth of her contribution, we salute her. Thank you very much Monique" (ICWES-XII, Complete Proceedings, p. 10; French version at p. 9).

Participants

The total number of delegates was 550, including 296 from Canada and 254 from 49 other countries. A large number of delegates came from the US (108), from Far East Asia (50, including 40 from Japan), English-speaking Africa (27), Western Europe (19), the UK, Australia, and New Zealand (16), Eastern Europe (10), French Africa (8), South-East Asia (6), Central Asia (4), and Latin America (3). Out-of-country attendance came up to 46.2% (see Appendix 3).

ICWES-XII, S&T Papers

The world had entered the twenty-first century relieved that the computer problems expected on January 1, 2000, at midnight, known as Y2K (Year 2000), did not happen. From a global *Information Society*, the world was running toward a *Knowledge-Based Society*. Increasingly conscious that human activity was affecting the world globally, it became necessary to consider many dimensions in any impact assessment—scientific, social, ethical, and political—to share this knowledge worldwide and to act on it. UNESCO explained the expression *Knowledge-Based Society in the Following Way* in a global report that would be published 3 years later (Bindé 2005):

> "Might we now have the means to achieve equal and universal access to knowledge, and genuine sharing? This should be the cornerstone of true knowledge societies, which are a source of human and sustainable development. [...] A knowledge society is a society that is nurtured by its diversity and its capacities. [...] The idea of the information society is based on technological breakthroughs. The concept of knowledge societies encompasses much broader social, ethical and political dimensions."

Invited to contribute to the main theme, *Women in a Knowledge-Based Society*, ICWES-XII authors shared their knowledge and expressed their hope for a better integration of women scientists and engineers' contributions in the solutions. They encompassed a very large variety of scientific, social, and educational subjects, aiming toward diversification in science and technologies. ICWES-XII invited a global audience of women and men working in NGOs, industry, academic, and governmental institutions from different countries and different sectors of economic activities.

The paper submissions were all peer-reviewed. The *Access* database created for ICWES-XII shows that 698 people from 49 countries submitted a poster, a paper, or participated in a panel. However, not all of them attended the conference: some people declined to attend, the peer review was unsuccessful, or some people from developing countries were not successful in obtaining funding to attend the conference or their visa was not approved. The organizers had obtained a grant from the Canadian International Development Agency to financially support the participation of 41 people from developing countries; unfortunately, an additional 68 people had asked for funding but could not be funded. In the final count, the program included 219 papers, posters, panels, round tables, workshops, and plenaries in both the S&T and gender categories; two symposia, 13 panels, 3 round tables, and 5 workshops were submitted. Three special tracks had been proposed on the themes of: women's studies, chaired by Ruby Heap, Canada; youth, led by Submaranian Saraswathy, India; and women and health, led by Monique Bégin and Miriam Stewart, Canada. The 31 poster presentations authors had all submitted a full paper for the proceedings.

Focusing on the S&T subset of content, ICWES-XII was organized into eight sessions and one symposium on climate change, as can be seen in Table 13.1. Out of 43 papers expected, 41 papers are present in the proceedings (ICWES-XII, Complete Proceedings). Global issues such as climate change and sustainable development were slotted into specific sessions. A plenary was presented by Elizabeth May (director of The Sierra Club of Canada) and Elizabeth Dowdeswell (founding president and CEO of the Nuclear Waste Management Organization) as part of the symposium on climate change (ICWES-XII, Complete Proceedings, pp. 1329–1330).

The S&T papers and abstracts were written by a total of 84 authors (32 women, 18 men, and 34 of unknown sex). Among them, 20 authors were from the host

Table 13.1 Distribution of S&T Sessions and Papers in ICWES-XII, Complete Proceedings

ICWES-XII S&T	Main topic: Women in a Knowledge-Based Society	# of papers expected	# of papers in proceedings
1	Engineering/Physics	7	5
2	Climate change	4	4
3	Chemistry/Biology	7	7
4	Information Technology	5	5
5	Technology and Health	3	3
6	Women and Health	6	6
7	Sustainable Development	5	5
8	Agriculture in Developing Countries	6	6
	Symposium on Climate change	1	1
	Total	*41*	*39*

Table 13.2 Distribution of S&T Posters in ICWES-XII, Complete Proceedings

ICWES-XII S&T poster sessions	Poster session titles	# of posters
1	Engineering/Physics	14 (16)
2	Climate change	3 (15)
	Total	*17*

country, one from the USA, five from Eastern Europe (1 from Armenia, 4 from Russia), 3 from French-speaking Africa (1 from Congo, 2 from Benin), 41 from English-speaking Africa (20 from Ghana, 20 from Nigeria, 1 from Uganda), five from Central Asia (2 from India, 3 from Pakistan), three from the region of Far East Asia (2 from Japan, one from Mongolia), and 6 from the United Kingdom.

In addition, 17 posters with full papers were presented, prepared by 47 authors (8 women and 39 of unknown sex). Among them, three authors were from the host country, 1 from the USA, 5 from English-speaking Africa (Nigeria), and 38 from Far East Asia (Japan). In Table 13.2, the first number under # of posters expected indicates the S&T-themed posters, while the number in brackets gives the total of posters for a given session.

A Glimpse of ICWES-XII's S&T Content

With a population of 6.3 billion in 2002, it had almost doubled since the first ICWES conference in 1965 (3.3 billion). This population explosion was rarely mentioned during ICWES-XII, but it is certain that the impact of humans' activities on the earth was an increasing concern.

As a first example, four papers were presented in the *Climate Change* session. One of these papers, "The Moderating Role of the Oceans in Climate Change" (ICWES-XII, Complete Proceedings, pp. 470–475), by Brenda Topliss and Sherry Niven (Bedford Institute of Oceanography, Canada), reported on global temperature changes predicted by coupled ocean-atmospheric models. The authors highlighted the role of oceans as a climate moderator: "The oceans provide a huge storage (or sink) capacity for both heat[1] and carbon [...] the oceans have the potential to delay the impact of greenhouse gas (GHG) emissions and thus affect climate change" (ibid., p. 470).

The three other papers in this *Climate Change* session offered solutions for issues specific to different parts of the world. Karla Solis (Centre for Environmental Strategy, University of Surrey, UK) and her colleagues presented a well-documented paper called "The Implementation of the Kyoto Protocol: International Climate Change Policies in Latin America" (ICWES-XII, Complete Proceedings, pp. 1080–1085). The authors underlined that gas emissions were increasing not only in developed countries but also in developing countries such as Costa Rica, Brazil, and Mexico. They insisted on the necessity to implement environmental policies and institutions that would be economical, feasible, and have long-term objectives. The focus was put on "Activities Implemented Jointly"[2] between developed and developing countries. For Latin America, these activities would affect sectors such as forestry and renewable energy. To provide an example, the Mexican government had already established policies focused on three main aspects: increasing energy efficiency, supporting renewable energy techniques, and switching to natural gas to replace high sulfur-content fuels.

Similarly, the engineer Joanna O. Maduka (Chairperson, Friends of the Environment, Nigeria) wrote a paper called "Renewable Energy and the Empowerment of Women in Nigeria" (ICWES-XII, Complete Proceedings, pp. 424–430). She explored the energy needs of rural women in Nigeria and Africa in general. This need is mainly for heating, cooking, lighting, and agriculture production and processing. The women relied on traditional woodstoves, which may bring deforestation; on blacksmith burning of charcoal, which is hazardous for health; and on open-air solar drying, which exposes them to dust, insects, and animals. Moreover, Nigeria produced over 20% of the world's total GHG from gas flaring associated with crude oil production at the time of the conference. To improve women's quality of life, the author suggested the development of renewable energy sources such as microhydro, biomass, solar, and wind.

The last paper of this session, "The Role of Nuclear Power in Combating Climate Change," by Elisabeth Cooke (Hons., British Nuclear Fuels Limited, Warrington, UK) (ICWES-XII, Complete Proceedings, pp. 601–604), advocated for nuclear electricity generation, especially in the UK, to ensure a secure and diverse electricity supply whilst, at the same time, reducing carbon dioxide emissions.

All but one paper in the *Sustainable Development* session and in the *Agriculture in Developing Countries* session were from black African countries (Nigeria, Ghana, Uganda, Benin, and Congo). The paper, written by Lamjav Batmunkh

[1]The oceans heat capacity is approximately 10,001,000 times that of the atmosphere.

[2]*Activities implemented jointly* is a concept used at the first Conference of the Parties (COP-1) in Berlin in 1995.

(Mongolian State University of Agriculture) and named "Gender and Work in the Herding Community" (ICWES-XII, Complete Proceedings, pp. 310–312), described women's work in herding. According to official statistics reported by the author, over 40% of the population in Mongolia was nomadic and engaged in herding at the time of the conference. Herd milking was the exclusive responsibility of women, and this created a situation where herd growth was limited to the number of animals the women could milk and process themselves.

An example of a topic similar to those presented in previous ICWES conferences was a session on *Information Technology*, which had four papers. Three of them were very technical. Durdana Habib (Communications Enabling Technologies, Islamabad, Pakistan) and her colleagues Shoab Ahmed Khan and Sohail Sadiq (Computer Engineering Department, National University of Sciences & Technology, Pakistan) reported on "Code Generation for Digital Communication Algorithms on Application-Specific Processors" (ICWES-XII, Complete Proceedings, pp. 1032–1035); and Grazyna F. Whapshott (Senior Lecturer at the University of Greenwich Medway School of Engineering, Pembroke, the UK) presented "Genetic Algorithms as Self Organizing Methods Applied to Digital Systems Design" (ibid., pp. 362–365).

In the third paper of this session, Moyra McDill (professor at the Mechanical and Aerospace Engineering Department at Carleton University, Canada) shared her thoughts on "Developing and Maintaining a Large FEA Research Code" (ICWES-XII, Complete Proceedings, pp. 209–213). Finite Element Analyses (FEA) are used to study physical applications such as heat transfer in solid mechanics by providing a way to solve a discrete system of equations. Some commercial solvers using FEA were well known at the time of the conference: ABAQUS, DYNA, and NASTRAN. Nevertheless, research-based FEA codes allowed better access to the embedded mathematics and physics. The author demonstrated that developing these homemade codes permitted active research at the forefront of the discipline. She described two examples: prediction of stresses in weaved repair welds and prediction of high-temperature behavior in a Ni-based single-crystal alloy. However, according to the author's personal academic experience, maintaining and expanding a legacy code is a non-trivial task, and it carries potential career challenges related to the search for funding and publishing results.

The *Women and Health* session included five papers, three of which were presented at the conference. One of these was written by Galina Nechitailo (professor at the Institute of Biochemical Physics RAS Kosygin, Moscow, Russia), a specialist in Biochemical Physics. Her paper, "Protection of the Health of Women Occupationally Exposed to Unfavorable Ecological Factors is a National Task in Russia" (ICWES-XII, Complete Proceedings, pp. 258–260), explained that under unfavorable ecology, such as stresses in the social-economic sphere, unbalance in nutrition, wide application of various xenobiotics,[3] the normal microflora of the

[3] "Xenobiotic is a term used to describe chemical substances that are foreign to animal life and thus includes such examples as plant constituents, drugs, pesticides, cosmetics, flavorings, fragrances, food additives, industrial chemicals and environmental pollutants." (Patterson et al. 2010, p. 851).

population is disturbed, bringing potential health problems. This has been observed in organizations with harmful working conditions and in chemical industry, in particular, where many women are engaged. The author reported that using bacterial preparations, jointly with preventive and sanitary measures, can reduce the level of infection, gastroenteric, urological, gynecological, allergic, and other diseases for women exposed.

Two papers were presented by Nigerian scientists in this session on *Women and Health*. The first one was "An Epidemiological Study on the Health Status of Sawmillers in Benin City, Nigeria" (ICWES-XII, Complete Proceedings, pp. 706–711), by Obehi H. Okojie (Department of Community Health, University of Benin, Benin City, Nigeria). The second one was written by Antoinette N. Ofili, et al. and called "Psychological Disorder and Job Satisfaction Among Nurses at a Teaching Hospital in Nigeria" (ibid., pp. 607–611).

ICWES-XII, Gender Papers

In 2002, awareness that the participation of women in S&T was low and that this situation was one of the main issues to be tackled concerning the status of women had largely grown. Many groups, associations, governmental agencies, universities, and representatives of economic sectors gathered in Ottawa to share their experiences and learn about these issues. They had many concerns and hopes concerning women in S&T and presented related research or activities.

Therefore, topics at ICWES-XII were not only related to women's contributions in S&T and Women's Studies in S&T but also, in large part, to advocacy in favor of women's successful inclusion in STEMM.[4] Considering that 178 submissions were scheduled at ICWES-XII (ICWES-XII, Complete Proceedings, pp. 11–41) and that 58 submissions were related to S&T (Table 13.1), 77% of the submissions were directly related to women in STEM.

The following classification provides a glimpse of the richness of ICWES-XII on these issues:

- Personal experiences and women's experiences in different groups such as industry, information technology, and various countries and sectors of activities;
- Inclusion: a panel on the transmission of knowledge by Aboriginal women (Keepers of the Circle) and submissions in French;
- Issues related to women in S&T, such as outreach programs, career development, balance between personal life and work, and scientific policies, to name a few
- Studies on different aspects of women in STEMM by researchers in social sciences, education, and women's studies

[4] STEMM is a recent acronym that stands for Science, Technology, Engineering, Mathematics, and Medicine.

- Activities, such as workshops or panels, organized by prominent associations involved in or for women in STEM

This last item attests to the effervescence around the question of women in science and engineering at the time of ICWES-XII, in particular in North America. Specific panels or round tables were organized by the following organizations: ASME (American Society of Mechanical Engineering, USA), IEEE (Institute of Electrical and Electronics Engineers, USA), ASCE (American Society of Civil Engineers, USA), AWIS (Association for Women in Science, USA), WEPAN (Women in Engineering ProActive Network, USA), SWE (Society of Women Engineers, USA), Global Alliance (USA), CCWEST (Canadian Coalition of Women in Engineering, Science, Trades and Technology, CANADA), WISEST (Women in Scholarship, Engineering, Science & Technology, Canada), MentorNet (Canada), and Hypatia Project (Canada).

Concerning the social sciences and education inputs at ICWES-XII, it is worth mentioning the exceptional participation of Professor Sue Rosser, a leading scholar in multiple areas, including women's and gender history, history of public policy, and history of technology. To these must be added her strong interest in biology and women's health, and in women in science and teaching methods. Such a large teaching and research map and agenda illustrated Sue Rosser's continuing commitment to the promotion and pursuit of her interdisciplinary interests across the fields of health, arts, technology, and science.

Also of interest is the participation of Prof. Judy L. Lupart (University of Calgary, Canada) and her colleagues and students, who presented seven papers from a research program to understand the issues of women in STEM:

- "Predicting Science Achievement in Males and Females: Perceptions of Science" (ICWES-XII, Complete Proceedings, pp. 487–489)
- "Gender Differences in Science Achievement, Science Related Career Choices, and Perceptions of Adult Roles in Society" (ibid., pp. 680–683)
- "Gender Differences in Expectations for Success in Three Subject Domains" (ibid., pp. 792–796)
- "Promoting Science in Secondary Students: What have Parents Got to Do with It?" (ibid., pp. 893–897)
- "Secondary Students and Gender Differences in Computer Attitudes, Skills, and Perceived Ability" (ibid., pp. 919–923)
- "An Exploration of Gender Differences in Computer Usage and Affinity, and the Relationship with Career Plans and Adult Life Choices" (ibid., pp. 928–932)
- "Impact of Performance Anxiety and Parental Attachment in Relation to Gender and Academic Achievement in Math, Science, and English" (ibid., pp. 1069–1073)

In addition to the creation of INWES, which is discussed in Chap. 14, the creation of three Canadian organizations (Institute for Gender and Health, NSERC Chairs, AFFESTIM) was reported at ICWES-XII, summarized as follows:

In the plenary on "*Women and Health*" (ICWES-XII, Complete Proceedings, pp. 1337–1338), the Honorable Monique Bégin explained that "[i]n a country like

Canada, women's health remained for a good 20–25 years [after the late 1960s] a holistic quest nurtured by the women's movement" (ibid., p. 1337). However, in 2002, private and public funding for mainstream medical research in women's health led to the creation of specific institutes. Miriam Stewart, who spoke in the same plenary as Monique Bégin, focused on the creation of the *Institute for Gender and Health* as one of the *Canadian Institutes of Health Research*. These institutes have research priorities on issues important for ICWES participants, such as health services for marginalized groups, gender and health, and breast cancer research.

The paper "The NSERC/Industry Chairs for Women in Science and Engineering: A National Program in Canadian Universities" (ICWES-XII, Complete Proceedings, pp. 1101–1105), by Professor Mary Williams et al., reported on the establishment of five senior academic positions in science and engineering faculties across Canada. The chair holders are university professors in S&T from five regions who devote half their time to promoting the participation and retention of women in S&T. This chair program of the Natural Sciences and Engineering Research Council of Canada (NSERC), still existing today, expanded in 1997 on the model of the national chair for women in engineering held by Monique Frize (1989–1997). The paper by Mary Williams presented a summary of the challenges, results, and lessons learned from the program's first 4 years. Among the elements of the conclusion can be read: "The challenge of raising the participation of women to current levels has been met through energetic recruitment and promotion programs. However, any further gains will need much deeper changes to the environment, the curriculum, and pedagogy, as well as changes to broad public perceptions, so that some of the women who are now choosing more liberal professions (medicine, dentistry, law, pharmacy, business) will begin to see engineering and sciences as fields that benefit society and where their feminine perspectives are valued" (ICWES-XII, Complete Proceedings, p. 1105).

Canada being a bilingual country, one panel was held entirely in French. Called "Des réflexions et des actions pour stimuler les femmes à poursuivre des carrières en sciences et en génie au Québec,"[5] (ICWES-XII, Complete Proceedings, pp. 1169–1171), a panel brought together various organizations from the province of Quebec working with women to promote their participation in trades and scientific professions (Marianne-Mareschal Chair, MOIFEM, ministère de l'Enseignement du Québec, ministère de la Recherche, de la Science et de la Technologie, NSERC/Alcan Chair for Women in Science and Engineering). This event launched the new *Association de la francophonie à propos des femmes en sciences, technologies, ingénierie et mathématiques*[6] (AFFESTIM), which today has reached the francophone population across Canada.

Out of a grand total of 236 different speakers presenting 141 papers across 32 sessions, a little over half came from North America (105 from Canada and 33 from

[5] "Thoughts and actions to stimulate women to pursue careers in science and engineering in Quebec" (authors' translation).

[6] "Francophone Association on Women in Sciences, Trades, Engineering and Mathematics" (authors' translation).

the US). For the other half, there were: five from Puerto Rico, 15 from Western Europe (1 Austria, 4 Belgium, 4 France, 3 Ireland, 1 Spain, 2 Sweden), 4 from Eastern Europe (2 Poland, 1 Romania, 1 Ukraine), 2 from French-speaking Africa (1 Ivory Coast, 1 Senegal), 31 from English-speaking Africa (7 Ghana, 2 Kenya, 3 Mauritius, 17 Nigeria, 1 Uganda, 1 Zambia), 1 from Egypt, 7 from India, 5 from Southeast Asia (1 Bangladesh, 1 Laos, 2 Malaysia, 1 Vietnam), 7 from East Asia (2 Hong Kong, 5 Japan), 15 from the UK, and 6 from Australia (Table 13.3).

Two poster sessions, with 28 presenters (20 women, 2 men, 6 unknown), coming from the host country (10), the US (3), Western Europe (2 France, 1 Germany, 2 Italy), Nigeria (1), India (1), Japan (6), the UK (2). In the table below, the first number under # of posters expected indicates the Gender-themed posters, while the number in brackets gives the total of posters for a given session (Table 13.4).

Three more papers appear in the proceedings but do not feature in the conference program:

- "On the Achievements of the Women Scientists of the Gauhati University" (Pranayee Datta, India)
- "A Comparative Study of the Dowry System in Different Regions of India" (Monalisa Mazumdar, India)
- "Gender-Based Digital Divide in the Web-Related Occupations in the Universities: A Cross-Country Study" (M. Suriya, India; Carole Farber, Canada; Britta Schinzel, Germany)

For the sessions on Women's Study, Sue V. Rosser delivered a plenary lecture called **"Using the Lenses of Feminist Theories to Focus on Women, Science, and Technology."**

In 1973, Rosser obtained a PhD in Zoology from the University of Wisconsin-Madison, where she had also received an MSc specializing in French and Zoology in 1969. She held various important positions throughout her career: Dean of the Yvan Allen College, the liberal arts college at the Georgia Institute of Technology in Atlanta (from 2009); holder of the endowed Ivan Allen Chair of Liberal Arts and Technology; director for the Center for Women's Studies and Gender Research (1995–1999); Professor of Anthology at the University of Florida-Gainesville; Senior Program Officer for Women's Programs at the National Science Foundation (from 1995). Sue Rosser has an impressive number of publications to her name; examples: *Biology and Feminism: A Dynamic Interaction* (1992) and *Women, Science, and Society: The Crucial Union* (2000).

The full paper is not featured in the conference proceedings, but here is the abstract:

> "The impact of feminist theories has profoundly influenced both the disciplinary and interdisciplinary questions asked, topics explored, approaches taken, and conclusions drawn from the teaching and research in the humanities and social sciences. Now the case studies and increasing numbers of well-document work examining the relationships among women, science, and technology suggest that examination using a broader spectrum of feminist lenses would be likely to reveal useful information. Feminist analyses of science and technology have typically fallen under the three categories of liberal feminism, ecofeminism,

Table 13.3 Distribution of Social and Gender Papers in ICWES-XII, Complete Proceedings

Social and gender session titles	# of papers expected	# of papers in proceedings
Women's Studies (1)	4	4
Computer Science – Attitudes	4	4
Initiatives by Associations (1)	5	5
Social Studies – Career in Information Technology	4	4
Women in Different Sectors – University & Research	5	5
Women's Studies (2)	5	5
Computer Science – Training	4	4
Initiatives by Associations (2)	5	5
Social Studies – Self Confidence/Career Choice	4	4
Women in Different Sectors – Industry	5	5
Globalisation	5	5
The Bold and the Brave	4	4
Women Experiences in Different Countries	5	5
Women in Different Sectors – Working Conditions/Safety	4	4
Youth Program – K to 12 (1)	5	5
Women in Different Sectors – Statistics	4	4
Social Studies – Gender Analysis	3	3
Outreach – Girls, Young Women	4	4
Family and Work Balance	6	6
Youth Program – K to 12 (2)	5	5
Women Experiences in Different Countries	5	5
Social Studies – Enrollment	4	4
Outreach – Girls and Information Technology	3	3
Personal Experiences	4	4
Youth Program – University Recruitment and Retention	5	5
Outreach in Different Countries	4	4
Social Studies – Education	5	5
Outreach – K to 12	6	6
Gender Equity & Policies	4	4
Outreach/Retention in Universities	4	4
Retention in Science, Mathematics and Engineering	4	4
Women and Technology	4	4
Total	*142*	*142*

Table 13.4 Distribution of social and gender posters in ICWES-XII, Complete Proceedings

Social and gender poster session titles	# of posters expected	# of posters with papers in proceedings
I	2 (16)	2
II	12 (15)	12
Total	*14*	*14*

> and most particularly, socialist feminism. Expanding the theoretical framework to include racial/ethnic, essentialist, existentialist, psychoanalytic, radical, lesbian/separatist, post-modern, post-colonial, and cyber-feminism uncovers subtle, rich insights to the dynamics of the co-evolution of gender with science and technology" (ICWES-XII, Complete Proceedings, p. 48; also at 1336).

One of the 14 panels of this conference was titled "**Aboriginal Women: Keepers of the Circle of Knowledge**." It featured four speakers in addition to the facilitator, Corinne Mount Pleasant-Jetté. Mount Pleasant-Jetté, a member of the Tuscarora First Nation and holder of a Bachelor of Education, taught at the Faculty of Engineering and Computer Science at Concordia University (Montreal, Quebec) and founded and directed the Faculty's Native Access to Engineering Initiative. She was invested in the Order of Canada in 1992 (Miall 2014).

Mount Pleasant-Jetté introduced the panel by presenting how indigenous peoples around the world and across periods had adapted to their environment by "building communities, nurturing children, and finding innovative, often creative solutions to everyday challenges" (ICWES-XII, Complete Proceedings, p. 1022). Clothing, crafting, and food were often the responsibility of women. "The value of these objects and activities meant the difference between survival and submission to the elements of nature. For many Aboriginal groups in Canada, the capacity to conceptualize, create, design and implement solutions enabled their populations to live in harmony with nature" (ibid.). Food and medicine came from local sources and and preserved during the winter; shelters were meant to integrate with the environment; food and water were used to their full capacity.

None of this knowledge was recorded in writing but kept and passed down through oral traditions: "the stories, the techniques, the artistry, and the detailed 'scientific' secrets of their societies remained the domain of the elderly men and women of the communities" (ICWES-XII, Complete Proceedings, p. 1022). Women played a key part in this transmission of knowledge and wisdom as "Keepers of the Circle."

Karen Decontie, an Algonquin from Kitigan Zibi (Quebec) and a structural engineer, worked for Public Works and Government Services Canada in Calgary, Alberta. She graduated from McGill University with a Bachelor in Civil Engineering and from the University of Calgary with a Master of Engineering in Civil Engineering ("Karen Decontie" n.d.).

In a paper titled "Keepers of the Circle," Decontie talked about the historic role of indigenous women within their societies and their current position in the field of engineering: "[w]omen are taking more of an active role in non-native society and demonstrating to their offspring that there can be a blending of traditional culture and living in a modern society" (ICWES-XII, Complete Proceedings, p. 1023).

Wanda Wuttunee, a member of the Red Pheasant Nation, was at the time an associate professor at the Department of Native Studies at the University of Manitoba (now a full professor). She graduated from the University of Calgary with a Bachelor in Communication, a Bachelor of Laws, and a Master of Business Administration. She pursued and obtained a PhD at the University of Manitoba ("Dr. Wanda Wuttunee" n.d.).

Wuttunee presented a paper on "Living Rhythms: Lessons in Aboriginal Economic Resilience and Vision." The objectives were: "(1) to ascertain the place held by Aboriginal wisdom in economic development theory; (2) to document the blend of approaches and economic development strategies currently employed in selected Aboriginal communities; and (3) to identify the role of Aboriginal wisdom in selected Aboriginal communities that are striving for economic development in tune with individual community rhythms" (ICWES-XII, Complete Proceedings, p. 1023).

Wuttunee stressed how Aboriginal scholars were challenging mainstream strategies and proposing alternative solutions grounded in traditional indigenous knowledge. Contrary to non-native approaches trying to turn away from tradition, Aboriginal methods were a combination of tradition and modern economic models. Wuttunee based her research on "Elements of Development," a model created by Aboriginal women that is based on 16 factors to measure economic success (Salway Black 1994). This alternative approach called for a re-examination of sustainable development with a renewed emphasis on the role of spirituality.

Priscilla Settee is a Cree activist. At the time of the conference, she held a Bachelor of Arts in Sociology from the University of Guelph, a Bachelor of Education from the University of Saskatchewan, and a Master of Education from the University of Manitoba ("Priscilla Settee" n.d.-a, b).[7] She was director of the Indigenous Peoples Program at the University of Saskatchewan.

Settee's talk, titled "Indigenous Contributions to Science, Environmental Impacts and Concerns," dealt with indigenous knowledge systems and their contribution to the well-being of nature and community. One important example was the concern over recent advances like genetically modified organisms.

Lillian E. Dyck, a member of the Cree Gordon First Nation, was a full professor in neuropsychiatry at the University of Saskatchewan. She graduated from the same university with a Bachelor of Arts, a Master of Science in Biochemistry, and a PhD in Biological Psychiatry ("Lillian Dyck" n.d.-b).[8] In 1981, Dyck became the first Indigenous Canadian woman to obtain a doctorate in science ("Lillian Dyck" n.d.-a).

Using the "Medicine Wheel of the Plains Indians," Dyck sought to (re)examine what science was, as regards both education and research practices. Her talk first explained what a Medicine Wheel was and how it could be used as an analytical framework. She argued that the low numbers of women and Aboriginals who chose to study science could be understood by the neglect of certain aspects of the Medicine Wheel. Dyck also addressed the issue of recruitment and retention of women in science education and careers.

---

[7] Settee would later obtain a PhD at the University of Saskatchewan in 2007, with a dissertation on indigenous knowledge systems. She is currently professor emerita at the University of Saskatchewan.

[8] Dyck would later be appointed to the Canadian Senate with a nomination by then Prime Minister Paul Martin.

Mary Braithwaite, "An Assessment of the European Union Research Funding Programme 'Improving Human Potential and the Socio-economic Knowledge Base'"

Mary Braithwaite (director of Tacitus) presented the framework, methodology, and results of a 'gender impact' study done in 2000–2001 by her firm and other experts. This assessment was carried out for the European Commission's Directorate General for Research, more specifically its Unit "Women and Science: Mobilising Women to Enrich European Research" (established in 1999) and the research funding program "Improving Human Potential and the Socio-economic Knowledge Base," with the aim of improving the "integration of the gender dimension in EU research activities and programmes" (ICWES-XII, Complete Proceedings, p. 482). Six other studies were conducted in parallel on other research programs within the Fifth Framework Programme (FP5) for research, technological development, and demonstrations.

The program "Improving Human Potential and the Socio-economic Knowledge Base" (IHP for short) was concerned with: "training and mobility of researchers (including the Marie Curie Fellowships); improvement of research infrastructures; dissemination of high-level research; raising public awareness; research projects and networks in the field of socio-economic knowledge; and development of science and technology policies, indicators and statistics" (ICWES-XII, Complete Proceedings, p. 483).

For this study, Braithwaite's team relied mainly on desk analysis, or desk research, of reports and documents dealing with FP5 and IHP. Regular exchanges, visits, and participation in meetings with Commission staff also took place. Within Braithwaite's team, frequent workshops and meetings were also useful to share ideas and information during all phases of the assessment.

The evaluation was grounded in the concept of gender as "a significant variable in understanding contemporary social, cultural, economic, and political changes" and "a dynamic factor influencing these changes" (ICWES-XII, Complete Proceedings, p. 483). Six main aspects of this notion were used:

1. "Sex counting": an analysis oriented towards statistics and based on biological sex
2. A certain specificity of the lived experience: for instance, the difference in the unemployment rate, or the struggles specific to single mothers
3. Gender differences and gender relations: socio-economic roles typically ascribed to women and men
4. Gender differences translated into gender inequalities: inequality of rights, control of resources, access to decision-making
5. Gendered structures: systems and processes that reproduce gender inequalities, with a focus on how gendered, often unconscious, biases operate on the meso- and macro-levels
6. Gendered concepts: the ways in which concepts such as "science," "productivity," and "work" are at their core gendered visions of society and the world.

The study used "gender mainstreaming" as a framework, which is "the current strategy, adopted and endorsed at European and international levels, for achieving gender equality" (ICWES-XII, Complete Proceedings, p. 484). This approach is threefold:

1. Promote the participation and integration of women in decision-making and higher levels of structures.
2. Make sure that all policies and programs recognize different profiles and backgrounds.
3. Make sure that all policies and programs are built in such a way that they are conducive to gender equality.

As Braithwaite explained, gender mainstreaming is "more than simply ensuring 'fair' or 'equitable' treatment of the needs and interests of women and men. It requires the specific orientation of policies and programmes so as to redress existing inequalities" (ICWES-XII, Complete Proceedings, p. 484).

This approach can be measured by ten factors:

1. Women's and men's experiences and interpretations of reality
2. Policy
3. Representative political structures
4. Institutional culture
5. Political commitment
6. Resources
7. Procedures and instruments
8. Information
9. Location of responsibility for gender issues
10. Staff development

According to Braithwaite, positive achievements of this assessment included an increase in the presence of women in the IHP program and as panel experts, the extension of age limits for Marie Curie Fellowships and Research Training Networks "to take into account time spent away from professional activity for family reasons and provision for childcare costs" (ICWES-XII, Complete Proceedings, p. 484), and an overall effort to promote the participation of women in science.

Important weaknesses were also identified in regards to gender issues within the IHP program. As project coordinators and partners, women and men were not equally successful. Braithwaite suggested that "a higher value is given to a seniority of the proposal coordinator or how well they are known and this would indirectly favor men given the sex imbalance at higher levels of scientific profession" (ICWES-XII, Complete Proceedings, p. 485). Another problem raised by Braithwaite's team was the disinterest in promoting gender equality beyond official documents and calls for proposals. Concrete plans of action were not required of project promoters. Most project proposals concerned with "themes where gender is known to be a significant variable" do not take gender into consideration. For instance, more than half of the proposals financed by the Key Action "Improving the Socio-Economic Knowledge Base" do not cite gender at all as a relevant issue. "Gender-blind" proposals were financed at 53% (first call) and 59% (second call, in 2000). This slight difference could be explained by the diverging presentation of gender in the two calls: in the first, a "crosscutting dimension" that should be included in the project design in all relevant aspects, and in the second, one example of possible research topics. Braithwaite's team found that when gender was cited in calls, it was not particularly well developed as a concept, and it mostly addressed "the special characteristics of women" and "gender differences and inequalities in specific research areas" (ICWES-XII, Complete Proceedings, p. 486).

Another element of interest was that the research proposals financed by the IHP program were quite vague on the human aspects of their topics. Terms like "employee," "citizen," "learner," and "public" were frequently used without further explanation of the categories and their specific characteristics. Similarly, studies of processes and structures generally neglect their human dimensions. A comparable weak treatment of gender issues could be expected.

While most Commission staff claimed to encourage and promote the participation of women in science, the study showed that in practice there was little actual understanding of the gender dimension. "No specific procedures or instruments exist for supporting the mainstreaming of gender in the management of the programme. Nor is the gender dimension integrated into the criteria and procedures for ex-ante evaluation and selection of proposals, nor into the requirements for project reporting" (ICWES-XII, Complete Proceedings, p. 486).

Braithwaite finished her report with a series of key recommendations:

- "Strengthening the policy framework on "gender and science," in particular to clarify and reinforce the gender mainstreaming approach in science and research.
- Making stronger reference to the importance of ensuring equality of participation of women and men in the guidance to proposals for all actions and calls.
- Requiring project proposers to provide information on how they intend, concretely, to ensure gender equality in participation.
- Establishing a more systematic and integrated system for encoding, collating and producing data on the participation of women and men.
- Improving information and data on gender issues in relation to mobility; science and technology, particularly concerning education and awareness-raising; and scientific discourse and epistemology.
- Raising the awareness and knowledge of staff and project promoters about community policy on gender equality and on gender mainstreaming as it applies to science, technology and research.
- Adapting instruments for managing the programme to take into account the gender dimension, in particular: ensuring that work programmes reflect community policy in their objectives and themes; including clear definition of key concepts and identification of the human populations concerned in the quality criteria for proposals; clearly mentioning equal opportunities and gender mainstreaming in all calls and guides to proposers; adapting the criteria and procedures for ex-ante evaluation and selection of proposals; and including gender aspects in projects and programme monitoring and reporting.
- Ensuring that all committees and expert groups are gender-balanced and contain an adequate level of expertise and experience in gender issues in the relevant domains.
- Clearly, locating responsibility for gender mainstreaming at "mainstream" levels among those responsible for managing the programme, and clarifying the roles and responsibilities of staff in relation to the implementation of Commission policy on women and science" (All taken from ICWES-XII, Complete Proceedings, p. 486).

Business Meeting

The main goals of the business meeting at ICWES-XII were to ask the delegates to approve the creation of the International Network of Women Engineers and Scientists (INWES)[9] and to discuss and approve modified guidelines on organizing the ICWES Conferences, a task that had not been completed at the Business Meeting of ICWES-XI in Japan. It was also critical that, if INWES was created, the assembly agree with a proposed list of Board Directors of INWES so that the new corporation could be established legally and develop its operations and management processes. Of course, the proposed directors were selected for their geographic representation and for their previous experience with ICWES and with major associations of women engineers and scientists.

Agenda: ICWES Delegates Meeting, Tuesday, July 30, 2002

Meeting Chair: Dr. Moyra McDill, ICWES-XII Convenor
Meeting Secretary: Ms. Alejandra Carmona

1. Introduction of the International Network for Women Engineers and Scientists (INWES)
   Motion 1.
   Moved by: Monique Frize, Conference Chair
   Seconded by: Claire Deschenes, Program Chair
   "That ICWES delegates support the formation of INWES."
2. "Adoption of ICWES Guidelines as the preliminary guidelines for INWES Conferences or, if Motion 1 fails, proceed to item III."
   Moved by: Monique Frize, Conference Chair,
   Seconded by: Claire Deschênes
   "That INWES adopt the ICWES Guidelines as the preliminary guidelines for the INWES International Conference."
3. Presentations of motions with respect to ICWES Guidelines resulting from the acceptance of Provisional Guidelines in February 2000.
4. Presentations of Notification(s) with respect to holding ICWES-XIII. Note: If Motion 1 passes, ICWES-XIII will be the international conference held by INWES.
5. Consideration of the need for a continuity committee and/or further discussion of INWES.

[9] Minutes of the following meeting may refer to propositions on the creation of the International Network of Women Engineers and Scientists (INWES) made the previous year in Merrickville (= INWES BM0), see Chap. 14.

ICWES 12 Delegates Meeting Minutes, July 30, 2002, 2:30 p.m., Ottawa

Meeting Chair: Moyra McDill, ICWES Convenor (Canada)
Recording Secretary: Alejandra Carmona (Canada)
Meeting Assistants: S. Oddy (Volunteer), S. McLoughlin (Volunteer)
Conference Chair: Monique Frize ICWES-XII

Delegates Present:

Australia – Sally Male (Women in Engineering Panel of the WA Division of the Institution of Engineers)
Canada – Dormer Ellis (ICWES ICC)
Czech Republic – Josee Iva Lekesova
Egypt – Prof. Azza Elleboudy
France – Marianne Rodot (Association française de Femmes Ingénieurs, AFFI)
Germany – A. Reinhard (VDI)
India – Suriya Thevar
Iran – Mahin Rahmani (Prof. of Physics Dept. Tehran University; ICC)
Ireland – Ita Richardson (Member of Women in Technology & Science (WITS))
Ivory Coast – Issié Yvonne Gueye (CIFISATS)
Japan – Sachiko Okuda (Chairperson, ICWES-XI Japan)
Kenya – Jennifer Nyambalia
Poland – Anna Szemik-Hojniak (U. Wroclaw)
Slovenia – Alenka Buria (Free-lance Journalist)
Switzerland – Anita Phaff (Swiss Association of Women Engineers, SVIN)
Taiwan – Li – Ling Tsai
Uganda – Wilbrod Birabwa (Solar Promotion & Women Eng. Tech. and Scientists)
UK – Jackie Longworth (WES, UK)
Ukraine – Olena Shubrarska
USA – Peggy Layne (SWE and AWiS)
Zambia – Agnes Mofya

Delegates Absent

Austria, Belgium, Congo
Lao PDR
Latvia
Lithuania
Malaysia
Peru
Philippines
Puerto Rico
Romania
Russia
Senegal
Sweden
Vietnam

Delegates Sending Proxy Votes
Bangladesh – Lutfor Rahman
Bermuda – Kimberley Stevenson
China – Li Jianying
Ghana – Peggy Oti-Boateng
Korea – Kwang Hwa Chung
Mauritius – Pritam Parmesseur
Nigeria –Ahbor Ighoroje
Pakistan – Durdana Habib
South Africa – Brenda Sudano

The agenda included a list of motions and was posted during ICWES for participants to learn about them. Handwritten (signed and dated) proxy votes were accepted.

Opening: The Convenor, Dr. Moyra McDill, welcomed the delegates and participants and explained that only delegates (1 per country) and previous ICWES conference chairs were permitted to vote. Each pre-selected delegate was given a colored card to facilitate voting. M. McDill then asked M. Frize (Canada, Conference Chair) to introduce INWES, as it was the first item on the agenda.

Monique Frize discussed the current limitations of the informal ICWES structure and the need for a more formal network. She discussed the characteristics of the network, such as financial details, a formal identity, name, and address; issues such as liability; and the initial steps taken to date, including suggesting the potential names of an Interim Board of Directors to help set up the organization.

1. Creation of INWES

Discussion Followed:

- **P. Layne (USA):** Felt better about INWES because of the insurance but was concerned about financial issues. She asked how all the responsibilities or committees are going to be divided and if the delegates need to make special considerations for the next conference ICWES-XIII.
- **M. Frize (Canada):** Replied that INWES will be incorporated and have limited liability; the directors and all the members will have the protection of an insurance policy. It is easier in Canada. It is important that the Chair of the committee for the ICWES conference be aware of this development. She may need to fundraise for insurance.
- **L. Mantel (USA):** Asked if INWES was going to have a status as an organization that can recognize donors so people who give donations can get taxes deductions?
- **M. Frize:** Replied that INWES will be a non-profit organization. This allows it certain privileges, e.g., providing statements of donation. That is going to be discussed with the lawyer.

- **S. Chung (Korea):** Proposed to further discuss the policies later in ICWES-XIII, so all members can understand every item and make a decision. It is better to have everything in hard copy.
- **M. Frize:** Replied that, first, it is necessary to create INWES. If it is created now, it will give more flexibility with fundraising; it will have an address and identity. These are helpful in raising funds. This meeting's main goal is to create INWES.
- **M. McDill**: Explained that we are here as delegates to an international conference. We are not part of INWES. INWES does not exist yet. She asked if the group supported, in principle, the formation of INWES.
- **I. Y. Gueye (Ivory Coast):** INWES will lead to answers on how to provide support to women, having its own identity, and fundraising. More countries want to organize ICWES if the organization is supported; countries such as Nigeria or Ghana could host a conference, and it would be easier for fundraising. She hoped INWES could support and help women from developing countries, and she supported the formation of INWES.
- **S. Till (Canada):** Was concerned about the credibility issue of having already chosen the Board of Directors. She wondered if it was possible to have an interim board that would include women from different countries.
- **M. Frize:** Replied that this would be discussed in item 1B.

Motion 1A: Moved by: Issié Yvonne Gueye (Ivory Coast); Seconded by: Gloria Bonder (Argentina): "That ICWES-XII delegates support the formation of the International Network of Women Engineers and Scientists (INWES)."

Votes For: 22 + 8 proxy; Votes Against: none; Abstentions: none; ***Motion 1A Passed.***

M. McDill congratulated the group for its unanimous support of the resolution and re-introduced M. Frize to discuss the Board of Directors.

M. Frize: Presented a list of proposed board directors, chosen for worldwide representation. The 13 directors will finish their work after the initial 12 months, although for continuity, some may serve 24 months. Then, the Members of INWES will elect up to 18 directors, and every organization that joins will have a vote. She pointed out that some companies and associations had already indicated their interest to join INWES, and some funding had been received from IBM and promised from UNESCO.

The proposed **Board of Directors** list was:

Sue Bind (UK)
Issié Yvonne Gueye (Ivory Coast)
Joanna Maduka (Nigeria)
Sophia Chung (Korea)
Marianne Rodot (France)
Anna Szemik-Hojniak (Poland)
Mahin Rahmani (Iran)
Gloria Bonder (Argentina)

Sachiko Okuda (Japan)
M. Suriya (India)
Gail Mattson (USA)
Claire Deschênes (Canada)
Monique Frize (Canada)

There was a discussion among a number of delegates and participants. J. Longworth (UK) supported the formation of the Interim Board. S. Okuda (Japan) was glad to hear about INWES's creation and about the possibility of fundraising. A. Elleboudy (Egypt) felt there was not enough representation from the Middle East. She wanted to represent the Arabic countries. A similar concern was voiced about representation from Europe.

M. McDill then suggested that rather than a formal motion, ICWES recommend to the Board that INWES consider the issue of Middle East representation and European representation.

Motion 1B: Moved by: Dr. Monique Frize Conference Chair; Seconded by: J. Longworth (UK);

"That ICWES delegates support the establishment of a preliminary Board of Directors as listed for a 16-month term."

Votes For: 23; Votes Against: 0; Abstentions: 1 (Slovenia); No proxy votes were included as this motion was not stated on the agenda. ***Motion 1B Passed.***

M. McDill returned the meeting to ICWES matters, reminding the group that ICWES could go on as it was. She reminded the group that the Provisional Guidelines (PG) for ICWES had been accepted by majority vote in February 2000. The guidelines were provisional because the Delegates Meeting at ICWES-XI had not been completed and some matters needed to be resolved. The first area of concern is to recognize that the ICWES Guidelines represent a wealth of knowledge and history that INWES could find useful. There were also three motions resulting from the acceptance of the guidelines. There was a period of overlapping discussion relating to the Area Representatives of ICWES and the Board of Directors of INWES.

Li-Ling Tsai (Taiwan) commented that there were many women in Taiwan willing to participate in this organization (INWES). She made a suggestion for an amendment to the definition of director on page 4 (INWES documents). She noted that the criteria differ for each country and organization. She also proposed changing 'INWES: a member consisting of organizations' for 'INWES: a member consisting of countries'; it would be more inclusive and would give more flexibility to the operation of the network.

M. Frize: Responded that the Board will consider that suggestion very carefully and that it will be part of the work of the initial board. Some things have to be changed.

M. McDill: Responded that these will be suggestions to the INWES board. Preliminary conference guidelines will be largely the responsibility of the ICWES-XIII organizers.

Azza Elleboudy (Egypt) commented that she did not like the use of the word Chairman in the ICWES guideline.

M. McDill replied that there was a motion to follow on the agenda on that matter.

W. Birabwa (Uganda) asked if every country was going to be represented in INWES and how it was going to be balanced.

M. Frize: Responded that every organization will have a vote but not necessarily have a director. There will not be 10 directors from a country, but there will be 10 votes for 10 associations that have joined. INWES will be meeting the next day, and all suggestions and questions will be discussed.

Motion 2: Moved by: Dr. Monique Frize (Canada); Seconded by: Dormer Ellis (Canada): "That INWES adopt the ICWES Guidelines as the Preliminary INWES Conference guidelines."

Votes For: 20; Votes Against: 0; Abstentions: 0; ***Motion 2 Passed***.

2. Motions resulting from the acceptance of the Provisional Guidelines

M. McDill stated that three issues were outstanding from the acceptance of the provisional guidelines by delegates of ICWES-XI:

Motion PG1: Moved by: Dormer Ellis (Canada, ICC, XI); Seconded by: Susan Bird (UK, ICC, XI); "That Elizabeth Laverick's contribution to ICWES should accompany the History portion."

Votes For: 22 (unanimous) + 8 proxy votes; ***Motion PG1 Passed***.

I. Gueye (Ivory Coast) requested that the History section include a note on the major modification of the Guidelines in 1987 (Ivory Coast ICWES), in which the Convenor was to be associated with the country holding ICWES.

M. McDill asked for a show of support for this. It was the will of the delegates to add this change to the History section.

Motion PG2: Moved by**:** Filipa B. Sotto (Philippines, ICWES-XI); Seconded by: Vera Verica Lukic (Australia, ICWES-XI): "That the word Chairman be replaced in the guidelines with the word Chairperson."

Discussion followed, with a member of the delegation speaking to the use of the world Chairwoman as being appropriate. Others preferred the word Chair.

Amendment to PG1: Moved by: Ita Richardson (Ireland); Seconded by: Peggy Layne (USA); "That the motion PG1 be amended to use Chairwoman rather than Chair."

Votes For: 9; Votes Against: 9; Abstentions: 0; No Proxy Votes counted.

M. Frize, the conference Chair, then exercised her vote against the amendment, stating that the word Chair is more inclusive and that men are invited to attend ICWES.

Amendment to PG1 Failed.

Motion PG2B: Moved by: M. McDill; Seconded by: (secretary could not hear); "Considered that the motion PG1 be amended to use the word Chair rather than Chairman."

Votes For: 15: Votes Against: 4; Abstentions: 3; No Proxy Votes counted; ***Amended Motion PG2B Passed***.

Motion PG3 was then introduced. M. McDill noted that the concerns of PG3 were addressed in the INWES documents.

Moved by: W. Birabwa (Uganda, ICWES-XI); "That the object of the guidelines to be modified include: to bring women engineers and scientists together; to exchange views to promote technology and find solutions to problems; to encourage more women to become engineers and scientists.

There was no seconder; ***Motion PG3 failed***.

3. Presentation of Notifications and Invitations to ICWES-XIII

M. McDill mentioned that Switzerland and Australia had been interested in holding ICWES-XIII, but that the only formal invitation submitted was from South Korea. She then read the principal portion of the invitation to attend ICWES-XIII in September 2005 in Seoul, South Korea, with the conference title to be "ICWES-XIII: Women Engineers and Scientists: Main force to reshape the future world."

South Korea, located between China and Japan, has four distinguished seasons, and September is the most beautiful month of the year. The area is 99,373 km^2, and the population is 46.1 million, with more than 10 million in Seoul. Seoul, being the capital city of Korea in present and old times, has many historical places as well as theaters, parks, sports facilities, and museums. It has 27 large-scale department stores and many traditional market places where you can enjoy shopping. There are 43 universities and more than 10 buildings with sophisticated conference rooms that are equipped with modern conveniences, in which many international conferences are taking place. The public transportation in Seoul is very convenient, with subways, and the international airport is only 52 km away. There are many hotels, with rates ranging from $300 to $30. Korea had successfully hosted the 2002 FIFA World Cup, the 1988 Olympic Games, and many other international events.

The ICWES-XIII theme aims at studying the impacts of science and engineering on society in the twenty-one century and finding out the roles and responsibilities of women engineers and scientists in the reshaping of the future society. We need to develop new leadership among women engineers and scientists and develop joint programs with women in social, educational, economic, and other studies. The conference will have special features like mentorship between the established and upcoming women scientists, supporting and encouraging the participation of women scientists in developing countries, a joint gathering of women CEOs, CTOs, and other scientists, and the creation of international contact points and databases to expedite the exchange of women scientists, information, and job offers between developed and less developed countries. The conference will consist of invited speakers, oral and poster sessions, workshops, discussions, panels, and laboratory and industry tours.

Motion PG3: Moved by: Dormer Ellis (Canada); Seconded by: Peggy Layne (USA): "The motion for accepting the invitation to ICWES-XIII in South Korea."

Votes For: Unanimous acceptance + 8 proxy votes. ***Motion PG3 passed***.

4. The ICWES Continuity Committee:

M. McDill noted that the ICWES Guidelines refer to the International Continuity Committee (area representatives) and that Motion 2 has now made the ICWES Guidelines the preliminary INWES Conference Guidelines.

Motion 3: Moved by: P. Layne (USA); Seconded by: J. Longworth (UK): "That the INWES Board of Directors replace the ICWES Continuity Committee."

Votes For: Unanimous consent: ***Motion 3 passed***.

5. Other Business:

(i) A number of delegates expressed concern with respect to visas. It is suggested that for participants from the developing world, it is important to have a personalized and sealed letter copied to the embassy and to the participant. It was also suggested that we ask the embassies what they would like to see in the letter of invitation.

M. McDill: Commented that for ICWES-XII, embassies were called, but they did not discuss files over the phone. In all, more than 300 personalized letters were written. Few participants in the end could not get their visas. Letters faxed to the embassies between 24 hours before the person went for their visas were most likely to be accepted, if between 48 hours before the person went, the visas were rejected.

(ii) M. McDill reminded the group that the ICWES delegates wished to formally suggest to INWES that the Board of Directors carefully consider appropriate representation for the Middle East and Europe. The delegates *accepted* this recommendation to INWES with a show of colored voting cards.

(iii) J. Longworth (UK) asked how and when INWES will sign up members.

M. Frize: replied that a month or 6 weeks are required to finalize the bylaws and that important fundamental things like memberships have to be clarified. In the meantime, forms for applying will be developed (this information is not in the bylaws). In the bylaws, the lawyers have to be paid every time that they are changed. Also, the policy manual is going to be developed, and the Board of Directors and its members can apply changes. By the end of the year 2002, hopefully an election with memberships can be organized for the next ICWES, with directors proposed by new members.

(iv) S. Male (Australia) suggested that the documents should be checked by lawyers in every member country. It was also suggested by a participant that electronic communication documents were needed.

M. Frize replied there would be a non-public website, and each of the delegates would have the address. The documentation cannot be given to all the organizations but will be mailed electronically.

(v) Thanks were expressed to M. Frize and M. McDill for holding the meeting. M. McDill moved (Motion 4) that the meeting be closed; it was seconded by all delegates. Motion 4 passed with unanimous consent.

Testimonials

Anne M. Perusek and Peggy Layne, "ICWES-XII: 'Women in a Knowledge-Based Society"

"On July 27, 2002, I joined over 600 women from around the world in Ottawa, Canada, for the 12th International Conference of Women Engineers and Scientists (ICWES12). The five-day program included sessions on technical topics, from climate change to globalization, and non-technical topics such as women's studies, social studies of science and technology, outreach to young women about science and engineering, retention of women in engineering schools, and encouraging women to succeed in their careers. An opening reception, banquet, and closing lunch provided opportunities to catch up with old friends and to meet women from more than 40 different countries."

"Ninety-eight delegates from the United States participated in the conference, presenting 37 posters, papers, panels, or plenary lectures. SWE members served on the program committee (Kathleen Harer), moderated panel discussions (Gloria Montano, Betty Preece, Roberta Gleiter, Judy Forbes, Marilyn Reeder, and myself), and presented papers (Judy Vance, Patricia Eng, Mary McCarthy, Marie Laplante) in addition to networking with colleagues from across the globe. Gail Mattson served on the organizing committee for the new International Network of Women Engineers and Scientists, and was elected to the interim board of directors of this new coalition that will serve as the permanent secretariat for the ICWES conferences and facilitate interaction between women engineers and scientists in the years when no conference is held."

"Dr. Sue Rosser, Dean of the Ivan Allen College at Georgia Tech and author of many books on women in science, gave a plenary lecture on feminist theories of women, science, and technology. Her remarks discussed the various perspectives of feminist thought, from liberal feminism through socialist, essentialist, psychoanalytic, and post-modern feminism. Each of these schools of thought approaches the relationship between gender and science from a different set of assumptions about the position of women in [the] world, and whether or not you agree with any or all of them, they can provoke thoughtful discussions of how the professions of science and engineering operate and why women continue to be underrepresented in these fields."

"One of the more unusual papers at the conference told the story of the Kate Gleason College of Engineering at the Rochester Institute of Technology (RIT). Kate Gleason was the first woman to enter Cornell University's engineering program in 1884. She did not graduate, but left college to work in the family business, Gleason Works, a manufacturer of gear cutting equipment, and later became the first woman member of the American Society of Mechanical Engineers (ASME). The Gleason family has a long relationship with the Rochester Institute of Technology and its predecessor institutions, so when the school embarked on a major fund-raising campaign in the 1990s, they naturally turned to the Gleasons for a major gift. The family readily agreed to donate $10 million, through their existing foundation, and the RIT College of Engineering became the first and only school of its kind in the United States named after a woman. The funds are being used to renovate the main engineering building at RIT, to endow a named professorship for a woman engineer, to support scholarships for women engineering students, and to support faculty research as well as ongoing programs for women and minority students. A statue of Kate Gleason now stands in the entry of the engineering building."

"Representatives of ASME, the Institute of Electrical and Electronic Engineers (IEEE), the American Society of Civil Engineers (ASCE), and the American Institute of Chemical Engineers (AIChE) discussed their organizations' initiatives to increase the participation of women and minorities in the profession and in their technical societies. ASME's Board on Minorities and Women has been in existence for more than ten years, while ASCE's Committee on Diversity and Women in Civil Engineering was established two years ago. All four organizations have international membership, and have established committees that are working to get more women into leadership roles."

"The Association for Women in Science (AWIS), the Women in Engineering Programs and Advocates Network (WEPAN), and the Society of Women Engineers (SWE) are all dedicated to supporting the professional fulfillment of women engineers and scientists in the United States. These three organizations participated in a panel discussion about ways they are working together to inform young women of career opportunities and to support women currently pursuing science and engineering careers. AWIS also sponsored a reception at ICWES-XII to introduce conference attendees to leaders of the three organizations, as well as representatives from the American Association for the Advancement of Science and the National Science Foundation."

"Speakers from the New Mexico Institute of Technology, Texas A&M University, the University of Michigan, Rowan University, and Western Kentucky University presented papers about retaining women engineering students through curriculum changes, residential programs, support networks, role models, and mentoring. Research at these and other universities documents the success of programs aimed at increasing the retention of female undergraduates in science and engineering. Many of the schools work closely with SWE student sections on campus to support and complement the administration's efforts. These studies document the continued need for support networks like SWE."

"In addition to the formal conference activities, we had plenty of time to enjoy Ottawa's multicultural restaurant scene, view the changing of the guard at the Canadian Parliament, shop for maple leaf souvenirs, practice our French, and take pictures of the monument to the Famous Five, who led the fight for equal rights for women in Canada. This was the second time that I have traveled outside the United States to attend an ICWES conference. It is easy to be so focused on our own lives, activities, organizations, and challenges that we don't take the time to think about the challenges faced by women in other countries, some similar to our own and others different. The wide variety of presentations and participants at ICWES-XII, all dealing with different aspects of the challenges still faced by women in science and engineering throughout the world, energized my own commitment to SWE's mission of encouraging women to reach their full potential in careers as engineers and leaders" (Perusek and Layne 2002, pp. 88–90).

Jacky Crook and Dianne Patterson (WES Members), "ICWES-XII"

"This was Jacky's first visit to ICWES but Dianne had been to the previous two conferences and reports that she is now 'hooked' and already looking forward to the next one... The following WES members received bursaries to attend: Sue Bird, Alenka Burja, Jackie

Carpenter, Jacky Crook, Urszula Dzvvonowska, Abi Jacobs, Jackie Longworth, Jane Magill, Dianne Patterson, Karla Solis, Brenda Sudano, Pam Wain and Grazyna Whapshott."

"Professor Monique Frize who organised this conference, has done an outstanding job in making this a prestigious event in Canada, and for women everywhere. We took over the Ottawa Congress Centre for 5 days where there was plenty of space for the meetings and banquets organised for the 550 plus delegates; Monique's reason for coming to ICWES was quoted as: "We are here to get energised" — and energised we were. This is a great opportunity to meet amazing women from all over the world who are inspired, and inspirational, by their passion for engineering and science. There are women just starting their studies, taking a family break, working at all levels in universities and industry, and who have achieved status as professors, coordinators of government bodies, and company directors."

"Monique opened the conference on the first evening. The keynote speech was given by Kathleen Sendall, senior VP of Petro-Canada who talked about the Famous Five. Not the Enid Blyton ones, but five women suffragettes of Canada who challenged the ruling that 'women are not persons' and therefore could not vote. They took the challenge to the highest court in the Empire — the House of Lords — and won."

"Eight women who were at the first conference way back in 1964 were presented with certificates. Then the entertainment started. Inuit throat callers were the first act, followed by a local band—French Canadian folk- rock fusion. A Celtic Dance troupe—although they had a lot more gusto than River Dance—finished off the evening."

"The next three and a half days were crammed full. There was an early start (8.30 most days but an even earlier start on Monday with a working breakfast at 730) with working lunches and evening functions. The conference sessions covered a huge range of topics, including world issues such as climate change. ethics, globalisation; education including outreach, retaining women, gender analysis; experiences from countries such as Poland, Africa, Egypt, Korea, technical papers on engineering, physics, biology, chemistry, IT and personal experiences on portfolio careers, family & career, gender blocks."

"There were 70 separate sessions, each with 3–5 presentations. The problem was deciding which session to go to. We usually had to choose one session out of a group of 7 – it was too hard, we would have liked to go to them all. Below are some highlights from the sessions that we attended:"

"**Keepers of the Circle of Knowledge**: This was a fascinating presentation by five remarkable aboriginal women of Canada. The tone was set when Wanda Wuttunee, with her daughter Drew (10), opened the session by singing a Cherokee song to 'Welcome the Sun.' While Western women still know about child bearing, the aboriginal women remain in touch with the earth and with the true female nature. We have many debates about choosing between career and children, or about 'having it all' — but these women know that what we really need to keep is our spirit. Lillian Dyck's analogy of modern-day engineering with the traditional Medicine Wheel was particularly striking."

"**Portfolio Careers in Engineering:** Melody Kratsios suggested that her upbringing with a family of brothers had a big influence on her career in the 'man's world' of engineering. This also applied to a large number of people in the room. Melody identified 5 skills that women need to work on: Ability to tolerate ambiguity and to avoid perfectionism. Learn to network more. Laugh at themselves, don't take everything so seriously. Go out on a limb—do something scary. Tenacity — 'No' just means 'Not just yet' — keep trying again. There was discussion on the best time to have children. The consensus was that you should do what is right for you, and that no time is good or bad. Having a child will always put a delay into a career, but this need not limit you in achieving what you want. The importance of networking was emphasized. This is something that women are (surprisingly) not very good at. Jackie Carpenter probably hit the nail on the head when she suggested that women tend to feel guilty about using friendships for business opportunities – whereas men do it all the time. This is one reason why conferences like ICWES are so important."

"**Peggy Oti-Boateng** gave a very interesting talk about the use of the jatropha plant in Mali and Ghana which is empowering the women to improve the economy in their villages. This plant is currently considered worthless because it cannot be used for food but the seeds can be converted into a good quality oil for running engines, for insect repellent, and for soap; the foliage can be used for fertiliser. Peggy wants to see more commercialisation of the product in order to improve the production effort (eg: more effective methods for seed collection). However, there is great concern about the 'men taking over' when they realise the full potential of the product."

"As we were given the conference papers on a CD, we are hoping to get these put on the WES website. One of the great joys was seeing so many women (over 550 from 40 countries) engineers and scientists from all over the world getting on like a house on fire. and hearing others' experiences."

"Our accommodation was about 30 minutes amble (it was steaming hot even early in the morning) from the Congress Centre, so we were able to vary our route back and forth and see some of ordinary Ottawa, and our hosts were local so we actually got the feeling of being in Canada rather than at a generic conference venue."

"**Jackie Longworth adds**: I focused on sessions to do with women's issues, organisational culture, and different countries' experiences in promoting engineering and science for women. The key encouragement for me was that WES is not alone in beginning to think it is time we shifted our tactics from seeking to change women to be more like men to seeking to change organizations to be more women friendly!"

"The very opening sessions on Sunday morning examined why women leave SET partway through their careers to a much greater extent than men do. The conclusion was that cultural constraints were a much greater influence than family constraints. Of particular interest was a Hypatia Project working with employees at the Bedford Institute of Oceanography (Canada) to identify issues and implement strategies affecting workplace culture and environment. The key accomplishment to date is 'a demonstrated willingness by staff and managers to work together to create a work-place that respects and supports individual difference.'"

"**Jacky recalls the bonhomie and hard work**: 'Whilst queuing up to register, and wearing a dolphin t-shirt, I got chatting to Suzanne. who spends a lot of time off the coast of Vancouver Island studying dolphins. She thought I might be a dolphin person too. Although I like living in Dunbar, I'm not sure it compares to the Pacific Rim Park! In one session I was chatting to the woman sitting next to me, and it turned out that she worked for the Open University too, but in the science faculty in Milton Keynes (I'm a regional academic in Scotland for the Technology Faculty). One of the more conference-experienced said she was taking a week's leave when she returned home to go through the papers. Now I understand. Three weeks after my return home when I sat down to write this, I looked at the programme as an aide-memoire (well, Canada is bilingual'). The rest of the papers are sitting in a heap waiting. The WES bursary covered my flight and most of the registration fee. I found the experience well worth the upheaval to my body clock, and would recommend others to apply for the bursary to get to the next ICWES. One really good point that came out of this conference is that there is going to be a full-time web page set up for ICWES. Watch our web page and journal for the URL.'"

"**Dianne's reflections**: 'It was great to see so many people at this conference (including several men). However, for me, some of the magic of ICWES was lost. When you don't see the same person more than once in three days, it is very hard to network and to build up new relationships. One person that I knew from previous conferences was registering at the same time as me, but then I didn't see her again for two days! Seven parallel sessions for eight of the presentation periods was far too many and involved many difficult choices. It must have been hard to accommodate everybody's papers, so I don't know what the solution is. A longer conference would have been nice—I was only just getting over the jetlag and managing to see some people more than once, then it was over—but this would increase the costs for everybody. I paid for myself at the last two ICWES as at that time I was working as a systems engineer in a 'well' paid job in industry. As I am now unemployed I am very grateful for the bursary from WES which enabled me to attend this conference in Canada. Women from academic institutions usually predominate at these conferences (probably because they can get funding and time off). It would be really nice to see more women there from industry and other non-educational establishments. Talk to your company, they may be happy to fund you or just to allow time off. Or talk to WES, they have a bursary fund to help members to attend. The next ICWES will be in Korea in 2005. Korea is a unique country just as ICWES is a unique experience—so put KOREA into your diary for 2005.'"

"**Notable quotes from Dianne's notebook:** "If there were equal numbers of women, as there are currently men, in engineering and science, then there would be no worker shortage. Currently there are roughly equal numbers in education but as the seniority of the positions increase, so does the disparity between the numbers of men and women" (Ruta Sevo, USA)."

"Oceans are like women scientists: they are forgotten about and they act as moderators. (Brenda Topliss, Canada)."

"Our society's view: we must put ethics to one side for now—first we must make money, then we can improve the world. However, the products of wealth are pollution and ill health. The faster we run, the poorer we get. (Willem Vanderburg, Canada)."

"Good facts are essential for good ethics—and good ethics are essential for good law. Humans bond and build their cultures and communities through their shared stories. There are two standard world views: the pure science view and the pure mystery view. But we should be seeing the world with a science spirit view (Margaret Somerville, Canada)."

"On seeing me as one of the first women to do engineering in Uganda, the tutors' comments were, "Has engineering suddenly become easy, or what?' (Emily Clare, Uganda)" (Crook and Patterson 2002, pp. 8–9)."

"INWES—Looking Forward to Korea 2005: Sue Bird Reports on Developments"

"Most WES UK members have heard of ICWES, the International Conference of Women Engineers and Scientists. This is staged at 3-year intervals, and has been going since 1964. The UK hosted ICWES in 1967 and 1991, and WES has been involved throughout its life. Those who have been to ICWES will know what a terrific event it is, and how friendships are made with people all over the world."

"ICWES is not an organisation. Events are organised by those in the host country. Recently the International Network of Women Engineers and Scientists (INWES) was created to continue the ICWES enthusiasm between conferences, and to help obtain funding and insurance for the next conference."

"The guiding light of this organisation is Monique Frize of Carleton University, Ottawa in Canada, where the last ICWES took place. She raised money from UNESCO and gathered together interested people from around the world. I attended this first meeting with Nicole Rockliff and Pam Wain was also involved."

"At ICWES-XII, an interim Board for INWES was set up with 14 people from all areas, both voluntary organizations and companies. I started on the Board as chair of the Communications Committee but my 'portfolio' has changed from communications to conference, and I am now ICWES convenor. Since then, we have held Board meetings in Paris and Daejeon, South Korea."

"WES was one of the first members of INWES – very fitting, I thought, for an organisation which has been so very involved with ICWES. We are starting slowly but things are happening and everything is starting to come together. You can find out more about INWES on www.inwes.org."

"The next ICWES will be held in South Korea in August 2005. The INWES board attended a women's science conference held in Daejeon in November. One of our ideas is that INWES will also support smaller regional conferences in the years between the three-yearly ICWES conferences. There is already a great deal of interest from people wishing to organize these conferences."

"The Daejeon conference was very successful with over 200 women delegates, mainly from the Far East but also from Asia, the Middle East and Africa. This success encouraged us all to believe that the next ICWES will be a very good one. We were pleased to meet the person who will be leading the Koreans organising ICWES, and will now be able to give support in a way which has not been possible before."

"I also visited Seoul for a couple of days and met contacts at the British Embassy and the British Council. These will be useful when we go to ICWES and I will keep you posted on any help they can provide."

"Korea was fascinating. In Daejeon there were few westerners, and very few English speakers, but in Seoul if you stood looking puzzled someone would always ask in English if they could help you. (Do you get that in London?) The people are all very pleasant and helpful. Prices are cheap – metro at about 35p a journey and very cheap taxis. Meals and hotels were reasonable – and the food was wonderful. So I hope you will think about attending ICWES-XIII, it will be a fascinating place to visit" (Bird 2004a, p. 2).

"INWES: Sue Bird Reports on the Latest Developments"

"A Board meeting of INWES was held at a conference on leadership organised by the African Women in Science and Engineering (AWSE) in Nairobi, Kenya in August. Once again, we had women board members from all over the world in attendance, and we worked very intensely for 5 days. INWES is growing slowly but surely, and as a result of this conference, we now have many individual African members. We were made very welcome by the organisers and the delegates, and we really had a wonderful (if tiring) time both professionally and socially. I did a joint presentation with Gail Mattson of the Society of Women

Engineers (SWE) from the US, on how to set up a new organisation for women in science and engineering. These suggestions were made from our experience of working with WES and SWE (Gail is a former SWE President) and also of being involved with INWES from the start."

"The session seemed to go well, and we learned a lot from the delegates. I managed to get a day off to go on a trip to a game reserve and see the Rift Valley, which was terrific, but really it needs much more time to do it justice. After the board meeting we have lots of work to do and ideas to sort out. The main issue we now have to address is fundraising. We need funds, but it is difficult to attract funds until we are doing something – the usual problem. The good news is that we have recently received some promises of funding so this will give us a good start" (Bird 2004b, p. 2).

Part II
INWES in ACTION!

Chapter 14
Merrickville 2001, A First Meeting

Meeting to Discuss the Creation of a Transnational Organization for Women in STEM Fields, Held in Merrickville, Ontario, Canada, May 26–29, 2001

The hope in organizing this global meeting was to examine how ICWES could be solidified by creating a corporation with a fixed address, limited liability, and an expanded mandate—that is, not just running conferences but being involved in several other activities that would help women in STEM fields around the world. Another goal was to respond to UNESCO's call for such an organization.

At the UNESCO World Conference on Science held in Budapest in 1999, where Claire Deschênes was one of the Canadian delegates, a resolution (Paragraph 90, in "Science for the Twenty-First Century") was passed encouraging the development of a "global network" of women engineers and scientists. After, through the support and encouragement of the Canadian Commission for UNESCO and a successful application for a grant from UNESCO Headquarters in Paris, Monique Frize, Claire Deschênes, and Moyra McDill organized a meeting in Merrickville (near Ottawa) on May 26–29, 2001, to examine ways to create such an international network. It was obvious early in the development process that a good approach would be to build this network on the already successful ICWES network; the International Continuing Committee (ICC) had successfully overseen the organization of 11 conferences since ICWES-I in 1964. The ICC was a committee of between 9 and 11 area representatives who met at each conference (usually every 3 years) to decide where the next conference would be and to help and support the organizers of the next event to ensure its success. Since the ICC had been an informal but global network of women in science and engineering for nearly 40 years, it was natural to build on this successful base. It was also timely; some of the ICC leaders were now deceased. It was important, at this time, to create a formal structure that would

M. Frize et al., *Women's Contribution to Science and Technology through ICWES Conferences*, Women in Engineering and Science,
https://doi.org/10.1007/978-3-031-33800-7_14

respond to the wishes expressed so long ago at ICWES-II during the business meeting and at several other meetings later. The Executive Committee of ICWES-II felt "this meeting should plant an acorn and let an oak tree grow" (ICWES-II, Complete Proceedings, p. 40). This image may well represent a wish that ICWES would one day be supported by International Network of Women Engineers and Scientists (INWES).

The connection with the World Science Conference resolution (Paragraph 90) happened because Claire Deschênes, Professor of Mechanical Engineering at Laval University and Natural Sciences and Engineering Research Council (NSERC) Chair for women in science and engineering in Quebec), was asked by the Canadian Commission for UNESCO to be a member of the scientific committee for the "European Regional Conference – Women in Science – Quality and Equality-Conditions for Sustainable Human Development," which was held in Bled, Slovenia (November 5–7, 1998). Claire made a presentation on the NSERC initiative on the creation of the chairs on Women in Science and Engineering. Since she had attended this pre-conference, Claire was asked to accompany the Canadian delegation to the World Science Conference organized by UNESCO/ICSU in Budapest (June 26–July 1, 1999). When Claire returned to Canada, she shared the exciting news about the UNESCO resolution with Monique Frize, and they decided to move forward with a plan to create an international network of women in STEM fields. At that time, Dormer Ellis, who had been a Canadian delegate at all but the first ICWES conferences and a continuous member of the ICC since 1967, suggested to Monique that the time was right to hold the next ICWES in Canada. Monique readily agreed and put together a committee in Ottawa to discuss the development of a proposal (a bid) to be submitted at ICWES-XI in Japan. The bid from Canada was accepted unanimously at the Business Meeting of ICWES-XI. Issié Gueye, from Ivory Coast, expressed the hope that there would eventually be a formal organization to ensure the continuity of the ICWES series. Monique responded that there was already a plan to examine how to create a formal organization, and she will have a proposal to create a network of women in these fields ready for the business meeting at ICWES-XII.

Upon Monique Frize's return to Canada from Japan in August 1999, she met with Claire Deschênes and Moyra McDill (Mechanical and Aerospace Engineering Professor at Carleton University) and a staff person (Gisèle Trubey) from the Canadian Commission for UNESCO to discuss an application for a grant from the UNESCO Headquarters in Paris, to be submitted through our contact, René Clair, who had met Claire in Budapest. A grant of $20,000 US from Paris and $4,000 CAN from the Canadian Commission were received in the fall of 2000. Moyra McDill, Claire Deschênes, and Monique Frize began to organize an international meeting to discuss the creation of the international network and proceeded to invite 20 women from around the world, several of whom had organized an ICWES conference in previous years. There were also representatives from major organizations like SWE, WES (UK), and the Japanese Society of Women Scientists. Invitations covered North America, Latin America, Asia, Africa, and Europe. Eighteen women confirmed their attendance. Two were not able to come at this time.

Why Merrickville? The town of Merrickville was founded in 1794 and has a history marked by mills and manufacturing. Now it is mostly known as a tourist destination with a variety of artist studios and shops. There are locks on the Rideau Canal where boaters can secure their crafts for some shore time. "Sam Jakes Inn, which opened in 1991, was one of the most popular spots in Eastern Ontario for weddings, conferences, and weekend getaways. Overlooking the Rideau Canal, it was renowned for regional cuisine and featured 33 guest rooms with period furnishings and a spa. It also had a meeting room large enough for 25 people. The stone structure was built in 1861 as the residence of Sam Jakes, a teacher, postmaster, and department store owner who served as Merrickville's second reeve" (Cook 2013).

This location was chosen to provide a pleasant environment with good meals, a sufficient number of comfortable rooms for the group, and a meeting room. Monique Frize visited the inn prior to booking and confirmed the dates early to ensure accommodation for everyone. The UNESCO grant could cover the accommodation for everyone and the travel expenses for the women who did not have an organization covering their travel expenses. It was very fortunate that several of the women did not need their airfare to be covered by the event budget. Another important aspect of the cost was the expense of a consultant to facilitate the three-day meetings. Engineer Elza Seregelyi (from Ottawa) did an amazing job of keeping everyone involved and on topic. She had a bell to call everyone back in after breaks. The meeting held plenary sessions to ensure everyone had the same opening information and small group discussions to develop the plans.

Participants who attended the meeting:

Renée Clair, UNESCO, Paris

Gisèle Trubey, Canadian Commission for UNESCO

Mitsuko Kazuno, Organization of ICWES-XI, president of Japanese Society of Women Scientists

Sue Bird, Organization of ICWES-IX, past president of Women Engineering Society, UK

Nicole Rockliff, WES (UK) president

Suzanne Brainard, Women in Engineering Proactive Network (WEPAN), USA

Gail Mattson, President Society of Women Engineers (SWE), USA

Kathleen Harer, past president of SWE, attended several ICWES conferences

Catherine Didion, Executive Director, Association of Women in Science (AWIS) USA

Marianne Rodot, vice-president, AssocFrance des Femmes Ingénieurs (AFFI), now FI, France

Efstratia Zafeiriou, German Association Women Engineers (DIB) and Greek Association of Women Engineers

Aba Andam, UNESCO Chair, Women in Science, Ghana

Gloria Bonder, UNESCO Chair, Women in Science, Argentina

Hiromi Matsui, president, Canadian Coalition Women in Engineering, Science, and Technology (CCWEST), now CCWESTT (and Trades)

Claire Deschênes, Program Chair, ICWES-XII, Professor Engineering, Laval University and NSERC/Alcan Chair for Women in Science and Engineering, Quebec

Moyra McDill, Convenor, ICWES-XII, Professor Engineering, Carleton University

Colleen Ennett, PhD Candidate, Systems and Computer Engineering, Carleton University

Monique Frize, Chair ICWES-XII, Professor Engineering, Carleton University and University of Ottawa, NSERC/Nortel Chair for Women in Science and Engineering, Ontario

Presentations at the plenary sessions: [Minutes of the meeting taken by Colleen Ennett]

The first talk was given by Renée Clair, UNESCO; she was directly connected to the goal of this meeting through the resolution (Paragraph 90) of the World Science Conference. Renée Clair, technical advisor on science for the National Commission of UNESCO in France, was instrumental in helping our group obtain a grant to organize this founding meeting. She spoke about the six international fora on women in science that she organized with Gloria Bonder (a UNESCO Chair in Argentina). Her main point was about the need to respond to Paragraph 90 of the World Science Conference Report, that is, to create an international network of women in engineering and science. She provided a copy of a booklet that was a summary of the six regional fora.

Renée explained how UNIFEM and other non-governmental organizations (NGOs) helped to develop the resolution concerning women at the World Science Conference organized by UNESCO on the creation of a network for women in science (Paragraph 90). The resolution was to promote education, recruitment and retention, research, collecting gender-desegregated data, and especially to creating international network of women in science and engineering. She mentioned that UNESCO is keen to encourage groups of women to start organizations of women in science and engineering, beginning with developing countries, and create an international network of solidarity, but not imperialist or monolithic; she insisted that it should be open-minded, flexible, and become a link; it should not be oppressive; these networks need a dynamic collective link to progress.

The next activity at this historical meeting was for everyone to speak about their hopes and fears about the future network. These points were recorded and listed below. They are included here as this information may be useful for anyone planning to create a new organization.

Hopes expressed were: To develop an international federation; to identify short-term goals; to focus on actions that each country can do and implement them; how to communicate between countries, remain practical, and focus on just a few actions; need a permanent structure to help networking; help organizers of ICWES meetings to save time and money; end up with an umbrella organization so we know who is around in the world and can communicate with them; have the spirit of ICWES but with a better structure; concrete ideas because we know this is an important subject;

a structure with equilibrium between flexibility and structure, global and local, specificity in women in science and technology with links to other initiatives related to women's issues, links to groups in other subjects that enrich results, bridges with other initiatives; ideas of structure from WEPAN's website, think of ways to penetrate the infrastructure of the all-male version; need an informal umbrella organization without money, but need money to have power; find a balance; build flexible and original ideas, respect the culture, language, and originality of each country, and be realistic about the culture of the region; focus on what we want without worrying about structure, networking, or sharing information—structure will come naturally; need something inclusive for everyone; effectively use our time; create something powerful for women globally; agree on something—what we want to do and how we want to do it; identify permanent "headquarters" of ICWES; find a formula that will work around the world where UNESCO exists; UNESCO hosts and mentors this group; find a common ground to build something positive; end up with an organization that can solve the problem; need agenda; this organization is a "tree" with "branches" that reach for the sky and "roots" to the individual cultures, religions, and languages; plant a seed for an idea for ICWES-XII, regardless of whether we leave this conference with something concrete; create momentum to continue the ideas that are developed over the next few days; look at the University Women's Organization as a model because they do not have much funding either (similar to us), need an international network, and need to find a way to incorporate the conference to reduce the liability of the organizers; identify who the representatives around the world are, solidify our international committees, create continuity and flexibility, "share," too few people are able to attend ICWES conferences, need a database of international women, newsletters, websites to connect, stability; build great conferences using structure, make it work without undue burden—time, money, etc., need "profile"/recognition, identify short—and long-term objectives.

Fears expressed were: Should not be a burden on UNESCO; no expectations for financial support from UNESCO; the structure created for structure's sake rather than for our needs; should be resilient and flexible; the structure may be difficult to create; a federation is big; do not just have an organization for the sake of having an organization; concerned that working on the suggested structure could result in nit-picking without any concise results; ending with nothing concrete or with something very abstract; be an organizational burden; no consensus on where to go; failure to criticize the draft model; begin a club of western women scientists and engineers; limit ourselves because we worry about money and because it's never been done before; do not spend a lot of time on structure; don't create a bureaucracy with lots of red tape; speaking English too fast for non-Anglophones; not able to convince UNESCO of its importance; if you talk about negative, then the result will be negative; difficulty with ICWES is the legal "liability"; liability of the ICWES conference is the responsibility of conference organizers. The floating committee for ICWES has not been organized with succession planning in mind, currently has weak links.

Our team (Frize, Deschênes, McDill) believed that it was helpful to have a round table expression of hopes to guide the group into possible ways to accomplish the mission, that is, to create an international organization of women in engineering and science; the fears would help to avoid pitfalls or unsuccessful models. In the end, this exercise did help guide participants towards a successful model.

Other presentations were selected to set the tone and identify important considerations in our task. The summaries are below:

Monique Frize: Main Issues North:
Monique Frize spoke about some of the current issues that limit the participation of women in STEM careers in industrialized countries (North).

One issue is the lack of self-esteem in teenage girls and the overconfidence in boys. By Grade 7 (age 13), children have decided whether their career will be related to science and mathematics or not. Young people make career decisions much earlier than we originally thought. Activities led by Monique and some of her students, such as "Pinocchio Noses" and team games, with urban and rural girls in the Ottawa/Carleton region increased girls' interest in science and engineering and their knowledge of opportunities in these fields; 88% urban girls and 96% rural girls enrolled in courses in science and mathematics in the next school year; 88% urban and 84% rural girls mentioned a career choice change following the activity. These strategies work well to open girls' minds to opportunities they had not considered, such as a career in STEM fields. Women are clustered in chemical, civil, and environmental engineering and not in high-tech areas with good job opportunities like electrical, computer, and mechanical engineering. On the question of what girls liked least about science: tests, taking notes, reading manuals, working with noisy boys. What girls liked best about science were labs, experiments, research, trying new things, and seeing use and benefit for society. Other factors limiting the participation of women include the absence of information on visible benefits to society and misconceptions of what engineering is (artifacts, symbols, and images). Another important point: the percentage of women scientists drops as career stages advance (undergraduates, PhD awards, postdocs, professors). There have been several intervening strategies in Canada: the National Chair for Women in Engineering (1989–1997), NSERC Task Force Report (1996), five regional NSERC Industrial Chairs (1997 onwards), and the University Faculty Awards (1990–1995), which pay part of the salary for women professors (tenure track). Finally, a new definition of success that better fits women's goals and a corporate culture that provides a better balance between work and life would be helpful. Increasing the number of women is one thing, but what is their impact? Redefine success in terms of the integration of women's approaches and perspectives. Do women make a difference? Yes, if there are enough women (a critical mass) and if they are comfortable in being themselves (that is, with feminine attributes).

Eva Rathgeber – SOUTH: Issues in Developing Countries: From 1992 to 2001, Eva was IDRC Regional Director for Eastern and Southern Africa, based in Nairobi, Kenya; she was in Ottawa at the time of the meeting. Summary of the talk:

Some facts: The Royal Society of the UK did not admit women Fellows until the 1950s; only 2.9% of women were Fellows in the 1990s. The situation is similar in developing countries. Research says women demand careers with a strong impact on society and nurturing social service. In developing regions, Latin America and Central America have the highest percentage of women scientists. Girls receive less intensive training in early years; girls are given answers by teachers, whereas boys are encouraged to figure it out on their own. Male attitudes can have a negative impact on girls in science. Girls are open to doing different things, but boys feel certain things are for boys and certain things for girls, and this influences girls. Girls are influenced not to go into science because they will not find a husband. Women are predominant in the natural sciences in Algeria (36%) and Namibia (35%). There is very little data about women in science in developing countries, but in Bangladesh (13%), Cuba (13.3%), Mexico (25.8%), Venezuela and Argentina (46%), Benin (12.5%), and Senegal (25%). Where are the women working? Sometimes lower levels of employment mean less opportunity for career advancement. Some experiences in the workplace: In the Kenyan study, both men and women preferred to have male bosses. Asked why? Men are hardworking, able to handle crises, more exposed to new ideas, always obeyed by everyone, have high esteem, and are good negotiators not women bosses because they are emotional, petty, easily distracted, hate women who are smarter than them, and are not always efficient because they think of family and children all the time. Women identify themselves with a male model of behavior. How can we change the way that science is done to allow space for women's ways of being without having to be "one of the guys" to be accepted? Girls do not have mentors in developing countries; there are few women professors in science; there are 19 women out of 450 full-time professors in natural sciences in 10 African countries. How can we make a difference? The current model—by those men and women who have gone through the system—is that anyone can do it if they remain focused, which means they have to accept the obstacles. Increasing the number of women scientists will not necessarily cause change.

Suzanne Brainard: Global Alliance for Diversifying the Science and Engineering Workforce (member of WEPAN): Summary of the talk:

Some international participation rates of women in the engineering workforce: Denmark 6%, France 5%, Ireland 2%, Germany 10%, Japan 0.5%, Mali 4%, Sudan 16%, Sweden 14%, the UK 5%, and the USA 11%.

Why form a global alliance? Most countries are facing shortages of engineers and scientists, decreasing enrollments in the USA and abroad; by 2005, women will be 62% and minorities will be 51% of the new entrants to the workforce; engineering accrediting boards worldwide are requiring graduates to have new skills (communication team building, etc.); and businesses need to look beyond the traditional pool of talent for recruiting. The goal of the Global Alliance is to diversify the global engineering and science workforce with particular attention to improving the recruitment, retention, and advancement of women. Partners: AAAS, WEPAN, and AWIS. Strategies: develop networks of institutions; identify and disseminate best practices; develop common standards for data collection; showcase prominent women scientists and engineers; develop and provide training and seminars; organize collaborative projects; publish occasional papers; research; and share information on the website. Future steps: follow-up meeting to discuss corporate and higher education best practices; project funded by the Engineering Information Foundation to create networks and support the efforts of women engineering associations in three African nations (strategies to get past the glass ceiling, etc.); planning of a forum at the next World Engineer's Convention in Shanghai, China, in 2004. Sections: about us, accomplishments, best practices, prominent women, notable quotes, current events, reports, special projects, bibliography, links (future expansion: report on corporate and higher education best practices, special section on Mali, Nigeria, and Egypt). It is not a legal entity at this time because money for special projects will come through partners; therefore, there are no liability issues because collaborators take responsibility.

Gail Mattson, President, SWE: Summary of Notes

We need to have a presence at mainstream international engineering conferences (i.e., World Federation of Engineering Associations (WFEO), World Engineer's Convention, WiTEC[1], etc.). We should do an exhaustive search for existing mainline national, regional, and international organizations related to science, engineering, and technology (SET) to contact. We need to find out how they are organized, their objectives, and how we should interrelate with them. For example, WFEO has regional, national, and specific associations as members. Use WFEO to find out about women engineers and associations in the member countries and how we can participate in their future events. Find out about existing and upcoming women in SET associations as potential members, such as WiTEC.

Renée Clair, UNESCO: Summary of notes:

There should be a special foundation with money from advising companies between UNESCO and around the world; it needs sponsors from Africa and Asia. There is a network of UNESCO chairs (academic contacts who work on women in science and technology): two in Africa, one in Latin America, two in Arab countries (in progress), and one in India (by 2004). There should be a network around the six regional fora to discuss what each group has been doing and future work.

Claire Deschênes – Claire showed and discussed a diagram of the structure of the International Federation of University Women (IFUW) as a potential model for our network.

Efstratia Zafeiriou – Efstratia described the board, members, committees, and associations—national and regional—in Germany and mentioned that networking

[1] WiTEC is a European association for Women in Science, Engineering and Technology (known as SET). It was formed as a network in 1988. There have been more than 10 years of networking and project activities related to women and SET. Then, it established itself as a non-profit European association (May 2001).

is done through conferences and the web. They have projects such as training, seminars, research, education, and statistics. Their funding comes from the public sector, industry, marketing, and advertising. They do advocacy concerning laws, politics, and governments.

Arlene McKechnie – Legal presentation on the incorporation process. Arlene, an Ottawa lawyer, provided ideas for the international corporation to be created and explained what it would do for us. She assumed our primary purpose would be to promote women in science and technology around the world. She recommended the following:

(1) we want this corporation to be independent of its members so that it can exist despite a change of members; (2) we need legal capacity that can enter into contracts, hire people, retain people, purchase insurance for directors; we should consider becoming a Canadian federally incorporated not-for-profit organization (same as "non-profit"), which does not insist that members or directors be any particular nationality or residence; and meetings can take place anywhere around the world, subject to Canadian law. The discussion that followed was based on the model presented by Claire Deschênes (IFUW).

The incorporation process takes 4–6 weeks. We need to have an annual general meeting (AGM) of members to make the organization accountable, and we are required to deliver the audited financial statement at an AGM. The types of organizations that could become members:

(1) NFA = National Federation or Association represents more than 50 people, larger, more organized organizations (i.e., SWE at a national level), which breaks down into state-level organizations and then local organizations; (2) other organizations that are smaller; (3) individuals where these organizations do not exist. These three groups could have different rights, fee structures, voting rights, and membership issues. At an AGM, each class of membership should have representation, but perhaps not all have the same voting rights.

Arlene asked: How will you run this organization? Use a board of directors that carries out the wishes of the body? How will you elect this board of directors? Usually, the members elect the board of directors, and then the directors elect the officers (president, vice president, treasurer, and secretary). Arlene recommended that we make sure the board of directors is representative geographically, linguistically, and by discipline (engineering, science, social science); each group can appoint some directors or elect directors, but we can end up with a large board that may be unwielding. We need to make it efficient and cost-effective, otherwise the organization may collapse; therefore, define an executive committee of perhaps no more than five (president, key vice presidents, treasurer, secretary). We need a number of committees (for example, finance, membership, international relations, women's studies, communications, and youth); these can be permanent or temporary committees. All committees report to the board, and the conference convenor could be on the board of directors.

Legally, INWES must have an AGM; however, ICWES has a triennial conference. Canadian requirements dictate the need for an annual meeting; thus, create a governance bylaw to define how you are organized and discuss annual meetings of members. For meetings that are between conference years, they are limited to including the following items: budget, audited financial statements, basic reporting requirements; members can attend by proxy; include "housekeeping" business; current list of members; appoint a new financial auditor. AGMs can be done physically (but a teleconference will do); it is important to make sure that the objectives of the corporation are being met. Board meetings can take place as frequently or infrequently as you need (can be teleconferenced); you may need to meet a quorum for board meeting votes. For example, officers meet every 3 months, board members annually, and members triennially.

How will you select directors, members, the rights of members, and the level of participation of convenors? Similar, identical issues in any country. Every organization must file whenever you change directors, change objectives, or make other major changes; income tax filing must be done every year. You should have directors and officers' liability insurance to protect the organization due to negligence or omission (accidental) by the director or officer; you can make this a mandatory requirement in your bylaw. Insurance should be obtained and maintained throughout the existence of the organization. You can obtain a supplemental policy for the duration of the conference/event, whereas directors and officers' insurance should be ongoing. You need to develop a procedure (policy) manual to ensure appropriate insurance coverage. Unincorporated entities are loosely organized and can get insurance that applies to specific events, but corporations can attain this more easily, and it is less expensive because they are more accountable. Are there province-specific laws? Typically, the provincial requirements are similar, but you are advised to incorporate federally if you are going to operate anywhere outside of Ontario; typically, federal requirements are looser than provincial requirements (less restrictive).

Some issues to consider: some members or groups will end up with more power or influence (good/bad); you can require half the membership to participate in votes, proportional votes based on regions (weighted), or majority votes, for example. You must maintain records in Canada.

[Authors' note: The presentation by the lawyer was most helpful, and many of the points she made were taken into consideration when INWES was created.]

Discussion of Structure and Funding

Participants commented on the desired organizational characteristics; they expressed the importance of being simple and flexible and respecting country differences. Some asked: Why create an organization in the first place? Is it worth the effort? Can it be effective? Who wants it? Does the ICWES conference drive the structure, or is it the opposite? It was stated that it must support ICWES conferences and expand the structure to support common interests beyond the conferences; in other words, it must be a complete organization with an expanded role. On funding sources, it was mentioned that proceeds from an ICWES conference should provide seed money for the next conference and should help to fund the operations of the 'federation'. Another comment suggested that the current ICWES structure be maintained, which consists of an organizing committee with regional representatives, the chair of the current ICWES, the last ICWES chair, and the convenor. The ICC ensured that a choice would be made at the business meeting at each ICWES on the location of the next ICWES. Usually, the current ICWES chair was a business meeting convenor. For all past ICWES conferences, no seed money was available, and each conference was responsible financially, which meant that a deficit would have to be covered by the organizers or by the host association. There was no database of potential participants, only a list of past attendees.

The next comments related to the types of members that this federation should have. Engineers, scientists, or anyone interested in these issues and fields? National organizations representing engineers and scientists? Include all regions of the world? Could individuals be members? Could other organizations be included, such as government, non-government, the United Nations, and Asia Pacific Economic Cooperation?

On the issue of the governing body, would it be based on geography or discipline? Geographical regions can provide one representative from engineering and one from science for board members. Would the term of the board members be the same as the conference: 3 years? Which month of the year? Would it be limited to 2 or 3 terms? Would there be members 'at large', to bring in other people to serve on the governing body who do not fit into the other categories? How many officers should there be? Five to provide continuity and serve with the same term as the conference? It would be good to have someone who has several years of experience; therefore, it could be someone in the second or third term. A suggestion was made that officers consist of a past president, a president, a vice president for the conferences, a secretary and a treasurer, and perhaps other vice presidents for advocacy, membership, and communications.

Other suggestions were to create a standing committee and ad hoc committees. Would the new organization use the fiscal year? Would terms be staggered? Would there be a regional meeting at a conference, at the beginning or at the end?

Characteristics and Role Desired for the New Organization

The discussions became more and more focused on the desirable organization/federation and on the decisions on officers for an executive committee, for the conference committees, for the governing body (by region), on the types of members, and on funding sources such as membership fees, conference fees, and funding of projects. Would funding be obtained from corporations, governments, foundations, UNESCO, individuals? Would there be fundraisers, administration fees for projects, partnerships in kind? It was suggested that terms not be too long to avoid burnout; for example, there was a possibility of being on the board for 9 years. The size of the board may be too large, especially if an engineer and a scientist are elected for each region. It was argued that geographical representation was important and that members of the board should be actively involved in issues.

It was suggested that discussing the structure before the content of the organization may be a problem and to keep in mind that the organization intends to promote women in SET; the focus should not be only on the ICWES conference, as that would fall into place naturally. The need to develop a "mission statement" for this future organization was expressed: to fill a need for an international alliance and promote the interests of women in SET around the world.

The next discussion was about an advocacy role for the new organization. Examples were provided: preparing recommendations or guidelines for governments on policies that would impact women in STEM. It could ensure the presence of women at key conferences and political events. Representatives at these events should receive some training to be efficient advocates. Training is also needed on how to prepare a portfolio to argue for adequate representation at mainstream meetings. A question was: what about the private sector and NGOs? This led to the suggestion that there was a need to liaise with industries and institutions, to develop a database of contact people, and to provide a means where people can learn who they can plug into and how to be effective in our actions. We want to be proactive in our strategies, whether it has to do with industry or government, to have an impact.

The group expressed the need for a comprehensive website with links to organizations worldwide as well as the identification of best practices and lessons learned. There should be a research bibliography, data, and statistics examining the workforce (corporate and non-profit), research instruments and evaluation, funding sources, fellowships, and study abroad for professionals and students; identifying career opportunities through prominent women as role models in different fields, to find out what they do; following current events, model programs around the world, and effective policies; chat rooms, blogs, and comprehensive list serves.

Other suggestions: curriculum transformation to make it more woman-friendly; mentoring programs; collection of relevant databases; the importance of having information that is not in a digital format to increase accessibility. Once priorities are identified, we can decide where to spend our time and funding—that is, how we support women who are already showing an interest in science and technology but lack the financial resources to choose this path. Dissemination of data and statistics is important to have an impact on agencies; standardization on the collection of gender-segregated data is available on the UNESCO website. It was noted that not every country collects this data. Also, it is important to know what research is being done and to link a team of researchers in one country to those in another country, a sort of "partner search" for projects. Knowing who has data and information enables us to pool resources and can lead to a concerted effort to put an item on the public agenda; looking at "critical mass," with this type of organization, we would be increasing our numbers and more organizations would take us seriously.

On the ICWES Conferences Series:

The group expressed the importance of the new organization being a legal entity, having influence and presence, being able to contract out, and having insurance against loss in case there is an airline strike or other obstacle to holding the event as planned. There must be a physical and an electronic address to reach the conference organizers, which can move every 3–5 years while maintaining the legal entity in one country. There is a need for rules and issues that arise with these rules, like money; for example, what is the sharing of investment in the cost of the conference and in the profit (if there is one) between ICWES and the local host organization? It is also a good idea to hire conference managers who have a lot of experience organizing events.

On Communication:

The purpose is socio-technical communication and networking. Goals: share, help, and support, formal and informal discussions and interactions; benefits of face-to-face interaction with the possibility of virtual interactions. Physical side: physical meetings, physical links (paper documents or data, CDs, USB keys), mailing one version to a local or regional society for local distribution from the international body. Regarding the electronic side: need a website (funding acquired with many corporate logos), with a search engine by topic and word, e-meetings of the board business, a blog with news flash, list serve (sign on and off), links to other associations such as AWIS, SWE, DIB, WEPAN, UNESCO, associations of women in universities, technical (i.e., IEEE), industry, governments, corporations; there should be a database with content, statistics, dissemination of information, a calendar of local, regional, national, international events, awards (other groups' awards, our own awards), links to projects; need a web-spinner to maintain the website and keep it up to date, which needs to be budgeted; whoever is maintaining it must receive instructions on what to keep and what to delete, and the content must be vetted by someone from the organization.

Another question was: How do we get women, who were pioneers in their field and who have stepped down from committees, to be involved again? These women may be influential in male-dominated organizations; we should involve them in our mission. How do we infiltrate organizations led by men since they control the funds? We also need women to contact politicians to increase our profile. Currently, most people in powerful positions around the world are men; therefore, we need men as champions, those who are already convinced. We need to be an effective lobby; to do that, we must find influential people and understand the processes to accomplish our goals. It is also desirable to find a role for young men who are open-minded and involve them in our cause; they are sometimes more flexible than older men. However, we are told that in Germany, the problem is backwards: young men are more difficult to deal with than older men, as they view women as obstacles to their careers. This also happens in other countries.

Responses to: "Is this something worth doing from your perspective?" and "Is it of enough value to motivate you to participate?"

- **Claire**: had an idea of what this could be before she came; now we are getting a better idea of what it is going to be. It is worth pursuing, but it needs to be more focused.
- **Renée** – The project is very exciting. From the UNESCO side, she is very enthusiastic to see some momentum to build the association.
- **Gail** – Enthusiastic about how much we have accomplished; good progress. Excited because they have talked about these issues in SWE. Looking forward to having some action items to hand to her international committee with SWE.
- **Catherine** – Enthusiastic. Wants this to be a bold, strategic move. Show that this organization is something different. Leap far!
- **Gloria** – Needs to figure out exactly what she needs from an international association. Needs an association that would strengthen the capacity of associations working on women in science, technology, and engineering worldwide through the exchange of information, networking, advocacy, and promotion of specific projects. A louder, stronger, more effective voice on science and general issues, in addition to women's issues. Influence the "mainstream". If these are achieved, she will be very interested. She has a regional agenda. If the organization makes sense for our regional agenda, then we should move forward.
- **Eva** – Must focus on broader issues (advocacy, science and technology policy, and practice) and speak with authority for our needs. Concerned that we were talking too much about putting together an organization without figuring out what it is supposed to do. We need two or three strong objectives as a rallying point to continue these discussions. Have a continuing electronic discussion after the meeting ends to keep the momentum going. Supportive and interested in this cause.
- **Marianne** – Exciting possibility to have a forum with material being exchanged (website, physical site). Not sure if we need this organization to achieve these goals. Look at what exists already—the current organizations do not work. It would be more efficient if it was regional. Exchange regional experiences to get more power from that. Liability for ICWES changes depending on the country organizing it. Large conferences are less productive than smaller ones.
- **Gisele** – Require a consensus on the feasibility of the network. Willing to help promote the network at an international level through UNESCO.
- **Hiromi** – Believes in diversity and inclusiveness. There have been differing interpretations of what is on the table. Important to communicate our ideas, which is difficult because of different backgrounds and cultures. Still a challenge to work through issues like why have it. Not excited yet, but she thinks she can be. Perhaps this is a job where we need to get some strategic advice from communication professionals regarding how we should proceed. We are all committed to the issues, but we need to listen to each other's input into what is being developed.
- **Nicole** – Fairly happy with the direction we are heading in, but there is still a lot of work ahead. She would be committed to the organization. There are some

mismatches in what we are looking at, so we are not yet in complete agreement. She feels quite positive.

- **Kathleen** – Discovered a whole list of common interests, regardless of background. It is easy to build from there. ICWES is a very effective meeting, important for sharing information. Would like to see us move ahead. It would provide more power for our issues. She is committed to the issues and the committee.
- **Mitsuko** – Had some problems understanding some of the discussions in English. The current structure of ICWES is very incomplete and would benefit from development.
- **Aba** – Brilliant idea. There will be difficulties. Bold idea for Canadians to start going. She believes it will succeed despite the challenges. Willing to commit.
- **Sue** – Originally afraid of leaving without achieving anything. Small groups were very effective. It is a fine start, but we need to construct it in a way that we can change it. The idea of this organization has been discussed for years. We are at a good point to expand on the ICWES conferences. Willing to support this new organization.
- **Efstratia** – Not yet excited. Should support, coordinate, and give guidelines to national and regional associations towards the realization of equality for both genders in the workplace. We need to articulate our objectives, long- and short-term expectations, a flexible organization with fewer people than proposed, and the need to hear each other (developing countries and developed countries). The role of engineering in ICWES is not satisfactory to her. Proposal to have 2 or 3 parts of ICWES: science, engineering, and gender. To make an attractive conference for engineers as well. Perhaps we will be excited tomorrow if these goals are met.
- **Monique** – She does not mind incorporating the conference separately if that is a barrier; it just costs more money. Let's find linkages and be stronger when we're together. This is just the beginning, not the end. There will be actions that take place once we have agreed on the issues.
- **Suzanne** – Does not think we have all the right players in this room to answer the question. We only have one or two developing countries here. Something needs to be done internationally to share information. The flavor of ICWES, with its technical part and gender part, is very effective. Potential is exciting. One more session is needed to determine who (what other organizations) need to be at the table for these discussions and how to bring them in.

Response to Suzanne's comment:

- **Gisele** – But at the six regional fora, all these women were there, and they said yes to these ideas; the result was paragraph 90.
- **Aba** – Do not need to have "everyone" at the table to build this network. In the beginning, when the ideas of the first women in science organizations were discussed, there were only a few women at the table. It took several years, but they worked to convince other organizations to come onboard and were successful. This same approach will work for the international federation as well. It is up to us here to decide whether we should move forward.

- **Gloria** – At one of the six fora in 1998, 250 women from 60 South American countries met; the regional agenda was important, and this is a piece to link to the international organization.

Discussion on the Mandate of a New International Organization: The group offered suggestions based on Gloria's comments to define what the organization should be and do:

Mandate: To strengthen the capacity of organizations working on women in STEM (science, technology, engineering, and mathematics) worldwide through the exchange of information, networking, advocacy, and promotion of specific projects in collaboration with relevant organizations. To be a voice and influence on scientific, engineering, and technological issues. To increase the presence of women in mainstream science, technology, and engineering organizations.

Sub-committees were created to continue with the "wordsmithing" of the mandate.

An interesting discussion ensued about what we want and do not want from this organization. Are we competing with other large women's organizations? What will be different about our organization? We need to be open, inclusive, and collaborative with other organizations, not exclusive. We need a way to connect with people around the world who are working on women in STEM issues. If another group is doing the same thing at the same time as us, then we should work together.

Modus Operandi: The next step was to identify the DO's and DON'Ts to focus even more intensely on what the organization should be and do and the pitfalls to avoid.

DO's: Be open, inclusive, and collaborative with other organizations; be responsive to events; react; be flexible; understand that it takes money and resources to make things happen; we cannot depend entirely on volunteers; be professional; show solidarity, unity, cohesion, strength, and a sense of community; each country should be heard, not only the strong countries from the north; work with committed individuals, not just "representatives" of organizations.

DON'Ts: Build an empire; expect to get everyone to join at once; make an old-girls' club; have too much bureaucracy; no patronizing of developing countries; arrogance.

Long-Term Objectives: In 5 years, what will we have accomplished? This should be measurable and specific. We are the place where people get information; we are the first point of contact. This does not mean that we do all the work; just that we are the network. We are effective and responsive. We develop a database of women in STEM that covers the entire world and links with all other organizations. We have a solid base of sponsors who are motivated by our work to ensure continued funding for this organization. We are recognized as an important and vital voice in several mainstream science and engineering organizations, and our opinion on many STEM issues is solicited; we are asked to participate on important committees. We have a great, effective website. We have a diverse membership with respect to age and geography to help develop succession planning. To have at least one member from every country in the world. To see a dramatic improvement in women's STEM data worldwide. All countries collect and analyze desegregated data and make it available publicly. Establish a worldwide program with Girl Guides to encourage participation in science and engineering. Help establish effective organizations for women in STEM in 20 new countries. Be financially viable, successful, and self-sustaining. Have an endowment to fund awards and scholarships to support individuals and organizations. To have more women win Nobel Prizes, to be recognized with awards in STEM, and to nominate women for existing awards. To see an improvement in the participation of women in powerful, policy-making positions. Advocacy and leadership training programs for individuals would be held at different locations around the world on a regular basis. One or more prestigious conferences organized by our affiliated organizations. Multilingual information is available for conferences, the website, and resources. Engineering becomes an established part of the regular curriculum for elementary and secondary schools (ages 5–18) worldwide. Facilitate projects as listed in Science twenty-first Century's Paragraph 90 (World Science Organization) and Beijing (Beijing +5).

Resolution to Form the Enabling Body:
We resolve to cause an enabling body to be formed to achieve the following objectives:

To strengthen the capacity of organizations working on women in science, technology, and engineering worldwide through the exchange of information, networking, advocacy, and promotion of specific projects; to be a stronger, more effective voice and influence in scientific, engineering, and technological issues; and to increase the presence of women in mainstream science, technology, and engineering organizations.

[Authors' note: These objectives will be achieved in collaboration with relevant organizations. The women in attendance are voting as individuals participating in a special workshop funded by UNESCO.]

Votes on the resolution: FOR: 18; AGAINST: 0; ABSTENTIONS: 0; the resolution was adopted unanimously.

Incorporation or Alternatives? The last step was to obtain consensus on the structure of the new organization. It is desirable to have a simple structure; limited dues because many organizations have limited funds; find other sources of funds; address liability issues for "board" and "representatives"; a written commitment by "members" to the body; a written contract for projects where the body funds projects with finances clearly outlined; use international association experts to help develop an example of an incorporated body for every potential "member" to review; a high-priority issue to get this written document out to all potential members for them to discuss and give comments; and give time to incorporate those comments into the final document before the ICWES-XII conference.

Define Types of Membership:

We need different types: individuals, NGOs, organizations, and associations of women in technical and scientific fields that are national and regional, intergovernmental, and governmental, academic institutions and centers, corporations, industry, and foundations.

Criteria for membership:

NGOs, organizations, and associations must support the goals of the organization with proof of being a legal entity. There should be a resolution concerning their participation in the new body; there should be an annual fee or commitment of resources (in kind). We do not want inequity with respect to participation due to money issues. The same would apply to all other types of organizations wishing to join our new entity; they must support the goals of the new body, its policy on projects, and be willing to share technical expertise and resources. There should be an annual fee for all membership types. An important question raised was: are there certain organizations or industries that we will not take currency from? Tobacco? Arms manufacturers? Pharmaceuticals? Another point is that individual memberships could result in competition with local membership organizations; thus, they should be as an exceptional case by demonstrating with proof that they live in an area where no organization exists.

Resource Requirements to Set up the New Organization:

A deadline was set for the end of March 2002 for a plan to be ready for discussion at the ICWES-XII Business Meeting, which was approximately 4 months later, where a vote would be held on establishing the new entity. Several steps need to be completed in just over 1 year: We must find a name for the enabling body, have a well-defined draft proposal that defines the body, identify a coordinator and start funds to collect and disseminate information from committees and consultations, organize the next meeting, and prepare a schedule to complete the tasks. A budget was defined to cover the incorporation and other founding expenses and another to cover the first year of operation if the network is accepted at the business meeting at ICWES-XII. The final step was to collect votes on potential names for the new organization: confederation (0); council (6); board (0); committee (0); federation (0); association (7); congress (0); conference (0); organization of (0); representatives of (0); gathering (0); union/united (0); society (0); collaboration (0); network (10); alliance (7) (French meaning); coalition (2).

Follow-up Activities: Committees were created with participants to continue the work after the end of the meeting:

- Communication (includes website): Gloria, Suzanne, Marianne, and Sue
- Structure: Kathleen, Monique, Mitsuko, and Nicole
- Membership: Efstratia and Kitty

- Advocacy (short-term: opening channels): Hiromi, Kitty, Aba, Efstratia, and Renée
- Funding: Monique and Kathleen

The committees met by electronic means, and the work advanced rapidly. A major action was a meeting between Gail Mattson, Monique Frize, and a consultant held in a meeting room at the Philadelphia airport for convenience. The goal was to discuss the creation of the founding document. SWE covered the cost of this important step in creating the new entity. On the flight between Ottawa and Philadelphia, Monique thought about the name and came up with the INWES, which was quite an obvious choice considering that a majority of participants in Merrickville voted for the word NETWORK. The new mandate in the founding document must go beyond the single function of ensuring the existence of the conferences. A few additional meetings with the consultant enabled the development of the By-Laws, Policy Manual, and Strategic Plan. When the document was ready, some weeks later, Monique met with a lawyer to incorporate INWES as a Canadian Federal Corporation. The charter and letters patent name three owners of all the shares: Monique Frize, Claire Deschênes, and Gail Mattson. The initial corporate address was Monique's house. We were ready to propose the adoption of INWES at ICWES-XII.

Chapter 15
INWES Between 2002 and 2008

This chapter describes the early development of INWES, a model of a successful transnational organization. Not only was INWES built upon a solid base, the ICWES conference series, and its International Continuing Committee (ICC), but it expanded its mandate greatly when it was created, even in the early years, to support women in STEM fields around the world. INWES also added regional events in years between the ICWES main conference. These regional events were either in the form of a conference, a symposium, or a workshop. This chapter describes regional events: a conference in 2003 (Daejeon, South Korea); a workshop in 2004 (Nairobi, Kenya); and ICWES-XIII in Seoul, South Korea (see Appendix 7 for a list of all INWES meetings between 2002 and 2008).

Another factor for its long-term success is that it was built on a very careful examination of what INWES should be and what it should do at the Merrickville Meeting in 2001. Moreover, the enthusiasm of the Board of Directors meant that a vast amount of volunteer time (in-kind contributions) was dedicated to build INWES. Agendas and minutes of Board meetings are shared to describe the various activities and evolution of the management and structure of the organization. Summaries of Board meeting minutes are provided.

2002: The First Meeting of the INWES Interim Board (BM1)

Ottawa, July 31, 2002

The first meeting of the Interim INWES Board was held in Ottawa on July 31, 2002, one day after the adoption by the ICWES-XII delegates of the resolution to create INWES. The main purpose of the first meeting was to establish the mandate and tasks to be accomplished by the Interim Board until the first election of INWES

M. Frize et al., *Women's Contribution to Science and Technology through ICWES Conferences*, Women in Engineering and Science,
https://doi.org/10.1007/978-3-031-33800-7_15

directors to be held in 2005 in South Korea. The business meetings are now called INWES Board meetings. A summary of the main discussion points of these meetings is included in the chapter.

Present: All directors nominated and who accepted to be on the Interim Board at ICWES-XII Business Meeting were present, except Marianne Rodot who had to leave prior to the meeting.

Agenda:
1. Election of Interim Officers.
2. Identify important tasks.
3. Select Committees and Chairs.
4. Identify the tasks to be done and a schedule to accomplish them.

Highlights of the Minutes: An Interim Board Communications Protocol was agreed to by all: At meetings, everyone is to speak slowly and clearly and allow for interpretation or clarification, if necessary. Emails: Respond within 1 week and provide information requested in the time period requested; (usually 3–4 weeks would be expected).

The officers elected were President: Monique Frize; Treasurer: Gail Mattson; Secretary General: Claire Deschênes. The committee chairs and members appointed were

- **Conference Committee**: Sophia Kwang Hwa Chung with the help of Mahin Rahmani; ICWES-XII organizers will provide help and guidance where needed
- **Communications Committee**: Sue Bird and help from others (Pam Wain)
- **Vigie/Clearing House Committee**: Co-chairs: Gloria Bonder and Claire Deschênes
- **Membership Committee**: Co-chairs: Gloria Bonder, Issié Yvonne Gueye, and Suriya Thevar
- **Nominations Committee** (to organize the first election): Joanna Maduka, Sachiko Okuda, Anna Szemik-Hojniak

Marianne Rodot will be contacted to ask which committee she wishes to join.

Tasks to be accomplished by the Executive Committee and other Committees (12–24 months):
- **Executive Committee**: Establish a fee structure and a process, forms, and criteria for joining INWES for the various types of members; complete the By-Laws, Objectives, and application for letters patent; obtain comments on By-Laws and Objectives from other organizations; set up the first website with critical documents for setting up INWES; develop the policy manual; organize the first-year meeting of the Interim Board (2003); get young people involved in INWES.
- **Membership Committee**: Review By-Laws and develop a list of potential organizational, corporate, and individual members.
- **Nominations Committee**: Develop the protocol for putting forward nominations and receiving nominations from the members of INWES; develop the

voting procedure; implement the suggestions made at the ICWES-XII Business Meeting regarding membership and representation for potential candidates for Board Director positions.

- **Conference Committee**: Revise the Guidelines of ICWES adopted at the Business Meeting of ICWES-XII. Organize ICWES-XIII; propose a fee-sharing structure between INWES and the host country for both raising funds for the conference and for sharing profits/deficits; involve treasurer in discussions.
- **Communications Committee**: Update the communications plan that arose from the Merrickville meeting; develop a press release that each country can use to inform about INWES. Identify what we want on the website, in collaboration with Gloria Bonder and Claire Deschênes.
- **Timelines**: The tasks are to be carried out between now and December 2002.

2003: Interim Board Meeting 2 (BM2)

Paris, France, May 1–2, 2003

The selection of Paris was an easy choice, as it represented the lowest overall cost of air travel for interim Board members and Renée Clair (UNESCO) invited the Board to meet at the headquarters on May 2. May 1, being a Holiday in France, the meeting was at the Ibis Hotel located near UNESCO. Highlights of the discussions and actions are reported below. The actual minutes were 15 pages.

Highlights: Present, in addition to the interim Board Members, were a few guests: Renée Clair (UNESCO), Karin Dűrmeyer (IBM Germany), Yolanda George (AAAS, USA), Diane Derby (Graphic Designer, US), and Monique Moutaud (AFFI, France). Sophia Kwang Hwa Chung (South Korea) could not attend.

Monique Frize summarized how the ICWES ICC was replaced by an INWES Interim Board and provided information on the Merrickville founding meeting. Monique presented a draft of the By-Laws, policy manual, and strategic plan developed with the help of Tecker Consultants (US), a Canadian lawyer, and the following persons: Gail Mattson, Moyra McDill, Pam Wain, Claire Deschênes, and Monique Frize. There was a brief update on INWES sponsorships and expenses.

Report from the Executive Committee: Monique informed the Board that the incorporation would be confirmed soon. The By-Laws and Application of Letters Patent for Incorporation were submitted in December 2002. As for the policy manual, Sue asked Pam Wain (UK) if she can do it and is waiting for a reply. Claire will send an electronic copy of the policy manual to the Board. The document will be adopted by email when ready. There was a discussion on the fee structure and process, forms, and criteria for joining INWES for the various types of members: Monique explained the fee structure proposed and used in the past few months to recruit members. The invitation to join proposed payment by cheque, money order, or bank wire transfer. Another means of payment will be found if the ones proposed are not practicable in some countries. Credit card payment was not considered at this time, as it involves some costs.

Monique presented the list of members who joined INWES to date: Individual: A. Reinhard; K. Bensouda, Corporate: IBM. Organizational: Nipissing University; Society Women in Science & Technology (Canada); WES (UK); SWE (US); Comité des femmes en ingénierie, Ordre Ingénieurs Québec.

Fact Sheet, News Release, Stationery, and Envelopes: Gail provided a kit to everyone and asked for comments. Monique reminded the Board members that the intent was to have official documents in three languages (English, French, and Spanish); later, other languages could be added depending on the volunteers found. Gail distributed a draft of a press release for immediate use by every Board Member. Comments are due to Gail by May 15. Karin will ask some women engineers and scientists at IBM if they agree to translate the web and flyer information into other languages.

ICWES-XIII and ICWES-XIV Conferences: Sue Bird agreed to become Chair of the Conference Committee; this new title replaces the title "Convenor." Sophia Chung, organizing ICWES-XIII, will be sending reports to Sue. Among her responsibilities, Sue will collect information from Board Members about upcoming events that they attend or organize and post them on the website.

Progress Report on ICWES-XIII (2005): Sophia is asking for $80,000 US from her government and is planning to raise additional funds from other sources. She should reach the breakeven point with approximately 300 attendees if her fundraising goals are met. Yolanda suggested that Sophia gets support from the Minister of Science. Monique mentioned that she has approached the Canadian representative (Marie Tobin) on OECD, who sits beside the Korean representative at these meetings. Marie agreed and encouraged him to help fund the event.

Europe was thought to be a good location to hold ICWES-XIV since the last events were held in Asia and America. Sue will approach different groups in Europe to encourage bids. On the matter of a fee-sharing structure (contract) between INWES and the host country, for both raising funds for the conference and sharing profits/deficits: Monique presented a draft of a contract, involving $5000 US seed money from INWES and recovery of up to $20,000 US if a surplus is made at the conference. She informed the board that some past ICWES conferences had a healthy surplus. Examples were ICWES-IX, XI, and XII.

Report from the Communications Committee: Sue Bird said that Helen Francis (Canada) and Pam Wain (UK) helped her to prepare this report. They developed a policy for press releases and information on INWES to be provided to the media at conferences or meetings. As Sue has taken responsibility for conferences, Yolanda offered to help with the Communications Committee. Yolanda agreed to prepare a newsletter.

Report from the Vigie/Clearing House Committee: (Gloria Bonder, Claire Deschênes, Marianne Rodot): Claire stated that the communication report and the Merrickville meeting minutes formed the basis from which the Vigie and clearing house plan was developed. The committee proposed additional ideas for the website such as policy studies, statistics, and STEM issues; a calendar of events (conferences); a Notice Board with career opportunities; grant opportunities; scholarships, awards, and competitions; formal information for members such as newsletters, educational tools, share of experiences, art and/or games related to women in science and engineering, testimonials and personal experiences; good practices, user-groups, chat groups, forums, and policy statements on science issues. Claire will incorporate the issues brought forward at this meeting such as the need for a webmistress, involvement of youth, the support of associations that need help in developing their websites. Karin was particularly interested to be part of this committee.

Membership Committee: (Gloria Bonder, Issié Gueye, and Suriya Thevar): Gloria reported that she sent emails to around a thousand persons and associations. She thought from the feedback that our message needs to be clearer. Issié emphasized the need for personal contacts in Africa, through regional seminars and provided information on grants, and how to use computers to network together. Issié will contribute to an African website with INWES information. The Membership Committee will develop a strategy for recruitment, considering the characteristics of various regions. It should consider the inclusion of young women and prominent women and be coordinated by the Secretary General's office.

Concept and Application for Individual Membership: There was a discussion on Organizational versus Individual Memberships. This was an important question; for example, should we allow persons who are already members of organizations to become Individual Members of INWES? The intent is for persons without any organization in their region to join as Individual Members; then they are encouraged to create an organization and join INWES as Organizational Members. INWES is developing a Manual on how to create an organization of women in STEM.

Corporate Members Benefits: Questioned about what benefits would make INWES attractive for corporations, Karin (IBM) mentioned: A link on the website; appear on the list of sponsors; opportunity for recruitment of staff; present a poster at conferences; information on retaining women engineers; opportunity to propose projects. Speaking of corporations, a question was raised by the board on who would be an acceptable Corporate Member and who would not be, from an ethical point of view. Karin will prepare a draft of benefits for corporations to be added to the communication policy.

Nominations Committee: (Joanna Maduka, Sachiko Okuda, Anna Szemik-Hojniak): Sachiko recommended adding the category of Honorary Members. According to the By-Laws, two categories are possible: members with a long service to ICWES or individuals who have contributed greatly throughout their life to the objectives that INWES stands for. Joanna presented an update on the process for putting forward nominations and to receive nominations from the members of INWES. A short list of nominees will be prepared by the Nominations Committee considering regional representation as the main criteria.

Voting Procedure: Gail mentioned that the total period for nomination and voting proposed by the Nomination Committee is probably too short (6 months before GM). The time frame should be extended at least to 9 months before GM and the other target dates extended accordingly, which was accepted. Call for nominations will be sent by the Nominations Committee to the Election Officer (Secretary General) who will check the validity of these nominations, compile numbers, send the list to all Board Members, set the voting procedure, and count the ballots.

Strategic Plan: Gail explained that basic ideas for the strategic plan were drafted at the Merrickville meeting and developed by a US Consultant (Tecker Consultants), a Canadian incorporation lawyer, and a sub-committee from Merrickville composed of herself, Moyra McDill, Pam Wain (UK), Monique Frize, and Claire Deschênes.

Review of Purpose, Vision, and Objectives:

Why INWES? Gail reminded the Board of the current statements on these points and a list of benefits of joining INWES:

Purpose: "Strengthen the capacity of individuals, organizations and corporations to influence policies in STEM worldwide and encourage education, recruitment, retention, support, and *advancement of professional women and students through an international network of* organizations and experts."

Vision: "To build a better future worldwide through full and effective participation of women and girls in all aspects of STEM."

Objectives: (1) Becoming an influential voice on STEM issues for the benefit of women, gender equality, and society. (2) Increasing the presence of women in mainstream STEM decision-making bodies. (3) Promoting exchange of information, networking, advocacy, and specific projects.

Benefits of Joining INWES: The most evident traits of distinction of INWES are:

- ICWES conferences have both technical and gender papers
- Leads to the development of new organizations, especially with the help of the manual being developed on creating a new organization of women in STEM
- First international organization of women in STEM with a long history, since 1964
- Building on existing networks and long-standing organizations (SWE, WES UK, etc.)
- Developed by women in STEM
- Includes career development of professionals
- Inclusive of developing countries
- Network of organizations for women in STEM
- Develop policies on STEM issues
- Includes multiple disciplines, technical aspects, gender, and education

- Includes professionals, students, and academics
- Includes Corporate Members
- Addresses women's issues for the entire life cycle
- Leading voice for the World Science Conference

INWES for 2003: Finalize materials and reports from all committees; complete the policy manual; develop a list of projects and prioritize them; translate main texts (By-Laws, policy manual, website, flyers) in French and Spanish.

Committee Goals and Chair for Each Committee: Conferences Committee: Sue Bird; Nominations Committee: Sachiko Okuda; Communications and press releases: Yolanda George; Membership Committee: Suriya Thevar; Executive Committee (president): Monique Frize; Finance and fund development: Gail Mattson; By-Laws and incorporation: Monique Frize; Policy manual: Sue Bird with Pam Wain; Election Office and Secretary General: Claire Deschênes; Vigie and clearing house: Claire Deschênes; Fact sheets, official documents: Gail Mattson. Gail will include a brief history of ICWES/INWES on the fact sheet. Everyone on the list above is to update the definition of their role and functions and update their action plan.

Fund Development Strategy and Budget (Gail): This agenda item was supported by documents circulated by Gail. The idea was to develop a fundraising strategy, considering ICWES history, and including long-term relationship (multi-years) with organizations, NGOs, etc. Focusing on international organizations will prevent regional conflicts of interest. The discussion began with previous experience with ICWES fundraising. The points brought forward by Board Members with experience were: a few people on the fundraising committee; approach of local or national agencies, some international (governments, companies, NGOs, and individuals); focused on specific projects and products; start 2–3 years before the conference; fund development plan with budget goals; identify benefits to donors; sponsorship package.

Lessons learned from ICWES:

- Had to sell products with recognition, profile, specific projects or events (cannot sell operating expenses)
- Uncertainty about reaching budget goals
- Have to follow up with donors by telephone or visit (personal contact)
- Initial contact, multiple media (letter, email, fax, telephone, visit)
- Need to know the fiscal year and donation strategy of donors
- A lot depends on personal contacts
- Make sure to deliver and follow-up (at the event itself) and postevent thanks
- Competitive allegiances are an obstacle
- Getting a grant from one major company helps to get grants from others
- Asking for money is stressful, difficult, surprised when the YES comes easily

What is different about INWES (from ICWES Fundraising):

- Multiyear support
- Long-term relationships with international organizations, companies, NGOs
- Need to develop files and keep reports
- How it would benefit the region
- Need to identify common goals between corporations and INWES
- Need to consider different departments of corporations, agencies
- Country-specific strategy
- Need statistics on membership (number of members represented by INWES)
- Focus on international organizations

Criteria for Region-Specific Benefits: These criteria have to be developed. Mahin will send the list of previous ICWES conference sponsors to Claire (from proceedings). Gail will send the SWE model of fundraising for corporations to Claire for distribution to everyone. Board Members will send a list of potential funders. Yolanda will send to Claire for redistribution criteria for region-specific strategy.

Listing of Potential INWES-Specific Projects: We need a director responsible for receiving projects. As this function grows, however, regional responsibilities will be needed. The idea behind this proposal is to give power to regions, an intrinsic INWES goal.

- ICWES conferences
- Website services
- Scholarships, internships, fellowships, grants
- Awards, prizes, and recognition of outstanding individuals and programs
- Vigie; clearinghouse
- Regional seminars (entrepreneurship, professional development, outreach for girls, share materials and best practices, consulting services, collaborations to optimize expertise).

A Three-Year Budget (Gail): It was decided to postpone the hiring of a part-time executive directorship (Management Company or shared director), but this will become important to consider in the future. Every Board Member will check in her country to see if such firms exist and check costs.

Other Issues: Joanna made a motion, seconded by Sue: "To adopt the By-Laws in the latest version submitted by the lawyer for incorporation." The motion was approved unanimously.

Issues to Think About in Developing Partnerships: Renée raised the issue that UNESCO (along with some governmental bodies) cannot possibly be on the same level of membership as other associations and NGOs such as universities. Therefore, we agreed that there is a need to develop different types of Organizational Members. Although the discussion was not completed and no decision was taken, there is a necessity to develop special partnerships. Questions and matters to think about include: Voting rights at the Annual General meeting? At Board meetings?

Sponsorship? Prefer projects to expanding programs? Partner provides expertise and receives expertise from INWES? Avoid duplication of efforts with other organizations. Political role? Provide connections for scientific policies. Limitation of liability? Responsibilities?

Collaboration with GASAT and Other Networks That May Exist: Monique is going to the GASAT conference (Mauritius Island, July 2003) and will discuss with their Board ways in which the two organizations can collaborate.

Draft Resolution for UNESCO (Renée): There will be a General Conference of UNESCO in Paris, on October 29, 2003. INWES should present a resolution to the UNESCO Commissions in countries represented by Board Members and then bring it to the General Conference. The idea is to include women's issues in all reports. Renée Clair will prepare a draft of the proposition to be presented at the next UNESCO General Conference. After revision by the Executive Board, it should be supported by the UNESCO Offices in the countries represented on the Interim Board.

Next Meetings (2003, 2004, and 2005): After the Paris meeting, the Board decided to hold meetings in various regions between the years with an ICWES conference. This would encourage networking in regions, ensure a solid connection between regions and INWES, and enable to hold face-to-face Board meeting each year. This could lead to the eventual development of INWES Regional networks.

2003: Interim Board Meeting 3 (BM3)

Pai Chai University (Seo-gu, Daejeon, South Korea), November 13–16, 2003

Although the 2003 fall Board Meeting was to be held by conference call, as decided at BM2 meeting in Paris, the organizers of the BIEN Conference in South Korea offered the INWES Board a travel grant to enable the Interim Board to meet during the BIEN event. This was very welcome and gratefully accepted.

Highlights of the Minutes:

Interim Board Members Present: Sue Bird, UK; Myung Hee Jung (Chair ICWES-XIII); Claire Deschênes, Canada; Monique Frize, Canada; Yolanda George, USA; Issié Yvonne Gueye, Ivory Coast; Gail Mattson, USA; Sachiko Okuda, Japan; Mahin Rahmani, Iran; Marianne Rodot, France; Anna Szemik-Hojniak, Poland; Suriya Thevar, India.

Interim Board Members Excused: Gloria Bonder, Argentina; Renée Clair (UNESCO); Karin Dürmeyer, Germany (IBM); Joanna O. Maduka, Nigeria.

Invited Observers: Nageen Ainuddin (Pakistan); Sabina Bakhtyaar (Pakistan); Misa Gratton (WISE-Ottawa President); Sultana Razia (Bangladesh); Heisook Lee Kim (South Korea).

Adoption of May 2003 Minutes (Paris Meeting): Sue proposed the adoption of the Paris minutes, seconded by Issié. A change was asked by Sachiko: there were 1400 ICWES attendees in Japan instead of 350, on page 5 (including engineers, scientists, and children). The minutes were approved unanimously with the change proposed.

Report from the Executive Committee: Monique mentioned that the by-laws are now official. The lawyer asked for the board members to specify which category or association they belong to. Translations in Spanish and French are almost ready. A bank account that will soon accept Visa and Master card payments has been opened at the Bank of Nova Scotia in Canada.

Fact Sheet, Stationery, and Envelope: Gail presented a recent version of INWES sheets, stationery, envelopes, business cards, and press releases. Comments are to be sent to Gail who will provide a draft of a press release about the Korea meeting; she wants a short biography for the website. Note: Myung Hee Jung replaces Sophia Chung on the Board.

Membership Update: INWES now represents around 200,000 persons. This includes ICWES Japan which joined after a meeting arranged by Sachiko Okuda and Misa Gratton in Tokyo, accompanied by Monique and Claire earlier this week. Issié's organization in Ivory Coast (CIFISAT) joined during the meeting in South Korea.

Recruitment Plan: A need was identified by the Board of Directors to verify which member will contact which company, for consistency and to avoid duplication. Monique will prepare a draft of a letter that will be customized by each board member, with Monique's signature if needed.

GASAT Joint Conference? Monique's update seems to indicate low interest from GASAT to hold their next conference with ICWES-XIII; they have not responded since her proposal in July 2003, and they gave her a very cold reception during their conference in Mauritius. The board agreed not to insist on having the GASAT event jointly with ours. One point made by Monique was that GASAT and INWES should hold their conference in different years to avoid competition for participants and funding.

Reports and Actions from Committees:
Finance The draft plan for fund development will be finalized during the Regional Conference (a workshop in Nairobi, Kenya) in August 2004. Directors' responsibilities will be: Learn about fund development via a series of emails; provide Gail information about your fundraising experience; nationalize and/or regionalize standard INWES presentation; make three contacts by 03-31-04 (letter—email—phone call follow-up and meeting if possible). Gail will send the website of the international money index to the Board.

Funding Programs/Projects: We need a process for proposing and deciding on INWES projects. A project proposal format was suggested that includes: Title;

Responsible persons; Description; Benefits; Schedule; Cost estimate; Deliverable/ result and report.

Donor Recognition Levels (Proposal): The recommendation was: Platinum: >$10,000; Gold: >$5000–$9000; Silver: >$1000–$5000; Bronze: $1000.

Possible Sources of Funds: Membership dues; UNESCO; United Nations; National Science Foundations; organizations of economic development; international development agencies; industry, science and technology organizations; international corporations; foundations—corporate and association; wealthy individuals; associations of engineering and science; international associations of women in similar missions (girl guides, business women); percent benefit from ICWES conferences.

Program Committee: proposed a list of activities for the INWES programs that were reviewed and prioritized by the board members as follows:

Essential: Web portal; establish archives for ICWES-I to ICWES-XII by scanning proceedings and photos; add professional/student development program at conferences.

Votes on Priorities: A manual on how to start up an association for women in STEM: seven votes.

Establish scholarships for students to study abroad and for presenting papers at international conferences: six votes. Establish an international summer camp program for high school girls: four votes. Award program: Establish an INWES award; and submit applications for other association awards on behalf of INWES Members: two votes. Establish a memorandum of understanding with the girl guides for STEM activities worldwide (or where they exist): three votes. Develop professional development programs for distribution worldwide, to develop research programs, and to start a business: four votes. Establish "sister" association program between associations in different countries—self-selecting and facilitated by INWES: six votes. Create a position paper on global climate change to present to governments, based on a resolution made at ICWES-XII Forum on Climate Change: two votes. Best paper contest with prizes by regions for high school girls on why they wish to consider a science or engineering career: six votes. Note: The priorities above may change according to sponsors' desires and preferences.

Conference Committee: Sue is the conference Convenor (now called Chair of Conference Committee); it was suggested that the new title should be: Vice-President (conferences), since this is a major activity of INWES. A tentative date and location for ICWES-XIII are Seoul, last week of August 2005. Myung Hee Jung (MHJ) will meet her organizing committee and send information by email on the Board proposal of the contract, update on the conference, insurance. Previous conference organizers (Sachiko for ICWES-XI and Monique for ICWES-XII) are asked to send a summary of their experience to MHJ. Sue will draft a memorandum of agreement between INWES and ICWES-XIII.

Next Regional Conference and Board Meeting: Monique proposed Nairobi, Kenya, for our next board meeting, to be held in August 2004. Prof. Joseph Massaquoi (from the UNESCO Commission in Kenya) put us in contact with Mabel Imbuga, Chairperson of the African Women in Science and Engineering (ICRAF, AWSE, awse@cgiar.org) who could help organize the meeting as well as a regional workshop. Their network covers 23 African countries. AWSE will use this opportunity to bring together women engineers, highlight their research, improve their networking, and interest them in community outreach programs. They proposed subjects such as HIV/AIDS, leadership skills, and peer counseling skills, empower women to reach out to their communities by teaching them mushroom growing skills, packaging of natural products for pest control, essential oil production for perfume industry, bee-keeping technology, poverty eradication and nutrition improvement, silkworm culture, and entrepreneurship. UNESCO will be approached to help support the meeting. A consensus was to have our next Board meeting in Kenya. Sue will prepare another memorandum of agreement for Kenya.

The proposal from Anna for a board meeting in Poland was also interesting. Anna should make a proposal that follows the process to be established by Sue. It should be completed and forwarded to the Board for discussion at the next board meeting.

Support for Regional Conferences: The question of support from INWES for regional conferences was raised by Suriya who spoke about her conference: "Women in the Digital Era: Opportunities and Challenges," to be held at Annamalai University, India, in December 2003. All board members are invited. However, no financial support could be provided at this time because of the short notice.

Communications Committee: Yolanda will develop a PDF format of the newsletter which will be posted on the website. Regional items and Suriya's conference will be good items to be included in the newsletter.

Annual Report Content: Yolanda proposed the content for an annual report: Message from the president; project activities with photos; Board members/committee members; awards; Honorary members' short biographies.

Board Meetings and Regional Conferences: ICWES-XIV: possibility in France? Monique Moutaud from F.I. (Femmes ingénieurs) will put together a proposal for an ICWES-XIV conference to be held in France in 2008. There is a possibility for Australia in 2011. SWE is also offering to hold an ICWES conference or a regional conference.

Motion Proposed by Sue and Supported by Sachiko: "Yolanda George is invited to become a director of the INWES Interim Board as a representative of AAAS (USA)": Motion carried unanimously. Yolanda accepted to become a new member of the Interim Board.

Motion Proposed by Gail and Supported by Anna: "The president will ask UNESCO (Renée Clair) if we can use the remaining funds from Paris to cover

Issié's airfare for her representation of INWES at the conference held in Tunis (Tunisia) on Engineering and Digital Divide (14–16th October 2003)." This conference was organized by the World Federation of Engineering Organizations (WFEO) to prepare for the World Summit on the Information Society (WSIS) to be held in: Geneva, 10–12 December 2003 and Tunis 2005. The motion was carried unanimously.

Web Portal: Monique Frize and Peggy A. Pritchard from Queen's University submitted a grant request to the Canadian Institutes for Health research (CIHR), to create a Web Portal that would become a "one-stop shopping" portal for women in science and engineering.

Membership Committee Report (Misa Gratton): The four Categories of the Members (Organizational Members, Corporate Members, Individual Members, and Honorary Members) are defined in the By-Law document. When an application for membership is received, the Policy Manual will govern the processing of the application. In the Policy Manual, Honorary Membership and criteria for nominations should be added.

Succession Planning and Youth Recruitment: The following items should be incorporated into the Policy Manual: Student and youth components in future ICWES; international summer camps for girls in STEM; exhibitors' participation in ICWES and regional meetings to provide hands-on experience.

Strategy for Recruitment: The "Lifetime Membership" was discussed to ease the payment process problems, especially for developing countries. Since INWES is a new organization, it was agreed not to consider the subject for the time being. As strategies for developing countries, the Committee suggested the following: focus on the reduction of the digital divide and provide opportunities for women in IT. Use UN agencies (e.g., UNDP) to gain funding for the participation of girls and women in ICWES and regional meetings. Include researchers and students in the Women's Studies programs.

The meeting was adjourned just prior to the closing ceremony of BIEN which all Board Members attended.

2004: Interim Board Meeting 4 (BM4)

Nairobi, Kenya, August 15–19, 2004

Board Members Present: Susan Bird, UK; Claire Deschênes, Canada; Monique Frize, Canada; Yolanda George, USA; Issié Yvonne Gueye, Ivory Coast; Myung Hee Jung, South Korea; Joanna O. Maduka, Nigeria; Gail Mattson, USA; Sachiko Okuda, Japan; Mahin Rahmani, Iran; Marianne Rodot, France; Anna Szemik-Hojniak, Poland; Suriya Thevar, India.

Excused: Gloria Bonder, Argentina; Renée Clair, UNESCO.

Invited Observers: Catherine Didion, USA; Misa Gratton (Canada); Jennifer Flanagan (Canada); Catherine Nyambala (Kenya); Aggie Mofya (Zambia); Teresa McNutt (Canada); Kong-Joo Lee and Jung Sun Kim (South Korea).

[Authors' note: Observers covered their travel costs and were very helpful with meeting arrangements and in bringing useful ideas to the discussions. The Board meeting in Kenya was held between workshop sessions. Several of the Board members and observers were speakers at the workshop. The idea was to provide experience to workshop participants on leadership skills, and several other topics that would enrich the discussions. Prior to the Board meeting, the Executive Committee met to review the agenda, prepare resolutions, and a finance strategy.]

The Board meeting objectives were to approve the minutes of the Board Meeting BM 3 in South Korea and review its action items. The discussions were focused on updating the membership list, budget and financial statement, a guideline for travel of directors, update from the Secretary including policy manual and official documents, update from the VP on conferences, creation of an Advocacy committee and an Advisory committee, develop and adopt Terms of Reference for all committees, the newsletter, nomination process, and the agenda for the first General Meeting of Members to be held at ICWES-XIII. It was noted that Karin Dürmeyer' responsibility was changed within IBM and so she could not represent the company at INWES anymore. IBM will send a new name. Jennifer Flannagan has been invited to this meeting, as she is an experienced professional in youth programs; she prepared a youth program for INWES. Jennifer was invited to become an Organizational Member of the Interim Board for ACTUA, thus filling an available organizational position.

- ***Motion*:** "Nominate Jennifer Flannagan as an organisational member of INWES representing ACTUA to serve on the board until the election." Motion carried unanimously. Jennifer left the meeting during the motion discussion and vote.
- ***Motion*:** "To adopt the minutes of the Interim Board Meeting in Paichai University (South Korea) as presented on the CD-ROM for the Nairobi meeting." Motion carried unanimously.
- ***Motion*:** "Change the conference convenor position to Vice-President Conferences." Motion carried unanimously.

President Report: Monique explained the main actions taken since the last Board meeting. A new CEO for INWES has been appointed: Kitty Didion has been hired to work part-time for INWES. Her main task will be fundraising. Monique obtained a grant from IDRC (International Development Research Centre) and from CIDA (Canadian International Development Agency) to support travel and conference fees for women from African countries to attend the workshop in Kenya. Monique discussed Youth Programs and the Web Portal, INWES finances, new members, and fundraising. By August 3, 2004, INWES had 41 members, including four corporations (two in USA, one in Poland, and one in Canada); 18 individual members from 12 countries; and 19 organizations from 10 countries.

Honorary Members: Monique suggested that some of the Honorary members form an Advisory Committee which may help to develop INWES. Some names suggested were: Dormer Ellis (Canada), Kathleen Harer (US), Mitsuko Kasuno (Japan), Marthe Achy Brou (Ivory Coast), Gloria Bonder (Argentina), Mahin Rahmani (Iran), and Nicole Bécarud (France).

Recruitment and Renewal (by Regions): Claire compiled an extended list of potential members based on ICWES-XII participants' list and the names provided by directors. Recruitment efforts since the last meeting were discussed by region: An INWES Japanese chapter has been started and already has a website. Sachiko had success with a silver sponsorship. Suriya contacted 20 corporations in India for participation at 1/3 of the regular price (developing countries), plus universities, including students. She proposed to start a chapter and develop a program. In Canada, contacts have been made by Monique with Claire's support and success has been reached with eight or nine organizations. With a few colleagues, Claire has started a French-Canadian Association: AFFESTIM (*Association de la francophonie à propos des femmes en sciences, technologie, ingénierie et mathématiques*) which is a new member of INWES.

Update from the Treasurer: Gail provided copies of the financial report for the period February 2003–October 2003. She arranged for SWE Magazine to report four times a year on INWES activities. They have already published a nice article for which many directors have been asked to make some comments.

Update from the Secretary: Claire reviewed the secretarial work since the third Board meeting in South Korea. A large effort was put towards preparing the meeting and to update the Policy manual. The Policy Manual format was simplified, and space was made for the various committees' "terms of references" discussed during this meeting. The Secretary General will prepare a final version of the Policy Manual for the General Assembly that will take place during ICWES-XIII in Seoul, August 2005.

- ***Motion***: "To adopt the Sponsorship program as it appears on the CD-ROM for the 4th board meeting with the changes listed above." Motion carried unanimously.
- ***Motion***: "To adopt the Benefits for members as they appear on the CD-ROM for the 4th board meeting with the changes listed above." Motion carried unanimously.

Monique and Jennifer will clarify the notions of donors and sponsors according to Canadian law and come back by email to the directors.

Update from VP Conferences: Sue met with the British Embassy in Seoul to help prepare ICWES-XIII and she suggested that the other directors do the same as a preparation for their trip to South Korea. She prepared policies for regional and INWES conferences.

Creation of an Advocacy Committee: The item was postponed until a Chair is identified. Terms of reference will be prepared after the present meeting.

Distribution of the Terms of Reference for All Committees: Terms of reference (TORs) are intended to guide the committees regarding their mandate, objectives, reporting authority, constituents, chair, meetings, quorum, and budget. Our target is to include the different committees' Terms of reference in the Policy Manual for adoption during the August 2005 General Assembly. Monique will prepare a form for conflict of interest to be presented at the next board meeting.

Archives for ICWES-I to ICWES-XII: The idea is to electronically collect information on past ICWES conferences. Claire has prepared a list of available proceedings. Mahin has sent to Claire many of them. SWE owns some also. Suriya proposed to scan ICWES proceedings and artifacts using the librarian electronic database she is setting up.

Small groups discussed a variety of matters including the plan to develop a manual on how to start an association of women in STEM. INWES would give seed money and other kinds of support to groups wishing to do this and have the new organization join INWES. It was decided that INWES will not have chapters, only organizations.

There will be an election of the Board during the General Assembly in August 2005 at ICWES-XIII. The process should start in November 2004. Claire prepared a draft of the procedure according to the By-Laws and Policy Manuel. The voting procedure for the corporate members and the individuals must be prepared separately from associations. The main responsibilities were clarified as follows: the nominating committee, led by Sachiko, will update the procedure draft as soon as possible and gather the slate of nominees with the secretariat office. Claire, as Secretary General, will act as election officer. Monique will be an *ex-officio* member of this committee. Some directors should stay on the Board for continuity purpose (ideally at least 2/3). They will have to be candidates on the slate since they were nominated and not elected. The nomination process will be launched by the committee this fall. Monique will collect the willingness of directors to remain on the board after 2005.

New:

Fund Development: Kitty presented the fundamentals of fund development. She emphasized that everyone on the Board must participate in the fund development program for INWES to be successful. She will send out materials including sponsorship levels in September and meet with the South Korean organizational committee for ICWES-XIII regarding their fundraising efforts.

Board Responsibilities, Surveys, Future Involvement: Directors will respond to the survey and send it to Claire on Board effectiveness.

Joining Other Organizations as Partners? The idea is to develop an alliance or a partnership with such organizations as the Institute for Science and International Security; IBM, Monique, and Claire will make the first contact with these organizations; Monique will check UNESCO and other organization membership possibilities.

Regional Conferences (2006?): Directors will send proposals based on Sue's Policy for Regional Conferences.

ICWES-XIV (2008 Pre-bid): Marianne Rodot presented a PowerPoint as a pre-bid for ICWES-XIV. The French Association of Women Engineers proposed that this conference take place in the 2nd part of July 2008, either at Lille (near Belgium), Nantes (Loire-Atlantique), or Tours (Loire). An official request for bids will be launched and will be considered at ICWES-XIII.

2005: Interim Board Meeting 5 (BM5)

Ewha University, Seoul, Korea, August 24–25, 2005

Interim Board Members Present:
Susan Bird, UK
Claire Deschênes, Canada
Monique Frize, Canada
Issié Yvonne Gueye, Ivory Coast
Myung Hee Jung, Korea
Yolanda George, USA
Gail Mattson, USA
Sachiko Okuda, Japan
Mahin Rahmani, Iran
Marianne Rodot, France
Anna Szemik-Hojniak, Poland
Suriya Mayandi Thevar, India

Excused:
Gloria Bonder, Argentina
Renée Clair, UNESCO
Karin Dürmeyer, Germany, IBM
Joanna O. Maduka, Nigeria

Invited Guests:
Catherine Didion, USA
Jeanne d'Arc Gaudet, Canada
Misa Gratton (WISE-Ottawa) Canada
Durdana Habib, Pakistan
Ruby Heap, Canada
Florence Hudson (SWE) USA
Mabel Imbuga, Kenya
Mitsuko Kazuno, Japan
Jung Sun Kim (ICWES-XIII), Korea
Kong-Joo Lee (ICWES-XIII) Korea

Maureen Lofthouse (New Board Manager), Canada
Teresa McNutt (WISE-Ottawa) Canada
Monique Moutaud (for ICWES-XIV bid), France
Haryoung Poo (KWSE), Korea
Betty Shanahan (SWE) USA
Crystal Sissons, Canada
Uduakobong Okon, Nigeria
Pam Wain (WES) UK

1. Opening

Monique opened the meeting by introducing all members and guests. The directors' list was circulated for update by all present.

1.1 BM4 Minutes

Motion BM 5.1. Proposed by Sue and seconded by Anna. "To adopt the minutes 'Interim Board Meeting in Nairobi Safari Club, Nairobi, Kenya', as presented in the document package distributed at the Seoul meeting."

Motion carried unanimously.

1.2 Agenda for Meeting BM5

Gail requested that item 2.3.3. "Finance, Fund development update" be deferred to the next day, August 25, to allow incorporation of the Fund Development Committee's discussions.

Motion BM 5.2. Proposed by Gail and seconded by Marianne. "To adopt the agenda with this modification."

Motion carried unanimously.

2. Reports from the Executive Committee
2.1 President's Report

Monique reported on events and activities since BM4, including the Leadership workshop in Nairobi, launch of the Arab Network of women engineers and scientists, and related reports from UNESCO and WFEO (World Federation of Engineering Organizations).

Regarding the current INWES network, it has 70 individual members, 6 corporations, and 20 organizational members, representing over 200,000 women worldwide. Maureen Lofthouse has joined as our Board Manager, but we need more volunteers.

INWES is faced with the major issue of a dramatic decrease in the number of women entering ICT or STEM fields and must develop positive strategies to counter this. We are grateful to our current sponsors but need more.

In response to the report, Issié noted an ongoing series of meetings on ICT held by the WFEO and suggested that INWES should be at the table for events such as this. Detailed discussion on these and other points was deferred to item 8, Strategic Planning.

2.2 VP Conferences Report

Sue reported on the development of policies relating to ICWES conferences and regional conferences, as well as a contract format and draft contract for ICWES-XIV. These are identified as items I1, I2, I3, and I4 in the package distributed to Board members. She invited review and comment. She commented that cooperation and sharing of information between INWES members involved in the organization of ICWES conferences was very important and that ICWES-XIII was a good start. She also reported on her visit to the Arab Network, noting that she had been very well received. It would be important to follow this up and seek more Arab involvement in INWES.

Motion BM 5.3. Proposed by Sue and seconded by Gail. "To adopt the four policies presented, namely (ref document designations in the distributed package):

I1: Policy for conferences
I2: Policy for ICWES
I3: Policy for regional conferences
I4: Contract template for ICWES-XIV"

Motion carried unanimously.

2.3 Treasurer's Reports
2.3.1 Budget Report

Gail presented a draft report of income and expenses for the current year, to July 31, 2005, which showed that INWES is overextended to date and cannot undertake more commitments until additional funds have been raised.

2.3.2 Financial Report

Gail provided the financial statements for the years ending December 31, 2003, and December 31, 2004, and walked the Board through them. She proposed deferral of any motion to adopt until the next day, August 25, to give directors time to review them, and this was agreed.

With regard to the financial reports, directors expressed concern that there was no recognition of in-kind contributions to INWES, which have been estimated to run as high as $100,000 to date, the greatest portion of this being directors' time spent on INWES matters. It was agreed that there needed to be a way of defining and recording these contributions.

Action: Gail will develop a form for identification by those contributing to the value of their in-kind contributions.

2.4 Secretary General's Report

Claire indicated that she planned to make a change from the teleconference to computer/Internet for meetings such as those of the Executive, which will save money. She will inform directors what software to install and how to use it.

UNESCO is preparing a report on gender and science, which will include data from all countries. Pam Wain is involved with writing this UNESCO report (IRSTG, International Report on Science and Technology and Gender). Claire is putting INWES's contribution together and needs everybody's input.

Claire advised directors that the term of her Chair ends September 30, 2005, after which she will no longer have the three research assistants who have helped her with INWES work.

In response to a question about charitable organization status for INWES, Monique told directors that the Canada Revenue Agency (CRA) has acknowledged receipt of the application and that their final response should be available in about 2 months. Recent changes in the law have sped up the process.

3. Updates from Committees and Programs
3.1 Advisory Committee

Monique thanked directors who will not be standing for re-election and commented that we would still be able to benefit from their expertise as members of the newly formed Board of Honorary Members, to be chaired by Kathleen Harer. We will also have advice from the Advisory Committee, which will be a subgroup of this Board.

3.2 Program Committee

Anna presented the Committee's Summary Report which identified the following topics reviewed during the past year:

- Terms of Reference for all committees
- Professional/student development programs
- Sister association program:
- Best essay competition
- Regional conferences

Regarding the last item, a proposal had been drafted by Misa for a regional conference to be held in Moscow in 2006, but there has been a delay in moving this forward. The Russian women explained to Anna that they have had considerable difficulty in finding a project coordinator in the European Union, but have finally found one in England. They have now applied within the EU framework six program for an 18-month SSA project "Women in Science." Poland represented by Anna (for the University of Wroclaw) is also included. The project consortium comprises eight partners from six countries (the UK, Sweden, Russia, Spain, Italy, Poland) with world-known experts in the area. Russia is committed to organize the international conference and Poland will be the host institution for a final workshop for 20 persons. Costs will be covered by the project. This project is a step towards developing a larger proposal within the EU framework 7 program. Anna would like to invite the INWES executive to the workshop. All the above possibilities depend on whether the project will be positively evaluated or not. The final decision may be known only in late December or early January 2006.

The possibility of such a conference in Brazil has also been discussed informally. It was suggested that the Board should invite bids for the 2006 regional conference.

Anna noted that in the Committee's document "Subset of Programs and Projects," it was suggested that the Board discuss the question of which programs and projects should be included in this Committee's work, and how projects could become active if they needed sponsors. Monique asked the Board to look at these questions during their discussions in small groups later in the meeting.

3.2.1 Archives for ICWES-I to ICWES-XII

Suriya presented her work on digitization of the proceedings for these conferences, which is complete except for ICWES-VIII, for which some material she has received is in video format (WES and other organizations are looking for proceedings). She described the process followed by her team and gave the Board a demonstration of the resulting archive and search engine. Directors were impressed by the quality of the material and congratulated Suriya on her achievements.

In discussion, it was noted that this will be a worthwhile research tool for many, and that may be nonmembers of INWES could be charged a fee for accessing it, to help defray the costs. This is to be discussed by the Web portal group.

Suriya was asked whether the search engine could allow users to extract subsets of information for particular purposes, e.g., specific research projects or ICWES-XIV promotion. Suriya responded that it had not been set up to do that at this stage.

3.2.2 Starting Up an Association in India

Suriya reported on her activities in establishing a new organization, the Indian Association for Women in Information Technology, which was registered officially in May 2005 and now has about 150 members. Work began with seed funding from INWES and now continues with local grant money. Suriya showed her list of current and planned projects and programs, which include training in both development and use of information technology. Suriya has been widely cited and recognized for this work and was again congratulated by the Board.

Suggestions as to how INWES might help further include help to link to a sister organization, with opportunities for information exchange and mentoring, and identification of this new organization as an INWES affiliate. Monique suggested that a formal letter from INWES congratulating the Association on its formation would be appropriate and useful.

Action: Monique will write letters indicating INWES's positive response to the start-up of the Indian Association of Women in Information Technology and to the digitization project.

3.2.3 Manual on How to Start an Organization

This has yet to be fully developed, but drafters can refer to Suriya's experience.

3.2.4 Sister Associations

To be discussed later in small groups.

3.2.5 Youth Program

To be tabled.

3.3 Nominating Committee

Sachiko presented her Committee's Report, in which the principal activities focused on the need to elect new directors this year. A "Call for Nominations for INWES Board of Directors" was issued in January and, after checking nominee eligibility and suitability, a slate of 18 candidates has now been prepared for election at the General Meeting on August 28. A slate of Honorary Members is also available for presentation at the General Meeting.

There was some discussion on the lack of candidates from certain geographical areas, namely South and Central America, Caribbean, Middle East, Australia, and New Zealand. Concern was expressed regarding the predominance of North American candidates on the slate, although it was recognized that there is value in continuity at this early stage of the organization. It was suggested that INWES should treat as a priority the need to bring in more candidates from other regions; ideas included using OAS contacts for the Central/South American region, making relevant contacts at ICWES-XIII, and making specific efforts to recruit and groom candidates for the Board. Monique asked that this be added to the work of the Membership and Recruitment Committee.

It was confirmed that, as well as the slate of 18, the Board will be open to accepting names of new candidates proposed from the floor at the General Meeting.

3.4 Conference Committee

The work of this committee was reported on by the VP Conferences, under item 2.2 above.

3.5 Communications Committee

Yolanda apologized that her committee's report was not yet ready but promised to have it for the Board very shortly. Key points are

- One objective is to get the logo used more widely
- The newsletter needs more material, especially pictures and regional items. Board members are asked to send in news of, e.g., conference attendance as it happens. Pictures and news from ICWES-XIII will be very welcome. The newsletter has been published once a year so far, but we are aiming for twice a year or more. Monique added that she wants to include other items such as book reviews.

Yolanda was thanked and congratulated by the Board for her persistence in collecting material and publishing the newsletter.

3.6 Membership and Recruitment Committee

This committee has not yet been active but will be started up at this conference. Yolanda has agreed to chair it.

3.7 Advocacy Committee

Monique reported that the committee is in the process of being formed; Claire will chair it and it may meet this week. Relevant issues will be identified in the gender and leadership workshops during ICWES-XIII.

3.8 Web Portal Committee

Monique reported that Phase 1 of development was completed on time in January, to improve the look of the site and to make it easier to navigate. All Board members' photos are now there.

Phase 2 was to be completed by the date of ICWES-XIII and will be done in the next few days. It will provide members and the Board with sign-on IDs and passwords so that they can access areas of the site not open to the public. The Board will have a further separately accessible area within this, as will admin-related users such as Claire, Cate Palmer (bookkeeping), and Maureen Lofthouse (Manager). When members renew this year, they will be provided with their IDs and passwords. Corporate and organizational members such as SWE will have access to all their members/employees through their corporate ID/password combination. Questions were raised about the limitation on a number of users with the same ID/password using the system at the same time.

Action: Monique will contact the software developer to find the answer re-multiple users.

The next phase of web development is to include other languages and any other items defined by the Web Portal Committee in its small group meeting later.

A contractor, Cathy Thompson, has been engaged to work on the website. New ideas should be sent initially to her at cphil087@hotmail.com and she will compile them for vetting by the Web Portal Committee.

3.11 SWE/IBM/INWES Partnership

The presentation made by Betty and Florence outlined this partnership, in which Florence is a loaned executive from IBM working with SWE on strategic development. Their goals are to expand SWE's outreach and awareness programs. They propose adding INWES into the partnership to add a global dimension to this work, with benefits such as a better understanding of cultural differences. An example of activities proposed is Global Women in Technology Awareness Week, at the time of International Women's Day.

Discussion points included:

- Questions regarding other International Women's Day/Week celebrations in the USA and the possibility of conflicting messages
- Yolanda can get information through the OAS on global surveys of women in technology being done in the USA and the UK
- Monique Moutaud has information about IIWE events for young people

- Could IBM's EXITE program be brought to other countries if questions of language and cultural appropriateness permit

More details will be elaborated in small group discussions later in the meeting.

4. INWES Development: Work in Small Groups

Groups discussed issues and reported as follows. In some cases, two stages of discussion were involved.

4.1 Programs Other Than Conferences

Participants: Anna, Misa, Mabel, Marianne, Betty
Points of discussion included:

- A proposal for the Eastern European Group
- Recruitment of Dr. Haryoung Poo to the committee
- Need for a new internal document on programs and projects to replace the "Subset of Programs and Projects" distributed to Board members
- Programs should be identified and customized according to Life Cycle and Region
- The importance of the awareness about and interest in STEM among girls. Issues relevant to several areas were discussed in detail, including K-12, university students and professionals, start-up and sister associations, scholarship programs, and career development
- An ICT for Women Program was suggested

More specific details can be found in Attachment 1 to these Minutes.

4.2 Nominating Committee

Participants: Sachiko, Issié
Sachiko produced revised versions of the form for nomination of candidates for election as Board Directors and the form requesting nomination and the candidate's CV and handed out copies for the Board's information.

4.3 Conferences

Participants: Sue, Monique M, Poo, Durdana, Mahin.
Sue reported that the group had discussed two key points: ICWES-XIV and regional conferences.

ICWES-XIV can draw heavily on ICWES-XIII experience and data and will meet with ICWES-XIII organizers during this conference to discuss. Information can also be transferred from ICWES-XII in Canada. Points of interest re ICWES-XIII include:

- Funding, including EU funding application requirements
- Regions of origin of participants and gaps
- Possible re-use of ICWES-XIII website, or creation of a template for use at all future ICWES conferences
- Accommodation reservations—what % in what kind of facility?

Regional conferences should target areas where INWES has not been influential to date, i.e. South America, Arab countries, and China. UNESCO will be a useful reference source.

4.4 Communications and Newsletter

Participants: Yolanda, Suriya, Florence, Maureen

Florence reported that the group had identified three general communications concerns, namely branding (look and feel); marketing material; and logo policies, including recognition of INWES when co-sponsoring an event. Further detailed discussion is needed to identify priorities and next steps.

4.5 Membership and Recruitment

Participants: Yolanda, Suriya, Florence, Maureen

Florence reported that the group had identified recruitment of corporate sponsors and members as top priority, which requires a value proposition, to demonstrate the value of INWES to the corporation and its employees. Suggested elements to include:

- INWES builds and nurtures an integrated global community of women in STEM
- Brings women into an extensive global network of developed and developing countries to leverage for growth and success
- Professional development
- Enhances collaboration capabilities and opportunities
- Can demonstrate the use of advanced technologies for training and communications, to increase skill levels and facilitate person-to-person contacts

Targets for recruitment should probably include large or growing multinationals and companies in high-growth countries such as China and India. Development of a specific target list by region is proposed. INWES considers it important to involve more youth in its activities.

4.6 Advocacy

Preliminary discussion participants: Monique F, Teresa, Claire

Teresa reported that the group had discussed the identification of issues of interest, such as ethical concerns. More issues will be identified during the Symposium on Gender and Leadership at this ICWES-XIII.

Second discussion participants: Claire, Jeanne d'Arc, Ruby, Crystal Sissons.

Claire circulated the Terms of Reference for this committee for review and comment—to be approved later in the meeting.

Key issues identified by the group were:

- Inclusivity
- Linking with relevant researchers
- Including science and technology issues in the discussion

Action: Claire will use the results of the survey to be conducted during the ICWES-XIII Gender and Leadership workshop to identify the main advocacy issues to be addressed by the Committee.

4.7 Web Portal

Participants: Monique F, Teresa, Claire

Teresa reported that the Beta test of Phase 2, as described in item 3.8 above, will start in September. Passwords will be issued for Board and Committee members starting October 1, and for the general membership in December, as members register or renew.

Next steps will include:

- Verify the ability to accept multiple logins with one ID/password
- Include more profiles of women
- Determine whether the portion of ICWES archives can be made public domain on the site
- Align information on corporate sponsorship and membership.

Marketing recommendations of the Communications Committee will also be addressed.

4.8 Finance and Funding Development

Preliminary discussion participants: Gail, Kitty, Pam

Gail reported that the group had identified four issues relating to funding:

- General membership dues (less administration costs)
- Programs and projects—specific plans and wish list
- ICWES multiyear effects and relationships
- Regional conferences

Overall funding strategy needs to be implemented around the world, and badly needs some descriptive material (pamphlets, Annual Report) on programs, projects, and other INWES activities and benefits, to attract sponsors. A project manager should be appointed for each project requiring funding, to provide information for the pamphlet and to answer any questions from potential donors. INWES needs to understand charitable status requirements and advantages in different countries, as well as restrictions related to the donor corporation's headquarters location. A list of funding sources should be developed, with research requirements and deadlines. Follow-up can be done via a software program.

Action: Gail will provide information on an appropriate software program for this purpose.

Action: All directors will check charitable status requirements in their countries and inform Gail whether it is necessary for fund-raising.

Action: Maureen will work on the Annual Report, in consultation with Monique and Claire.

Second discussion participants: Gail, Kitty, Yolanda, Pam

Gail reported that the Committee had noted many needs from earlier discussions, including the sister organization program, starting new organizations, and various projects. An aggressive drive to raise new funding from corporate sources is needed but requires specific actions in order to proceed. Successful completion of this plan could bring in new funds before the end of calendar 2005.

- Charitable status has been applied for in Canada and is under review, but non-profit status (5013C) in the USA should also be applied for. However, this costs $2500 which we do not have, so cannot be done immediately.
- Planned projects and programs should be developed into handout/flyer material and project managers named
- Conflicting information on corporate sponsorship vs. membership on the website needs to be resolved.
- Status of INWES organizational needs should be updated so that specific requirements are known.

Actions: Gail will present a resolution to the sponsorship vs. corporate membership conflict issue. Gail will also update INWES's financial status and clarify the organization's financial needs.

4.9 Manual on How to Start Up an Association

Not discussed.

4.10 Sister Association Program

Participants: Sue, Durdana, Mahin, Maureen

Durdana reported the group's findings that emphasized the value of linking with sister organizations to help with starting/running an organization, to exchange fund-raising ideas, and to make useful personal contacts. Organizations could send delegates to each other to make presentations about their own situation. INWES can play a role by establishing contacts and suggesting a framework for such linkages. A form could be provided at the INWES desk to enable organizations to identify others they would like to partner with, and discussions could begin at this ICWES-XIII. Sue would remain the key contact at INWES, to advise and monitor these linkages as they develop. Paired organizations should be asked to report back to INWES on a regular basis.

4.11 Youth Outreach

Not discussed

4.12 Scholarship Program/Essay Contest

Participants: Okon, Issié, Marianne

Marianne identified the objective as the promotion of STEM disciplines to girls through contests. The target audience would be students at presecondary and secondary schools (up to age 18) and contest organizers.

The role of INWES would be to:

- Help develop girl-friendly contests (look at existing contests, good practices/experiences, etc.)
- Support contests, by endorsement, sponsorship, rewards

It would be important for INWES to provide a letter of support to indicate, where appropriate, that the contest and organizer have INWES approval.

4.13 Professional Development

Participants: Anna, Misa, Monique M, Teresa

Teresa reported the group's proposed first steps toward the development of a program specific to INWES:

- Compile an inventory of readily available programs/workshops/courses from its member organizations.
- Select a list of "quick hits" in terms of target populations and areas of INWES priority.
- Arrange access to courses/workshops/coop programs, etc., for items in the "quick hit" list.
- Develop building blocks to success, using items available from Step 3.

Key areas of interest to be included, such as coop programs, manager/executive programs, information sessions, and mentoring/networking were discussed. More detail can be found in Attachment 2 to these Minutes.

4.14 Policy Manual

Not discussed in group sessions. See item 5 below.

4.15 SWE/IBM/INWES Partnership Committee

Participants: Monique F, Florence, Betty, Suriya, Mabel, Sachiko

Florence reported that the Committee had decided not to spread itself too thin but to focus on one specific project, namely the International Women's Day event (Global Women in Technology Awareness). Pilot countries and cities proposed were

- Kenya: Nairobi, Mombasa, Kisumu
- India: Annamalainagar (Suriya's city)
- Canada: Toronto
- Japan: Tokyo
- USA: TBD
- Pakistan, TBD
- Ivory Coast, Abidjan

SWE would provide marketing materials, including a global press release and a basic agenda and would create and host a website for this material. Local organizations would translate and customize, and identify their budget and possible local funding sources. This will be done with INWES collaboration and contacts. Any proposed use of the INWES logo requires INWES approval.

Action: The SWE/IBM/INWES team will provide sample content materials in September 2005.

4.16 Guidelines for Conflict of Interest and Ethics

Monique noted that the proposed Guidelines had been distributed in the package for the Board; these are based on the guidelines of the Canada Foundation for Innovation. The new Board will be asked to adopt them.

4.17 Guidelines for Hiring Personnel

These have not been worked on, as INWES has not hired anybody, but has used contractors. It was suggested that separate guidelines should be developed for contractors. Discussion on other guidelines would be tabled at a future Board meeting.

4.18 Starting an Association in Nigeria

Okon requested time to present her project. She is working to set up a new organization for women in engineering and science but needs additional support and financial resources. In particular, she would like INWES to send her an official letter of support, to assist her in getting the association officially registered in Nigeria. She is also looking for support to provide motivational rewards for girls who get involved, such as scholarships, etc. Monique asked her to send in an official request, with all the required information, for consideration by INWES. At the moment there is no funding to provide this kind of support, but if her request is approved and if funding comes in, she can certainly benefit.

5. Policy Manual Update

5.1 Terms of Reference

Terms of Reference were presented to the Board for approval for the Program, Nominating, Conference, and Advocacy Committees. In discussion, the following changes were proposed:

Program Committee

The version presented was not the latest one. Anna will send the latest version to the Board for consideration.

Nominating Committee

It was noted that the Chair of this committee was to be the immediate past president of INWES. Until there is one, it is suggested that the Chair be identified by the Executive Committee.

Given that teleconferencing will not be used once we convert to the use of the Internet, it was suggested that references to teleconferencing be removed and that the budget be reduced to $500.

Motion BM 5.4. Proposed by Sachiko and seconded by Marianne. "To adopt the TOR of the Nominating Committee as amended with these changes."

Motion carried unanimously

Conference Committee

Editorial changes are required to replace "Sub Committee" with "Committee" throughout and to replace "Program" with "Conference" throughout.

Under "Constituents," the sentence "A balanced regional representation shall be maintained," should be revised to read, "A balanced regional representation shall be attempted," this being a more realistic goal.

The budget should be revised from $1500 to $500 and reference to "teleconferencing" be changed to "email" or "Internet."

Motion BM 5.5. Proposed by Sue and seconded by Claire. "To adopt the TOR of the Conference Committee as amended with these changes."

Motion carried unanimously.

Advocacy Committee

Under "Objectives," the word "lobbying" should be deleted.

The fourth paragraph under "Objectives" should be modified to read "The committee will monitor and develop position papers on women in STEM, and will report on their accomplishments and also on challenges they face…etc."

The budget should be revised from $1500 to $500 and reference to "teleconferencing" be changed to "email" or "Internet."

Motion BM 5.6. Proposed by Claire and seconded by Monique F. "To adopt the TOR of the Nominating Committee with these changes."

Motion carried unanimously.

Action: Claire will amend these three TOR documents, for the files.

5.1 Update of Policy Manual

Work is not complete. To be tabled at a future Board meeting.

5.2 Continuation of 2.3.2 Financial Statements

This item is a continuation of the Treasurer's presentation on August 24. Directors were asked if they had any comments on the reports presented.

In discussion, it was noted that the letter from Monique to the auditors and their cover letter with the statements had not been provided to the directors.

Action: Gail will provide the Directors with the letters relating to the review of the financial statements for 2003 and 2004.

Motion BM 5.7. Proposed by Gail and seconded by Sue. "The Board accepts the Financial Reports for 2003 and 2004 and recommends them to the General Assembly for approval."

Motion carried unanimously.

6. ICWES-XIII Conference

Dr. Myung Hee Jung presented information on the plans and results to date for ICWES-XIII, including organizers, sponsors, and participants (about 700 from 53 countries). The Association of Korean Woman Scientists and Engineers (KWSE) is the main organizing body, with support from many organizations in research, the private sector, and government. A full timetable over 4 days offers plenaries and eight parallel technical sessions, with over 650 presentations. The budget is estimated at $819,000 US and sponsorships are expected to cover the full amount. Profit sharing with INWES has not yet been discussed. Seed money for ICWES-XIV is to be included in the ICWES-XIII budget.

7. Next Meetings
7.1 Regional Conferences

The possibility of holding regional conferences in each of the 2 years between ICWES conferences was discussed. Brazil had been suggested for 2006. It was suggested that a work session on this topic could be held during this ICWES-XIII.

Monique commented that members should be advised of this at the General Meeting on Sunday, August 28. It should also be on the agenda for BM6 Monday, August 29. Monique suggested that, if suitable interested organizers can be identified, we aim to confirm South America for 2006 (although Russia is still open for 2006) and that bids be open for 2007. However, we will be guided by the response at the General Meeting, where Sue will present an item on regional conferences. Suriya suggested that India should be the host for 2007. Monique invited her to put in a formal bid.

7.2 ICWES-XIV Conference

Monique Moutaud presented the proposal from the Association Française des Femmes Ingénieurs for ICWES-XIV to be held in Lille, on July 20–23, 2008, with the theme "Role of Women Engineers in policy making in a global and changing world." A handout with some preliminary details was provided. In discussion, some concern was expressed about reference to engineers and not scientists in the theme, and also that "policy making" might not send the right message.

7.3 Next Board Meeting (BM6) and 2nd General Meeting (GM2)

The next Board Meeting will be held on Monday, August 29, 2005, from 9:00 a.m. to noon.

The date of the 2nd General Meeting remains to be determined, following GM1 (to be held on August 28, 2005).

8. Strategic Plan and Program Development

Deferred to BM6.

Motion to adjourn. Proposed by Anna, seconded by Marianne. Carried unanimously (August 25, 3:50 p.m.)

BM5 Minutes—Attachment 1

4.1 Programs Other Than Conferences

Participants: Anna, Misa, Mabel, Marianne, Betty

Anna and Misa introduced a few items, e.g., a proposal for the Eastern European Group, recruiting Dr. Haryoung Poo to the Program Committee and drafting a document on program and projects. This was followed by discussions on the following points:

(a) After some discussion, it was agreed that an internal document on programs and projects should be available, not the Subset of Programs and Projects distributed in the BM5. The "subset" misses some key background information and it may cause problems leading to duplication or omission of program-related items. The internal document will be useful to the Program Committee as well as to any member interested in participating in programs and projects. (After the final BM5 minutes, Misa will update Draft #2 and circulate it for comments.)

(b) It was agreed that a program or project will be defined using the two variables mentioned in Draft #2, namely, Life Cycle (K-12 or equivalent age, undergraduate, graduate, entry-level professional, mid-level professional, senior-level professional and retirement and unemployment) and Region (as defined by INWES). In terms of needs, a program or project should be customized over the Region variable. For example, some projects might be applicable over multiple regions while others might be Region specific.

(c) The group identified the following items to be important, in particular, the awareness about and interest in STEM among girls.

1. For K-12, in particular for ages 12–18, raising interest in and awareness about STEM-related subjects (Math in particular) is important. In Africa, raising girls' interest in Math is considered a priority.
2. For undergraduate students, graduate students, and professionals, INWES could encourage the development of internship, assignment, and exchange programs.
3. The initial work was led by Marianne on the manual on how to create an association. It needs to be revisited and additional work must be carried out before a completed version can be distributed.
4. A similar case would apply to the sister association program, i.e., follow-up is required.
5. Scholarship programs need to be developed for students.

6. The group suggested an ICT for Women Program. This would include
    - Articulating benefits of promoting women in ICT to private and public sector organizations.
    - Making women aware of professional opportunities through appropriate means, e.g., workshops, career seminars.
    - Offering skills development courses and workshops at professional levels, e.g., entry (technical), middle (ICT management oriented), and senior (leadership).
7. The group suggested career development programs for students, entry-, mid-, and senior-level professionals. They would include re-employment of laid-off workers and leaving and re-entering the profession due to family needs.

BM5 Minutes—Attachment 2

4.13 Professional Development

Participants: Anna, Misa, Monique M, Teresa

Professional development programs have been developed by a number of organizations, including supporters and member organizations of INWES. Therefore, the group would suggest the following steps for the INWES to develop its own program:

Step 1: Compile an inventory of readily available programs/workshops/courses from its member organizations, e.g., AWSE in Africa, SWE in the USA, and associations in the UK and France. The data should contain target populations including Geographic and Life Cycle variables.

Step 2: Select a list of "quick hits" in terms of target populations and areas of INWES priority.

Step 3: Arrange access to courses/workshops/coop programs, etc., for items in the "quick hit" list.

Step 4: Develop building blocks to success, using items available from Step 3.

Some of the "key areas" identified:

- Coop programs, workshops, awareness courses for students entering the job market
- "Survival Toolkit" type workshop/courses for the junior-level professionals
- Leadership/Manager/Executive training for mid- and senior-level professionals
- Mentoring and networking, both formal and informal
- Information session for leaving and re-entering the job market
- Soft skill areas, e.g., negotiation, communication, assertiveness, presentation (generic courses/workshops should be tailored for women in science and engineering).

2005: First General Members' Meeting (GM1)

(=AGM1) Seoul, South Korea (at ICWES-XIII), August 28, 2005, 3 p.m. to 5 p.m.

The first Board of Directors' election was on the agenda of this meeting, to replace the Interim Board. Below are highlights of the minutes of this meeting, which replaced the ICC Business Meetings that had been held since 1964; this was a historic transition.

Reports from the Executive Committee: Monique Frize, President, explained the role of each of the Officers and reminded everyone that INWES was created in Ottawa, Canada during ICWES-XII in July 2002. Monique gave a brief overview of the Board meetings held to date.

The Executive Committee met almost every month by teleconference. Pam Wain, Shirley Malcom, and Claire Deschênes have been actively representing INWES at UNESCO meetings pertaining to their upcoming International report on science, technology, and gender (IRSTG). In October 2005, Monique Frize and Suzanne Brainard will be in Puerto Rico with 30 minutes each to present at the World Federation of Engineers Organization meeting; this was originally an all-male committee dealing with the eradication of poverty amongst other issues, but Monique asked for INWES to be included on the committee. Suzanne Brainard (Global Alliance) was able to present on behalf of INWES in Shanghai at the UN conference on gender. Maureen Lofthouse has been appointed as the new board manager, replacing Catherine Didion. Monique explained that INWES is not an umbrella organization, INWES is its membership. Some of INWES corporate members are IBM, Global, Nortel, Bechtel, SNC Lavalin Inc., and L'Oréal Paris has just applied for corporate status. Monique concluded that there is an urgency to address women in engineering issues, as enrollments at university have dropped down to the 1989 levels; similar trends are noted around the world.

VP Conferences Report (Sue Bird): Sue's committee has the following members: Issié Gueye, Myung Hee Jung, Marianne Rodot (France). Sue explained that each time this group meets, it assesses the current conference and regional events, and decides which future conferences to support. Representatives have attended some events including the launch of the Arab Network.

Current issues include the bidding process for ICWES and the regional conferences and how to carry this out. Sue announced that Lille, in France, has been chosen for ICWES-XIV.

Secretary General Report (Claire Deschênes): Claire has been holding this position since July 2002. Maureen Lofthouse and Misa will now be assisting her. The position includes the preparation of agendas, documents, invitations, recording minutes during meetings, and drafting of resolutions as well as maintaining corporate documents, ensure regular meetings with the president and executive, by phone as necessary; Cate Palmer does registration for the memberships separately. The website will soon allow registration online. The Policy Manual has been updated; it *is* a living document that will continue to grow. Claire is developing what the "benefits" are for members and sponsorship packages to recruit new members. Claire also creates contracts for new organizations asking for a grant.

Programs Committee (Anna Szemik-Hojniak [chair] & Misa Gratton [Vice Chair]): Committee Members: Monique Frize, Jennifer Flanagan, Haryung Poo. This committee was established in South Korea in 2003. Some of the initiatives include the establishment of the groundwork for scholarship programs, summer camps, award programs, professional development programs. Fundraising must now be done to make these plans a reality. Other ideas include having professional student development at future ICWES conferences.

Nominating Committee (Sachiko Okuda, Chair): Committee Members: Joanna Olu-Maduka, Anna Szemik-Hojniak, Claire Deschênes. The mandate includes the preparation of the slate of nominees for the Board of Directors. Nominations closed on the website March 28, 2005, and there was a good response. The idea was to create a policy for selecting directors, ensuring they are the right people and well-suited to play key roles on committees. Ideally, a director should have attended at least two ICWES conferences and have an adequate knowledge of the issues of concern.

Web Portal Committee (Monique Frize): Committee: Claire Deschênes, Misa Gratton, Teresa McNutt. Adlina Hamid, a student at Carleton, designed the website. The mandate of this committee is to create an information center for INWES. There are three phases in the evolution of the site: Phase I: Creation and new face (completed in January 2005). Phase II: Creation of sign-in for (1) directors, (2) committees, (3) members, while also maintaining a public access site, with rich information; members will have access (in December/January). Phase III: Construction of a web portal, including statistics. December membership renewals will get an ID and password (individuals for themselves and organizations for all members).

Programs and Projects: The digitalization of ICWES-I to XI was completed by Suriya Thevar. Some proceedings are missing (ICWES-VIII). This allows papers to be available from 1964 to 2005 except for the 8th; ICWES-XII and XIII are already in digital format; author title indexes have been created as well as indexing by name and country. Another project is to continue developing the manual on how to create associations; people can join as individual member and then get the manual. Jennifer Flanagan, an expert in youth initiatives, has created excellent strategies (brochure) for involving youth and the parents of girls. We also need men who care about our initiatives to be involved as well. Another important aspect is to be represented in mainstream policy-making bodies to ensure our voice is heard. Career development via videos, etc., can be added to this list. Sue has excellent models to share. Another area worked on has been outreach; in 3 years, INWES worked to have a voice in as many international groups involved in STEM as possible such as UNESCO, WFEO. Another area of focus is career development.

Advocacy Committee (Chaired by Claire Deschênes): Key questions of the Committee: (1) What are the concerns of women in STEM? (2) What should INWES do? The Gender and Leadership Symposium held at ICWES-XIII collected responses to these questions from the 100 attendees. Answers were prioritized.

Elections of the Board of Directors: Claire Deschênes made a motion: "To have Marianne Rodot act as Election Officer as Claire is on the nomination slate." Seconded: Sue Bird. 0 against, 0 abstentions. Motion passed unanimously. Maureen Lofthouse and Teressa McNutt were the vote counters.

Corporate Directors: Three candidates for three positions: Janet Higgins (NORTEL—North America); Melody Kratsios (SNC-LAVALIN, North America); Deborah Landers (IBM—North America): Elected by acclamation. **Individual Directors:** three candidates for three positions: Suriya Thevar (India—Central Asia); Anna Szemik-Hojniak (Poland—Eastern Europe); Unesu Ushewokunze-Obatolu (Zimbabwe—English-speaking Africa). The vote: 14 individual members in favor; two abstentions. **Organizational Directors:** 12 nominees for 12 positions: Sue Bird (WES—Western Europe); Claire Deschênes (AFFESTIM—North America); Monique Frize (Pathmakers—North America); Yolanda George (AAAS—North America): Misako Gratton (WISE Ottawa—North America); Issié Gueye (CIFISATS—French-speaking Africa); Durdana Habib (WESTIP—Southeast Asia); Mabel Imbuga (AWSE—English-speaking Africa); Gail Mattson (SWE—North America); Yvette Ramos-Aivazian (AFFI—Western Europe); Akiko Tsugawa (INWES Japan—Far East Asia); Myung Hee Jung (KWSE—Far East Asia). The vote: 13 in favor; 0 against; 0 abstentions; approved.

By-Laws: They were adopted by the Board in Paris, France. Today was the first general meeting, so it was important to adopt them formally at this meeting. They may be changed as the organization evolves. Motion: Adoption of INWES By-Laws, moved by Claire Deschênes, Seconded by: Issié Gueye: 0 against; 2 abstaining; motion passed.

Review of Financial Statements (Gail Mattson): There have been two financial reviews, one in 2003 and 2004. INWES was officially established in 2002, but did not get incorporated and establish a bank account until 2003, therefore, it was in 2003 that INWES officially started to function financially as a nonprofit corporation. All the information has been reviewed by the Board of Directors, recorded and approved, and presented at the General Meeting. All funds are presented in Canadian dollars. No questions or issues were identified. Future reviews will be done for the 2005–2006 and 2006–2007 financial years.

Appoint Auditor (Accountant Firm) for 2006: Motion: To name Raymond Chabot Grant Thornton LLP as auditor for 2005, 2006, and 2007. Vote: 0 against; 0 abstaining; motion passes unanimously. **Authorize Board to fix the remuneration of the auditor (accountant firm):** Motion passed.

Next Regional Event 2006 and 7th Board Meeting: Bids will be submitted in the next couple of months.

ICWES-XIV Presentation by Marianne Rodot: PowerPoint presentation and brochure available in draft form. Date: July 20–23, 2008 in Lille, France at the University of Science & Technology. Title: "Role of Women Engineers and Scientists in a Global Changing World." Main organizers will be the French Organization of Women Engineers (AFFI) in partnership with INWES. The goal is to be a European effort, so AFFI is looking for partners.

2005: Board Meeting 6 (BM6)

Seoul, South Korea, August 29, 2005

Highlight of Minutes:
Board Members Present: Susan Bird, UK; Claire Deschênes, Canada; Monique Frize, Canada; Misa Gratton, Canada; Issié Yvonne Gueye, Ivory Coast; Durdana Habib, Pakistan; Mabel Imbuga, Kenya; Myung Hee Jung, Korea; Gail Mattson, USA; Anna Szemik-Hojniak, Poland; Suriya Mayandi Thevar, India; Akiko Tsugawa, Japan; Yolanda George, USA.

Excused: Janet Higgins, USA; Melody Kratsios, Canada; Yvette Ramos-Aivazian, France; Unesu Ushewokunze-Obatolu, Zimbabwe.

Observers: Florence Hudson (for Deborah Landers), USA; Kyeongsook Kim, KWSE; Marianne Rodot, France; Board Manager: Maureen Lofthouse, Canada.

Monique congratulated Board Members on their election at the General Members' Meeting on the previous day, August 28, 2005. Draft Minutes of Board Meeting BM5 of August 24–25, 2005 were circulated for review. Comments and corrections were provided to the Board Manager for incorporation into a revision to be circulated to the Board with the Minutes of this meeting.

ICWES-XIII Budget: Myung-Hee provided clarification to questions regarding how much contributions were in kind, and how much was in cash, in addition to the seed money provided ($5000 US), which will now be available for ICWES-XIV. In-kind contributions are also included as well as the organizing committee's time, travel, and supplies in Myung's report. There were unexpected extra charges such as the Daily Bulletin. There were also unexpected contributions such as the Gala Night expenses, so the budget is expected to balance, but with no significant overage. Myung Hee Jung was thanked by the Board for an excellent conference.

Election of Officers: The following slate of officers was nominated: President, Monique Frize; Vice President Conferences, Sue Bird; Treasurer, Gail Mattson; Secretary General, Claire Deschenes.

Motion proposed by Anna Schemik and seconded by Misa Gratton: "*To elect the officers as identified in the above slate.*" Motion carried unanimously.

To support the Secretariat and the Board, Misa Gratton accepted to serve as Associate Secretary General and Maureen Lofthouse as Board manager, two volunteer positions. The Board accepted gratefully!

Board Responsibilities and Ethics Guidelines: Monique handed out documents on "Governance" and "Code of Ethics for INWES Directors" and asked directors to study and abide by them. Directors will be asked to sign the Code of Ethics document to indicate their agreement to comply. Monique emphasized that Directors are expected to familiarize themselves completely with INWES background and organizational documents, such as the By-Laws and Policy Manual, and to contribute actively to the work of the Board and its Committees. In addition, Directors are asked to play an active role in finding new members, particularly corporate members, and sponsors. Directors are expected to each bring in at least two new members, preferably corporate.

Regarding Board expenses, Monique emphasized that INWES has not paid and does not pay for Board members' travel and other expenses, whether to Board meetings or other events in view of our limited budget. Members are expected to draw on other sources for this. An exception is made for developing countries and countries in transition. Note: Several organizations have covered the cost of travel for their representative and some funds were provided at times by organizations like UNESCO, KWSE in South Korea, ICWES-XII surplus, etc.

Assignments to committees were made as shown below, on the basis of members' expressed interests. Board members were asked to find others from outside the Board who can participate actively in the committee work of INWES.

Advisory	Monique (lead), Kathleen Harer + other invited members
Advocacy	Claire (lead), Mabel, Issié
Communications	Yolanda (lead)
Conference	Sue (lead), Akiko, Myung Hee
Finance & Funding Devt	Gail (lead)
Membership & Recruitment	Yolanda (lead), Durdana, Suriya
Nominating	Issié
Program	Anna, Misa (joint lead), Mabel
Web Portal	Monique (lead), Suriya, Misa, Teresa McNutt, Claire

Participation in International Events: Gail noted that INWES is invited to participate in many international fora/events but cannot afford to accept indiscriminately. A guideline is needed for decisions on such participation. Directors are asked to identify what events they have been involved in that could represent important opportunities for INWES.

Prioritization of Programs and Projects: Gail led an exercise in which directors were asked to list their planned/wished-for programs and projects and then vote to determine priorities.

1. Participation in other organizations/advocacy (3 trips/year): $10,000
2. Manual on starting an association: $1500
3. Seed Money to start an association (3/year): $1800
4. Youth participation at ICWES conferences: $8000
5. Essay competition: $16,500 and Math competition (Ivory Coast) $5000

Top priority has been given to items currently included in the budget, namely Web Portal, Regional Conferences, and ICWES-XIV. Cost of the other projects is not in the current budget. Fundraising is needed to support them.

Fund Development: Gail distributed a document entitled "INWES Fund Development Fundamentals," which emphasized the need for every Board Member to be actively involved in fund-raising. The six basic elements of fund development were identified as: Projects must be concisely and convincingly defined; identify potential donors; donor research; donors want to see their funds effectively used; for continuity and stability, need different types of sponsorship; successful fund development is based on long-term relationship building.

An outline plan of Fund Development actions to be taken in the next 6 months was provided for Board Members to develop their own action plan and identify and engage new sponsors.

Regional Conferences: Some discussion of possible bids for 2006 was held at BM5, with candidate countries of Russia, Brazil, and India being tentatively identified. Monique advised that a formal request for bids would be placed on the INWES website, with a closing date of November 1, 2005, for 2006 bids. Bids for 2007 would also be invited.

Next Board Meeting: Next Board meeting will be held at the 2006 Regional Conference. The bid selected was Lille in France, to connect with ICWES-XIV organizing committee (May).

2006: Board Meeting 7 (BM7)

Polytech Lille, Salle du conseil, Lille, France, May 23, 2006, 9 a.m. to 15:30

Board Directors Present:
Nageen Ainuddin, Pakistan (replacing Durdana Habib)
Susan Bird, UK
Claire Deschênes, Canada

Monique Frize, Canada
Yolanda George, USA
Misa Gratton, Canada
Issié Yvonne Gueye, Ivory Coast

Mabel Imbuga, Kenya
Myung Hee Jung, Korea
Melody Kratsios, Canada
Debbie Landers, USA
Marie-Thérèse Malezet, France (replacing Janet Higgins)
Gail Mattson, USA
Aguri Nakano, Japan (replacing Akiko Tsugawa)
Yvette Ramos-Aivazian, France
Anna Szemik-Hojniak, Poland
Suriya Mayandi Thevar, India
Unesu Ushewokunze-Obatolu, Zimbabwe

Board Manager: Maureen Lofthouse, Canada

Invited Guests:
Michèle Baron
Nicole Bécarud
Samia Nasr Bishara
Mario Cervantes
Catherine Chochoy
Renée Clair
Dawn Fitt
Anne-Marie Jolly-Desodt
Tahani Lefebure
Paula Leventman
Marina Marchetti
Claudia Morrell
Monique Moutaud
Robyn Sandekian
Nathalie Vast
Pam Wain

Opening: **Monique Frize** opened the meeting by welcoming guests and directors and introductions were made. The agenda for the day was reviewed. Monique explained the purpose of this meeting which is to exchange information and seek partnership opportunities. Monique gave a brief synopsis of INWES for our guests and outlined the nature of the organization, its membership, its reach, and its role. One of the major factors INWES wants to address is the fact that women in STEM are not being reached and that there has been a major decrease in the number of women studying or gaining entry in science, technology, engineering, and mathematics. The proportion of women in Engineering in Canada has now reverted to levels seen a decade earlier. Programs that INWES created to address this major issue were outlined.

Presentations:
All guests were invited to present information about their organization and potential opportunities for collaboration with INWES.

UNESCO (Engineering): Tahani Lefebure presented for Tony Marjoram who could not attend.

ICWES-XIV: Monique Moutaud spoke of plans for ICWES-XIV.

CWIT: Claudia Morrell described the activities of the Centre for Women in Technology.

WFEO: Tahani Lefebure spoke of WFEO being an international engineering organization that operates in more than 90 nations and has 90,000 members. Their new project, "Empowering Women in Science and Technology" arose from the general assembly of 1999. Discussions are ongoing with INWES regarding potential collaboration. (For example—the Tunis conference 2007.)

Engineers Without Borders (EWB): Robyn Sandekian gave an overview of EWB activities at the University of Colorado at Boulder. The program attracts a broad diversity of engineering students but also does outreach to K-12. Robyn described an example of their work for developing countries introducing the use of water filters and this seems to be particularly attractive to women and minorities and may help to attract them to engineering. Link: www.teachengineering.com for grades 3–4.

European Union (EU): Marina Marchetti spoke of Science and Society and the Gender Equality Policy within EU. (See www.cordis.lu/science-society/women, and http://europa.eu.int/comm/research/science-society/women-science/women-science_en.html.)

OECD: Mario Cervantes gave an overview of OECD with a look at the current status of women in various regions. OECD is an intergovernmental organization that is funded by the various governments involved. Its goal is to promote sustainable economic development and its members include 30 of the most developed countries, and the organization is now trying to include other countries. OECD has created a new gender tool for researchers and policy-makers, the *Gender, Institutions and Development Database* (GID-DB). This database presents an analysis of obstacles to women's economic status and covers 162 economies and has 50 indicators on gender discrimination. It also includes institutional variables that range from intrahousehold behavior to social norms. Link to this site at: http://www.oecd.org/dev/institutions/GIDdatabase.

UNESCO: Renée Clair UNESCO provided funds for the Merrickville meeting where ideas to develop INWES were first discussed. Renée was present as UNESCO representative. She spoke of her vision for INWES. She made a number of interesting comments stemming from this vision and her ongoing experience in International Development. The following are a few of the key points:

- Politics is now a major component of STEM issues (in science and in its structure). For INWES to be a highly effective organization is no longer enough. There is a need to push global strategy to get fresh ideas of what science and technology are. There is a need to be more political in thoughts and strategy.
- Young people are not interested in old science and technology and in traditional ways of dealing with women in STEM issues. We need new reasons for science and technology in today's civilization to get them interested—especially women. A new vision is needed.
- The need for a strong international organization to join forces as separate organizations is not helpful (fractured work). To be part of globalization, INWES needs an international vision. Be ambitious for yourselves and for your partners.
- A strategy that reinforces national networks (associations, governmental bodies, national Unesco commissions, research chairs, NGOs for Women in STEM, programs for promotion, retention…) should be adopted. The strategy involves prioritising at a national level with project roots in the country as they understand the needs of their own population and can develop a national network. Connect all these national to the international dimension through INWES. Challenge for this meeting: find two or three good projects according to this vision.
- The role of UNESCO regarding gender in science and technology may be funding although there is not much money for women in STEM available these days. It might make recommendations, provide statistics, newsletters, contact lists. It can act as a catalyst through its various chairs on women in science and technology.
- Also important for INWES is to think "Gender mainstreaming" (for example: *water* and women in STEM).

Renée concluded by stating that the challenge for all is to understand the dimensions and the need to find two or three significant projects that we can undertake together toward reaching our goals.

Discussions in Groups:
ICWES-XIV Preparation—Monique Moutaud, Sue Bird. Monique M. should participate at the next Helsinki meeting to meet and invite others to participate with INWES. Need to link with European countries. Need translation at the opening session and maybe some other plenaries. Suggest we give the floor to some African speakers. Develop slogan/keywords for conferences i.e., change the world, new opportunities for women, challenge, etc. Possibility to circulate logos for response from directors.

WFEO, INWES, UNESCO Partnership, and Joint Projects—Monique Frize with Tahani, Mario, Claudia, Maureen. The WFEO conference will be held in

Tunis in 2007. All 90 members (countries) have been invited. Conference consists of lectures and workshops and a training workshop for N. Africa and the Middle East on leadership, promotion, advancement, etc. Case for INWES participation: If really intended to deliver ideas on developing the role of women in science and technology especially N. Africa, Middle East—then INWES is definitely interested.

Should approach funders together.

Board to decide whether to send delegates or the whole Board. A joint project could be on statistics presented in a joint session, with inputs from OECD, UNESCO, etc. Monique Frize started to work on gender in S&T and disaggregated data. We need to bring to the table all of the institutions involved with data development. For new areas of data such as working conditions, etc., we need to work towards setting standards so all data that are collected and used consistently. We need to post the INWES Web portal tool. Check the OECD tool currently being developed and the UNESCO Global Report on Women in Science.

Outreach Strategy to Reverse the Declining Trend of Women's Participation in Science, Engineering, and Computer Science—Gail, Claire, Debbie, Melanie, Anne, Robyn, Marina, Catherine, Dawn. The following are points raised by the Outreach group: Focus on awareness at the top level (of corporations, media…) with a message of technical relevance. Establish/provide INWES members with a database of best practices that are appropriate for dissemination and translate them. Money will be needed for the database and for translation. Identify one specific project, something that is cool and uses technology that is performance-based. Success leads to success (i.e., WAGGGS World Association of Girl Guides and Girl Scouts). Matchmaking of various organization programs and activities such as with EU program and EWB projects. Schedule the events. Work with corporations and associations to find good role models.

Guidelines: INWES role is to enable women in regions to do outreach in their areas. Winning strategy is probably to reach girls and boys equally.

Proposal for follow-up action: To organize a high-level meeting with corporate directors, communications committee to develop outreach activities. Date: September 9, 2006.

2006: General Assembly Meeting 2 (AGM2)

Held Online (Through Email), May 22–25, 2006

Two By-Laws clause changes were accepted. Our official documents need to be updated accordingly.

Action: Claire will organize AGM3 Electronic meeting in late fall.

Partnership with UNESCO: Monique F. talked about putting forward a proposal for an official partnership with UNESCO. A discussion followed on the obligations and advantages of INWES:

Obligation: keep UNESCO Director General informed in the field of competence and contribute to the annual report.

Advantages: UNESCO Director-General will provide documents concerning gender and we shall be kept informed; UNESCO executive committee can invite INWES to participate in their meetings, and we can be invited for consultations. If we are the most competent for certain tasks, there could be funding. UNESCO National Commissions could be a point of entry for us in many countries. It could affect our charity status request positively.

Action: Monique F. will send a proposal to UNESCO Director-General in Paris. This proposal needs to be supported by at least three Regional Commissions. The Canadians will support the request. Directors who are in contact with their National Commission for Unesco can help by asking for support from them. Monique F. will provide a draft letter for each director to customize their request.

Update from Newly Appointed Executive Director (Márta Ecsedi): Márta presented the Executive Director's Report. She talked about the membership database. Potential members in the database will be sent a letter. She suggested inviting organizations in universities to join INWES and organizing a deans' conference, maybe for ICWES-XIV.

Action: Márta will work on a conference of the Deans for ICWES-XIV with Yolanda and Suzanne. Gail and Monique F. to meet with deans at the SWE conference.

Action: Directors will send Márta contact information for potential memberships by the end of October.

Update from Committees and Programs (Yvette): As Program Chair, Yvette presented her vision for INWES programs. She asked that we continue working on the project forms and she will gather the information. Budget items should include web portal costs.

Program and Project Catalogue (Yvette): we need to make money. The expectations of sponsors, our program ideas (what we want to do), and members' expectations, all may be different but we need to find commonalities between them and make all three aspects work together.

The following summarises Yvette's suggestions: increase the number of women engineers in corporations; conference to showcase top women engineers; list we have is too long; focus on few projects (max. 10); emphasize objectives and beneficiaries. Four subgroups should be organized during this meeting: Group 1: work on three projects for the charity status (essay contest, best practices, Gender Tool). Group 2: work on one project for the directors (welcome and orientation kit). Group

3: work on three projects for capacity building for INWES members (train the trainers). Group 4: work on a plan for benchmarking. Outcome: to write an executive summary for projects and programs: What is the background? What are the main objectives? What is the strategy proposed to achieve these main objectives? Who are the key beneficiaries? Who are the potential financial partners? What are the main activities and budget? Key Milestones.

Various Presentations by Members, Followed by Partnerships (WFEO/UNESCO/NSF/Brazil 2008): Monique F. mentioned the World Engineers Conference 2008 in Brazil earlier and Yvette talked about it here. This conference occurs every 4 years. There will be five people picked out of 15–20 to be on the organizing committee. Yvette will be one of these five for Brazil. She will work with WFEO organization and an MOU should be signed between INWES and WFEO this fall.

Action: Monique F. to send a letter of intent on how INWES will support WFEO in cooperating on various projects where we have synergies. Gail will provide a template from SWE.

OECD 2006: The report from the Ottawa Workshop in September 2006 was made available.

Advisory Committee (Monique F.): Honourary members have asked to join this Committee to provide advice on membership recruitment and fundraising. Kathleen Harer chairs this committee. She reported to Monique that members are not responding; they are not very motivated and/or it is difficult to reach them by email. She suggested: Operate by mail and focus on only one or two topics instead of talking about the general mandate and on one or two projects.

Action: Monique F. met with Nicole Bécarud. Monique M. will follow up with Nicole and find a way to honor the ICWES founders at ICWES-XIV.

Communications Committee (Yolanda): Yolanda prepared a short Newsletter for this meeting. She offered to rework it in a longer version by the end of the summer.

Action: Directors should send articles for the next Newsletter by August 15, 2007. Monique F. will send a new picture of herself. A list of organizational members should be added to the newsletter.

Yolanda offered to print (at AAAS) the next Annual Report, to be prepared by Helga (student volunteer) with input from the Executive.

Action: Since we need to archive conference records, ICWES-XIII CD will be sent from Myung Hee (South Korea) to Suriya for file organization and to Monique F. for INWES website.

Nominating Committee (Issié): Elections will occur at GM4 at ICWES-XIV in July 2008. During BM5 in Seoul, the board decided to review the application form. The Committee was composed of Anna, Sachiko, Claire, Issié, and Monique F. as ex-officio. At the BM7 in Lille, a suggestion was to enlarge the committee for at least: 1 organizational director, chair; 1 individual director; 1 corporate director. New Committee members (Issié, Anna, Claire, and Monique F.)

Action: Nomination kit to be ready before the end of August. Post the nomination kit to our website by the end of 2007. January-March: reception and review of the nominations. March 20, 2008: deadline for nominations which are to be submitted to Chair Nomination Committee and Secretary General.

Develop the Five-Year Review Process (Claire): Claire summarized a draft document to help prepare a five-year review.

Action: Yolanda to document suggestions to improve the content of the five-year review. Claire and Monique will link outcomes to goals and identify gaps. Monique reviewed ongoing projects:

Gender tool: She informed that IDRC (International Development Research Center, Canada) is interested in receiving a proposal for funding the development of the tool.

Best practices: The Global Alliance website presents many Best practices. A manual on how to start an organization exists already.

2007: A Challenging Year for INWES

In September 2006, after the meeting in Lille, Claire Deschênes and Monique Frize spent a whole day at an OECD meeting in Ottawa. We were part of a four-person panel on women in engineering.

We prepared an application for a grant from CIDA to support women traveling from Eastern Europe and from developing countries to attend the symposium in Poland planned for July 2007, the next Regional event.

In April 2007, Claire and Monique met Isabelle Blain at NSERC. We took the opportunity to inform Isabelle about the upcoming INWES symposium in Poland and hoped that she would attend or send a representative to speak about the NSERC–CWSE chair program. We informed Isabelle that we were filling out an application for INWES to become an official NGO partner of UNESCO. At this time, the financial situation of INWES was worrisome. CIDA had rejected our initial request for funding for Poland, but the CIDA representative said we could re-apply a month later, in May. We modified the proposal according to the comments we had received. We also asked SWE if it could lend us $15,000, until our finances improved after the Polish symposium. SWE generously agreed and did not charge any interest on the loan.

In May 2007, Claire and Monique prepared the Board meeting in Poland and the third INWES AGM. Anna Szemik-Hojniak received a grant from the Polish Ministry of Science and Higher Education, which helped to cover the expenses of the symposium. Anna sent Claire and Monique 30 essays from women who were applying for financial support. The essay competition was again chosen as our way to select women receiving a travel grant; it was important to include as many women as possible from countries in transition and developing countries. The CIDA grant of $31,700 was confirmed at the end of May. The winners of the essay competition were notified so that they could apply quickly for their visas and make travel arrangements. We calculated that we could subsidize travel costs for 7 women from developing countries, 4 from Warsaw, and 11 from Eastern Europe. This was our third successful application for funding from CIDA; the generous responses over the past 5 years had enabled many women from Africa and other regions to participate in global discussions and share the information with others, hopefully helping to bring about change for girls and women in their countries.

Tribulations in Tunisia: A Major Battle (Excerpt from Frize Memoirs, 2019): In June 2007, a serious situation occurred, potentially threatening the existence of INWES. Some of us on the INWES board heard rumors that there was a plan to create a parallel international network; this action was led by the woman who had organized the event in Baltimore in 2005. One of the documents created by this group used our INWES global map design, without our permission, to lay out their plan. Their document was titled *Declaration of Agreement in Support of Women of All Ages Aspiring or Working as Professional Engineers*. Few people were aware that the group planned to create the international network of women in STEM at a WFEO global colloquium in Tunis. The official main goal of the colloquium was to empower women in engineering and technology. The WFEO colloquium had several partners: UNESCO, SWE, the Center for Women in Information Technology (CWIT) in Baltimore, Femmes Ingénieurs (FI, France); INWES was listed as a partner.

The program mentioned a workshop to be held the day before the beginning of the colloquium to discuss leadership, entrepreneurship, and networking skills; however, these topics were changed at the last minute, and the workshop was used instead as a forum for discussion of the "Declaration" document to create the network; specific persons backing the proposal were invited to participate. INWES had not been informed of this and had not been involved in the development of this document, which seemed to have, as a main objective, to create an international network of organizations of women in engineering and technology.

Since INWES was supposed to be a partner of WFEO for this event, six directors of INWES, including Monique Frize, came to Tunis to participate in the colloquium. When we arrived in Tunis, we heard rumors about the pre-event workshop, but no copy of the document was provided to us. The six INWES Directors decided to meet and discuss how to obtain a copy of the document and find ways to counter the establishment of a competitive organization. During the colloquium itself, the document was updated by a small group (no one from INWES); the plan was to present the document for adoption at the closing ceremony. An INWES Director, Pamela Wain, found a copy of the most recent version of the document which had been forgotten in the photocopy machine; we held a caucus to figure out how to react while finding ways for INWES to work with WFEO and its organizations on common issues. There was a way to accomplish both goals by making some small, but important changes, to the document. As the president of INWES, Monique Frize spoke with the president of WFEO, Kamel Ayadi; he was very supportive of our suggestions, and he amended the document to reflect our discussions; he agreed that the new body to be created would be confined to WFEO members, thus would be an internal organization within WFEO, and would not be a broad network like INWES. He also asked Monique to be one of the signatories of the document, renamed "'The Carthage Declaration.' Declaration". Monique signed just before Kamel presented the document to the 400 participants from 60 countries at the closing ceremony. The main person involved in the development of the previous document said she would not sign the Carthage Declaration if Monique Frize was a signatory. The response from the President of WFEO was that this was fine. She signed.

It was good that WFEO developed a plan for engaging its member organizations to support existing local committees on women in engineering and encouraging the creation of these committees in countries where there were none. These committees were forums for discussing common issues for women in the profession and sharing best practices. In a speech to the attendees of the colloquium, Monique Frize stated: "INWES and its members can be a source of expertise for this new WFEO project on gender; we hope that the new committee will find it useful to work with INWES once it has been established." A memorandum of understanding (MOU) was signed between INWES and WFEO a few months after the Tunis event. There was a positive outcome for everyone" (Frize 2019, pp. 308–310).

2007: Board Meeting 8 (BM8)

Wroclaw, Poland, June 29–30, July 1, 5–6, 2007

The symposium was a great success, thanks to the huge efforts of Anna Schemik-Hojniak. There was even a financial benefit provided to INWES. Minutes of the Board meeting:

Thursday June 29, 2007

Directors
Bird Sue (VP Conference)
Deschênes Claire (Secretary general)
Frize Monique (President)
Ecsedi Márta (Executive Director)
George Yolanda
Gueye Issié
Habib Durdana
Kimura Ryo
Imoto Ikuko (rep. Akiko Tsugawa)

Jung Myung Hee (ICWES-III Chair)
Mattson Gail (Treasurer)
Ramos Yvette
Szemik-Hojniak Anna
Ushewokunze-Obatolu Unesu

Excused
Imbuga Mabel
Kratsios Melody
Landers Deborah
Thevar Suriya

Observers
Brainard Susanne (USA)
Heap Ruby (Canada)
Ross Helga (Canada)
Lee Kong Joo (South Korea)
Leventman Paula (USA)
McNutt Teresa (Canada)
Moutaud Monique (ICWES-XIV Chair)
Wain Pam (UK)

Tour de table: Everyone introduced themselves. This was followed by a Press Conference with Polish media. The agenda was revised slightly at the beginning of the meeting and adopted. Motion by Yvette, seconded by Yolanda. The minutes of BM7 were also adopted. Motion by Yvette, seconded by Unesu. The action items were reviewed.

Action 1: Directors involved will do a formal follow-up with the potential partners with a copy to the executive committee: EWB, OECD, UNESCO, EU Platform, WISE.

Action 2: Yolanda will inform Monique F and Gail of the pricing of printing of INWES Progress Report (500 and 1000 copies).

Action 3: Board needs to provide contacts for Corporate Members to Márta.

Action 4: Yolanda will provide writing tips for proposals for grants or other fund-raising activities; INWES will post them on the website.

Action 5: Instead of the Tour de table, the Secretary will add time to future board meeting's agenda for directors to report and will provide a template for review. She will also change "guests" to "experts" on the agenda.

June 30, 2007

Update from the Treasurer: Gail read Financial Reports from 2005 to 2006. INWES had to borrow $15,000 from SWE to support expenses for lawyers, accountant, and others; this will have to be reimbursed by the end of December 2007.

Motion by Gail: Authorize payment of the invoice for 2006 Financial statement to Kelly Huibers McNelly Chartered Accountants. Motion passed unanimously.

In-Kind Contribution: There was a request at BM7 in Lille for financial records concerning directors' in-kind contribution to INWES. Forms were provided. Directors present completed the in-kind contribution forms provided by Gail during this meeting.

Charity Status in Canada and the USA: The discussion was on how a charity status can help to raise funds.

Action 6: Directors will seek information through their country's appropriate Ministry on NGO, Not for Profit, and Charity status, will summarize requirements and advantages for each and will report to Gail.

Update from the Secretary-General: Board structure discussion: Elections at ICWES-XIV at AGM4 and the new Board is to meet at BM10.

Board, Committees, Chairs, and Volunteers Update: Claire presented Board of Directors changes: IBM and Nortel did not renew their participation in INWES for 2007, director resignation will follow. Misa Gratton resigned from the Board last May. Claire presented lists of members in good standing and members who did not pay their dues. INWES should seek new memberships. Durdana volunteered to be the Membership Chair.

Action 7: Send Terms of Reference for the Membership committee and potential members lists to Durdana who will Chair the committee. Issié and Monique F will work on this as well.

July 1, 2007

Conference Committee Report: Monique Moutaud presented her report on preparations for ICWES-XIV. Yolanda and Gail offered to print material for ICWES-XIV such as flyers, large posters, and a full-page advertisement in SWE magazine. Awards are being negotiated with L'Oréal.

Board Support for Fund Development: Gail made a presentation on Fund Development, the main point being that it is the Board's Responsibility to help with funding.

Action: Each director needs to get three more organizations into INWES.

July 5, 2007

Report from work in small groups. Participants: Issié Gueye (Chair), Anna Szemik and Yolanda George. They suggest changing the nomination procedure for director (Appendix E of the Policy Manual and used at AGM1) to: Simplify the application form; request to attach relevant CV to this new form.

Motion 1: Moved by Issié seconded by Claire: To enlarge the Executive Board with a new position for VP of Programs.

Rationale: It will be easier to get money with this title; there is a need to increase action on existing programs and create new programs. The VP would help make this happen by following up with people and reporting to the Board and the President. This VP could have an assistant (paid position) for 25% of the job.

Arguments for the motion: VP title is needed when talking to people to empower them in external meetings. We need to empower our people with titles to get things done.

Arguments against the motion: A VP does not necessarily have to be on the executive committee; adding a vice president has major implications such as reaching the quorum for executive meetings. The executive director supports all the different functions of the committees.

Vote Motion 1: In favor: 6; Against: 0; Abstentions: 2. Motion carried.

Motion 2: Moved by Issié: seconded by Durdana: To enlarge the Executive Board with a new position for VP of International Partnerships.

Rationale: To interface with international organizations that have or may have MOU with INWES (WFEO, CWIT, TWOWS, Women's Forum, Global Alliance, Oxfam, GASAT …) as well as corporations and institutions.

Arguments for: INWES is involved with so many external things it's important to get someone who gets involved with international partnerships. Primarily the President has been doing that on top of all her many duties. Therefore, it would be useful for somebody else to take over this task.

Arguments against: President usually does external relations. Developing international partnerships is the role of the president (or president-elect).

Vote Motion 2: In favor: 3; Against: 4; Abstentions: 0. Motion not carried.

Monique suggested adding a new category of member called Institutional Members.

Rationale: Some institutions do not fit well in our membership structure at the moment, for example, universities or AAAS. Considering the goals and objectives of INWES, it is important to collaborate closely with institutions sharing the same goals as INWES

Discussion: The addition of Institutional members must go through bylaws change. How can we fit institutions in? How would be the voting rights not to have an over-representation of institutions against organizations? Some institutions such as the National Science Foundation could be a member without voting rights. Could UNESCO have a representation without votes? Adding a new category of members would allow changing INWES membership fee structure and would give more flexibility. At the moment, organizational membership price is based on the number of members the organization has.

Action: The executive will develop a rationale and a motion on this question for AGM3.

Motion 3: Moved by Gail, seconded by Issié: To create an award of recognition for an outstanding woman in engineering or science to be presented tri-annually at ICWES.

Rationale: Recognition and Celebration

Arguments against: Concerns that awards go to people who only focus on careers, not on women who have done something to help other women.

Argument for: This award will help us advertise conferences.

Amendment 1 to Motion 4 by Sue, supported by Issié and Gail: To create an award of recognition for an outstanding woman who has significantly improved the situation for women in STEM, to be presented tri-annually at ICWES.

Vote on Motion 3: The amended motion 3 was approved unanimously.

Motion 4: Moved by Gail, seconded by Claire: That the Executive Committee looks into the structure of the Board of Directors and the Executive Committee and associated changes to the By-Laws by December 2007 for consideration by the Board.

Vote Motion 4: In favor: 8; Against: 0; Abstentions: 0. Motion carried.

Action: Claire will prepare changes to the by-laws for AGM3.

Develop Templates for Specific Projects (in Small Groups)
The agenda was changed with a new reorganization of work in small groups:

Group 1 (***Gail***, Monique F., Marta, Claire, Sue): Work on three projects for Charity Status. (Best practices; Essay competition; Gender resource tool.)

Group 2 (***Pam***, Monique M., Durdana): Directors welcome and orientation kit

Group 3 (***Ruby***, Issié, Unesu, Yvette, Teresa): Work on three projects for capacity building for INWES members; "Train the trainers"; training for executives of corporations on gender issues; Forum for Deans on gender issues.

Group 4 (**Yolanda**, Myung-Hee, Paula, Suzanne, Anna, Ikuko, Ryo, Kong-Joo): Work on a plan for Best Practices or benchmark. International perspective for "Best practices benchmarking."

Topic: Creating a sustainable forum in academia and private sector in STEM

Objective: Increase participation of women in academia, the corporate world, nonprofit work in STEM sector.

Senior-level members of industry organizations need to be involved in INWES objectives. Increasing the number of women in STEM will increase the number of women that are attracted back to the industry. Strategy: Offer benefits like ICWES-XIV (networking) and do a follow-up after the conference. Beneficiaries: INWES, women in S&E societies, deans, CEOs, Institutions, and Industry. Benefits to participants of the Forum. Leaders of academia and private sector will share experiences. Women from the organizations could give speeches at their organizations.

Broadening the base for innovation by including women. Finance: The organizations and institutions would pay a fee to be a member of the Forum. Min: 1000 USD. Follow-up: In Dec. 2008 at WFEO WEC (Women in Engineering Conference) in Brazil. This initiative should be done before December 2008 (Unesu lead). Will need assistance from members to get this done (Guidelines).

A discussion followed around organizing an event at ICWES for corporation CEOs and deans together or only for one group. Will this type of event attract French Deans (Monique M. expressed some doubt). They have different issues and solutions but would the presence of industry be attractive to deans as it is elsewhere? Should it be a large event or a small forum?

Action: Group 4—Deans and CEOs (Unesu): Unesu is going to develop the proposal and include Monique M. and Monique F. in the discussions. After the end of the summer, she will circulate the proposal for a pilot program. Follow up at Lille in July 2008 and at WEC 2008 in Brazil in Decembre 2008 (Suzanne).

Report from Group 3: Charity Status: Participants: Márta, Gail, Pam, Monique F.

Gail presented a Charity Committee Report; recommendation that we establish another entity that will qualify for charity status (educational, research Institute).

Motion 5: Moved by Monique, seconded by Gail: to establish an Education and Research Institute.

Discussion: Canada considers that there is a vast difference between Charity and non-profit organizations, but the USA only has nonprofit. In order to be a nonprofit in the USA, however, Charity status in Canada must be obtained. Once this happens we can act more liberally in the USA. The attorney suggested to start with three best practices projects. INWES still has the large list of projects chaired by Yvette which will be presented to corporations that do not need a tax receipt while the other projects will be marketed by Márta to corporations that need a tax receipt. We could campaign separately to fundraise for the institute projects and for the other INWES projects. Charitable fundraising would take a percentage for administration fees and INWES would still benefit from these fees. We will later look at how our charitable status could reach reciprocal agreements in other countries. The attorney recommends against a foundation because they are limited in what they can do. Institutes are more flexible. Best practices under the institute must be of a charitable nature; we cannot include training our members. But it could include INWES best practices.

Vote on Motion 5: The motion passed unanimously.

The process will be the following: write Letters patent to incorporate the institute and then apply for charitable status for the institute.

Action: Group 3—Charity Status: Gail to investigate how much overhead can be charged for projects. Proceed with the Letters Patent application (Monique F, Claire, Gail, Márta).

Motion 6: Moved by Gail, seconded by Monique F.: Hire Jane Burke-Robertson, a lawyer, to help for Letters Patent and with the application for charity status for the Education and Research Institute.

Vote on Motion 6: The motion was passed unanimously.

Group 2: Partnerships—UNESCO, WFEO: Participants: Monique F, Monique M, Claire, Teresa, Paula. The UNESCO document prepared by Claire was presented.

Motion 7: Moved by Claire, seconded by Sue: "The Board of INWES supports the application to UNESCO to become an NGO partner of the operational relations type."

Vote on Motion 7: The motion was passed unanimously.

Action: Group 2—UNESCO Partnership: The proposal document will be completed with projects (Claire and Monique F.). Monique will write a letter to the Director General of UNESCO and request a support letter from the National Commission of Canada. She will send it to all the directors to hand to their respective National Commissions if one exists. Newsletters will also be given to National Commissions (Monique). Claire will give directors steps to follow such as making an appointment with National Commissions. All directors are to send letters, documents, and newsletters to their National Commissions.

Discussion on WFEO meeting in Tunis (6–8 June 2007): Monique F., Monique M., Issié, Pam, Nadia, and Yvette were present. The original intent for INWES in Tunis was to do a training session for the participants. A declaration was developed and was adopted. The idea of establishing a standing committee for women in engineering was circulated. There will be a members' meeting in India in December 2007 where a decision will be made. After we get this information, we should think about what we would like our interaction to be. The Ivory Coast women federation is part of WFEO. Five women will be invited to help in organizing Brazil conference (WEC). INWES participation should be discussed in detail and decisions made clear to INWES directors and ICWES-XIV organizers.

Group 5: Essay Competition: Participants: Claire Deschênes, Paula Leventman.

Summary of the one-page draft proposal: Aim: enhance educational opportunities in STEM fields for young women in all of the 11 regions of INWES. Target age: Ages 14–17. Prizes each year: Winners $1500 each, three Winners in each region. Financial Partners: Science and Engineering organizations and corporations. Set up a committee in each region consisting of two or three INWES members. Applicants will be asked: Background information; to describe why they want to follow a course of study toward a career in STEM; an Essay biography of a women scientist or engineer from their region.

Action: Group 5—Essay Competition (Claire and Paula): Send Claire and Paula suggestions (All). Claire and Paula to write up a more detailed copy of Scholarship description and rules using suggestions by the end of September. Monique M. and Okon to provide C. Deschênes with examples.

Discussion and questions arising: Need to define judging criteria. Is this too much money for a teenager? Should we give the money as part of school fees? Should we only look at High School Seniors? Ask who they want to become when they grow up. Since we have three per region, can we hand out the different scholarships to different ages maybe one for university and two for younger students? Should we continue with INWES official languages, or can they submit in their own language with translation in the region? To be translated before the Board vote.

ICWES and Regional Conferences:
Action: A clarification of relationship between INWES and ICWES will be worked through email by Sue, Monique M., Myung, Monique F., Claire. A follow-up will be done at an executive meeting.

INWES Operations: Address of INWES is the home of Monique F. (corporate headquarters). Ottawa UNESCO will be asked if they can provide office space for INWES. Corporations in Ontario need Ontario accountant. Monique F. recommendation: look for new headquarters for INWES in Ottawa.

Action: Monique F. is working with UNESCO to find a new permanent address for INWES headquarters.

Next meetings and events (Monique F.)

Proposals are India, South Pacific (places where we have not been yet); women in Puerto Rico still interested (Gail). Need a lot of people in Brazil 2008 to build contacts. Could be good timing to have it there 2009. Peru: One member has attended three conferences.

Action: A call for proposals will be put by Sue for the next Regional Conference (2009) on INWES Web as soon as possible. Bidding should be done before ICWES-XIV so we can advertise at ICWES-XIV.

The next Board Meetings and AGM: at ICWES-XIV.

2008: General Assembly Meeting 3 (AGM3)

Held Online (Through Email), May 24–27, 2008

Proposed Motions

1.1 Motion AGM3-1 Proposed by Gail Mattson, seconded by Claire Deschênes. To appoint for 2008–2009 the accountant firm Kelly Huibers McNeely, Chartered Accountants[1] to produce INWES Financial Review or Audit.

[1]Coordinates are: Michelle E. Huibers, CA, Kelly Huibers McNeely, 5992 Hazeldean Road, Stittsville, ON, K2S 1B9, Phone +1 (613)836-5626, Fax: +1 (613)836-2540.

1.2 Motion AGM3-2 Proposed by Monique Frize, seconded by Gail Mattson. To authorize the Board to fix the remuneration of the Accountant for the Financial Statement.

Results of Voting

A delegate from each organization was invited to vote on the two motions.

28 organizations were invited to vote, totalizing a possibility of 47 votes.

On this amount, 18 delegates answered, representing a quorum (more than 40% of organizations).

The total of votes received was 27 for each motion. Both motions were carried unanimously.

Related Documents

- INWES GM3 Executive Report
  - President report
  - VP Conference report
  - Secretary report
  - Program committee report
- INWES GM3 Financial Report 2007
- INWES GM3 List of Directors 2007–2008
- INWES Organisational Members in Good Standing 2007–2008
- Newsletter 2007 (INWES Regional Symposium in Poland and news from INWES members) http://www.inwes.org/PDF/newsletters/newsletter2007.pdf
- Report from Anna Szemik-Hojniak (Regional Symposium Chair in Wrocław)
- Report from the Membership Committee
- INWES Extracts from By-Laws for GM

2008: General Assembly Meeting 4 (AGM4)

Polytech'Lille, Amphitheatre Appert, Lille, France, July 17, 2008, 1:30 p.m.

1. Opening by the President

Verification of Quorum and Introduction

Quorum reached—8 individuals, 13 organizations (27 votes on 49 potential votes)

40% is 20 votes.

Introduction: INWES needs volunteers. Individuals who have paid will vote for individual positions on the Board. Organizations in good standing are entitled to vote on motions and to vote for director positions for organizations.

The agenda includes seven motions that are recommended by the Board of Directors. INWES has the legal responsibility to have a financial review. The accountant firm who produced the review was a much better firm this time and the previous firm one was charging double. We vote to allow our accountants to make the next review and to let the Board fix their remuneration.

In 2007, the main event held by INWES was a regional meeting that was organized in Wroclaw, Poland, in July 2007. Several Executive meetings were also held through the Internet using Skype software.

Note: if you are part of an organizational member of INWES, you are part of INWES and entitled to attend the members' general meetings.

Adoption of AGM3 Agenda and Minutes

AGM4—Motion. Adoption of the agenda The Secretary proposed the agenda, Virginia Connolly seconded.

Agenda unanimously adopted.

AGM4—Motion 0. The president proposes to adopt the AGM3 minutes prepared by the Secretary provided with the meeting documents. Minutes adopted

2. Report from the Executive Committee

2.1 President Report

The main elements are the use of Skype for the meetings (executive) helped INWES to save a lot of money. Last year, we spend $300 on teleconferencing calls.

The president reported on activities linked to UNESCO. In April 2008, INWES became an NGO of the operational type. INWES received support letters from Ivory Coast, Canada, and Poland National Commissions for UNESCO for this nomination. This will help INWES to be at the table of NGOs, to share our experience with them, to be in the loop, and to benefit from various links established by UNESCO. There will be a Meeting of the NGOs in September in Paris. INWES will send at least one representative out of three possible persons to this major meeting.

Creation of the INWES Institute

INWES previous application to be accepted as a charitable organization in Canada was refused because INWES was not considered charitable. Therefore, with the help of a lawyer who recommended the creation of a separate institute specifically for projects of a charitable nature, Monique Frize, Claire Deschênes, and Gail Mattson founded the INWES Institute (three people and three proposals to keep it simple). In November 2007, they incorporated the Institute. INWES reapplied for Charity Status and is now waiting for a final answer (the application was made in January/December and it takes about 6–8 months before receiving a response).

With SWE support, Gail applied for charity status in the USA. However, there is an agreement between the USA and Canada regarding the Charity Status and if the application is successfully accepted in Canada, we can apply in the USA for 501c(3) status—with this status it will be easier to have US corporations join INWES and get funding.

Regarding the accountant firm, for the review, they charged less because they believe in INWES vision and mission; they only charged 15% of the normal price that would be charged.

The president developed a MOU between INWES and WFEO.

Monique Frize presented the INWES organizational structure. The network has a very horizontal organizational structure; there is a link between individuals, organizations, and corporations.

Therefore, the three parts need to have an agreement about the next steps, the next directions. One goal of INWES is to encourage individuals to create new organizations. The Board is made of 18 directors: three individuals, 12 organizational, and three corporation representatives.

A MOU has been signed with Engineers Australia (Representative Marlene Kanga), which will be hosting ICWES-XV in Adelaide, Australia, in July 2011.

2.2 VP Conference Report

A regional symposium and a board meeting were held in Poland in July 2007. It was the first time to our knowledge that Eastern Europe was talking about women in STEM and gender, where we met with decision-makers and exchanged information. INWES obtained a grant from the Canadian International Development Agency (CIDA) to help delegates from developing countries and Eastern European countries to attend the workshop.

ICWES-XV—In July 2011. We were proud to announce that the bid was accepted from Adelaide, Australia. We will put the bid for the next regional event on INWES website soon hoping to have it in an area where we have not been and where there is a need (maybe Mexico or Porto Rico).

2.3 VP Program Report

The Catalogue of projects that potential sponsors and corporations can fund proposed different projects developed by working groups within INWES.

Benchmarking—Paula Leventman is working on Best practices with Claire Deschênes.

Gail Mattson, Monique Moutaud, and Monique Frize visited 43 companies during the last SWE meeting to do fundraising. The Math Techno Contest is a project from Issié Yvonne Gueye. Okon Uduakobong proposed with APAGESTE (Nigeria) to lead a pilot Scholarship Contest test case. If it functions well, the project will be extended to other regions.

Fundraising is difficult. The need for Charity Status in the USA to be able to have more fundraising in that country has been expressed several times.

There will be a training kit and session on Saturday for the new INWES directors (capacity building for the trainers).

2.4 Treasurer Report (Financial Statements 2007 and Budget for Next 3 Years)

INWES has been reviewed by the accountant. Note: Two-thirds of INWES organizations are scientific organizations. The Treasurer, Gail Mattson, presented the Financial Statements to the Assembly.

Action: The financial statements 2007 will be posted on the website.

Charity status in Canada vs. fund development: INWES hopes to get confirmation soon because we need to add the certificate from Canada to have a non-profit in the USA, to do fund development. We need to clarify our programs and projects and present them to international sponsors and corporations. We need to identify potential sponsors: there are 200 companies who have expressed interest in Women in STEM (Meeting in Baltimore November 6–8, 2008 for the follow-up).

In-kind contributions. The in-kind donation from INWES directors and members is very important. For example, AAAS has printed 375 colored newsletters and many copies of the progress report (which is a 25-page colored document). Thanks to Yolanda Gorge for this in-kind contribution. A special form has been developed for everyone who supports INWES, containing in-kind contributions and tracking of volunteer time. It is very important and helpful for fundraising and for the INWES Institute.

2.5 Secretary Report

The Secretary-General, Claire Deschênes, worked with the Executive Committee on the Charity Status application and, to prepare the documents to apply as NGO (Non-Governmental Organization) at UNESCO, to prepare the Executive committee, BM, and AGM meetings and minutes, lately on the Website with her assistant Pascale Dubé, on membership renewal, Scholarship Competition program, Women Study Workshop presented at ICWES-XIV, and general internal communications. Claire supported part of this work with her own funding as in-kind contributions.

3. Reports from Committees and Programs

Advisory committee: In the last 6 years, the initial Advisory Committee was composed of honorary members, mainly the founders of ICWES. Note: Mahin Rahmani was present at the AGM. They were consulted on the future of INWES. New members will be sought for the committee.

OECD—They no longer have a committee on gender.

Asian Network

Myung Hee Jung and Kong Joo Lee presented the Asian Network, developing phases 2 and 3 (within 3 years). This regional group will be linked to INWES. The aims are to have regional meetings and to increase the membership of INWES. They already have a sponsor.

Membership and Recruitment Committee

Durdana has been Chair since the meeting in Poland and did a very good job at developing recruitment ideas to increase the INWES voice and documents. She found new members from Pakistan, Monique Frize brought some from Canada.

She prepared a motion to be presented to the board to create a new category of Members University/Institute with three representatives on the board and three votes each at the Members' meetings. This motion was carried unanimously at Board Meeting 9 (Lille, July 12–15). The Universities and Research Institutes make a good pool of members for INWES since they need to recruit more women students, and they have funds allocated to outreach for women in STEM especially in Science and Engineering faculties.

Symposium in Poland

The "INWES Regional Symposium on Women Scientists and Engineers of New EU Countries and Eastern Europe: Strategies for a highly Skilled Global Workforce" was held at the University of Wroclaw, from July 2 to 4, 2007. It welcomed 75 persons from 22 countries. It was organized by the University in close cooperation with

INWES and funded by the Polish Ministry for Science and Education and Canadian International Development Agency (CIDA). It was the first symposium of this kind organized in Eastern Europe.

Post-symposium recommendations were:

- Create responsible policies for education and research that promote female potential in the sciences.
- Increase mobility of scientists in all of Europe.
- Improve the communication between scientists and engineers, for women and men.
- Integrate East European and Russian women in European science.
- Increase the proportion of women scientists from Eastern Europe with respect to international positions on gender equality.
- Foster start-up of local and regional networks.

Special thank is expressed to Anna for making this event a great success and from her to INWES president and directors for their important contribution (in-kind).

Web Portal Committee
It is the first tool for communication within INWES. It has a lot of information but the technology is 6 years old and different from the technology it could be today. Problems remain for example with the membership registration form and profile of the member organization. It is now necessary to move forward and to redesign the site. It has been updated by Helga Ross, and since February, by Pascale Dubé.

Wiki
An idea brought by Monique Frize is to create a Collection of Best Practices on Outreach and professional development through a Wiki. People will be able to send their good ideas. The web portal committee will integrate them.

Gender Resource Tool
There are 21 indicators defined and presented on maps in software developed by Monique Frize and her students for developing countries. It gives statistical information, following a specific color code on women's and girls' life and education. For example, one can find the mean age for a first baby, by country, mean age of marriage, etc. Once we have the funds, the project is to integrate it into the website.

4. Voting on Motions
4.1 Proposed Changed to INWES By-Laws

It is proposed that a new University/Institute Members category be created. This category would consist of Universities or Institutes interested in or engaged in the teaching or other application of science, technology, engineering, or mathematics. University/Institute Members would be entitled to receive notice of and attend all meetings of Members and have three votes.

Motion AGM4-1-A. Proposed by Claire Deschênes, seconded by Sue Bird.

To create a new category of University/Institute Members with three representatives and three votes at Members' meetings.

Clarifications:

The intention is to provide three votes at AGM, where the number of votes for organizations goes with the level. Not every organization has a director and at Board Meetings (BM), each person has one vote. At BM, the number of directors INWES should have is still to be discussed. Should we have for example 21 directors (three more than now) on the Board or leave it as it is?

- Who will be the three representatives at AGM (three people from the new category of members or one person having three votes)? The size of the organization goes with the number of members and the fees go with it. Every university or institute would be treated as a large organization with three votes.

A discussion followed on the votes at AGM. Following the By-Laws, organizations are the only ones to have a vote on motions. Institutes and universities would now also have votes. The intention behind is that INWES is built mainly for organizations, but INWES wants a dialog with large institutions and universities. The idea is not to engulf organizations' voices.

An amendment to the motion was proposed to allow no vote at members meetings for the new category of members.

Motion AGM4-1-A Amendment 1. Proposed by Claire Deschênes, seconded by Sue Bird.

To create a new category of University/Institute Members with three representatives.

35 votes for, 3 votes opposed, 3 abstentions.

Motion Carried

Motion AGM4-1-B The secretary explained that two of the motions that were adopted at AGM2 to change the by-laws need to be adopted again at a face-to-face members' meeting according to our lawyer. They are presented again at AGM4 as Motion AGM4-2 and AGM4-3.

Motion 2

Whereas INWES By-laws specify that the term for the Elected Officers of the Board is 2 years, we introduce a motion to change this clause to a term of 3 years, in order to align elections with ICWES conferences.

Motion AGM4-2. Proposed by Claire Deschênes, seconded by Marlene Kanga.

To modify INWES By-Laws clause 5.1:

From:

> "5.1 BOARD ELECT OFFICERS
>
> The Directors shall, as often as may be required, elect from among their number, and may remove, the President, President Elect, Secretary and Treasurer and one or more Vice-Presidents. No person may be nominated for or elected or appointed to hold more than one office. The President Elect, Secretary and Treasurer shall each hold office for a term of *two years*."

To:

"5.1 BOARD ELECT OFFICERS

The Directors shall, as often as may be required, elect from among their number, and may remove, the President, President Elect, Secretary and Treasurer and one or more Vice-Presidents. No person may be nominated for or elected or appointed to hold more than one office. The President Elect, Secretary and Treasurer shall each hold office for a term of *three years*."

Motion carried unanimously.

Motion 3

Whereas INWES By-Laws Clause 6.4 specifies, in order to ensure rotation, that Three Organizational Directors, one Corporate Director, and one Individual Director shall be elected at each Annual General Meeting. Here again, since it is difficult to meet face to face often, it would make sense to ensure this rotation happens every 3 years, at the AGM held at ICWES conferences.

Motion AGM4-3 Proposed by Claire Deschênes, seconded by Unesu Ushewokunze-Obatolu.

To replace the following motion:

From:

"6.4 ELECTION BY ROTATION

At each annual members' meeting, three (3) Organizational Directors, one (1) Corporate Director and one (1) Individual Director shall be elected from among the persons nominated in accordance with these By-Laws. Subject to the By-Laws, a Director so elected shall hold office until the third annual meeting of Members immediately following the meeting at which she is elected provided that if an election is not then held, they shall remain in office until the dissolution or adjournment of the meeting at which their retirement is accepted and a successor is appointed."

To:

"6.4 Election

Ideally, a minimum of one third of Directors in each category (Individual, Organisational, Corporate) will be replaced at each election. The elections will normally be held in conjunction with ICWES conferences. A Director so elected shall hold office until the third annual meeting of Members immediately following the meeting at which she is elected provided that if an election is not then held, they shall remain in office until the dissolution or adjournment of the meeting at which their retirement is accepted and a successor is appointed."

It is argued that INWES is supposed to have a general meeting and a board meeting every year. But, interpreting our By-Laws, if the meeting cannot happen or if a director cannot be replaced, the past director stays in place so the original clause does not cause a problem. Also, if someone is elected twice, she cannot be re-elected for a third term.

Three against, three abstentions.

Motion carried.

A question arose about President-Elect (existing position in our Policy Manuel) and past president (does not exist so far).

4.2 Accountant Firm and Remuneration

Motion 4

Motion AGM4-4. Gail moved, Nadia seconded. To appoint for 2009–2010 the accountant firm Kelly Huibers McNeely, Chartered Accountants[2] to produce INWES Financial Review or Audit

Go back to the firm with a better price and did a better job.

Motion carried unanimously.

Motion 5

Motion AGM4-5. Moved by Gail, Trish seconded. To authorize the Board to fix the remuneration of the Accountant for Financial Reviews or Audit.

Motion carried.

Motion 6

Motion AGM4-6. Proposed by Gail, seconded by Roseni Dearden. "INWES board proposes to change the fees for medium organizations from $200 to $250 and for large organizations from $300 to $400 from January 2009."

Motion unanimously carried.

Motion 7

Motion AGM4-7. Proposed by Gail, seconded by Durdana. "INWES board proposes to add a very large level of organisation (more than 5000 members) with four votes and a fee of $1000 annually from January 2009."

Motion carried unanimously.

5. Elections of the Board of Directors (AGM4 Slate of nominees for Director Positions)

Vote for election—unanimously carried.

Election

Motion: That Màrta Ecsedi and Pascale Dubé be named officer for counting the votes for the nomination of directors 2008 and report to AGM4.

Proposed by Monique Frize, Seconded by Suzanne Brainard. Motion Carried.

After the Votes

Motion: To destroy the votes after being counted and reported by the two independent counters.

Motion—1st Nadia (proposed)
2nd Roseni Dearden (seconded)
Motion carried.

---

[2]Coordinates are: Michelle E. Huibers, CA, Kelly Huibers McNeely, 5992 Hazeldean Road, Stittsville, ON, K2S 1B9, Phone +1 (613)836-5626, Fax: +1 (613)836-2540.

6. Next Regional Meetings
6.1 Regional Events 2009 and 2010
7. ICWES-XV (2011) Bid and Winner Presentation

2008: Interim Board Meeting (BM9)

Lille, France, July 12–15, 2008

Directors were: Sue Bird, Claire Deschênes, Monique Frize, Issié Yvonne Gueye, Durdana Habid, Myung Hee Jung, Gail Mattson, Yvette Ramos, Anna Szemik-Hojniak, Unesu Ushewokunze-Obatolu, Màrta Ecsedi.

Observers and experts: Ha Do (Aus), Pascale Dubé (Can), Nadia Ghazzali (Can), Marlene Kanga (Aus), Kong-Joo Lee (S. Korea), Petra Mayerhofer (Ger), Teresa McNutt (Can), Monique Moutaud (Fra), Angelika Reinhard (Ger), Marie-Hélène Therre (Fra), Pam Wain (UK).

Directors Excused: Melody Kratsios, SNC Lavalin (Corporation), Canada.

Directors Absent: Suriya Thevar Mayandi, Individual, India, Yolanda George, American Association for Advancement of Science, the USA, Mabel Imbuga, African Women in Science and Engineering, Akiko Tsugawa, INWES Japan.

1. Tour de table: Three Slides for Each Board Member

Sue—UK: PowerPoint presentation

Durdana—Pakistan: She worked on the new membership category, as established in Poland. An Invitation Kit was made by Yvette, Issié, Màrta, and Durdana to invite Universities to join INWES. Two new members have joined: le Cercle des Femmes Ingénieurs et Scientifiques du Congo and the Center for Advanced Research in Engineering, in Pakistan. She works on developing the Asian Network.

Unesu—Activities in Zimbabwe: To increase the visibility of women in STEM issues, she succeeded in getting support from key industries for women in STEM, such as Ariston, which produces goods for houses. The National Academy of Sciences, a key institution, will accept future initiatives by INWES. Unesu met the Chief Secretary of the Ministry of Sciences and Technology who was very receptive to the initiative of Unesu for women in STEM. Southern African Development (also decision-makers) is expected to participate at the national level to have an impact at a regional level. The African Union raised the number of women in STEM from 30% to 50%. The Women's Coalition had a meeting for the entire country; STEM is one of their priorities, with agriculture and education. They want to create more careers in STEM, to improve efficiency. The Minister of Agriculture also supports Unesu and sees women in STEM issues as a priority. There is a wish for a change at a political level, as the situation is really difficult in Zimbabwe; 50% of people in

Zimbabwe live below the poverty level. The main livelihood is agriculture, which has a need for infrastructure and technology. There is also a need for change in the social perception of equality because there is not enough access for women to good lands, and good places for agriculture.

Claire Deschênes—President of AFFESTIM (Quebec): AFFESTIM has now become a major organization of the Ministry of Science and Innovation; with this formal nomination, it will reach more people and help to bring funds to the organization. AFFESTIM has launched, at ICWES14, a collective book written by French Canadian Researchers on the leadership of women in STEM "Le leadership des femmes en STIM," PUQ 2008, ISBN 978-2-7605-1565-9. There will be a workshop in Moncton, New Brunswick, about feminism and sustainable development (October 8–10, 2008). For the 400th anniversary of Quebec City, there will be a workshop for women in STEM on September 10. The exhibition "Physique de Femmes," which is composed of portraits of French women in Physics, will be presented at both events with the addition of Canadian Women portraits.

Myng Hee Jung—KWSE (South Korea): The ICWES-XIII conference was an opportunity for KWSE to be Science Communicators and to promote science to the public. They want to make a Database for women's issues. The seventh president of KWSE—Mi Sook Won—was recently elected. The organization was funded 50,000 Won through WIST. There are many leaders in STEM in South Korea; some of them have given leadership education workshops for juniors. There is a necessity to achieve equality in STEM in South Korea. Kwang Hwa Chung—president of the Governmental Institute and Myung Jung Hee received an award in STEM. There is a need for many good leadership positions in Korea and for globalization through an Asian Network. $45,000 was obtained this year to help with ICWES-XIV and for the Asian Network. There is an increase in the number of meetings of Asian women scientists and there are many programs for leadership training for students. The Asian Network is very important for South Korea, as is the database for STEM and Women in STEM. The Network would include India, Japan, South Korea, China, Pakistan, West Asia and the Philippines. Much information about this network was provided by South Korea.

Monique Frize—Pathmakers (Canada, Ontario): Led by Monique Frize, women students from universities go to schools to present Engineering and Technology careers to grade 1–12 students with hands-on activities. In Ottawa, each year, it reaches hundreds of girls, resulting in more young girls inspired by women models considering a career in STEM. It was supported by the Ministry of Research & Innovation of Ontario.

Anna—Poland Symposium: Anna presented a report from the symposium that was held in Wroclaw during the summer of 2007. The high education system is similar to Germany. It was a very successful symposium for the women at Anna's University and her Faculty became more interested in the issue of women in STEM. Anna spoke to the Director of the national committee for UNESCO. Because of their family/career conciliation, women's careers advance slowly and their situation is much more difficult than that of men. She devoted much time to her *habilitation* (promotion to Professor process) at the university.

Gail Mattson—Treasurer Report (USA): As Treasurer, Gail worked to keep INWES account information up to date. She worked to establish a separate institute for charity status in Canada and assisted in the application process. She prepared a sponsor package for INWES that includes general information, future projects, and issues. This package was presented to many international companies at the SWE 2007 National Conference. It needs the charity status in Canada to get the non-profit status in the USA, to get more funds from these companies. She worked on developing the INWES award ideas, (persons from INWES that were involved in the women in STEM issues) and the Distinguished Service Award.

Issié—Nomination Committee (Ivory Coast): As Chair of the Nomination Committee, Issié developed the slate of nominees for directors positions to be voted on at AGM4. She also checked many points in the By-Laws concerning this election: She clarified whether the interim board members nominated at ICWES-XII could run again. She brought up the possibility to nominate a President-Elect at ICWES-XIV. She emphasized the need for a balance in the representation of women coming from different regions of the world. Regarding the creation of a new category of members within INWES for universities, she expresses concerns that this could lessen the impact of the organizations of women engineers and scientists within INWES and on the representation of engineers versus scientists. She was active to help a Congolese group to start an association, and to get a support letter from the Ivory Coast National Commission for UNESCO to help get NGO status for INWES. As a member of the ICWES-XIV Organizing Committee, she supported the Logistics Chair Yvette Ramos. In her region, she launched a special program in Power Engineering.

Yvette (West Europe): Yvette is Logistics VP for ICWES-XIV and Chair of the INWES Program Committee. She has been active in involving Portuguese groups in INWES and convinced women from the Bahamas to join INWES. In March 2008, she went to Venezuela where she met Orlanda Castro, a woman scientist, to discuss issues of women in STEM. This could be a future link for INWES in South America. Yvette is also involved in a UN agency specialized in meteorology. She met the Arab women's organization and was invited to Tunisia in March 2008. She has links with the European Platform for Women in Science, with WEE and Marie-Hélène Therre. She will participate in WEC2008 from December 2 to 6, in Brazil. For ICWES-XIV, she is responsible for the logistics committee and social events. She also helped with the organization of the volunteers.

2. Work in Small Groups (Sue)

2.1 Develop Institutes and Universities Categories of Members and Their Position on the Board and Prepare the Motion (Durdana): A motion was tabled by Durdana to develop an Institute and University category of membership for University/Research Institutes.

Motion 1: Proposed by Durdana, seconded by Unesu: INWES will create a new membership category and add it to its By-Laws.

In her opinion, Durdana thinks that INWES should phase out Individual Directors during the next 3 years. At this point in time, there are many countries that still do not have an organization and can only interact through individuals. However, these individuals need to be encouraged to form organizations before the next Board election and the Board directors can be brought back to 18 as follows:

- Organizations: 12
- Corporations: 03
- Institutes/Universities: 03

Since it is difficult to have organizations from every country, Durdana suggests keeping the individual member category for at least 2 years. She also proposed to keep the fees for the new membership category at the same level as the fees for organizations. If we raise the fee for the organizations, this new one will have to be adjusted. A discussion followed about the necessity for INWES to have a "University" category and how to integrate it into the INWES structure: Universities and especially STEM faculties need to recruit women in STEM and this goes in our direction. They are concerned about the issue (Universities have lost about 3% of women in engineering, everywhere except in France). It is one of the roles of universities to have and get men/women interested in gender in STEM issues; all women engineers need to be linked together and they need to have a voice on the Board; optimum size for the Board: an increase in the number of directors will make it more difficult to manage, but more directors could have roles in committees. Large boards tend to be downsized today but smaller groups lose their voices in reduced boards; there exist large size differences between universities so fees need to be defined accordingly (richer vs. standard vs. countries in transition); it is necessary to get more corporate funding; communication mechanisms on the Board are of two types: individuals and organizations both have a voice, but only organizations are allowed to vote. University members would have a voice, but no vote; what about at AGM? The vote was postponed until later in the agenda.

2.2 Develop New Directors Training for July 19th AM (Pam, Claire): No paper documents will be provided to the new directors, but a CD will be prepared and discussions on various items will be held with them: Budget, five-year review, Code of Ethics, Lessons learned. Pam and Claire are to define the content of the training document.

2.3 Develop a Declaration (Sue, Monique M.): Extending the Asian Network idea, Sue Bird tabled the following declaration to be announced at INWES-XIV closing ceremony about the creation of regional INWES Networks.

Motion 2: Proposed by Sue, seconded by Gail: ICWES-XIV supports the creation of INWES Regional Networks, to aid the operation's communications and recruitment for INWES, and to enable women in the area to realize their full potential in

STEM. INWES will initiate at least two Regional Networks within the next 3 years. ICWES-XV will enable Regional Networks to meet, exchange information, and report to the membership.

Discussion: The name "INWES Regional Networks" is confusing. Maybe it could be named "Alliance" or "INWES Regional Network-Platform." People from Canada and USA could bring people from South and Central America. In Tunisia, there is a new organization of Women Scientists and Engineers, started by UNESCO. A discussion arose about ICWES-XIV to support the creation of INWES, not to be confusing for the journalists for example, or should it be INWES, who represents thousands of women, or the INWES Board of Directors?

Votes: two for; five against; Motion not carried.

Amendment 1 to Motion 2: Proposed by Monique F, with additional text from Gail, seconded by Anna:

INWES and ICWES-XIV support the creation of INWES Regional Networks, to aid the operation's communications and recruitment for INWES, and to enable the women in the area to realize their full potential in STEM. INWES will initiate at least two Regional Networks within the next 3 years. ICWES-XV will enable Regional Networks to meet, exchange information, and report to the membership.

Votes: 8 for; 1 against; Motion carried.

2.4 Develop INWES Distinguished Service Award and Recognition for Women in STEM: (Gail, Issié, Márta): Two kinds of recognition were proposed:

1. recognizing outstanding support for women in STEM
2. recognizing the work done within INWES.

The group suggested that INWES develop three different awards:

Internal to INWES: an award to recognize women that have done a lot of work for INWES, i.e. a distinguished service award.

External to INWES: an award for a woman from around the world who has worked to further the cause of women in STEM.

An award for a company having put something in place, or done an action, or create an environment to help women in STEM.

A nomination package will be put in place with three criteria. A resumé and three letters of reference should be provided.

Action: The information about INWES Awards and the nomination package will be added to the Policy Manual and the website. A "Certificate of Appreciation" will be given to all retiring directors along with an Honorary membership.

2.5 ICWES Conference Series and Regional Conferences: ICWES-XV, 2011 (Anna, Yvette, Marlene Kanga, Ha Do, Jung Myung Hee, Kong-Joo Lee, Durdana): Engineers Australia presented a very good bid and they will hold ICWES-XV in Adelaide in 2011. They already have excellent support and are connected with the previous organizers. They will engage women from Australia and New Zealand and get the whole region involved. ICWES-XV will benefit from funding from Engineers Australia. Monique Frize suggested holding a regional meeting to prepare for the large international conference, for example in 2009, as we did in Korea and Lille.

Regional Conferences (2009 or 2010): No bids have been submitted yet for a regional conference in 2009 and 2010. Suggestions include French Africa (Ivory Coast), the Middle East, South America, the UK, Washington, USA. The location should be a place where the project has already started with the ability to have no deficit, to raise funds, and to have the support of the Mayor or of a University. Other suggestions were to help with the payment of registration by members in developing countries, consider the possibility of having a lifetime membership; see how to combine a Board Meeting with the regional meeting. Improve communications between committees and members.

Action: The VP Conferences will post a call for bids on the website. A mass email to members will be prepared to see if there are bids for the next two regional events; a solid proposal is needed with good possibilities for grants.

Action: All small groups will prepare a written report and send it to Secretary (Claire).

3. Agenda and Minutes

3.1 Adoption of the Agenda: Moved by Sue and seconded by Anna. Gail suggested adding a presentation on SWE (item 9.2). The agenda was changed, and a motion adopted unanimously.

3.2 Adoption of BM8 Minutes June 29–30 July 1, 5–6: The minutes were sent before the meeting. Motion: Moved by Claire, seconded by Gail. To adopt the minutes of BM8. The BM8 minutes were adopted without change.

4. Report from the Executive Committee

The PowerPoint that was prepared by Pascale and Claire with the Executive Committee and other Committee inputs was presented. Summary of the main points:

4.1 President's Report: The President and the Executive Committee have had many Skype meetings. The regional workshop in Poland, for which INWES obtained a Canadian CIDA grant for the travel fees of many delegates, was very successful and informative. The Committee has done a lot of work to make INWES an NGO with official relations with the UNESCO. The UNESCO Director-General informed INWES in April 2008 that this status was now official. UNESCO asked our opinion on one of their next program budgeting and we sent a reply emphasizing women in STEM through the Canadian Commission for UNESCO.

Creating a separate organization: "INWES Education and Research Institute" (ERI) was a necessity in order to obtain charity status, because a not-for-profit organization is not considered automatically as a charitable organization in Canada, contrary to the USA. The INWES Institute was incorporated in November 2007, and the charity application was sent afterward. An answer is expected at the end of the summer. As designed, the Institute has three founding members, a board minimum of three persons and it is expected to report to the INWES Board of Directors (if this is allowed by the charity laws in Canada). It will start to develop three programs. The objective is to include more programs and more people from other countries in the future. As soon as the Institute gets charity status, we will apply to obtain 501c status in the USA and we will look for funding for programs and their management.

Monique F. negotiated an MOU with WFEO to collaborate with their new Standing Committee on Women in Engineering and other projects. The discussion was a very complicated matter.

The President, with the VP Conferences, Sue Bird, also negotiated an MOU with Australia for ICWES-XV. Monique F. wrote two chapters for UNESCO's WFEO Engineering Report to be published in 2008/2009: one chapter for UNESCO was on women in engineering—she wanted to look at the engineering profession from a different point of view—the second was on ethics.

The President created, with one of her students, an organizational structure diagram to explain in a simple manner the links between INWES, ICWES, and the Institute. Engineers Australia (Marlene Kanga) welcomed help from INWES, to set ground rules for the ICWES-XV conference.

4.2 Update from VP Conferences: Poland regional conference fulfilled its aim, to get people to know the situation in Eastern European countries; CIDA was very supportive again! For ICWES-XV, there has been a very good and professional bid from Engineers Australia. A request for bids for the next regional events for 2009 or 2010 was to be sent out ASAP. It is necessary to find someone interested in hosting this where there has not been a regional conference or an ICWES before. The Vice President encourages someone to run the regional event next summer. The locations proposed are Mexico, Puerto Rico, Venezuela, and the Bahamas. It would be interesting to hold it in South America to increase INWES presence in this area and to have a Board meeting at the same time.

4.3 Update from Treasurer: INWES had a professional, independent 2006–2007 financial review by chartered accountant Kelly Huibers & McNeely. The treasurer presented the balance sheet for 2006–2007. It is a requirement in Canada to have the "statement of cash flows" page, showing cash flow from the US and Canadian dollar accounts. Board members were asked to complete the In-Kind Contribution form and give it to Gail. Items to be included are travel, accommodation, phone, and time at your hourly rate.

Motion: Proposed by Gail, seconded by Sue: To authorize payment for Financial Statement Review in 2007, to Kelly Huibers McNeely Chartered Accountants. The vote was unanimous.

4.4 Update from Secretary General: The Secretary follows the Canadian Corporation act. She worked on the preparation of AGM2, AGM3, AGM4, BM9, and BM10. In Poland, she helped Anna to obtain the money transfer of the CIDA grant. There was a delay in the international transfer through two banks (one in the USA and one in Poland) so it didn't arrive on time. INWES borrowed $30,000 US from the University, to give the promised travel funds to the women coming from developing countries and from the eastern European countries and Russia. This lesson shows that there is a need to get registration money in advance to resolve cash flow problems. The Secretary updated the Policy Manual so it would be complete before she retired from this position. She added the ethical guidelines, the revised nomination kit, a reference to the new VP Programs, and all the Terms of Reference. Based on the MOU for ICWES-XV, a typical contract or MOU, to be signed between INWES and future ICWES organization committees, will be added to the Policy Manual by the next Secretary. A discussion on the position of past president followed. This position does not appear in our by-laws or in the Policy Manuel at the moment. However, our by-laws provide the opportunity to create such a position if it is useful to the organization, similar to the creation of VP Programs. Monique explained that she reapplied for a director position since our official documents were not clear about the Past President position, but she would like to act as Past President. She offers sharing her experience and is not running as president for a new term. Màrta observed that Past Presidents can stay on the Board as a Past President with voting capability or as an observer. Monique F. observed that a past president of the Executive committee would bring stability to the Board. Petra remarked that if we want to change the by-laws, we have to tell the members in advance. However, according to the By-Laws, we can create this position exactly as we created the VP Programs position. The by-laws can be modified later. Monique suggested that the new Board decide about the position. Issié suggested that Monique is free to present herself or not as a director. But this added to the confusion, as it is illogical to elect a Past President.

Motion: Proposed by Issié, seconded by Yvette: If the President is re-elected, the President would have the possibility to stay or not. If the position of Past President is desired, the new Board shall propose a Past President position and vote to make it a part of the By-Laws.

[Authors' note: This was a confusing situation, as persons voting to elect directors would assume that the President would naturally become Past-President since this position could be easily created, and so would not be inclined to vote for her. It is the authors' view that the position of Past President should have been created at BM9 and the current President would have removed her name from the slate of nominees. This would have left 18 persons for 18 director positions.]

4.5 Review Action Items Since BM8: Claire reviewed the action item list with the board.

5. Develop Templates for Specific Projects in Small Groups

During BM8, in July 2007, INWES program development work organized and generated a vision. The rationale for the programs and projects catalog was to package the catalog in a nice way and find commonalities between the expectations of sponsors, members of INWES, and INWES program ideas to make them work together. The objective was to get sponsorship from the public and private sectors for the projects. Working groups developed specific projects; each established the expected results and target.

- Group 1 worked on the Essay Competition, the Collection of Best Practices, and the Gender Tool.
- Group 2 worked on a Director Welcome and Orientation Kit.
- Group 3 worked on three projects for capacity building for INWES members (train the trainers).
- Group 4 worked on a plan for benchmarking INWES—for a scholarship competition in Nigeria.

As of January 2008, the following has been completed: Slate of projects used for fundraising during SWE 2007 National Conference. Proposal received from Issié Yvonne Gueye from Ivory Coast for a project Math Techno Contest (or Miss Techno Contest). Proposal underway to benchmark the Essay Competition in Nigeria in 2008 with Okon Uduakobong as Chair. Proposal to create an Asian Women in STEM Network (a regional network, which could be a model for the rest of the world).

After the plenary, the Board was again divided into small groups:

5.1 Web tools and content (M. Frize, Pascale, Yvette, M. Moutaud): 5.1 Sub Project 1 (Gender resource Tool); 5.1 Sub Project 2 (Web Phase IV).
5.2 Outreach (Claire, Nadia, Issié): 5.2 Sub-project 1. Test case Essay competition in Nigeria. 5.2 Sub-project 2. Test case—Miss Techno Contest.
5.3 Help create new associations (Durdana, Gail, Màrta).
5.4 Regionalization (Asian, European Networks) (Myung Hee, Kong Jo Lee, Sue, Petra).
5.5 Advocacy Program (tabled).
5.6 Career development workshop and delivery.
5.7 Sustainability of networks and organizations (tabled).
5.8 INWES Institute (M Frize, Gail, Claire met at a special meeting of the Corporation).

Web Tools and Content: Monique Moutaud hired a firm to develop an efficient website (all-in-web.fr) for ICWES-XIV. INWES has a limited budget, but needs to take back the control of its site. The main problems at the moment are mass emails, import of Excel data, and registration. This will help the communication within INWES. It would be interesting to discuss various issues online and to have feedback online. There must be fundraising to expand the Institute, to do the project step by step. The website should be revamped with a professional package. We need to think about how much the web has evolved. We need an action for the portal and it needs to be evaluated by experts. It would be helpful to have a communication committee. The Web portal committee proposed to look at the different professional packages and make a recommendation to the Executive. When INWES has funding, it will move on with the Web portal project. The authority should be given to the Executive Committee.

Motion: Moved by Monique F., seconded by Durdana: That the Executive Committee of the Board be authorized to act on the proposal from the web portal Committee to upgrade the website with a professional package. The motion was accepted unanimously.

Outreach (Small group: Claire, Nadia, Issié):

Test Case: Essay competition in Nigeria: Proposal: Next November 2008, a test case could take place in Nigeria. Okon (APAGESTE—Association Promoting African Girls in Engineering, Science and Technology Education) will do it if INWES approves the project. It will be in a local girl's high school in her state in Nigeria. Nadia emphasizes the need to add an academic excellence component to the criteria for allocating the scholarship. Gail underlines the importance of evaluating the project at the end. Special thanks to Paula Leventman for her support both in writing the proposal and in partial funding.

Action: The Treasurer will consider this project in the budget 2008–2009.

The full project is to extend this scholarship competition to other areas of the world later using the test case in Nigeria as a promotion means and for fundraising.

Test Case: Miss Techno Contest: Proposal by Issié: Develop an excellence center for girls in school, get the best girls in STEM. To have girls help girls; it is a mentoring, tutoring project to give opportunities to girls who have no funds to get to higher education levels. The name of the contest should be revisited (the word "miss" might be offensive to some).

Action: Issié will write a budget and time frame for partial funding and give it to the Treasurer. Perhaps it can be considered for funding in the budget 2008–2009.

Help Create New Associations: (Small group: Durdana, Gail, Márta): A document on "How to start an association" already exists. Durdana mentioned that all INWES geographic regions should be represented. She underlines the necessity to give local help, to provide seed money, ask for reports, etc. The document on starting an association should be updated, especially for using at the WFEO December meeting in Brazil.

Action: Update the manual on starting an association (for seed money, reports, etc.).

Petra underlined that we must mobilize the members to participate on committees and we must provide good communications. Monique F. emphasized that a Communication Chair is needed and asked Petra if she is interested. Petra accepted to chair the Communication Committee.

Monique F. mentioned that INWES needs a process to vet people through the membership committee. Durdana agreed to take this on.

Asian Network and Regionalization (Sue, Myung Hee, Unesu): Myung Hee reported on the state of development of the Asian Network. A regional network is an amalgamation of INWES members in a region. Claire underlined that INWES has 11 regions defined in its official documents, coming from the ICWES tradition.

Action: The aspect of the 11 regions of INWES will be reviewed with the possibility that the regions will be redefined. The By-Laws and the Policy Manual will have to be updated.

To create regional networks, meetings of regional organizations are to be set up. The Asian Network will serve as a model for other regions. Other possible regional networks suggested: Africa (French and English); the Caribbean, South America, North America; Middle East. The WFEO Declaration in Tunis goes in the same direction. Yvette indicated that 3 years is enough time to make a strategic plan.

Action: Check the possibility of developing regional networks. Nadia will check if the Arab Network is still alive and if an Arab Regional Network could be started. Links between INWES and these regional networks have to be worked out. Anna suggested using ICWES conferences and regional events to link women in a specific region together and with INWES.

INWES Education and Research Institute (ERI) (Claire, Monique, Gail): A decision was made to wait on the answer from the Canadian government on the status of the Charity application, which is expected 8 months after February 2008. Then, if positive, fund development kits can be developed for our three projects.

New Category of Member: The second part of the discussion on the new category of members was held at this point in the agenda. A revised motion was drafted in a small group (Durdana, Unesu, Nadia, and Angelika):

Motion 1 Amendment 1: Proposed by Durdana, seconded by Unesu, Nadia, Angelika: It is proposed that a new University/Institute Member category be created. This category shall consist of Universities/Institutes interested or engaged in teaching/research and development in STEM. The position of the membership category on the Board is as follows:

The number of directors representing universities/institutes of education, research, and development shall be three (3). Each member in this category shall be entitled to three (3) votes at the AGM. The annual fee for this category shall be the same as for Organizational Membership level 3. There would be two Individual Directors, 10 Organizational Directors, three Corporate, and three universities/institutes Directors.

Votes: nine for; one abstention; motion carried.

Preparing ICWES-XIV Events and Conference Committee Report (Small group: Monique Moutaud, Sue): Report from the Organizing committee (Marianne Rodot, Laurence Broze, Yvette Ramos, Marcelle Rey-Campagnolle). Two European organizations are involved in addition to CNISF: WES-UK and DIB-Germany. There is a total of 16 persons on the organization committee and 50 persons are on the program committee (experts evaluating the abstracts). Management of volunteers, selection of abstracts, management of industrial and touristic visits, are specific tasks in this organization. There were: 277 abstracts received, 201 accepted, 58 rejected, and 18 waiting for evaluation, divided into eight different categories: 122 oral presentations and 46 posters accepted, from 56 countries; 40 travel awards were allocated out of 104 requests. The communication was made through the website, dissemination, package for participants, press release, flyers in French and English and Sponsorship documents for potential sponsors. The Logistics committee is composed of 15 persons with partners and 80 volunteers from 25 countries. Sponsors are EADS foundation, Google, UNESCO, and Areva. Local support was obtained from: Polytech'Lille, Ville de Lille, L'Oréal, Chambre régionale CCI-Grand Lille, Audi, Transpole, Lille3, GE Energy. There were 370 attendees from 58 countries, 30 accompanying persons, 50 students, 80 volunteers, 30 organizers for a total of 560 participants.

Activity for Honorary Members (Monique M): No special activity was organized at ICWES-XIV, but Nicole Bécarud from France and Mahin Rahmani from Iran were present.

Preparation for AGM4: Adopt new Institutes and Universities categories of members (Durdana); Nomination of directors and discussion on the number of directors (Issié); Re-adopt the AGM2 2006 motions (necessary to change By-Laws); various representations, meetings, and INWES visibility during ICWES-XIV

The display for the INWES booth was discussed (banners, newsletters, general information, nominations for director position CVs) and a list of volunteers was circulated among directors to ensure a presence at the table booth.

Five-Year Review Process (MF, Claire): Summary for Five-Year Review: Desired State and Evaluation; Future Plans. The President and Secretary explained what they prepared for the five-year review. Each director can help complete this review.

Action: Board Members must examine the five-year review evaluation process and send their comments after the meeting.

ICWES-XV (Marlene Kanga): ICWES-XV will be held in Adelaide, Australia and will be hosted by Engineers Australia (EA, 83,000 members). DVDs were sent to schools. They want to include every state of Australia and to develop national programs, strategies for women engineers. They want to attract, retain, support, and celebrate. Between December 2006 and March 2007, there was an increase of more than 15% in student members in Engineers Australia. The new President has a vision for Engineers Australia and three women are on the EA Board.

ICWES-XV: The Adelaide ICWES conference will be held around July 10–13, 2011. A website was set up: www.icwes15.org.

Action: Pascale Dubé will put a link to the ICWES-XV website on the INWES website home page and on the upcoming events page. INWES will give support to sponsors and fundraising. Peer review will be used for abstracts submitted. The South Australia Women in Engineering strategic planning wishes to diversify the Engineering profession (young, retired, managers, international engineers) and ensure that the leaders in engineering will champion the goals.

Next Regional Events (2009 or 2010): In 2009, it would be a good time to go to South America. Gail has contacts in Porto Rico, Monique M and Claire in Mexico, Yvette in Venezuela and more will probably be made in Brazil at the WFEO meeting this fall.

2010—There will be a meeting in Argentina in 2010.

2012—Middle East?

A note was made that the host country must be safe and politically and socially stable; there must be no dangers for travelers.

Action: The VP Conferences will send a bid to see if there are other bids for regional meetings. In conjunction with the ICWES or regional workshops, we usually have a BM. Otherwise an annual face-to-face BM must be organized if possible.

Responsibility and Role with UNESCO: There will be a meeting of NGOs in Paris next September where INWES should participate (allowed three representatives), Monique Moutaud has agreed to participate for INWES.

New Categories of Members and Revision of Organizational Members' Fees: A motion to add a category of large organizations and increase their number of possible votes at members' general meetings was discussed. A clarification was given: there are currently three levels of organizations, fitting their size and their registration fees, and increasing number of votes (from one to three). No vote is allocated for individuals and corporations, because the main goal of INWES is in favor of gathering organizations. Large organizations would be willing to pay increased fees, so more votes should probably be allocated to them. This motion would be effective for the membership renewals that will occur in January 2009. A discussion arose about the fiscal year of INWES that starts in January and that does not fit very well with INWES reality since many registrations come about during the regional events and ICWES meetings. However, it does not seem possible to change this situation for reasons related to the Canadian fiscal law. The possibility of creating a membership fee for a lifetime or for 3 or 4 years in advance was also discussed. This would facilitate many members, especially from developing countries who experience difficulties sending their payments. Also, with the professional package for the website, it should be made possible to pay the fees via the Internet with a credit card. A note was made to check if the Web can be changed so a member does not have to refill the whole registration form each year.

Motion: Moved by Gail, seconded by Durdana: INWES Board proposes to increase the fees for medium organizations from \$200 to \$250. And for large organizations from \$300 to \$400 from January 2009. Votes: nine for—Unanimous.

Motion: Moved by Gail, seconded by Sue: INWES Board proposes to add a very large level of organization (more than 5000 members) and a fee of $1000 annually from January 2009.

	Actual fee structure[a]	New fee structure[a]	Number of votes
Individuals	$30	$30	
Corporations	$2500	$2500	
Organizations small	$100	$100	1 vote
Organizations medium	$200	$250	2 votes
Organizations large	$300	$400	3 votes
Organizations very large		$1000	4 votes

[a]Note: 1/3 for developing countries

Vote: Passed unanimously.

Wrap Up and Review of Action Items: The organizations cannot vote for more people on the board than the By-Laws state. At BM10, there will be three vacancies; the New Board could add three people temporarily. There are 19 candidates and 12 to vote for at AGM. We could ask the candidates who are not elected if they want to serve on a committee. INWES should involve all would-be volunteers as much as possible.

Action: The Secretary will prepare a list of the main action items and resolutions for BM10.

Reminders: Board members have to revise the five-year review and send their comments. Each committee has to revise its specific TOR. Each director has to revise her action items. A special thanks is delivered to Marlene Kanga for attending this meeting.

Motion: Moved by Sue, seconded by Anna: To close BM9. The vote was unanimous.

2008: Board Meeting 10 (BM10)

Polytech'Lille, Pasteur Room, Lille, France, July 19, 2008, 9 a.m.

New Board
Sue Bird (President)
Nadia Ghazzali (Secretary general)
Gail Mattson (Treasurer)
Pam Wain

Teresa McNutt
Maki Iwakuma
Issié Yvonne Gueye
Jung Myung Hee
Yvette Ramos
Anna Szemik-Hojniak
Petra Mayerhofer
Yolanda George
Unesu Ushewokunze-Obatolu
Monique Moutaud
Monique Frize (Past president)
Claire Deschênes
Marta Ecsedi
Pascale Dubé

Excused
Durdana Habib

Missing
Francisca Nneka Okéké

1. Opening (Monique Frize)
2. Welcome to New Directors

Every director is expected to lead a committee or a program.
Motion. Gail proposed, Anna seconded. Add a position of past president to serve on the executive committee, commitment for at least 1 year.
10 votes. Motion carried unanimously.

By-Laws committee
Motion. Gail proposed, Nadia seconded. The motion would be to have a President-Elect.
This President-Elect should be ready to replace the president in case of incapability. It must be someone who has not already a role.
Not consolidated enough—at that moment we can elect a President-Elect.
It would be important to have a designated person to replace the president in case of incapability, someone who has already been on the executive.
1 against, 10 for. Motion carried.

Election for the positions.
Directors can nominate others or themselves.
Each person will be asked if she accepts to serve in the position they were chosen for.
Francisca is missing and Durdana is excused.

There is a necessity to get more membership and promote AGM.

Action: Fact sheets and newsletters left are to be distributed to promote INWES.

3. INWES Operations

3.1. WEB Development and Maintenance

3.2. Legal Matters

3.3. Marketing and Communications (Yolanda)

Yolanda negotiates with the organization committee of the WEC.

Newsletters left are for the meetings and ICWES14.

The SWE Conference will be held on November 6–8, 2008, in Baltimore, Maryland. (mid-October)

Scientific Newsletter—it would be interesting to modify, redesign it. Newsletter could be based on the website design.

It could be changed with the web portal.

It is better to prepare it on time for the SWE Conference.

Newsletter: on ICWES14; Meeting coming up in Pakistan; Asian Network.

Planning for the newsletter by Yolanda: send texts by October 1st, send to print about October 15th.

Current events can be directly put on the website.

Take the material to different conferences

We can invite members from around the world to send news checked before being sent on the website.

3.4. Human Resources (staff, contracts, etc.)

Action to be done soon.

They have a generous offer from AAAS; they could provide a student part-time, at no cost.

Professional help

Action: Have a look at what can be done, regarding the budget and the resources (Jennifer, Pascale…).

Partnership

Issié—INWES could have a permanent delegate to international relations; someone to link with international organizations.

We could have someone manage the new status with UNESCO—opportunity to find some funds with UNESCO.

The meeting for NGOs in relation to UNESCO is in September. The goal of this meeting is to avoid conflicts between NGOs and share practice.

International relations with NGOs: Sue, Gail, and Monique M. went to a meeting with members of WFEO and Marie-Hélène Therre from FI.

Provide nominees for the Women Standing Committee (new)-manage in France

Every standing committee with WFEO is hosted by a country member, depending on the budget.

There are eight standing committees in WFEO.

Action: Look at the standing committee, look for INWES members who could stand on those committees.

We could support an INWES member to stand on this committee.
Gail proposes two motions.

Motion 1: Have the executive committee look into how to handle international relationships and report on them

Motion 2: Appoint Issié as our representative for the standing Women committee of WFEO.

FI (but also represent INWES)
Yvette nominated by CNSF to stand on the committee.
Sue: has reservations; the president can be an international representative, different people can be good at different jobs
To change the executive to have a representative from Africa?
It is not a good idea to extend the Board.

Motion 1: Gail proposes. Nadia seconds it. Having the executive committee look at how to handle international relationships and report it to the Board.

Gail: New relationships with WFEO, think to have a discussion on it, new board.
Action: Have a discussion and look at the issue.
Motion carried unanimously.
We could start to evaluate the possibility to have a representative work with the WFEO—it can be a model.

Motion 2: To appoint Issié as the INWES representative on WIE of the Standing committee of the WFEO.

10 for. Motion carried unanimously.

Addition to Motion 1: Pam proposed, Issié seconded. Have the executive committee look at how to handle regional representation within the INWES Board.

1 against, 10 for. Motion carried.

3.6. Ethics Signature

Action: Send an Ethical guideline to Francisca Nneka Okéké and other members who haven't signed it.

4. Election of the Officers to the Executive Committee

Position
Sue Bird—President
Gail Mattson—Treasurer
Nadia Ghazzali—Secretary General

5. BM9 Minutes

Note: the new president chairs the meeting.
Status on Motions and **Action items from BM9**

Action: Secretary to correct the list of members- modify it following alphabetic order, add corporations, no institution/university indication.

Action: Each committee will revise its section of the Policy Manual.

Action: Budget 2008–2009.
Gail will finish the budget for 2008 and will send it to the Board. And also give a proposal for 2009.
Action: Update the manual on starting an association (seed money, reports, etc.).
Action: Check possible networks with the 11 regions of INWES (historical regions) and update the last if necessary.
Nadia will check for the Arab Network.
Links between INWES and these regional networks have to be worked out.

Committee's Assignment
Nominating Committee: Issié, (Francisca), Nadia, Sue.
Fund development: Gail, Marta, Nadia.
By-Laws (new): Pam (Màrta will be happy to help and give advice.)
Membership: Durdana.
Communication: Petra, Yolanda (strong link between communication and web portal).
Programs: Yvette, Unesu, Maki, Teresa.
Web portal: Monique Frize, Petra.
Conference: Monique Moutaud, Anna, Kong-Joo.

Action: Nomination committee needs to clarify, modify the nominating form.

Chairs need to write their priorities and look at the TOR.

Review of the budget for 2008–2009 and 2010–2011 (Treasurer)

Action: Group projects give their inputs for 2009 to Gail.
Miss techno Contest already provided.
The Essay competition has to provide the input for 2009.
Committee chair
Give 2009 budget by December.
Paperwork for INWES Institute
Essay competition -in Nigeria

- Scholarship awarded to two young students

33 × $500

Total: $80,000—Estimated Budget
Package for sponsors and corporate

9. Organizing Next Regional Conferences
10. Next Board Meeting

New directors training: July 19, 2008, 1:30 p.m.:

1. General information and background
2. Review of documents (By-Laws, Policy Manual, past board meeting minutes…)
3. Board responsibilities and ethics guidelines

Authors' Recommendations for INWES Directors

Compile a list to compare Merrickville's conclusions on which of the characteristics and functions have been achieved, and which ones were not done and whether these were still desirable. This process would help INWES to note its achievements and what other goals should be worked on in the years to come. A rationale could be developed on why certain of the original goals are not feasible or are not a good fit with today's environment and conditions.

An important question to include in these conclusions regards membership in INWES. For example, how does INWES attract women who were pioneers in their field, and who have stepped down from committees, to be involved again? These women may be influential in male-dominated organizations. They should be encouraged to become involved. There were also interesting wishes such as: We have a diverse membership with respect to age, geography, to help develop succession planning. To have at least one member from every country in the world. Help establish effective organizations of women in STEM in 20 new countries. The original intent to have Individual Members was to accommodate women who had no organization of women in STEM in their geographical area. They would join for a 3-year term, then hopefully create an organization in their location and join as Organizational Member. It was worrisome that individual members could be in competition with local membership organizations, thus they should be considered as an exceptional case by demonstrating with a proof that they live in an area where no organization exists. It was the wish of the creators of INWES and of several of its elected boards to consider Individual memberships to be a temporary (short-term) solution for women who were isolated.

Chapter 16
ICWES-XIII: Seoul, South Korea, August 26–29, 2005

Theme: "Women Engineers and Scientists: Main Force to Reshape the Future World"

Organizers

The Association of Korean Woman Scientists and Engineers (KWSE)
ICWES-XIII Chair: Myung Hee Jung
Organizing Committee Chair: Kong-Joo Lee

KWSE was founded in August 16, 1993. Its objectives were to foster education of women in science and technology; to facilitate research collaborations among women through information exchange; to support the rights and elevate the status of women engineers and scientists; and to establish networking among women in science and technology.

Sponsors of ICWES-XIII

Korea Science Foundation
Korea Science and Engineering Foundation
Dongwon F&B Co., Ltd.
L'Oréal
AUDI
Ministry of Commerce, Industry and Energy
SAMSUNG Electronics

M. Frize et al., *Women's Contribution to Science and Technology through ICWES Conferences*, Women in Engineering and Science,
https://doi.org/10.1007/978-3-031-33800-7_16

Korea Water Resources Corporation
National Institute for Supporting
Women in Science and Technology
Korea Industrial Technology Foundation
Korea Hydro & Nuclear Power Co., Ltd.
POSCO, SK Telecom, and more

Participants

There were 743 registrations from 49 countries (509 from Korea and 234 from other nations). Out-of-country attendance amounted to 31.5% (see Appendix 2). A large number of students participated: 179 from Korea and 18 from other countries. The organizers managed to provide 53 travel grants for persons from 27 countries and 51 free registrations went to 29 invited speakers from Korea and to 22 speakers from elsewhere. Accompanying persons were 10 from Korea and 17 from other nations. A very large number of abstracts were accepted: 644 in total, out of which 433 were for posters, 183 for oral presentations, and 28 from invited speakers (numbers taken from "Final Report" presented at INWES BM7, see Chap. 15).

Opening Ceremony

ICWES-XIII was the largest women scientists and engineers' symposium in the world. Held at EWHA, participants included Monique Frize, the president of INWES, and Norah K. Olembo, Secretary General of Africa Biotechnology Stakeholders Forum. Under the theme "Women Engineers and Scientists: Main Force to Reshape the Future World," this year's ICWES-XIII aimed to establish a global network among women scientists and engineers to exchange information on research and education. Professor Lee Kong-Joo (Industrial Pharmacy), the Chairperson of the Organizing Committee said, "It is important that this conference is held here at EWHA, despite all the construction going around the campus, because ICWES-XIII can be a wonderful chance for students to see what their lives as scientists could be like in the future" (Kong-Joo Lee, "Opening," ICWES-XIII). Lively discussions in eight areas of specialization (Information Technology, Nanotechnology, Biotechnology, Energy Technology, Environmental Technology, Space Technology, Education and Ethics, Gender and Leadership) and numerous workshops took place during the conference.[1]

[1] Ewha Voice: Women Write the Future of Science—Ewha Voice. Accessed March 22, 2023.

Business Meeting

Business meeting Minutes: These are now INWES board meeting minutes and have been discussed in Chap. 15.

Testimonials

Asha Gopinathan

Asha Gopinathan from GenSci-e-Tech in Trivandrum, India received a WES bursary to attend ICWES in August 25–29 in Seoul, Korea. She reports on her experiences:

"I nearly didn't make it to ICWES, as my airline ticket did not arrive until shortly before my departure, so I could not get my visa until the afternoon I was to leave. Some high wire drama resulted—but all ended well."

"Incheon Airport is about an hour away from Seoul City. The organisers had set up booths at the airport where I first encountered the yellow T-shirt volunteers. These boys and girls (mostly students) were extremely helpful throughout the conference."

"Upon my arrival at the university hall of residence, I met many women from different parts of the world. We were all given two towels with the ICWES logo embroidered on them along with our room key. Throughout the conference, I noticed the thoughtfulness and attention to detail that the organisers had shown."

"I spent my first day exploring the campus and previewing my talk—hoping that my CD-ROM would not fail (it didn't)."

"At the opening ceremony the chief guest was the First Lady of Korea and we met Drs. Myung Hee Jung and Kong-Joo Lee—the moving spirits behind this event. It was followed by a multimedia show showing the history of ICWES and contributions of women in science and technology. It ended with a song sung by the students—'Come and shake us into action'."

"I had to shake myself into action soon enough as my talk and the session I was to chair followed this event. Despite some technical glitches, my talk ('Reaching for the Stars: The Indian Institutes of Technology and Girls') went well and there were even a few questions. The session I chaired ('Sisters in Science and Engineering: Creating Warming Climate') included papers by Pam Wain and Roseni Dearden. This was the first time I had met Pam with whom I have been in touch for almost four years."

"After this I could relax till the afternoon of 28 August when I gave my poster session ('Women in Science and Engineering: A few case studies from Indian Women')."

"I attended many sessions ranging from ethics and education to space technology. It was enlightening to meet women engineers and scientists from countries where I assumed women did not work in such areas. There were women from UAE to Ukraine, Turkey to South Africa; Pakistan and Nigeria; both sent a large contingent."

"Many of the problems faced around the world are similar—fewer girls entering STEM and then the effects of the leaky pipeline as they go up the career ladder. Obviously, many of the strategies to attract more women into STEM were paying off if one looked at the numbers of women from so many countries. But what about keeping women in these fields, happy, doing what they want to do and moving into positions of ICWES leadership and power. Will things begin to change once more women enter and form a 'critical mass'?"

"In the evenings, dinners hosted by different groups provided such variety of fare that one simply could not stop sampling. The final dinner was hosted by the Mayor of Seoul. This and the First Lady's presence at the opening ceremony signalled a warm climate towards women in technical fields from all over the STEM in Korea. Entertainment included dancing, music and even a NANTA performance (a mime performance using cooking utensils). Evenings were an occasion to parade in one's finery and be photographed. The Nigerian delegation had dresses made for all of them in purple for this occasion!"

"The conference ended with the conference organisers—the Association of Korean Women Scientists and Engineers (KWSE) telling us about their organisation. It was started in August 1993 with a membership of around 150 women. They have grown in numbers and organisational abilities over the decade as could be seen very clearly from this conference. Every last detail had been thought out including sign language interpretation as well as regular translators. Even though very often language was an issue, it never became a barrier."

"My memories of Korea are very warm. What I have gained most from ICWES is a chance to meet so many women working in STEM from around the world. I plan to write and talk about the conference to as many schoolgirls and women in STEM as possible. I already have been offered a photo exhibition of my photographs taken in Korea (not just the conference) at the Youth Hostel in Busan, Korea. I am very glad to have met WES members—Pam, Betty, Roseni and others who were very welcoming. I hope we will continue to meet from time to time. Thank you to WES for making this possible" ("ICWES-XIII," 2005, p. 3).

Sue Bird

"I arrived in Seoul five days before ICWES-XIII, which gave me two days sightseeing and a couple of days at the INWES board meeting before the conference started. The Board Meeting and ICWES-XIII were held at the Ewha Women's University in Seoul. This was in a good position, only a few stops on the reliable and cheap metro from the centre of Seoul, and close to a busy and vibrant shopping area. There were two disadvantages to the Ewha campus. The first was that it was built on the side of a mountain, and each time we emerged from one building to go to another, it generally meant either a very steep climb or easy slide, or a wait for a bus. The second was that it was dry!"

"INWES felt that this conference was very important, as it is the main event in three years where we can let people know about us and what we do. We discussed this at the board meeting, and had a fairly obvious presence at the conference with a stand that was staffed most of the time."

"The conference was, as always, a terrific opportunity to talk with people of different countries, races, outlooks and ideas. The conference organisers had raised enough money to bring 50 people from developing countries—a wonderful total. ICWES conferences would be nowhere near as successful if they did not include people from all over the world, and many women do not have the money to travel. Incidentally, WES sponsored a delegate from Ivory Coast and one from India, both of whom are INWES board members."

"I spent much time talking to people about INWES and how contacts could be set up and maintained. Many people joined INWES at the conference, and we are now improving our website to facilitate communication. All WES members are also members of INWES."

"Pam Wain, Betty McCarthy and I spent some tiring hours looking around the acres of markets, and also visiting the museum and art galleries. We sampled the delicious Korean food both at the conference and outside, and were shown some examples of Korean culture."

"At the end of the meeting, a new INWES board was elected—a third of which are new faces. I am still on the board as vice-president (conferences) for the next 3 years. ICWES-XIII was a great success."

"We look forward to ICWES-XIV in Lille, France" (Bird 2005, p. 3).

Chapter 17
ICWES-XIV: Lille, France, July 15–18, 2008

Theme: "A Changing World: New Opportunities for Women Engineers and Scientists"

Organizations Involved

ICWES-XIV was organized by Conseil National des Ingénieurs et Scientifiques de France (CNISF), in partnership with CNRS—National Centre for Scientific Research—(Mission pour la Place des Femmes au CNRS) and European Associations of Women Engineers and Scientists:

- in France with: Association Française des Femmes Ingénieurs (FI); Femmes et Sciences (F&S); Femmes et Mathématiques (f&m);
- in Germany, with: Deutscher Ingenieurinnen Bund e.v. (DIB);
- in the United Kingdom, with: Women's Engineering Society (WES).

Participants and Travel Grants: The organizers provided 40 grants ($50,000) and UNESCO provided 8 more ($15,000) to developing country participants. There were 481 delegates, 58 from France, and 423 from 55 other countries. Out-of-country attendance was 87.9% (see Appendix 2). More details are provided in Monique Moutaud's conference report (extracts are found below).

Description of the Conference (by the Chair, Monique Moutaud)

"The conference included scientific and technical presentations as well as sessions on gender, management, and leadership and methods to improve the position of women in Science, Technology, Engineering and Mathematics (STEM). Women engineers and scientists gathered from all over the world enjoyed the opportunity to exchange information on academic,

M. Frize et al., *Women's Contribution to Science and Technology through ICWES Conferences*, Women in Engineering and Science,
https://doi.org/10.1007/978-3-031-33800-7_17

industrial, small, and medium enterprise research. The main objectives of ICWES conferences are:

- To encourage women scientists and engineers from all over the world to meet and share their research and experiencein order to work out an international vision to everyone's benefit.
- To disseminate and strengthen the role of women engineers and scientists in society by their contribution to the cultural change in professional or academic research environments."

"The program was built with the contribution of each participant submitting papers to the sessions: plenary lectures, oral presentation, posters. The conference focused on issues like Management and Leadership, Diversity and Ethics. Moreover, key scientific and technical topics have been discussed including opportunities and challenges in sustainable development, waste management and water resources, climate changes, information and communication technologies, and new biotechnologies. Other important features of the conference were a Youth Forum, a formal workshop to meet 'Inspirational Role Model', and the opportunity to learn more about mentoring. To achieve these objectives, we have set up an organizing committee in January 2007 with a project team and collaborative tools to facilitate the communication between all the members of the team. The management of the project is assumed by a steering committee which is composed of the chair of the conference and the chairs of each committee (program, administrative and logistic, marketing and communication, finance). Referring to that organisation, the work plan was made according to a policy manual for ICWES provided to each ICWES chair by INWES. (International Network of Women in Engineering and Science). Moreover, the ICWES-XIV chair was present at the last three conferences and gained a lot of knowledge from Canadian and Korean colleagues. So, the work plan was set up, taking into account all recommendations. The project was structured in four committees. Each committee could progress in [its] activities in parallel and the coordination can periodically deal with the interactivity between them to achieve the results and respect the milestones targeted. To highlight the results obtained on the work progress during the preparation period, we would like to explain that all activities [defined] in committee are articulated around the website with a public side addressed to the future participants and a private side addressed for the organizing committee, called collaborative platform which allowed to share all the useful information of the project between the project team even if we are not physically close to each other. Each work package could benefit of a specific space shared with all the members of the global team. On the point of view of the organizing committee, all the achievement of the activities was facilitated by the use of the collaborative platform to put all deliverables of the project at the disposal of each one of the team. On the point of view of the participants, all the operations could be made with the website and general information or guidelines could be found on the same way. The different items developed were:

- Home page with welcome address and sponsors;
- Conference information with an overview of the event and the organizing committee;
- Program: scientific, social, and touristic, with CV of the eleven keynote speakers and the different steps to build the scientific program."

"Call for abstracts with submission guidelines, online submission form to propose an oral session or poster; afterwards we posted:

- The list of accepted abstracts and travel awards;
- Registration form to the conference and to industrial visits and social program;
- Registration form to visits, tour and accommodation in hotels or dormitory on the campus;
- Exhibition and Sponsorship;

- General information to easily [reach] Polytech' with access maps and other useful information;
- On the left side of the website, one specific item emphasized the news of the moment. During the last year of preparation, the Program committee and the Communication and Marketing committee represent a heavy [load] in terms of volunteer time as all the tasks to produce have to respect a coherent and precise communication between the website and all other supports to realize in French and in English and the energy to spend was the double of the previous [estimation]."

Opening, by Monique Moutaud, Chair ICWES-XIV

"It is a great pleasure for me to welcome you on behalf of the French scientific and engineering National Council, CNISF to the 14th ICWES Conference taking place in Lille. We are deeply honoured that Mr. Nicolas Sarkozy, President of the Republic, accepted the High Patronage of this conference. We are also highly honoured by the patronage of Mr. Jean-Louis Borloo, Ministre d'Etat, Minister of Ecology, Energy, Sustainable Development and Town and Country Planning, and also by that of Mrs Valérie Pecresse, Minister of Higher Education and Research, who has labelled this conference the first event to take place under the umbrella of the French Presidency of the European Union. We also wish to express our gratitude to Mr. Domino of the Regional Chamber of Commerce and Industry and Mr. Cluzel of the Chamber of Commerce and Industry for hosting the opening ceremony".

"We are very happy to see among the audience some loyal friends who have been attending this conference for many years now, and to greet new participants from European countries who joined us for this event. I would like to thank your excellencies for having come all the way to support some of your elite scientists participating in this conference. This year, ICWES-XIV is proud to welcome over 500 people from 60 countries. To highlight the diversity of the regions of the world represented at ICWES-XIV and its increasing popularity, I would like all the participants from Asia, to stand up, so we can welcome them. Now I will ask all participants from Africa and the Middle East, to please stand up; now comes the turn of our participants from North America; and now the participants from Central and South America; and finally, all the participants from Europe."

"We hope you will enjoy the diversity of the scientific and Gender sessions proposed to you during the next three days. No doubt, the exchanges generated by such an international gathering of women experts in a diversity of scientific fields will be a source of inspiration and bring about new initiatives and opportunities. A number of renowned women scientists, who are honouring us with their presence, will be giving thematic lectures during this three-day conference. You will also be able to meet outstanding role models who have achieved highly successful and rewarding careers. We believe that the collaborative work carried out during this conference will open new opportunities to women scientists and engineers at the international level. We also hope that this conference will help women from developing countries where gender issues need most to be addressed to improve their position in the scientific and engineering fields by 'raising the awareness of their government to gender equality. Three years from now, we will be happy to hear you tell us at ICWES-XV how the knowledge acquired during ICWES-XIV was translated into projects for your communities of women scientists and engineers."

"And now we would like to thank the major French institutions whose support has been instrumental in making this conference possible. I will acknowledge here the Directorate-General of Research of the European Commission and the Nord-Pas-de-Calais Regional Council, as well as the Ministry of Higher Education and Research, and the Ministry of

Ecology, Energy, Sustainable Development and Town and Country Planning. We would also like to thank UNESCO for their generous awarding of travel grants to some of the participants. Finally, we would like to acknowledge the University of Lille 1 for their warm welcome and for hosting our meeting at Polytech'Lille. We are also greatly indebted for their active participation in the organization of this event to five European associations of Women Engineers and Scientists, to the Mission for the position of women at the CNRS, as well as to the board members of INWES, who are a member of the organizing committee. We would also like to acknowledge the generous investment in time and energy of the volunteers whose efficient input has contributed towards making this event a success. Before giving the floor to the various personalities who have honoured us with their presence today, I would like to express my deepest gratitude to Mrs Anne Lauvergeon, chairwoman of the Directoire of AREVA, for giving us some of her precious time to attend this ceremony in Lille. Mrs Lauvergeon's strong commitment to helping women access the highest level of decision making in both the public and private sectors should be an inspiration to all of us. We are very grateful that she has come today to tell us how important it is for women engineers and scientists to attract a greater number of young women scientists in high-tech companies. I declare ICWES-XIV OPEN and I give the floor to Yvette Ramos, Chair of the LOGISTICS committee, who will introduce our guests involved in the major issues that will be developed during the three-day-program of ICWES-XIV. Welcome again to all and thank you for your attention" (Moutaud, "Opening", ICWES-XIV).

Welcome from Monique Frize, President of INWES

"It is a great pleasure for me to welcome all of you to this 14th conference in the series of ICWES . The first one was in NY City in 1964, so there is a long history, 44 years, preceding this 14th conference in Lille. INWES is very proud to see the successful organization of this conference by an excellent team in France, under the leadership of Monique Moutaud, and the sponsorship of the CNISF (le Conseil National des Ingénieurs et Scientifiques de France). As in each preceding ICWES, you will find a high standard of scientific and gender papers, with a wonderful blend of cultural events. You will meet old friends and make many new ones. I wish you all an experience to be remembered and thank you again to all who made this possible!" (Frize, "Welcome", ICWES-XIV).

Report from the Chair, Monique Moutaud

"We are very proud of the success of the event which could not be possible without the professional work of all the volunteer people involved in the organization."

"That success rests on the following aspects to emphasize:

- First of all, the diversity of countries represented. We never reached such a number of attendees coming from 60 countries on 5 continents in other ICWES. The European countries were well represented with 230 attendees coming from 21 countries. The European partnership of the organizing committee was the key to dispatch the information with good tools. Africa was represented by 100 participants coming from 17 countries. Asia was represented by 83 participants coming from 11 countries. America was represented by 78 participants coming from 6 countries and the

Middle East was represented by 9 people coming from 5 countries. We welcome 500 women and a few men.

- Then, the high quality and diversity of the scientific program included 130 oral sessions, 11 plenary lectures and 50 posters (accepted abstracts) completed by a Youth forum, round table on 'Inspirational Role Models' with outstanding women like Claudie Haigneré (scientist, first French women astronaut, politician), Valérie Manning, Vice President EADS... Among the famous lectures, we must [name]:
  - Judith Glover whose presentation dealt with the speech and stereotypes developed by companies on hiring very qualified women in electronic and telecommunications fields;
  - Pauline Gagnon who talked about the place of women in Research Laboratory of Particle Physics with the biggest detector ever built;
  - Nicky le Feuvre whose speech deal with 'The feminisation of scientific and technical professions: an issue for the evolution of gender relations in contemporary societies';
  - The Future of the Internet and telecommunications presented by Martine Lapierre (Alcatel-Lucent);
  - The view of the European Commission on the situation of women in science by Johannes Klumpers, Head of the scientific culture and gender issues unit".

"Were also put on the table a lot of subjects about the sustainable development and the civil society like climate changes but also about innovation management in big companies or the direct impact of new economy and TIC on the environment of women workplace" (Moutaud, "Final Report ICWES14", p. 4).

Business Meeting

The Board Meeting Minutes: These have been discussed in Chap. 15.

Testimonials

Pat Battams, "From the Editor..."

"Nearly a month has passed since I returned from ICWES-XIV in Lille. I can echo the views expressed elsewhere in this issue that it was a fantastic experience to be in the company of 450 women scientists and engineers from all over the world. It was inspiring to learn their determination, past and present, to succeed in their chosen career".

"It was impossible to report on everything that took place as there were five parallel sessions at most times. It was difficult to choose which to cover. There was a good turn-out of WES members amongst the speakers and delegates. We took the opportunity to display our new recruitment literature, explain our activities, such as MentorSET, and to publicise our own conference."[...]

"There was much talk at ICWES about flexible working" (Battams 2008, p. 2).

Pat Battams, Ruth Carter, and Betty McCarthy, "Four Action Packed Days in Lille"

"Pat Battams, Ruth Carter, and Betty McCarthy report on a busy programme of events at ICWES-XIV":

"Imagine the buzz, 450 women scientists and engineers from all over the world meeting for four days at Polytech'Lille, the engineering school in Lille, northern France."

"While other ICWES meetings have had an emphasis on the work that women do as scientists and engineers, this one also had a very strong set of gender themes, such as attracting girls and women into scientific and technical careers, women as leaders, entrepreneurs and role models, gender perspectives on scientific careers, mentoring, and work-life balance. The issues of male bias in the content of science and engineering curricula were tackled too."

"Conference kicked off with a presentation by economist Pietra Rivoli from Georgetown University, USA considering attitudes towards globalisation. Economists tended to believe that if people were better off, they would favour globalisation but this ignored the value we place on fair play. A broad multidisciplinary approach was needed to solve global problems. Women were skilled in networking and mutual support. This was the sort of bridge-building that was needed to solve the problems facing society."

"Problems that women face in pursuing academic and research careers seem to be much the same throughout the world. Speakers from all countries reported on a high representation of women at lower grades but few reaching the top positions. Disappointingly the situation was much the same in the eastern European countries where up to 50% of PhD students in science and engineering were female. In Latvia, Lithuania and Estonia, a considerable female pay gap existed in research jobs – up to 34% in Latvia. In Poland women's progress to senior levels was slow mainly due to the expectations of Polish society about women's domestic role."

"In Australia the number of engineering students had dropped from nearly 16% in 2001 to 14% in 2005, and women were leaving engineering careers faster than men. This was despite a major increase in carers' leave and flexi-working that women were not using because they believed that this would cast them in a negative light. Similar results were found in a study of retention rates carried out by the US Society of Women Engineers. SWE has widely publicised these results amongst US corporations, through congressional briefings and in the media. SWE's corporate partners had expressed alarm at the rate they were losing women engineers. SWE's solutions include corporations working with women early in their careers to map out career paths and leadership opportunities."

"The proportion of women working in information and communications technology (ICT) is also decreasing. This industry appeared initially attractive to women, yet fewer than 1 in 5 managers are female and women are most highly represented in the technician grade. There was a perception that women were better at 'soft skills', such as communications, team working and presentation, and these were not as highly valued as analytical and technical skills at which men were perceived as being best. If ICT wished to attract more women back to the industry it needed to tackle gender stereotyping, offer varied career paths and a gender aware progression system."

"The UK is not doing well in the comparative tables on the participation of women in engineering, despite the best efforts of WES and the UKRC. In 2004 the UK was one of only four European countries which still had less than 20% women graduating in engineering and less than 4% of engineering apprentices are women. In a paper: Engineering: a future for women? Pat Morton set out our own UK-specific difficulties and suggested how holistic approaches to student recruitment and support, career and workplace issues might be sustained."

"**Seeking Solutions.** Milada Williams explained one UK approach to retaining the 50% of women engineers who don't work in engineering. Mentoring schemes, such as MentorSET, were instrumental in helping women return to technical jobs as well as helping those who had reached a crossroads in their careers."

"Women overlook the influence that they could assert through engineering academies, according to Moyra McDill from Canada. The International Council of Academies of Engineering and Technological Science (CAETS) has members in 24 countries – the Royal Academy of Engineering in the UK. Academies are widely consulted by government bodies and other decision makers. Women are poorly represented amongst fellows often because women tend to ignore these bodies as being 'old pale males."

"Self-knowledge and understanding the difference between male and female behaviour and values were key to women breaking the glass ceiling, according to Silvie Klein-Franke from Germany. Women needed to understand their own leadership skills and challenge organisational culture. 'Don't count on men, count on yourself and your networks', she said. Korea is a great example of female empowerment. The Association of Korean Women Scientists and Engineers (KWSE) was set up in 1993 with 50 members. There are currently 1217 members, 800 of whom are PhDs. KWSE has been influential in pushing women's employment in technology because the Korean government is keen to become a knowledge-based supplier. Since KWSE organised ICWES-XIII, it has set up programmes for promoting science to the public and developed leadership training programmes. It is now setting up an Asian network."

"Leading international technology companies desperately want to recruit more women engineers, according to the round table discussion with HR professionals from some of these companies. They need diversity to generate more diverse ideas in areas such as problem-solving and forward planning."

"The companies were recognising the importance of inviting young women for hands-on job experience and going into schools. They were also realising the importance of different branding and promoting differences in workplace culture to appeal to women. Interesting initiatives being explored included career progression to an expert or executive expert status as an alternative to a managerial role – believed to be an attractive proposition for some women. Also recognition that people's careers move at different rates could enable women's careers to slow down when raising a family and later speed up again."

"In the ensuing discussion the company representatives were told that they needed to listen to their women employees and give them the tools to run their working lives according to the needs of family life."

"An area in which female employment is growing is in 'technology transfer (TT)' – companies involved in bringing academic research to the marketplace. This was often a career option for women scientists and engineers returning to work after a career break. The work offered fascinating projects and high flexibility to combine work and family commitments. A study amongst women working in TT in Germany showed the downside – low wages and limited career prospects. It was speculated that the high number of women working in TT is undoubtedly due to the networking and communication skills that they are able to develop, but also unfortunately because the prestige and rewards are not sufficient to attract men. There were many presentations offering an analysis of the absence of girls and women in STEM, together with descriptions of some very interesting initiatives to address the abysmal situation in most countries. It was clear that analyses nowadays do not reveal anything new about the barriers to women's participation in non-traditional fields, but they do serve to remind politicians, managers in education and industry and a variety of other budget holders that much remains to be done."

"Some of the most impressive, multi-pronged programmes to attract and retain girls in STEM came from the more affluent countries. For example, scholarships from foundations or campaigns directed at high school girls in the USA, Australia and Switzerland. In less developed countries there is strong awareness of the problems and what needs to be done, but a shortage of resources to implement initiatives that have been designed."

"Climate, Water and Sustainability. 'There will be no climate justice without gender justice,' said Ulrike Roehr. It is only recently that the gender aspects of climate change have come to be recognised. Climate change affects men and women differently. In poor countries women have less access to energy and are more vulnerable to climate changes because of their poverty, lack of education and responsibilities for providing food and clean water. Elsewhere energy production is male dominated and women tend to be excluded from energy planning. Men tend to favour technological fixes whereas women look to lifestyle changes."

"Five years ago at the United Nations Framework Convention on Climate Change (UNFCCC) a network was set up, Women in UNFCCC, and at the UNFCCC conference last December papers by women on gender perspectives were adopted."

"In many parts of Asia and Africa, clean water is not available from a tap at home, but collected from a distance source. Even when it is collected, it needs to continue to be kept clean during storage until it is used in order to prevent contamination and the spread of disease. In the session on water management, we heard about the use of traditional water filtering methods and in particular, using fabrics, to safeguard the quality of water that is collected and then stored at home. That water 'husbandry' is mainly the responsibility of women was clear from this and other descriptions of small-scale water supply and management in Japan, Mexico and Ukraine. Another paper raised awareness of water as an increasingly scarce resource by suggesting ways that less water could be used in garment manufacturing."

"Amongst the low technology solutions described in the sustainable development session, were phytoextraction – plants which remove metals from contaminated soil; compressing shredded bamboo into a material to replace plastics packaging and adobe housing that is three times cheaper to build. An unexpected benefit of adding fish farming to small farming is malaria reduction as fish eat mosquito larva."

"Added Value. "But ICWES-XIV was not all work, although the crowded programme left little time to see the city. There were plenty of networking opportunities during coffee and lunch breaks, and two lavish evening receptions at the grand buildings of the Lille Chamber of Commerce and Industry and the Musée des Beaux Arts, where there was also private access to an exhibition of Goya drawings. And then there was the gala dinner."

"The real value of an event of this type and size is undoubtedly the opportunity to meet like-minded women from all over the world and to affirm the existence of a global community working towards common goals. You leave with restored energy and a sense of renewed commitment, looking forward to ICWES-XV in Adelaide, Australia in 2011" (Battams et al. 2008, pp. 8–10).

Betty McCarthy, "Dear Editor", Writing About Her Impressions of ICWES-XIV

"I am writing this letter in the hope that future ICWES Conferences will be properly organised. The organisation of ICWES-XIV broke down under the pressure of the number of delegates. In my opinion if WES members had not helped out to a level beyond the call of friendship, the event would have been a disaster."

"Before:

(a) Grazyna Whapshott (who was temporarily immobile) had to review more than her fair share of abstracts. She got her reviews in on time unlike others.
(b) No welcome packs were sent out before the event either by post or email so there were troubles with registering on the Tuesday. I spent over an hour walking around the university site trying to find out where to register, only to be told it was in the centre of Lille.
(c) There MUST be plenty of signs to aid navigation especially where the talk's location and canteen are either ends of the site. Margaret Ajibode ran around the site putting up hand-made signs to rescue the situation.
(d) A more serious concern is that proper respect should be given to those who have submitted posters. These, mainly students, should be allocated places where their work will be seen not just odd rooms sprinkled over the site.

During:

(a) Talks should not start at 08:30 in the morning – there are too many people arriving late who interrupt the keynote speakers.
(b) Talks should not be in odd rooms sprinkled all over the building.
(c) Proper facilities should be provided for the disabled, such as all three lifts working, toilets truly accessible – not just a label stuck on the door. Not having the restaurant over the other side of the site, and having to travel back and forth to the centre of town.
(d) Speakers in the various sessions should be respected and chairs be given some guidance (such as, that they are not there to give a 15-minute presentation on their own subject after each speaker but to introduce the speakers and to ask a relevant question if the audience are reticent).
(e) The gala night was a disaster – coaches turned up at least half an hour late so we missed the conjuror doing table tricks.
(f) Respect should be given to the diets required by the delegates – some have strict religious rules and a printed menu should have been provided at the gala night so that delegates could check the meal placed in front of them very slowly. As a vegetarian I was catered for even if it was not very nice but others could not identify the meat supplied.
(g) Dancing should be possible at the gala event as some cultures really appreciate this activity.
(h) The evening was supposed to end at 10:30 but about 11:30 an escape committee was formed to go back by metro. As the word got around so many wished to go, a coach was called up for us.
(i) The conjuror did some rather old cabinet tricks very badly – loud music does not cover bad technique.
(j) When the successful INWES members were announced, Pam Wain, who was one of those elected, was not there as she had taken a delegate, who was sick after the fish course, back by taxi."

"In short, I remember the organisation of Korea with fond memories even if it was half-way up a mountain" (McCarthy 2008, p. 11).

Chapter 18
Development of CIWES-ICFIS

A Canadian Charitable Organisation to advance the Education and Success of Women in STEM

The charitable organisation *INWES Education and Research Institute* (ERI) was created in Canada in 2008. Maintaining this name was confusing and not ideal, as it was discovered that Canadian Charity Laws require that the Institute be at arm's length with noncharitable organizations like INWES. This requirement changed drastically what was permissible, and the ERI could not report to the INWES Board as originally planned. It had to be completely independent. The Canadian rules also made it very complicated for a charitable organization to provide funding to non-charitable corporations like INWES. Thus, the ERI had to revise its goals and plans and redefine itself to respect the laws. This led to the need to hold a strategic planning meeting with partners and supporters of the ERI. This meeting was held on June 24, 2019. Prior to discussing this meeting, it is important to mention the main activities of the ERI between 2007 and 2019.

The first project funded occurred prior to ERI obtaining Charity Status: it was an Essay competition organized by the Association for Promoting African Girls in Engineering Science and Technology Education (APAGESTE Nigeria), led at the time by Uduakobong Aniebiat Okon. APAGESTE, an NGO registered in Nigeria, has been an INWES member since 2005. The association aims to encourage and support African girls and young women in their pursuit of education and participation in STEM.

The competition, first offered in 2008 in the state of Akwa Ibom as a test case, targeted girls between 16 and 19 who were looking to undertake a STEM study program in college or university. Financial partners to APAGESTE were, in addition to INWES, NNPC/Mobile Exxon Unlimited, Bethesda Family Clinic, and Audeo Nigeria Ltd.

Participants should have taken several STEM classes in their senior secondary year (at least five in physics, chemistry, biology, mathematics, agricultural science,

M. Frize et al., *Women's Contribution to Science and Technology through ICWES Conferences*, Women in Engineering and Science,
https://doi.org/10.1007/978-3-031-33800-7_18

computer science, introductory technology, electrical electronics, mechanical technology, woodwork, automobile technology, welding, and textile technology). The application consisted of a motivation letter explaining why the participants wanted to follow a study program and pursue a career in STEM and a short biography of a women scientist or engineer coming from their region. The award ceremony took place on November 27–28, 2008. Three prizes were given ($500 US, $200 US, and $100 US), as well as six consolation prizes ($50 US each). The whole project amounted to $2310 US.

After obtaining Charity Status in 2008, two important workshops were organized with major outcomes.

In 2011, a workshop was held at the University of Ottawa on April 29–30 to examine the barriers that continue to limit the participation of women in the engineering profession, update our inventory of best practices, and build more bridges to increase women's participation in engineering programs and consequently, in the profession. The *Canadian Committee on Women in Engineering* (CCWE) official report released in April 1992, made 29 recommendations designed to break down stereotypes in early education, universities, workplaces, and professional associations. Now, 20 years later, it was time to bring together, yet again, key stakeholders from these communities of interest to review what had taken place over the years.

The **CCWE^{+20}** workshop brought together 75 key players from various backgrounds to discuss past and present conditions that contribute to a low participation of women in the engineering profession. After 2 days of intense discussions, workshop participants developed 25 recommendations aimed at improving the image of engineering for girls and young women as well as the climate and curriculum to make engineering programs more appealing to women. These recommendations were also directed at developing good practices in workplaces, within professional associations, granting agencies, and research organizations with a view to increasing the participation of women at all levels. Each organization and individual should focus on recommendations that best fit their own particular environment and resources. Collaboration between individuals and organizations to make this happen will be encouraged.

Two immediate next steps were established at the close of the workshop. They were (i) review of the recommendations by participants and selection of 3 top priorities for media release and (ii) make the *Recommendations* and *Final report of CCWE^{+20}* available on the INWES-ERI website this summer. The workshop project was a huge success, and participants were unanimous in recommending that this process be repeated in 2 years. It was also suggested that the *INWES-Education and Research Institute* set up a permanent committee to continue connecting people and to monitor progress in the implementation of new recommendations, including set timelines, metrics, and lead organizations identified for implementation (www.ciwes-icfis.org).

In 2014, a two-day workshop was also held at the University of Ottawa to explore the creation of a national archive of women in STEM fields: "Creating the Memories and Celebrating the Legacy of the Bold and the Brave: Building the Archives of Women Scientists and Engineers in Canada." Gathered were historians, archivists, women scientists, and engineers to learn how they can contribute to one another's fields, with the common objective of accomplishing the first step in the greater initiative of building the archives of women scientists and engineers in Canada. Together women in science and engineering and leaders in the archiving community could recover, research, write the history, and preserve the archives for current and future generations.

The workshop brought together 45 participants from Ottawa, other parts of Canada, the USA, and the UK. The event program was balanced with presentations in which speakers shared their expert knowledge and time for working groups to discuss what they had learned and answer key questions. The discussions from the working groups were the basis for establishing strategic priorities during a plenary session planned for day two. Four of the five current chair holders of the NSERC Chair for Women in Science and Engineering and four former chair holders were participants in the event. Four former chair holders, who were not able to participate in person, prepared videos in which they spoke of their experience and contributions as chair holders and their ideas on building the Archives of Women Scientists and Engineers in Canada.

Organizers were Professor Catherine Mavriplis, Faculty of Engineering, University of Ottawa and NSERC/Pratt Whiney Chair for Women in Science and Engineering, Ontario; Monique Frize, Distinguished Professor, Faculty of Engineering, Carleton University, and President, INWES Education and Research Institute (ERI); Professor Ruby Heap, Faculty of Arts, Department of History, Associate Vice-President, Research, University and Ottawa and Director (Research), INWES ERI, Professor Claire Deschênes, Department of Mechanical Engineering, Laval University, and Director INWES ERI (Treasurer).

Many secondary goals were identified such as to pay full tribute to women scientists and engineers' accomplishments and their contribution to women's history, to the history of science and engineering and to Canadian history; to provide critical historical insights into the current status and role of women scientists and engineers; to help establish important connections with gendered practices in the past and present-day challenges and policy issues; to develop a sense of pride in the achievements of women scientists and engineers, and to provide girls and women with inspiring role models that could lead them to pursue careers in science and engineering.

Seven outcomes were a direct result of the workshop's planning and execution:

- Establishment of a Task Force to continue the discussions.
- Four strategic priorities for the task force to act upon were identified.
- Links to be established with external organizations (SWE (US), key organizations to involve, key journalists of interest, the Wellcome Library (London, UK), and the WISRnet (UK)).
- Awareness (Strong Twitter presence, post-event articles, videos produced by former and current NSERC CWSE, podcasts and recordings, CWSE-ON event webpage and event blogs) to be created.
- Book plan: Current and former NSERC Chairs planned on writing a book to document their work and experience.

- Tour of the Canadian Women's Movement Archives, University of Ottawa was carried out.
- Presenting Sponsor visibility benefits.

Sponsors included the NSERC/Pratt & Whitney Canada Chair for Women in Science and Engineering for Ontario, the INWES—Education and Research Institute, the NSERC Chairs for Women in Science and Engineering National Network, the Social Sciences and Humanities Research Council of Canada, the Canadian Commission for UNESCO, the British High Commission, the National Science and Engineering Research Council of Canada, the Institute for Science, Society and Policy, the Faculty of Engineering of the University of Ottawa, the Faculty of Science of the University of Ottawa, IBM, and the University of Ottawa.

The Canadian Archive of Women in STEM initiative is led by the University of Ottawa Library. The partnership between the Canadian Institute for Women in Engineering and Sciences (CIWES-ICFIS) and Library and Archives Canada is now in its sixth year. The main focus of the initiative remains the portal index, which provides links to records held across Canadian archival organizations that relate to women in STEM: this includes the records of women and the records of the organizations that supported them. There are now 482 entries in the portal—representing 482 archival fonds/collections related to women in STEM in Canada. The availability of this robust tool provides a central location for the records of women's contributions to STEM in Canada, helping to ensure that these are more easily known and accessible.

The University of Ottawa Library Archives and Special Collections has also added several new fonds to their holdings. In time for Canadian Women's History Month, the Archives and Special Collections also hosted an exhibit (virtual and in-person) that celebrates women and work. As part of the exhibit, the records of engineers Claire Deschênes, Monique Frize, and Claudette Mckay-Lassonde were showcased.

CIWES has also added information to their website about the Canadian Archive of Women in STEM initiative. The University of Ottawa Library provides a guide to help potential donors navigate the archival donation process.[1] This guide was developed with the assistance of archivists at the University of Toronto, Wayne State University (SWE Archives), and Women's College Hospital Archives.

Strategic Planning in 2019

The Board of Directors of the ERI decided to hold a strategic planning meeting with partners on June 24, 2019, at the University of Ottawa Library to assess what ERI was when it was created and the boundaries set by Canadian charity laws and to identify what ERI should become. Present at the meeting were the five Board

[1] https://biblio.uottawa.ca/sites/biblio.uottawa.ca/files/how-to_guide.pdf. Accessed Mar 21, 2023.

members of ERI (Monique Frize, Claire Deschenes, Ruby Heap, Peggy Layne, Marina Bokovay) and representatives of NSERC, Engineers Canada, the Canadian Commission for UNESCO, Libraries and Archives Canada, the University of Ottawa Library, and of the NSERC Chairs program.

ERI was originally created because INWES could not obtain charity status. However, rules in Canada are restrictive and we cannot easily provide funding to non-charity organizations; NGOs are not qualified donees, and the process to provide some funding to these is very complicated and requires much overseeing by the ERI Board, which would be quite time-consuming. Since ERI could not easily provide funding for INWES or for any of their partners who were not a charity organization, we needed to examine what new role the Institute could play in Canada and elsewhere.

The objectives and hopes for the meeting outcomes were as follows: identify gaps in which ERI can play a role; identify activities that ERI can do; partners and the board are to identify ideas and projects that address knowledge, the status of women in these fields and the profiles of women who work in this field; identify boundaries—perhaps start with Canada or North America, and see how we can share more broadly later on; what boundaries are going to limit or support what we want to do—who we are and who we want to be; find a more suitable name—our partners are far more numerous than just INWES; define an election process and board structure—we have not had elections to date; evaluate the past; and keep a focus on academic perspective and involve collaborators to participate in our activities and vice versa.

The main objectives of INWES-ERI from the 2010 annual report were as follows: award scholarships to students wanting to study STEM; carrying out research in developed and developing countries concerning fields of STEM; acting as a resource center for data and information on women in STEM; developing and maintaining a website. The INWES-ERI was governed by a Board of nine directors, and activities were planned through five sectors: programs and projects, research, education, communications, and fund development. However, we needed to reexamine the objectives as some of them were very expensive to carry out yet provided little benefit. An example was the idea of providing scholarships; this idea requires a lot of funding, a competition management team, and provides a benefit to very few persons. So it was eliminated.

Looking back at past activities, there were two very successful workshops. In the annual report for 2017–18, we note a change in ERI primary objectives to advance the education of women in STEM by (1) creating the archives of Canadian women in STEM and (2) developing and maintaining a website to raise public awareness. These objectives were supported by the participants of the June 24, 2019, meeting.

Name Change

The Board members met the next day (June 25) to finalize the plans. It was decided that the activities and projects should, in the next 5 years, be focused on Canada to be expanded later on to the USA and elsewhere. Thus, it made sense to change the name from INWES Education and Research Institute to the Canadian Institute of Women in Engineering and Science (CIWES) and its French name: Institut canadien des femmes en ingénierie et sciences (ICFIS). The application to change the name was made to the Canadian Government and confirmed shortly after. A new logo was designed and is now used for the letterhead and website. The most important challenge remains to find funding for projects. A Finance Committee was created and is developing a plan to define important projects and develop fundraising strategies to implement these projects.

Conclusion

The Book Project

A unique adventure began in 1964, the creation of the ICWES conference series, so it is more than timely for the story to be told: not only why and how it was created, but also how it was sustained and how it grew over the nearly 60 years and gave rise to the not-for-profit organization INWES in 2002, and the charitable organization INWES Education and Research Institute in 2007, now CIWES-ICFIS since 2019.

It was a great luck to be able to collect the ICWES material from Mahin Rahmani (Iran), Issié Gueye (Ivory Coast), testimonies from SWE magazines and WES UK Journals publications (*U.S. Women Engineer*, *SWEMagazine*, *SWEMembers-at-Large (MAL) Newsletter*) and WES Journal, and personal testimonies from attendees such as Kathleen Harer and Dormer Ellis (Appendix 8).

The main intent of this book was to tell the story of the ICWES conferences, so that the readers may have a glimpse at the rich content of the archives located at the University of Ottawa Library, what they say about the participation of women in STEM all over the world during the second half of the twentieth century and the beginning of the twenty-first century. We believe that this book will be useful to both scientists and engineers, to historians, to organizations interested in women in STEM, and to educators in various interdisciplinary fields such as feminist studies.

Continuity of ICWES Through the Creation of INWES

The women leaders who created the ICWES conference series were already aware that, in the longer term, a formal organization would need to be created not only to provide more stability to the process but also more financial security through

M. Frize et al., *Women's Contribution to Science and Technology through ICWES Conferences*, Women in Engineering and Science,
https://doi.org/10.1007/978-3-031-33800-7

insurance and more focused fundraising. The organizing committee, at the Business Meeting of ICWES-II, in 1967, stated: "plant an acorn and let an oak tree grow." This led to a lively discussion by delegates who concluded that "a bulletin to disseminate information on women engineers and scientists was highly desirable although it would probably cost a great deal. A centre to help keep the contact between supporting organizations was highly desirable but probably prohibitive in cost." This became a reality in 2002, 35 years later.

At this point, it is important to recommend that current and future INWES boards of directors carefully read Chap. 14, on the creation of INWES in Merrickville, Ontario, Canada, which describes what was expected from a new INWES organization by the 20 persons who created it, and what exists now in reality. Chapter 15 is also important as it contains many of the important aspects to consider to ensure the continued success of INWES.

CIWES-ICFIS

Although it was originally planned that the institute would be closely affiliated with INWES, the Canadian Charity Laws forbade it and thus the institute became independent and created a new set of goals that would be helpful to Canada, to INWES, to other NGOs, and eventually to many other Women in STEM organizations around the world. The main achievement of the institute so far was to create a successful model on preserving the contributions of women in STEM fields and thus help researchers to include women in science and technology in their writing about the history of science and technology. Another achievement was to collect and preserve precious documents in a database that enabled the authors to write the extraordinary story of ICWES, as well as the exciting story of INWES and CIWES.

S&T and Gender Content Highlights

In producing the first in-depth investigation of the ICWES conferences content, our goals were, as stated in the introduction, to provide an overview on both the emerging or recurrent concerns about S&T, as well as concerning the feminist views on issues regarding women in STEM.

The main themes (Table 1.1) allowed contributions in both categories. Not counting the special events, posters, and plenaries, the ICWES database contains 765 papers and abstracts, of various types, going from scientific reports to research quality papers. Of them, 479 were classified in the S&T category (73%), and 286 in the gender category (37%).

The main themes of the 12 conferences were so broad, as shown in Table 1.1, that the proceedings contain contributions from virtually all areas of knowledge (physics, mathematics, engineering, technology, medicine, biology, geology, social sciences, history, art and culture, ecology, etc.) The contents' important testimony is about the contribution of each of the authors in S&T. The scientific content of the ICWES database is sufficiently large to indirectly reveal part of the history of the development of science and technology in the second half of the twentieth century.

S&T ContentHighlights: ICWES-I to ICWES-XII

Focusing on the S&T category only, we see that the scientific issues that interested the participants evolved in time. The organizers of the first ICWES in 1964 decided that their conference would focus on basic human needs: shelter, clothing, urbanization, food and nutrition, transport, water, and pollution. ICWES-I was held in a period of great hope for peace and prosperity. At the same time, they were preoccupied with the fast increase of the world population. Indeed, it almost doubled during the period covered by these conferences: from 3.3 billion in 1964, it reached 6.3 billion in 2002, leading to problems of pollution and overpopulation. This concern for basic needs will spread to all 12 conferences, representing roughly a third of all contributions in S&T.

As Annelli Hattari, a Finnish participant to ICWES-I, wrote in her contribution: "Nevertheless, the triumphal march of development has proceeded at the cost of pollution of our environment. [...] What have we done, and what must we do, to prevent the pollution of our environment? Attention has been drawn to this regrettable fact only within recent years" (ICWES-I, Complete Proceedings, p. 143). From this statement of pollution in humans' surroundings, the concerns presented by ICWES contributors shifted over time to the concept of *sustainable development* (ICWES-IX, 1991a) and the global issue of worldwide *climate change* (ICWES-XI, 1999).

With time, however, we see in the contributions to ICWES the emergence of other topics of interest. One of them is the important development of computers and computer applications. The word *computer* is mentioned in ICWES-I in the context of future requirements for manpower, various potential usages, and the predictable explosion of information to store. With time, ICWES testified to the participation of women in the development of computers and their languages, computer software and graphics, numerical analysis, measurements and control, computer-aided design, data storing, and so on. Communications also interested many ICWES participants, including telecommunications, satellite applications, and education and training.

A clear feature of ICWES conferences, as well, is that they were interested in issues present in the host countries or the regions at the time of the conference. For example, ICWES-II held in Cambridge in 1967, looked at how to produce "Enough for everyone," ICWES-VIII in Abidjan in 1987 focused on science, technology, and development, and ICWES-VI, which was held in India in 1981, invited contributions on both space technologies and rural problems.

In fact, due to the extraordinary development of sciences and technologies during this 40-year period, we see the arrival of papers with increasingly complex issues reflecting this evolution. This complexity is revealed in some paper titles; an example among others is this paper proposed by Renaye Peters, from Australia (ICWES-XI, Complete Proceedings, p. 89) titled "Integrating ecologically sustainable development and information technology for more efficient building design." Three topics, perfectly relevant together, appear in this title. With time, the ICWES scientific content became increasingly multidisciplinary and multifactorial.

Moreover, aside from the timeline aspect, multidisciplinary and the regional preoccupation already mentioned, the presence of *transversal issues* appears clearly in ICWES contributions, such as technological development, environmental issues, communications, women's studies, and technology transfer. This prevents making a simple classification of the papers included in the ICWES database. In addition, it is not possible to draw conclusions on women's specific role in S&T issues without analyzing simultaneously the whole context of the history of science in the second part of the twentieth century. But we believe that the book provides critical information that can be built on by others in multifactorial research.

Gender Content Highlights: ICWES-I to ICWES-XII

The women contributing to ICWES were certainly proud to present their work in an international recurrent series of conferences, aware of being pioneers among all the scientists and engineers of the world. ICWES interestingly provided a safe noncompetitive environment in that aspect. However, the desire to network among peers in one's own field of expertise was not the motivation of the women who participated to ICWES conferences, as can be seen in regular international scientific meetings.

These scientists and engineers wanted to network with other women in similar situations, learning from role models, sharing ideas to mitigate the low participation of women in STEM, and coming back home with ideas to implement and support regional groups of women in STEM. They wanted to become a strong international force, as revealed by the mission of INWES: "*Strengthen the capacity of individuals, organizations, and corporations througheducationand professional development to encourage the participation of women in Science, Technology, Engineering, and Mathematics (STEM) worldwide through an international network of organizations and experts.*" In so doing, the ICWES database also reveals part of the history of the contribution of these women scientists and engineers to the women's liberation movement, and to how feminism influenced their thinking.

One of the goals was related to the "women's liberation movement." It is interesting to recall that the early ICWES conferences generated insights into the emergence of second-wave feminism in the 1960s in the Western world, and on the activism of women engineers and scientists in various countries. This, however, was different for the Russian women who contented to be fully equal in practice to their male counterparts in science, technology, and engineering, thanks to the support of the Soviet State.

One strong example about women's issues in the proceedings was presented by Evelyn Harrison (ICWES-III, Complete Proceedings, p. 200), in 1971, who had great hopes for the advancement of women. She announced the advent of the contraceptive pill: "Some viewers on the American scene believe we may be entering into a new phase of family relationship—the keystone of our societal structure—where the roles of men and women are converging and the old male dominated, female serviced, family life is being replaced by one of companionship among all members of the family. Medical science, through the advent of the 'pill', has released the American woman from sexual dominance and gainful employment despite its inequities is affecting the power structure of the family unit as the woman contributes significantly to its financial base."

The concern for gender issues was present from the first ICWES conference, and the interactions between women and S&T issues grew from ICWES-I to XII. For example, many papers presented at ICWES-XI (1999), in Japan, were aimed directly at women's quality of life, health, education in S&T, and the role of women scientists in S&T. These concerns were embedded in the program of the conference, as illustrated by the following session's titles: "Human impact," "Quality of life," "Education and ethics," and "Women in science."

In Canada in 2002, an important trend was reversed: more papers were presented in the gender category (77%) than in the S&T category (23%). Indeed, the 131 papers on gender issues represented 48% of all the papers and abstracts of all the 12 conferences. Many of these contributions are directly related to the participation, retention, and advancement of the women involved in STEM careers. This probably meant an increase in the concern for women in STEM, including in education and governmental institutions, in North America in particular. Moreover, Ruby Heap and the organizers strongly called for the participation of researchers in feminist studies.

Suggestions for Future Research

This book builds on the collective memory of women in STEM. It provides relevant information on the internationalization of their community and creation of regional networks through INWES. It assesses continuity and changes in the dialogs between North and South, East and West. It attests to the development of knowledge and technologies by women in a short but exciting period, as well as their concerns for negative aspects brought by S&T development to the human populations and the environment.

But there are still so many aspects to explore for researchers of the future. These include key questions such as

- What is the specificity of women's contributions to S&T, as seen through the ICWES database during the period covered (1964–2008)?

- How did the broad and complex issues concerning the participation and retention of women in S&T evolve over time?
- To what extent did ICWES conferences serve to illustrate and document the reciprocal relationships between science and society?
- How were scientific interests distributed between regions at the beginning of the ICWES Conferences, and how did they change over time through an international dialog?

Appendix 1: Selected Biographies of ICWES Founders, Leaders, and Organizers

Marthe Achy Brou

Conference chair at ICWES-VIII (Abidjan, 1987).

Marthe Achy Brou, née Biley in 1949, graduated from the École Polytechnique Féminine (EPF) de Sceaux, in France. She was the first woman engineer in her country and had a brilliant professional and political career as Congresswoman, Deputy Chairwoman of Congress of Ivory Coast, former mayor of Grand-Bassam (1985–1990)—the historical capital city of Ivory Coast, Deputy Chairwoman of the Economical Council of Ivory Coast, and Minister of Social Affairs and Solidarity.

Achy Brou was an honorary member of INWES and was President of ICWES-VIII in Abidjan, in 1987. She was also involved in the setting up of the INWES African Regional Network. She was ICC convenor after ICWES-VIII and honorary president of CIFISATS. She passed away on November 10, 2011, in Abidjan ("Marthe Solange Achy Brou", n.d.).

Anna E. Amour

Conference secretary at ICWES-III (Turin, 1971).

Born in Turin, Anna Amour graduated from the Polytechnic of Turin in 1933 as an electrotechnical engineer. After graduating she joined ENEL (Italian Company of Electricity), where she held various roles: from the statistics office she moved on to the technical secretariat, then dealing with corporate planning in the form of suggestions and technical proposals. She was among the founding members and national secretary "for life" of AIDIA (Italian Association of Women Engineers and Architects) in 1956. Acting on an idea by the engineer Maria Artini, she set up a network of women graduates in engineering and architecture at the Milan and Turin Polytechnics; the other founding members of this association were engineers Emma Strada, Ines Deltetto Noto, Adele Racheli Domenighetti, Laura Lange, Alessanfra Bonfanti Vietti, and the architect Vittoria Ilardi. Amour was also a long-time Member-at-Large (MAL) of SWE and became a fellow in 1985 ("Amour Anna Enrichetta", n.d.). Anne E. Amour has passed away, date unknown.

M. Frize et al., *Women's Contribution to Science and Technology through ICWES Conferences*, Women in Engineering and Science,
https://doi.org/10.1007/978-3-031-33800-7

Nicole Bécarud

Conference president at ICWES-V (Rouen, 1978).

Nicole Bécarud obtained her undergraduate degree in engineering and a PhD in chemistry (École Supérieure de Chimie de Paris). Bécarud worked as chief of the Scientific and Technical Unit of the French National Association of Food Industries (ANIA). She was president of the Cercle des Femmes Ingénieurs at least by 1978. She was also a prominent member of Association Française des Femmes Diplômées des Universités (AFFDU) from 1965; she was vice-president in 1983–1988 and president in 1995–1999. She was part of INWES from its inception.

Bécarud was named Knight of the Légion d'honneur on March 19, 1985 (for her work at ANIA) and officer on July 13, 2001 (namely for her work with AFFDU). She also organized, in 1994, the first Women's Essay Fair, which excluded, on purpose, literary fields typically ascribed to women, such as novels and poetry. On April 12, 2010, Bécarud was named Knight of the Arts and Letters for her contribution to the journal *Diplômées* ("Hommage à Nicole Bécarud", 2020; Auzac 2021). She passed away on April 8, 2020.

Sumati Bhide

Conference president at ICWES-VI (Bombay, 1981)

Sumati Bhide was specialized in cancer research, namely, on the preventive effects of turmeric, worked at the Indian Cancer Research Institute (CRI), and directed the Bharatiya Vidya Bhavan's Swami Prakashanand Ayurvedic Research Centre (SPARC). She was the first president of the Indian Women Scientists' Association (IWSA) and acted as regional editor for Asia for the journal *Ethnomedicine* ("Sumati Bhinde (1932–1999)", n.d.).

Aileen Cavanagh

Gave the opening address at ICWES-I (New York, 1964).

Aileen Cavanagh completed her undergraduate degree in physics and chemistry in 1951, Boston University, her Master's in Business Administration in 1978 at the University of Massachusetts and was a Fellow Life Member of SWE ("Aileen Cavanagh", n.d.). Cavanagh worked in several companies, starting her first engineering job at Tracelab, developing a device to measure radiation. She spent a short time at Raytheon and then was hired by Bell Telephone Laboratories in 1955 where she worked at MIT's Lincoln Laboratory. She became active in IEEE's Engineering Management Society and was president of SWE in 1963–1964. She was named SWE Fellow in 1980. In 1967, she joined the New Jersey Department of Community Affairs, in Trenton, New Jersey where she started technical and management programs and helped local governments to use computers. In 1972, Cavanagh was a consultant and taught management courses at Bentley College, University of Massachusetts, and Northeastern University. In 1982, she joined MITRE Corporation but returned to consulting from 1986 to 1988 for small firms involved in military contracts. In 1988, in Atlanta, Georgia, she worked for Horizons Technology, Inc. Cavanagh passed away on September 27, 1992, in Atlanta ("Aileen Cavanagh", 2018).

Isabel Hardwich

Program chair and secretary at ICWES-II (Cambridge, 1967).

Isabel Hardwich, born in 1919, obtained a degree in physics and took on an engineering apprenticeship as part of a new mixed scheme at Metropolitan Vickers in Manchester. She worked in their electron microscope division and the photometry lab, while teaching at schools and colleges. She married in 1945 and continued working with MV (now Associated Electrical Industries), switching to recruitment and training of their women engineers, until 1969. She worked as a chief clerk at the Open University in Manchester for another 10 years. She joined the Illumination Engineering Society in 1947.

Hardwich came to the Women Engineering Society in 1941 and was elected President in 1961–1962. Her husband, John Norman Hardwich, also joined WES. After her death, the Isabel Hardwich Award was set up as the highest award possible for service to WES. Hardwich was also named SWE Fellow in 1986 (Baker 2019b). She passed away on February 19, 1987.

Beatrice Hicks

Director at ICWES-I (New York, 1964).

Beatrice Hicks, born in 1919, obtained a BSc in chemical engineering in 1939 from the Newark College of Engineering (now New Jersey College of Engineering) and an MSc in physics from the Stevens Institute of Technology in 1949. She also took courses at Columbia University's School of Continuing Education.

A registered engineer, Hicks, was the inventor of the gas density switch. She first worked at Western Electric, becoming their first woman engineer in 1942. She then went on to work at Newark Controls Company (Bloomfield, NJ), a company founded by her father, where she served as vice-president, chief engineer, and president.

She was named *Mademoiselle*'s Woman of the Year in Business in 1952 and received several honorary degrees and awards. She joined the National Academy of Engineering in 1978. She was a member of the Defense Advisory Committee on Women in Service between 1960 and 1963. In 1950, Hicks became the first president of the Society of Women Engineers, for which her husband, engineer Rodney D. Chipp, was a valuable ally. Thirteen years later, she was presented with the SWE Achievement Award, for her remarkable contributions to science and society. Hicks passed on October 21, 1979 ("Beatrice Hicks", n.d.).

Michiko Ishii

Chair at ICWES-XI (Chiba, 1999).

Michiko Ishii, born in 1933, graduated from the Tokyo College of Pharmacy. Ishii had a fruitful political career with the Liberal Democratic Party (LDP) and its Miyazawa faction: Deputy Minister for Labour and for the Environment, Deputy Director for the party's Environment Division, Minister of State, and Member of the House of Councillors Committee on Education. She was by 1999 Director General of the Environment Agency (Hudson et al. 1991, p. 117).

Suzanne Jenniches

Conference chair at ICWES-VII (Washington, DC, 1984).

Suzanne Jenniches graduated from Clarion State College (now Clarion University of Pennsylvania) in 1970 with a BSc in Biology. Jenniches first taught biology in high school in Maryland and in 1974 was working at Westinghouse as a computerized test engineer. She took evening classes at the John Hopkins University Whiting School of Engineering and obtained, in 1979, a master's degree in Environmental Engineering. She pursued her education in International Affairs (Catholic University of America, Washington, DC) and Management Development (Harvard Business School).

Jenniches held many technical and managerial positions (supervisory engineer of robotics development for electronics manufacturing, operations program manager, manager of Systems and Technology Operations, vice-president, general manager) at Northrop Grumman Government Systems Division, both locally and at international sites. Jenniches was in charge of projects resulting in the B-1B electronically scanned antenna and the APQ-164 offensive radar systems. She was also director of the Northrop Grumman Electronics Systems International Campaign and the Diversity and Inclusion Council for the Sector ("F. Suzanne Jenniches", 2010).

Jenniches joined SWE in 1979 and was president in 1988–1989. She became Life Member in 1990 and was elected Fellow in 1993. She was presented with the Achievement Award in 2000. ("Suzanne Jenniches", n.d.).

Elizabeth Laverick

Conference chair at ICWES-IX (Coventry, 1991).

Elizabeth Laverick, born in 1925, grew up in a family of chemists. She was the only girl of her class for the Higher Schools Certificate at Dr. Challoner's Grammar School. She obtained a degree from Durham University in Physics and Radio and a PhD in Physics with a doctoral dissertation on "Dielectric Measurements at Audio Frequencies Using a Differential."

She married in 1946 a fellow engineering student, Charles Laverick, and both were hired by GEC Stanmore (Marconi Defence Systems Ltd.) 5 years later. Laverick worked as a microwave engineer, namely, on guided weapons systems. In 1954, she was hired at Elliot Automation to work on microwave instruments, eventually becoming General Manager of their Radar Systems sector. She collaborated in the development of the Early Warning airborne system (Nimrod). In 1971, she became the first woman to hold the position of deputy secretary for the Institute of Electrical Engineers. Retiring in 1985, Laverick came to the Court of City University and took on consulting jobs in advanced manufacturing in electronics.

Laverick joined the Women Engineering Society in the late 1950s, held the presidency in 1968–1969, was named Honorary Treasurer and edited the Society's journal for several years. She was awarded with the Order of the British Empire in 1993 and an honorary fellowship from the University of Manchester Institute of Science and Technology (UMIST). Laverick passed in 2010 (Baker 2019d).

Naomi McAfee

Conference chair at ICWES-VII (Washington, DC, 1984).

Naomi McAfee graduated from the Western Kentucky University with a BSc in physics. She first worked as a mathematician for the Westinghouse Defense and

Electronic Systems Center in their Reliability Engineering Section and then in projects on airborne missile control and radar systems and communication satellites. She was the first woman to become supervisory engineer for Westinghouse, led all reliability, maintainability, and safety engineering activities for their Defense and Space Center and was in charge of the research group who developed the television camera system used on Skylab.

McAfee joined the Society of Women Engineers in 1961, was elected president in 1972–1974, and was named SWE Fellow in 1980. She became the first woman to hold office in McAfee is a Fellow of the Society of Women Engineers and served as its national president from 1972 to 1974. She was the first woman elected to office in the American Society for Quality Control. She was president of the Federation of Organizations for Professional Women in 1978–1979, sat on the board of the Engineers Joint Council, between 1975 and 1979 ("Naomi McAfee", 2003).

Pat Patterson

Conference chair at ICWES-VII (Washington, DC, 1984).

Patricia A. Patterson, born in 1937, obtained degrees in Special Education from Saint Augustine Academy, Lakewood, OH, and Coppin State College, Baltimore, MD and master's degrees from Loyola College (Special Education) and John Hopkins University (Technology Management). She worked as manager in the Radar Division at Westinghouse. Upon retirement, Patterson worked as a special education teacher at Joppatowne High School, Joppa, MD.

Patterson joined SWE in 1977 and was elected president of the Baltimore Local Chapter in 1986 ("Patterson, Patricia A.", 2004).

Ruth Schafer

Operations chair at ICWES-I (New York, 1964).

Ruth Shafer worked at Eastern Division Manager for Overhead Heaters, Inc. in the 1950s. She designed a pump for oil-fired furnaces and flues. Later she worked at Gibbs and Cox, Naval Architects and Engineers. Ruth held several roles in SWE and raised money which allowed their headquarters in New York to open. She was the New York Section Chairman, national Nominating Chairman, Development Chairman, NY Section Representative on the National Board and Executive Council, Treasurer and Chairman of the Employment Committee, and was instrumental in organizing ICWES-I as the conference's Operations Chairman. Ruth was also a member of the British society (WES). She made donations to WES in the 1960s and hosted WES members visiting New York. An amazing side to Ruth Shafer's story is that she was running auctions at the end of each national SWE conference, auctioning off objects left in the "Lost and Found" as well as joke objects donated by members. She brought the tradition to the International conferences which provided much needed funding. The funds raised at ICWES-I were donated to WES to ensure the holding of ICWES-II. Ruth was the first recipient of the SWE's Certificate of Recognition and a member of the American Society of Mechanical Engineers and the American Society of Heating, Refrigerating, and Air Conditioning Engineers. She died in New York of cancer after 6 years of illness on May 19, 1972 ("Ruth Schafer", n.d.).

Zsuzsa Szentgyörgi

Conference chair at ICWES-X (Budapest, 1996).

A prominent Hungarian engineer, Zsuzsa Szentgyörgyi, graduated in electrical engineering with an MSc and a PhD from the Technical University of Budapest. She first worked as a designer of industrial automatic control systems; she then turned to computing and language programming for the Research Institute for Computing and Automation. In 1985, she moved on to the Ministry of Industry and served as Director General overseeing collaboration between scientific research and industry. Szentgyörgi was also a fellow of the Hungarian Academy of Sciences (MATE) and worked as Deputy Secretary General of the Scientific Society of Measurement and Automation (Szentgyörgyi 1999, p. 57).

J. Cicely Thompson

Conference chair at ICWES-II (Cambridge, 1967).

Cicely Thompson, born in 1919, graduated from Girton College, Cambridge, in Mathematics in 1940. She first worked for the Leicester Electricity Service, on turbines and a power station extension, in the Construction Department of the Fulham Power Station, for the British Electricity Authority (BEA), SE Division, in their power generation system, and in 1956 with the Associated Electrical Industries John Thomson Group (AEIJTG). She designed several nuclear power stations, namely, for the Central Electricity Authority and for the Italian government.

She joined the Institution of Electrical Engineers in 1958 and became IEE Fellow in 1986. She was named Member of the British Empire in 1980 for her contribution to the nuclear industry. Thompson joined the Women's Engineering Society (WES) in 1947 and was elected President in 1963 and in 1965. She was awarded with the Isabel Hardwich Medal for her outstanding service to WES in 1990 and became honorary member in 1993. Thompson was also presented with Senior Membership for the Society of Women Engineers (SWE) (Baker & Bonfield 2019, p. 15). She passed away on February 3, 2008.

Izabela Zachwatowicz

President of the coordination team at ICWES-IV (Cracow, 1974).

Izabela Zachwatowicz was born in Paris in 1922. She graduated from the Faculty of Architecture of the AGH University of Science and Technology in Cracow in 1948. She worked on the design of several award-winning buildings in Warsaw, like the Grand Theater in Warsaw (1951) and the Central Broadcasting Station of Polish Radio (1958), and the Provincial and Municipal Library in Rzeszów (1964).

Zachwatowicz joined the Association of Polish Architects (SARP) and sat on the board between 1957 and 1959. She took part in the Warsaw Uprising in 1944. She was president of the Educational Society of Cecilia Plater-Zyberkówna. Zachwatowicz passed in 2015 ("Izabela Kunińska-Zachwatowicz", n.d.). She passed away on March 1, 2015.

Appendix 2: The Evolution Over Time of the Regions Used by ICWES to Divide the World

Since the second ICWES, the world was divided into 11 regions, in order for leaders of the conference's series to ensure good regional representation, not only for attracting participants but also to decide where the next ICWES should be held. An analysis of participants by region for each ICWES provides an interesting picture which can be used in the future to optimize regional participation and collaboration. The regions defined by the ICWES International Continuing Committee (ICC) were redefined over time. Initially, each country had a representative who would help recruit participants, advertise the conference in their country and region, and help organizers in ways that they could. Ideally, conference locations were selected under the following conditions: (i) a good plan was presented at the Business Meeting of a current conference; (ii) the organizers were backed by a solid Association or Society. The ICC then followed the development of plans to ensure a successful event.

It was at the Business Meeting of ICWES-V in France, in 1978, that a more formal ICC was formed, with the current conference Chair named as the next convenor (chair of the Business meeting at next event and chair of ICC for the planning of the next conference). The regions were divided as follows: North America; Latin America (including South and Central America, Caribbean, Mexico); Europe (West and East); Africa; the Middle East; Asia; and Far East Asia.

However, at the Business Meeting at ICWES-VII, in 1984, the regions were redefined as: North America; Latin America; the United Kingdom, Australia and New Zealand; Western Europe, Eastern Europe, French-speaking Africa; English-speaking Africa; Indian subcontinent; Far East (including Philippines); and the Middle East. The convenor chosen at this conference was Lydia Pickup.

The definition of regions changed again at ICWES-IX in 1991: North America; Latin America; the Middle East; India; Arabic-speaking Africa; French-speaking Africa; English-speaking Africa; Western Europe. No one was found to represent

M. Frize et al., *Women's Contribution to Science and Technology through ICWES Conferences*, Women in Engineering and Science,
https://doi.org/10.1007/978-3-031-33800-7

Eastern Europe, so this region was not included in this ICC. The convenor, Elizabeth Laverick, was from the United Kingdom.

Later, after the creation of INWES, 11 regions were defined; note that the current analysis includes Israel within the Middle East:

(i) North America
(ii) Latin America (incl. South and Central America, Caribbean, Mexico)
(iii) Western Europe (including Turkey)
(iv) Eastern Europe (including Russia)
(v) French-speaking Africa
(vi) English-speaking Africa
(vii) The Middle East and North Africa
(viii) Central Asia (incl. India, Pakistan, Nepal)
(ix) Southeast Asia
(x) Far East Asia
(xi) United Kingdom, Australia, and New Zealand

Appendix 3: Participation at ICWES-I to ICWES-XII

ICWES #	Year	Place	Total persons	# host country	# not local	NA	Latin A	W EU	E EU	Fr Afr	E Afr	Mid E	C Asia	SE Asia	FEA	UK NZ AUS	# coun-tries
I	1964	New York, USA	486	391	95	Can 4	25	16	3	0	1	5	3	4	8	24/2	35
II	1967	Cambridge, United Kingdom	307	122	185	100 + 2	7	41	6	0	9	3	4	2	11		35
III	1971	Turin, Italy	233	45	188	66 + 2	15	39	8	0	5	6	3	3	7	34	32
IV	1975	Cracow, Poland	618	470	148	43 + 1	12	24	27	0	3	7	4	5	6	14/2	30
V	1978	Rouen, France	146	43	103	34 + 1	8	14	6	3	6	4	3	5	1	18	30
VI	1981	Bombay, India	394	339	55	United States 10		F 14								14	17
VII	1984	Washington, DC, USA	1204	1150	53	7										9	13
VIII	1987	Abidjan, Ivory Coast	221	166	55	11+ 3	0	6	2	14	7	2	6	2	0	2	24
IX	1991	Warwick, United Kingdom	466	305	141	35 + 4	5	32	3	9	34	1	5	2	6	5	42
X	1996	Budapest, Hungary	82	17	65	14 + 3	0	11	22	2	3	1	2	0	9	15	21
XI	1999	Chiba, Japan	349	203	120	23 + 3	5	10	10	3	4	8	7	9	26	12	38
XII	2002	Ottawa, Canada	550	296	254	108	3	19	10	8	27	3	4	6	50	16	49

M. Frize et al., *Women's Contribution to Science and Technology through ICWES Conferences*, Women in Engineering and Science,
https://doi.org/10.1007/978-3-031-33800-7

Appendix 4: Elizabeth Laverick's Retrospective of the First Nine ICWES Conferences

"ICWES One to Nine" (ICWES-X, Complete Proceedings, pp. 8–12)

Introduction

In the "Sixties," with the increasing interest in nondiscrimination and equal opportunity in the Western world, conferences concerning the role of women in the scheme of things were not unusual, and the position of women in science and engineering, where women professionals had little visibility except in the USSR, was (and still is) a favorite subject for discussion and analysis. However, as Aileen Cavanaugh, then president of the United States' Society of Women Engineers said at the official opening, in 1964, of the First International Conference of Women Engineers and Scientists held in New York, (ICWES-1), such debates, "had taken place for far too long without the benefit of facts, and most often by people—including both men and women—who had never met a woman Engineer or Scientist." So, when I was invited to address the Japanese Forum of Women Engineers on the subject of "Women Engineers and Scientists Worldwide" I naturally turned to the published reports and proceedings of that Conference and the following eight such Conferences, which have taken, place between 1964 and 1991. I was fortunate enough to be able to attend seven out of the nine, and preparing this gave me great satisfaction as I looked back and tried to put into perspective what I have learned from those Conferences about women engineers and scientists worldwide.

M. Frize et al., *Women's Contribution to Science and Technology through ICWES Conferences*, Women in Engineering and Science,
https://doi.org/10.1007/978-3-031-33800-7

Historical Background to ICWES

The First International Conference of Women Engineers and Scientists took place in New York City while the great World Fair was in progress and the World Fair organizers declared June 15, 1964, to be Women Engineers and Scientists Day. The World Fair made a great impact on the Conference: various exhibits were visited on the first day of the conference including General Motors Futurama and Behind the Scenes and Second Mobil Exhibit. The second day was devoted to technical tours including Bell Labs and the new Research Center of IBM. The third day we got down to business. The Conference had taken as its theme "Focus for the Future—Developing Engineering and Scientific Talent." Its aims (to quote Aileen Cavanaugh yet again) were "to look at the future technological needs of a peaceful world, define those areas of endeavour in which our scientifically inclined and inventive youth of today- boys and girls alike can find their opportunities for service to their fellow men of tomorrow."

> "The Conference brought together more than 400 delegates from 32 nations to exchange information and to create new patterns and new plans for the activities of women engineers and scientists. In her keynote address, Dr. Lillian Gilbreth, a renowned management consultant who, with her husband, pioneered the study of time and motion, which became so important in Industrial Engineering at that time, said we have come together to focus our attention on the future needs of the world. We shall try, through pooling our diverse experiences, to estimate what those needs will be and then to consider ways in which we can best contribute to those needs."

The technical program consisted of six sessions. The first session considered the requirements for engineering and scientific manpower to meet the challenge of sustaining the expanding population of the world, and the technical services needed in support thereof. The following three sessions established the specific types of scientific and engineering talent needed, describing the state of the world in the major areas of human need, namely, food and water, clothing and shelter, heat and power, communications and transportation. Session 5 covered the current status of women engineers and scientists and information was published about Australia and New Zealand, Europe, Iran, Japan, North America, the Philippines, South and Central America, Syria, Turkey, and the USSR. Session 6 was in the form of a symposium on "Developing Engineering and Scientific Talent," at which various eminent people were invited to contribute statements as a basis for discussion. There is no report of the discussion as such, but a summary statement emerged from the conference which pointed out the many ways in which science and engineering were contributing to improved world living conditions, and the possibility of achieving the outstanding potential of the future by cooperation. Also emphasized was the inadequate supply of trained personnel to implement future world needs.

Finally, the conference resolved:

1. To encourage each participant to report about the conference in her home country or region.

2. To encourage women to increase their participation in the professional societies in their countries.
3. To encourage women to enter the fields of engineering and science and improve their qualifications not only during their student days but also throughout their professional life.
4. To maintain the central file of women engineers and scientists used for this conference and enlarge it as much as possible.

So ended the First International Conference of Women Engineers and Scientists. It was attended by a delegation from the United Kingdom of some 22 engineers and scientists, who were so inspired that on their return they persuaded the Women's Engineering Society in the United Kingdom to host a second such conference. This took place in the University City of Cambridge in 1967 the technical theme chosen was "The Application of Technology to World Food Problems."

The inaugural address was given by Dorothy Hollingsworth OBE, a chemist and Head of the Food Science Advice Branch of the Ministry of Agriculture, Fisheries, and Food. Her address was entitled "Nutritional Goals in a World Context," and in it she traced the growth of thought on the application of nutritional knowledge to the problem of feeding populations. Experiences in the United Kingdom since the beginning of this century were used to illustrate certain general principles. International action in matters of food started with the League of Nation's concept in the 1930s of "the marriage of health and agriculture." This was followed by the setting up in 1945 of the Food and Agricultural Organization of the United Nations for the purpose of raising nutritional levels and standards of living worldwide. Twenty years on and it was clear that the difficulties were even more daunting than anticipated, and a food crisis of staggering proportion was predicted by 1985 unless the rich nations drastically increased aid to poor nations for food and population control programs. This address provoked much discussion and formed a stimulating introduction to the ensuing sessions on "Enough for everyone," "The Use of Power in Agriculture," "Increasing Food Production," and "Future Trends."

Again, there was a second theme—"The Woman Professional Engineer" and Jo Webb, an American chemical engineer and author gave the opening address, entitled "Science versus the Humanities—a harmful dichotomy." She put forward the hypothesis that science and the humanities are not separate and unrelated but are subclasses of philosophy. She traced the division between the rational and the intuitive approaches to an understanding of life, through the ages to the present day. She expressed the view that changes in education to emphasize the importance of both modes of thinking were imperative, as was the need for, and satisfaction to be found in, social responsibility combined with professional commitment. Jo Webb also reported on a survey, which she had carried out in which she invited 34 people from different countries to complete a questionnaire about women engineers. She pointed out that professionally qualified women engineers were in a striking minority in every country in the world one in three in the USSR, one in one thousand in the USA. Her questionnaire was an attempt to collect information about women engineers on a common basis, something that the first ICWES had failed to do.

Most importantly, during the conference, a meeting of delegates was held to discuss the possibility of further cooperation between women engineers and scientists. In addition to delegates representing Supporting Societies from France, Italy, Japan, USA, and United Kingdom, there was one delegate from each country, and a few interested observers. After a lengthy discussion, the major points, which emerged, were as follows:

1. There was a unanimous desire for the continuation of International Conference of Women Engineers and Scientists.
2. However, it was agreed that, although an international secretariat to organize the Conferences was desirable, the cost would be prohibitive. Each conference would therefore build upon its predecessors and self-help, and enthusiastic volunteers would provide the necessary workers. Even so a host organization would be very necessary.
3. A Conference representative in each country was desirable who would organize her own voluntary supporting and fundraising subcommittee.
4. Three years was a reasonable gap between conferences.
5. The Conference venue should move round the world.
6. Conference participation should be truly worldwide, and a great effort should be made to secure the attendance of delegates from all countries.

A third conference was mooted for 1970, and a small international committee was entrusted with the task of ensuring the continuity of the conferences. Its members were drawn from Africa, Asia, Europe, Latin America, the Middle East, and North America. The chair was Isabel Hardwich from the United Kingdom, the Honorary Secretary of the ICWES-II Organizing Committee.

So ICWES was well and truly launched.

ICWES-II was the first conference to introduce "The Bringing of Greetings Ceremony." On the eve of the conference delegates were invited to attend in national costume, and each country delivered greetings to the host country. This was followed by a "get to know you" party. Of course, we English are at a great disadvantage when it comes to national costume and rely on the Scottish and the Welsh to represent the United Kingdom. However with countries such as Austria, Brazil, Ghana, India, Iran and of course Japan, to mention just a few; it was a most colorful evening, and the UK delegates were put in the shade.

ICWES-II vied with ICWES-I in providing a wide range of visits, some technical and some cultural (taking Jo Webb's comments on unification to heart). As we were in the City of Cambridge, using the facilities of the University, we were able to tour some of the historic colleges and enjoy such differing pleasures as an organ recital in King's College chapel, and a demonstration of Morris dancing—a ritual English folk dance symbolizing death and rebirth, performed by groups of white-clad men wearing bells and carrying sticks and handkerchiefs. The host Society also gave its own party in the historic town of Bury St. Edmunds where, in the Athenaeum, the eighteenth century assembly room where Dickens used to give readings, we "treated" the delegates to a full-blown demonstration of the Scottish pipes, and haggis for supper. An additional attraction was the post conference tour of the Irish

Republic. All in all, the conferences are a golden opportunity for delegates to learn more about the heritage of the host, as well as getting to know each other.

The ICWES Series

So, ICWES-I and II were a great success involving 374 and 313 delegates, respectively, 87 from 32 overseas countries at ICWES-I and 191 from 35 overseas countries at ICWES-II.

The third ICWES was held in Turin, Italy, in 1971 this time with 233 delegates, 188 from overseas. They discussed "Planning for Progress" and "Women's Professional and Family Duties." The opening took place in the Royal Palace, Palazzio Madama. The Inaugural address given by Professor Andrea Ferrari Toniolo examined and analyzed five points relating to the deep links existing between technical and human progress. A total of 69 papers were presented on various topics such as energy sources, communication and transport, computer technology, housing, town and community planning, industrial production, and human engineering. The second theme, "Women's Professional and Family Duties" attracted a further 28 papers.

It was at this conference that the meeting of delegates, in addition to choosing Poland as the host country for ICWES-IV, discussed the format of the conferences. Clearly some delegates felt that time should be allowed for more detailed consideration of the topics, for example, by working groups. Some would have preferred less time to be spent on the papers (which should preferably not be too specialized) and more time in discussion. However it was appreciated that the large number of papers could provide a greater opportunity to learn how other countries resolved their problems also that some organizations would not support delegates if the theme were too narrow. In practice each of the three conferences was quite different from one another.

Again, the program included the possibility of visiting places of historic interest and beauty as well as of technical interest. Particularly memorable was the visit to the historical motor museum which was followed, after a luncheon provided by the Fiat Company, by a trip round their manufacturing facility where the UK delegates were particularly interested to see women and men working together on the production lines. This was something you would not have seen in England at that time.

Four years later found us in Cracow, Poland—the first time for many of us "behind the iron curtain." Again more than 30 countries were represented, and there were 148 foreign delegates in a total of 618. The themes were "New Techniques in the Service of Mankind" and "Contemporary Sociological Problems of Women Engineers and Scientists." The opening address was delivered by the Scientific Secretary of the Polish Academy of Sciences and President of the Council of the Polish Federation of Engineering Associations, Professor Jan Kaczmarek. He told us that in Poland "women accounted for 46% of the labour force. They constituted 39% of manpower with University education, 54% with secondary school vocational education and 70% with general education. 50% of medical doctors were women, 91% dentists and 83% pharmacists. They account for 31% of the University teaching staff. They are active in light industry, food and agricultural industry, as well as in printing, electro-technical and chemical industries. They playa more and more important role in the scientific life of the country. They are awarded about 25% of all university degrees." Not surprisingly, 50 of the 106 papers were given by Poles.

At the delegates meeting, where 32 countries were represented, the chairman put forward the aims of ICWES as *First*, to bring together professional women engineers and scientists worldwide. *Second* to encourage the use of technology for the betterment of life, and *Third* to consider how to increase the participation of women in this. These aims were adopted, and although slightly differently worded are still the aims today. In discussing the host country for ICWES-V, it was recorded that "There was a general feeling that Japan would be an ideal venue." (There were six Japanese delegates present at ICWES-IV.) However, the Japanese representative regretted she was unable to commit the Society of Japanese Scientists. In the event, no definite offer to host ICWES-V was forthcoming. After the conference an offer was received from France and accepted by postal ballot among the delegates.

Rouen, the ancient capital of Normandy situated on the River Seine, became the venue for ICWES V, and it was at this conference that the meeting of national representatives confirmed the aims of ICWES and laid down guidelines which included the composition of the delegates meeting and its purpose (mainly to decide the host for the next conference). They also regularized the ICWES continuity committee set up at ICWES-II.

Their responsibilities are to encourage organizations to put forward proposals to host the next ICWES, and once the host is agreed to assist in the dissemination of its details. These guidelines can only be amended at the delegates meeting. Some minor amendments have been made in ensuing years. Finally, the offer to the meeting by India to host the sixth ICWES in 1978 was unanimously accepted.

The theme chosen for ICWES-V—Technology and Freedom—included a session on "Information of women on scientific and industrial careers and their training in these careers," as well as sessions entitled "Industrialized production—wherefore?," "From the transfer of technology to industrial cooperation" and "Responsibilities of the Research Scientist." At the close, it was emphasized that engineers and scientists have not only a responsibility by their professions but also to the public. This responsibility is of the utmost significance to women scientists and engineers, who are in a better position to convey their knowledge to a seldom well-informed public and to help all women to realize they should not be unconcerned with a world ruled by technology but have a part to play in it.

1981 took us to Bombay to discuss "Science, Technology and Society." The conference was opened by Indira Gandhi, the Prime Minister. It was a great thrill to meet one of the few women Prime Ministers in the world. There was a large delegation from India, mainly from academia and research institutes and many of the papers they presented were highly specialized and concerned with very specific research problems. Only 18 papers were offered by delegates from overseas, of whom there were 55 in all. Many of these took the opportunity to extend their stay following the conference and learn more about this fascinating continent.

In 1984, ICWES returned to the USA, to Washington DC with its museums, galleries, historical monuments and of course the White House. The theme was "Technology—An International Bridge"; its aims were to reduce the gaps in understanding and knowledge among the peoples of the world. Not only was it run in conjunction with the US Society of Women Engineers' National and Student Conferences, but the extensive program of talks was arranged in three parallel sessions covering 15 topics, ranging from defense, through Technology and Society to Transportation and Our Planet Earth. There were also sessions on Undergraduate Education, and Women in Engineering and Science, and for those papers, which did not fit into any of 14 topics there was a "Pot pourri" session. Industry was very much to the fore at the conference, running recruitment evenings and participating in an exhibition. Professional development workshops were available most afternoons, proving a great attraction to the younger delegates. Technical and other tours abounded usually at least six to choose from every afternoon. We were spoiled for choice...

The year 1987 took us to the Ivory Coast—the theme "Science, Technology and Development." The Conference was opened by the Head of State, preceded in procession by a group of Kings and Chieftains resplendent in beautiful robes and carrying golden carved staffs and maces. The theme was explored at plenary sessions, and working parties met to discuss "Science and Women," "Communication and Transfer of Technology," "Science and Development," and "Sciences and Techniques." Recommendations from the working parties emerged at the final plenary session. Visits were made to the Port Authority, palm nut plantations, and a power station. The conference ended with a buffet banquet, with African dancers and fireworks, and an enormous ICWES iced cake.

And so to ICWES-IX—hosted by the Women's Engineering Society (United Kingdom) at the University of Warwick in Coventry, England in 1991. The theme was "Communication" and topics included Transport, Satellites and Telecommunications, Basic Sciences, Information Transfer, Technology Transfer, Education, Career Development and Gender issues and Demographics, i.e., the education, training, employment, and status of women engineers and scientists worldwide. It attracted 466 delegates, including 139 from 41 other countries.

On the first evening a "Bringing of Greetings" ceremony was held followed by a party in the form of an Old English fayre, which included stalls providing such Old English delicacies as Game pie, Syllabub and Mead (a honey-based drink), and entertainment from Morris Dancers and a Jazz band. We were honored by the "visit" of the famous Lady Godiva who in the eleventh century rode naked through the marketplace of Coventry in order to persuade her husband to reduce the taxes he had imposed on the town.

Needless to say a variety of tours, technical and social, were made available, and, learning from the Washington experience, a career guidance and training seminar was held for the younger delegates. With a view to increasing the number of girls interested in engineering, and to tell our delegates more about our country, the opportunity was taken to hold an exhibition in parallel with the conference with stands from industry, government, and educational and professional organizations. A series of educational events was also held during the week to which local schools were invited and in which delegates were also able to participate. These included a WISE (Women into Science and Engineering) bus, (a traveling exhibition offering hands-on experience), part of a project organized by the Engineering Council and the Equal Opportunities Commission to tell schoolgirls about Science and Engineering. Also, a lecture for schoolchildren organized by the Institution of Mechanical Engineers was given by the Post Office. This was attended by some 1000 pupils, who also visited the exhibition.

The delegates meeting, held on the Thursday, appointed the new ICWES continuity committee. They agreed that subject to their scrutiny of detailed proposals, the Nigerian offer to host ICWES 10 in 1994 be accepted. However, because of political instability in Nigeria, the continuity committee was not able to confirm the acceptance of the Nigerian offer Fortunately Hungary stepped into the breach and offered to host ICWES-X in 1996.

So what can we learn from the ICWES series so far in terms of its stated aims?

First the figures of attendance and of countries represented show that our first aim, that women active in the fields of engineering and science shall know each other worldwide, is being achieved (see Table 1). More needs to be done to promote

Table 1 International Conference of Women Engineers and Scientists (ICWES)

	Year	No. of delegates	No. of countries
ICWES I USA (Focus for the future)	1964	374(87)	32
ICWES II United Kingdom (Application of Technology to World Food Problems) (The Woman Professional Engineer)	1967	313(191)	35
ICWES III Italy (Planning for progress)	1971	233(188)	32
ICWES IV Poland (Contemporary Sociological problems of Women Engineers and Scientists) (New Techniques in the Service of Mankind)	1974	618(148)	30
ICWES V France (Technology and freedom)	1978	146(103)	30
ICWES VI India (Science, technology, and society)	1981	394(55)	17
ICWES VII USA (Technology—An international bridge)	1984	1204*(54)	9
ICWES VIII IVORY COAST (Science technology and development)	1987	221(51)	21
ICWES VIX United Kingdom (Communication)	1991	466(139)	42

( ) = Overseas
* = Combined with SWE students conference

the conference worldwide, and to help us all to keep in touch. However, a lot has been learned also 10 terms of getting to know more about each other, and this I will discuss again under the section "The status of women engineers and scientists."

Second, the themes chosen have been wide ranging and, in line with our second aim, have enabled us to consider and discuss areas where technology can be used for the betterment of life. Not only has this encouraged a large number of papers, and attracted large audiences, but also the papers have provided an impressive record of the growing technical contributions made by women in these particular areas over the last 27 years.

This impinges on our third aim—that the participation of women in engineering and science shall be increased. In addition, many of the papers in the sociological sessions discussed the different ways in which countries are tackling the problem of the shortage of women engineers and scientists, and the waste of talent that this implies.

Appendix 5: Prof. Rita Levi Montalcini's Opening Talk at ICWES-III

"Women Scientists and the Woman's Liberation Movement" (ICWES-III, Complete Proceedings, pp. 10–13)

October 1964, 3 months after the first International Conference of Women Engineers and Scientists, which took place in New York, June of the same year, another conference was held in Cambridge at the Massachusetts Institute of Technology on the topic: Women and scientific professions. At variance with the conference in New York, with another one which followed in Cambridge, Great Britain in 1967, and with the third which opens today, the symposium at M.I.T. did not include scientific nor sociological papers but was of a pure speculative character, with three well defined objectives as stated in the introduction: the first was to acquaint girls seriously interested in a career in science and technological fields with the mythical and actual difficulties that they may expect to encounter and to convey to them that these are not immutable, and that the satisfactions and rewards of such careers are high.

The second aim was to bring together outstanding men and women, active in the fields of education and industry, to focus their attention on the concerns of women preparing for careers in science and technology, in the hope that they may be stimulated to suggest new approaches and discover new solutions to the existing problems, through discussions with the students and each other.

The third aim was to attract the favorable attention of industry, other educational Institutions and the public at large, to the present successful contributions of women in these fields and the desirability of decreasing the present barriers that now prevent maximum utilization of the abilities of qualified women in these areas.

The conference took place in a dispassionate atmosphere, well-fitting to an academic Institution such as M.I.T. and the characteristic of a time which was still not ripe for considering radical solutions to a problem which has plagued mankind or at least half of it and to be exact that part of mankind which consists not of men but of women, for such a long time, that it became customary to accept it as an evil which

M. Frize et al., *Women's Contribution to Science and Technology through ICWES Conferences*, Women in Engineering and Science,
https://doi.org/10.1007/978-3-031-33800-7

is a part of the nature of man. Likewise other similar abuses which dishonor mankind such as the exploitation of men by other men, the oppression of minorities on the basis of religious and racial prejudices, the infamy of slavery were accepted by the victims with a passive and hopeless resignation. And yet, in spite of the subdued tone of the conference, which is well apparent in reading the proceedings which were published in form of a volume one year later, one feels surprised and saddened by realizing that what has been said 7 years ago is still true today in spite of the many events which took place ever since, particularly in these last years, in the hot climate which developed soon after women of all countries joined together under the flag of Women's Liberation Movement.

Perhaps the most impressive of all reports presented is that of the great nuclear physicist Wu Chien-Shiung, who, together with two young associates, had a few years earlier proved that the fundamental law of conservation of parity in quantum mechanics was not valid. Dr. Wu started her report by quoting what had been said 100 years earlier by Matthew Wassar, the founder of the Wassar Women College. The generous endower of that College said that since women have received from the creator the same intellectual constitution as men, they should have the same right and responsibilities as men to add to the scientific and cultural progress of the world.

> "I am sorry to say that we have made relatively little progress since then, said Mrs. Wu. What has contributed to this lack of progress? I sincerely doubt that any open minded person really believes in the faulty notion that women have no intellectual capacity for science and technology. Nor do I believe that social and economic factors are the actual obstacles that prevent women's participation in the scientific and technical field. The main stumbling block in the way of any progress is and always has been unimpeachable tradition."

In support of her contention that women are indeed equally qualified as men for this career and that the lack of women in science is a terrible waste of potential talent, she proudly quoted women's achievements in nuclear physics. It was the well-known discovery of radioactivity by Madame Curie and her husband that made people realize the existence of the nucleus. Madame Curie discovered and identified several chemical elements and received not one but two Nobel Prizes. "No man in history, commented Dr. Wu, has yet equaled that honor and distinction." Her daughter, Madame Irene Curie Joliot, working in team with her husband, also was awarded a Nobel Prize for their discovery of artificial radioactivity. Coming to times closer to us, another woman, Lise Meitner, contributed greatly to our understanding of the alpha and gamma radiation and together with her nephew, Dr. Frisch, she gave the first explanation of a phenomenon first described by Hahn and Fermi and named by her "nuclear fission." Another woman, Mary Geppert Meyer, was awarded the Nobel Prize in physics for her important contribution to the nuclear shell model.

> "Never before, said Mrs. Wu, have so few contributed so much under such trying circumstances."

To the list, we must add the name of the same Mrs. Wu and of two other women who were awarded the Nobel Prize: Dr. Gerta Gori for her important discovery in intermediate carbohydrate metabolism and Dr. Doroty Hodgkin for uncovering the structure of Vitamin B12 by physicochemical most sophisticated techniques.

It is of interest to recall that 6 of these 7 women were able to achieve their work, in spite of their family responsibilities as wives and mothers, and it is somewhat amusing the way how one Italian newspaper announced the news that Dr. Hodgkin had been awarded the Nobel Prize in chemistry: "Grand-mother awarded the Nobel Prize."

Dr. Wu ended her report in quoting some depressing figures published in 1962 by the American Association of University Professors on "Women in Top level Teaching and Research." The survey based on the faculty members in ten leading high endowment and ten high enrollment universities, showed that women make up about 10% of the assistant and associate professors, but only 5% of the full professors. Actually in the physical, biological, and social sciences, only 1% of professors are women.

I asked and obtained similar statistic data from the most prestigious United States Universities such as Harvard, Stanford, Berkeley, M.I.T. and Yale, and it is quite distressing to see that the situation in fact in these 9 years has remained practically unchanged.

With the wisdom which reflects the philosophy of her country of origin—Dr. Wu was born in Shanghai in 1913 and came to the United States in 1942—the scientist ends her report in a mildly optimistic vein. "The underutilization of women... is thus severe. However there seems to be a glimmer of hope on the horizon. Women are currently making substantial contributions at the lower ranks in virtually all fields. There is a good chance that these women are today providing a solid base for the greater and surely necessary use of women at the highest levels tomorrow."

I am sorry to say that I do not share this optimism. The structure of human society does not as a rule—change for better or worse in a step like fashion, but only when they are overthrown by cataclysms which all of a sudden break down barriers which were supposed to be unbreakable (not only barriers raised by the law but also less tangible and far more powerful barriers rooted in the tradition) and sweep away, as a river in flood the peaceful rhythm of human life.

The Women's Liberation Movement viewed in this light appears to the observer as a true revolution. It started in the United States and from there it spread to all other Countries and has now taken such a proportion and direction as to disconcert even the liberally minded persons of both sexes and to appear as repulsive to those who feel deeply attached to the century-old tradition of a patriarchally ruled society.

It is certainly not this the place to discuss the positive and negative features of this movement which calls forth such violent reactions loaded with emotional overtones in its supporters as well as in its opponents. I suppose it is needless to state that I belong to the first category.

What can be the effect of this movement on the activity of that still too small number of young women who have selected a career, considered to be a typical men's career such as that of engineer and scientist? I am afraid that the advantages that we can hope to obtain for women of the young generation who have made this choice, will be slim if the today scientists, satisfied with the concessions which will certainly be granted in the near future and will make available more academic positions for women, will loose interest in the situation still prevailing in other fields, first of all that of politics, which today as yesterday, and thousand years ago, has been the indisputed man's privilege and field of activity. Women—and among them also scientists—never showed too much interest in politics, sharing in the general belief that this field belongs to man. How much this prejudice is still widespread, I had occasion to verify, much to my regret, in discussing the problem with colleagues and women scientists who had gained international reputation. Many of them declared their total indifference and almost annoyance toward this movement, stating that they had never suffered from discrimination, and in any way they saw no relationship between political activity and their own scientific activity.

I do not share this viewpoint and I hope that many in the audience feel in the same way. Human activities should not be divided into closed compartments—unfortunately it happens—and the excitement and pleasure which are part of a scientific career should not obnubilate the sensibility to other problems nor absorb all energies of men and women which should be—at least in part—available for other causes.

The women's problems, as I see them, are not so much or at least not only to make use of the reservoir of intellectual energies untapped for centuries, as to give to women the full measure of their intellectual talent and dignity as human beings.

In the same way, the Negro's Liberation Movement—which shares so many aspects in common with Women's Liberation Movement—has not been conceived by its more enlightened leaders, such as Martin Luther King, as a rebellion to physical oppression, but rather as directed to give back to the colored man his human dignity. "If a man hasn't found something he will die for" said Martin Luther King, shortly before being assassinated, "he isn't fit to live."

Today the Women's Liberation Movement does not certainly ask from its followers any act of heroism, and this quotation would be out of place, were it not for its profound significance which brings into focus the universality of human values which have neither color not sex. If women will succeed in solving the problem of pursuing the scientific career of their choice, without at the same time loosing sight of ethical and social problems which are today most urgent and acute and encompass far more than the relationship between the two sexes, they will have accomplished a much more important task than if they will concentrate all their efforts in gaining recognition in scientific fields.

Women's sensibility to social injustices—as proved by their heroic and relentless fight against slavery in the past centuries—while being of no use in understanding and solving mathematical equations, is instead of tremendous value in the field of human relations. In this area, women—more compassionate and understanding than men—can play a leading role without neglecting other scientific and biological assignments, that they should face in association and not in bitter fight with the other sex, no matter how profoundly justified is their resentment.

In closing in this conciliatory vein, I should, however, not like to convey the wrong impression that I do believe that differences can be met and injustices repaired by acceptance from the party used to accept.

A revolution needs not to resort to violence, but indignation, one of the most powerful and effective forces which move human actions, as proved by the tremendous impact that it played in exposing the infamy of slavery, is well appropriate in this early stage of the Women's Liberation Movement. Tomorrow, when women will have regained faith in themselves, and men will have recognized the advantage of their changed role, both will be able to work together in full harmony, to the great benefit of mankind, which—as all of you know, but as one is tempted all too often to forget—consists of men as well of women.

Appendix 6: Elizabeth Laverick, on the Organization of ICWES-IX

"Organizing ICWES-IX" (ICWES-IX, Conference Report, pp. 7–11)

The Organization

Our first action was to set up a company—we called it ICWES-IX—limited by guarantee, (thereby protecting the Society's own funds) to undertake what became a considerable business venture, ICWES-IX. Almost 100 (12%) members of the society volunteered to be guarantors. Next we applied for the company to be registered as a charity. It took about 6 months to achieve charitable status, our main aim being to "advance the education of the public in the study and practice of engineering and science for women" with powers to "promote and organize the Conference ... and to foster international cooperation to ensure the continuation of the exchange of engineering and scientific information between women scientists and engineers."

The original Executive Committee was expanded to include representatives from the Institute of Physics and the Royal Society of Chemistry, both of which had women's groups. The secretary of WES also joined the Committee acting as a focal point for the overseas corresponding members, and as a valuable link with the society. Although officially, according to the ICWES guidelines, information about the conference should be disseminated through the Continuity Committee appointed at ICWES-VIII and representing nine world areas, we had decided in addition to set up our own network of overseas corresponding members in those countries where we already had contacts. This proved a very sensible step to take.

Six subcommittees were set up, chaired by members of the Organizing Committee and responsible to the main committee for finance, fundraising, travel and social events, conference technical program, exhibition, and public relations. Nine patrons were appointed mostly from industry and academia.

M. Frize et al., *Women's Contribution to Science and Technology through ICWES Conferences*, Women in Engineering and Science,
https://doi.org/10.1007/978-3-031-33800-7

Every 2 months or so for 3 years the 11 members of the Executive Committee gave up their Saturday and travelled to London from all parts of the United Kingdom to meet and bring each other up to date on their separate activities and plan their next steps. We monitored progress and expenditure against budget. We had fun and we worked hard!

Fundraising

The most immediate problem was to raise some money to enable us to operate and to appoint professional conference organizers to carry out the administration. This we did with the help of some of our patrons and a voluntary fundraiser, Alan Gunson, whom we appointed. His first action was to prepare a fundraising leaflet which gave the background to ICWES, details of the proposed conference and its aims, and the reasons why we needed funding. Next we held a meeting followed by lunch hosted by the Rector of Imperial College, London, Sir Eric Ash, one of our patrons, at which we expanded on our needs and our fundraising strategy to a group of eminent supporters who had agreed to act as our advisers. This acted as a launch pad for our campaign which, over a period of 18 months, raised more than £100,000, thereby enabling us to subsidize the conference fees, and to offer £20,000 in bursaries to assist delegates both from home and overseas who would otherwise not have been able to afford to attend. The British Council helped a large number of these overseas delegates with their travel as well. A certain number of prizes were offered, mainly by the Institutions for various categories of posters and papers. Bearing in mind that we were busy going into recession to say nothing of fighting the Gulf War, this was a very creditable achievement.

Public Relations

As with all conferences, the importance of promoting the event cannot be overestimated. Communication is an essential element in this respect, and of course doubly important to succeed as this was also the theme of the conference. Getting our message out to women engineers and scientists worldwide presented problems, particularly as the names and addresses of delegates of previous ICWES were not readily available. Attempts were made to publish the dates in all relevant "diaries" and "event lists," and information was sent out to a large number of organizations—engineering Institutions, professional bodies and the like. Many of the overseas corresponding members played a vital role in promoting the conference in their own countries.

A series of press releases were issued as plans developed, a News sheet, "UPDATE" was produced at approximately six-monthly intervals and used to inform the overseas corresponding members, sponsors, patrons, and all interested parties of the state of play. A special postcard, showing views of Coventry, was overprinted with an invitation to attend the conference and this was distributed widely as was the First Announcement inviting papers. Articles were prepared for publication in relevant journals, but the uptake was disappointing.

However, by May 1990 over 400 people from 35 countries had expressed interest in attending ICWES-IX. We felt we were getting through.

Technical Program

As always this was a "chicken and egg" situation. Having chosen the theme of "Communications," we wrote down all the topics we could think of under that heading and sent out our First Announcement requesting offers of papers and registration of interest. So we began to draw up our own data bank. Based on the returns we received we were able to highlight six major areas: Transport, Telecommunications and Satellites, Basic Sciences of Communication, Media, Education, and Technology Transfer. Next we had to locate chairmen and key speakers in each area and, with over 80 offers of papers and posters resulting from the issue of the first announcement, the long slog started—getting in titles and abstracts, vetting them, notifying those successful with a deadline for receipt of the final papers, and identifying gaps which had to be filled. We were determined to have the proceedings available in printed form for issue at the conference to all full-time delegates.

As time went on it became clear that two additional categories of papers were emerging, and we were able to offer sessions on Human Communications and on Gender Issues. It was also decided that we would ask all our overseas corresponding members for help in providing demographic papers on the present status and trends occurring in education, training and employment of women engineers and scientists in the different countries, and this information would be published separately after the conference. The hope is that this will be a permanent feature of future ICWES's, and that gradually a worldwide demographic picture will emerge.

Technical Visits and Social Program

It has become a feature of ICWES's that a number of technical visits and social events are arranged for the interest and enjoyment of delegates and ICWES-IX was no exception. Bearing in mind the depth of the recession at the time, local industry was surprisingly cooperative and hospitable and 23 individual technical tours were available. Each delegate was entitled to two such visits with the result that the size of visiting parties was in some cases embarrassingly small, and WES members nobly volunteered to provide transport by private car in some instances where a bus could not be justified. However, in the event, an interesting time was had by all.

Socially too we were well catered for with visits to local places of interest, and the usual run of parties, receptions, etc., although at the last minute the barbecue which the society had offered to lay on was replaced by a visit to the Coventry Marks & Spencer. There we had the opportunity to sample M&S wines and various delights from their food department. Talks about the design of their corsetry and also of the stores themselves were unfortunately cancelled. Instead, delegates were allowed to shop in comfort. This was much appreciated particularly by our overseas visitors.

Exhibition and Schools Lecture

We decided at an early stage in our planning that we would organize an exhibition in parallel with the conference at which industry was invited to take stands demonstrating the career opportunities available to women as well as featuring their wares and illustrating the theme of the conference. The exhibition was also open to the public and schools who were to be invited to attend a special lecture on an engineering subject. We were fortunate too to be offered the use of a WISE bus, one of several which, as part of the Women into Science and Engineering campaign organized jointly by The Engineering Council and the Equal Opportunities Commission, tour the country showing girls what engineering is all about.

This was indeed an ambitious undertaking, requiring its own budget and administration. Thanks to the extremely hard-working sub-committee, supported by the conference administrators, who also had expertise in this area, and in the teeth of recession, this activity added an extra flavor to the week and put a few pounds in the bank.

Registrations

No matter how excellent the program or how much work WES members put in voluntarily, the financial outcome of the conference depended on the number of registrations. Looking on it as a business with a turnover of the order of half a million pounds, you can imagine we were sitting on tenterhooks, waiting for the registrations to come in. In spite of the generosity of our donors and sponsors, a week's conference does not come cheap. In addition, delegates were not necessarily given time off from work to attend and indeed could not, in some cases, afford the time away from their jobs. Everyone in industry was under pressure, and redundancies were the order of the day. Delegates from overseas, of course, had the additional problem of heavy travel costs.

We had decided to offer a generous reduction in fees to those who registered 3 months in advance. However, only 200 out of the 400 required (if we were to break even) took advantage of this offer. My hair started to go grey with worry as the numbers crept up very slowly. I went out and bought a wig! Were we going to have to go back to our guarantors with begging bowls in hand? We cut our budget as far as we dared. We agreed to dispense with the translation facilities which would have cost the earth and for which only one request had been made in answer to our specific question—a mistake as it turned out, as in practice it became very clear that this was a very necessary facility albeit for a small group of delegates.

In fact, it was not until the second day of the conference that we realized that we would break even as a large number of delegates just turned up and registered on the spot, even from countries as far away as Nigeria and the Ivory Coast. Future ICWES organizers take note. Mind you, over those last 3 months everyone involved made strenuous efforts to contact all possible sources of delegates, particularly those overseas.

Appendix 7: All INWES Meetings Between 2002 and 2008

Year	ICWES	Meeting	Location	Dates
2001		BM0	Merrickville	May 26–29
2002	**XII**, July 27–31	BM1	Ottawa	July 31
2003		BM2	Paris	May 1–2
		BM3	Seoul	Nov 13–16
2004		BM4	Nairobi	Aug 15–19
2005	**XIII**, Aug 26–29	BM5	Seoul	Aug 24–25
		AGM1	Seoul	Aug 28
		BM6	Seoul	Aug 29
2006		BM7	Lille	May 23
		AGM2	[email]	May 22–25
2007		AGM3	[email]	May 24–27
		BM8	Wroclaw	June 29–30, July 1, 5–6
2008	**XIV**, July 15–18	BM9	Lille	July 12–15
		AGM4	Lille	July 17
		BM10	Lille	July 19

KEY

AGM Annual General Meeting

BM Board Meeting

M. Frize et al., *Women's Contribution to Science and Technology through ICWES Conferences*, Women in Engineering and Science,
https://doi.org/10.1007/978-3-031-33800-7

Appendix 8: Dormer Ellis

Dormer Ellis, an engineer who is now in her 90s, attended all the ICWES conferences from II to XII, as well as ICWES-XVI. She wrote a letter to Monique Frize in March 2021, describing her experiences in attending the ICWES conferences:

> "For over three decades, I had regularly contributed to several series of international conferences, including those of the International Federation of Business and Professional Women (BPW) and ICWES, where almost all the participants were women, and also predominantly male conferences [...]. There were always a few men at ICWES, as a husband or employer sometimes wanted to hear a certain paper, but speakers and chairs were always women. There was never a special programme for men at ICWES like the concurrent Wives' Programme of fashion shows, musical concerts or visits to art galleries that was customary at UPADI conventions. [Authors' note: ICWES-XII was an exception with a program for partners organized by Peter Frize.]"

> "I participated in ICWES-VIII in Abidjan; ICWES-IX in Coventry, England; ICWES-X in Budapest, Hungary; ICWES-XI in Chiba, Japan; and ICWES-XII in Ottawa. (I will not discuss these in turn as information must be available from other, younger, still-living attendees.) When my husband had a stroke and I became his caregiver, I no longer attended conferences abroad as he was not well enough to either live at home or came with me. After a twelve-year absence from ICWES meetings, I travelled to Los Angeles in 2014 for ICWES-XVI, which was being held in conjunction with a national SWE convention. These annual gatherings grew bigger every year and thousands of Americans attended. ICWES delegates were a tiny minority at joint sessions. I was registered for both ICWES and SWE. I attended a SWE Life Members reception, and the SWE Fellows' Tea as well as ICWES sessions. I enjoyed meeting old friends at both conferences but would have liked ICWES to have been free-standing. Things did not go well back home during my absence and I have not attended *any* international conferences in the last seven years. I have fond memories of all the ICWES friends that I made and truly enjoyed and learned from every one of the twelve ICWES conferences that I was fortunate enough to attend."

Dormer Ellis was on the ICC from ICWES-II to ICWES-XII. She presented a paper at all the ICWES conferences she attended and was very helpful in providing translation for some of the delegates who struggled with the main language of the

M. Frize et al., *Women's Contribution to Science and Technology through ICWES Conferences*, Women in Engineering and Science,
https://doi.org/10.1007/978-3-031-33800-7

event. She was well-respected and made many friends over the three decades. She was instrumental in encouraging us, Claire and Monique, to organize ICWES-XII where INWES was born. Her experience expressed above was generally what everyone who attended an ICWES felt.

Bibliography

Conference Proceedings and Reports

ICWES I, Complete Proceedings = Proceedings of the First International Conference of Women Engineers and Scientists, New York City, 15–21 June 1964

ICWES-II, Complete Proceedings = Proceedings of the Second International Conference of Women Engineers and Scientists, Cambridge, England, 1–9 July 1967, 3 vols

ICWES-III, Complete Proceedings = Proceedings of the Third International Conference of Women Engineers and Scientists, Turin, 5–12 September 1971, 2 vols

ICWES-IV, Complete Proceedings = Proceedings of the Fourth International Conference of Women Engineers and Scientists, Cracow, 8–13 September 1975

ICWES-V, Complete Proceedings = Proceedings of the Fifth International Conference of Women Engineers and Scientists, Rouen, 4–8 September 1978

ICWES-VI, Complete Proceedings = Proceedings of the Sixth International Conference of Women Engineers and Scientists, Bombay, 8–12 September 1981

ICWES-VII, Complete Proceedings = Proceedings of the Seventh International Conference of Women Engineers and Scientists, Washington, DC, 17–24 June 1984

ICWES-VIII, Conference Booklet = Conference Booklet, Eighth International Conference of Women Engineers and Scientists, Abidjan, Ivory Coast, September 20–27, 1987a

ICWES-VIII, Proceedings = "Conference Report", "Discours", "Final Resolutions", "Constitution des commissions", "Synthèse des commissions", "ICWES Guidelines", 8th Congress of the International Congress of Women Engineers and Scientists (I. C. W. E. S.), Abidjan, Ivory Coast, 20–27 September 1987b

ICWES-VIII, Recording = Eighth International Conference of Women Engineers and Scientists, Abidjan, Ivory Coast, 20–27 September 1987c. [Speech audio recording]

ICWES-IX, Complete Proceedings = Proceedings of the Ninth International Conference of Women Engineers and Scientists, Warwick, 14–20 July 1991a

ICWES-IX, Conference Report = Ninth International Conference of Women Engineers and Scientists, Warwick, 14–20 July 1991b: Conference Report

ICWES-X, Complete Proceedings = Proceedings of the Tenth International Conference of Women Engineers and Scientists, Budapest, 8–10 October 1996a

ICWES-X, Minutes = Minutes of the ICWES-X Delegates, Tenth International Conference of Women Engineers and Scientists, Budapest, 8–10 October 1996b

M. Frize et al., *Women's Contribution to Science and Technology through ICWES Conferences*, Women in Engineering and Science, https://doi.org/10.1007/978-3-031-33800-7

ICWES-XI, Complete Proceedings = Proceedings of the Eleventh International Conference of Women Engineers and Scientists, Chiba, 24–27 July 1999
ICWES-XII, Complete Proceedings = Proceedings of the Twelfth International Conference of Women Engineers and Scientists, Ottawa, 27–37 July 2002

Society of Women Engineers (SWE): *U.S. Women Engineer*; *SWE Magazine*; *SWE Members-at-Large (MAL) Newsletter*

L. Arnold, Reflections on ICWES. SWE MAL Newsletter, May/June 1996, 7–10 (1996a)
L. Arnold, Special report: Reflections on ICWES 10. SWE MAL Newsletter., November/December 1996, 5–6 (1996b)
E.C. Barth, Women engineers of other nations: Their career access, professional concerns. U.S. Woman Eng. **28**(2), 3–5 (1981)
"ICWES 11: Lasting Impressions", SWE Mag. **45**(5), 17–22 (1999)
B.A. Krenzer "BK", An international conference: Impressions. U.S. Woman Eng.**28**(2), 2–3 (1981)
B.A. Krenzer "BK", ICWES VIII – A West African experience. U.S. Woman Eng.**34**(1), 6–8 (1988)
P. Layne, Report from ICWES-X, October 8–10, 1996, Budapest, Hungary. SWE Mag. **42**(6), 10–11 (1996)
N.J. McAfee, Report from ICWES business meeting. U.S. Woman Eng. **30**(4), 11 (1984)
A.M. Perusek, 'Focus on the future': A look at the first ICWES. SWE Mag. **44**(2), 14–15 (1998)
A.M. Perusek, P. Layne, ICWES-XII: 'Women in a knowledge-based society'. SWE Mag. **48**(5), 88–94 (2002)

Women's Engineering Society (WES): *WES Journal* (from the IET Archives and WES)

Vol. 9

S. Ferguson, Report on the New York conference of women engineers and scientists. WES J. **9**(14), 17–22 (1964)
V.J.K. Milligan, Impressions of the New York conference. WES J. **9**(14), 11–17 (1964)
"Second International Conference of Women Engineers and Scientists 1967", WES J. **9**(19), 13 (1965)
J.C. Thompson, The President's message: International engineering. WES J. **9**(14), 1 (1964).; 13

Vol. 10

D.M. Cridland, Bringing of greetings. WES J. **10**(6), 5 (1967)
M. Neal, Second international conference of women engineers and scientists. WES J. **10**(4), 1 (1967)
L.E. Obeng, Impressions of the conference banquet. WES J. **10**(6), 24–25 (1967)
B. Rahm, Need for a third conference. WES J. **10**(6), 38 (1967)

K. Saruhashi, Postscript from Japan. WES J. **10**(6), 28 (1967)
R. Wallin, Some recollections of the banquet from Sweden. WES J. **10**(6), 26 (1967)
R. Winslade, A message from the president. WES J. **10**(6), 2 (1967)

Vol. 11

D.M. Cridland, Fashion show. WES J. **11**(18), 8 (1975–1976a)
D.M. Cridland, Performance of folk songs and dances. WES J. **11**(18), 11 (1975–1976b)
D.M. Cridland, Final buffet supper. WES J. **11**(18), 20 (1975–1976c)
M. Dalski, Conclusions. WES J. **11**(18), 20 (1975–1976)
S.A. Fairweather, Reception at the banquet hall etc. 'TV Ardowska'. WES J. **11**(18), 8 (1975–1976a)
S.A. Fairweather, Conclusions. WES J. **11**(18), 20 (1975–1976b)
I. Hardwich, Fourth international conference of women engineers and scientists. IV ICWES: Krakow, Poland, 8–13 Sept., 1975. WES J. **11**(18), 6 (1975–1976)
P.L. Hodges, President's message. WES J. **11**(3), 1–2 (1971)
G. Howard, The magic of Poland. WES J. **11**(18), 7 (1975–1976)
"ICWES-IV", WES J. **11**(14), 18 (1974–1975)
"ICWES-IV", WES J. **11**(15), 2 (1975)
R. West, An opinion of the purpose and achievements of TICWES. WES J. **11**(3), 22–23 (1971)

Vol. 12

"Fifth International Conference of Women Engineers and Scientists", WES J. **12**(10), 1 (1978–1979)
I. Hardwich, International conference of women engineers and scientists. Newsletter, July 1976. WES J. **12**(1), 3 (1976)
I. Hardwich, Meeting of national representatives. WES J. **12**(10), 11–12 (1977–1978)
G. Skinner, Commission 5 – Development of information systems. WES J. **12**(10), 10–11 (1978–1979)
"VICWES Bursaries", WES J. **12**(6), 2 (1976–1977)
M. Watkins, Inaugural session. WES J. **12**(10), 5 (1977–1978)

Vol. 13

"6th International Conference of Women Engineers and Scientists", WES J. **13**(2), 7–10 (1982)
H. Blount, Personal impressions of the conference. WES J. **13**(10), 4–5 (1984)
I. Hardwich, Business meeting. WES J. **13**(2), 10 (1982)
"International News – VII ICWES", WES J. **13**(9), 6 (1984)
R., R, 1984 SWE Student Awards Banquet, 20 June 1984. WES J. **13**(10), 3 (1984)
G. Skinner, Impressions of ICWES. WES J. **13**(2), 9 (1982)
M. Watkins, Impressions of India. WES J. **13**(2), 10 (1982)

Vol. 14

J. Brookes, ICWES 9 update. WES J. **14**(6), 17 (1989a)
J. Brookes, ICWES 9 update. WES J. **14**(7), 20 (1989b)
E. Laverick, The ICWES phenomenon. WES J. **14**(15), 8–11 (1991)
M. Maple, Social aspects – Random jottings. WES J. **14**(1), 19–20 (1987)
J.C. Thompson, M. Maple, Eighth international conference of women engineers and scientists Abidjan, Ivory Coast 20–27 September 1987: A report. WES J. **14**(1), 19–20 (1987)

Vol. 15

E. Laverick, [Report from ICWES 10]. WES J. **15**(15), 3 (1996)
E. Rhodes, Destination Budapest – ICWES 10. WES J. **15**(16), 11 (1997)

Vol. 16

J. Carpenter, Renewable resources and icwes11. WES J. **16**(7), 21 (1999)
J. Crook, D. Patterson, ICWES 12. WES J. **16**(19), 8–9 (2002)
D. Ellis, [Impressions of icwes11]. WES J. **16**(7), 22 (1999)
M. Frize, [icwes12]. WES J. **16**(7), 22 (1999)
I. Ghose, [Impressions of icwes11]. WES J. **16**(7), 21 (1999)
P. Goodwin, [My personal experience]. WES J. **16**(7), 19 (1999)
Y. Gueye, [Impressions of icwes11]. WES J. **16**(7), 20 (1999)
J. Maduka, [Impressions of icwes11]. WES J. **16**(7), 20 (1999)
K. Solis Garcia, [Impressions of icwes11]. WES J. **16**(7), 22 (1999)

Vol. 17

P. Battams, From the editor.... WES J. **18**(1), 2 (2008)
P. Battams, R. Carter, B. McCarthy, Four action packed days in Lille. WES J. **18**(1), 8–10 (2008)
S. Bird, INWES – Looking forward to Korea 2005: Sue Bird reports on developments. WES J. **17**(3), 2 (2004a)
S. Bird, INWES: Sue Bird reports on the latest developments. WES J. **17**(6), 2 (2004b)
S. Bird, [ICWES 13]. WES J. **17**(10), 3 (2005)
A. Gopinathan, ICWES 12. WES J. **17**(10), 3 (2005)

Vol. 18

B. McCarthy, Dear editor. WES J. **18**(1), 11 (2008)

General Bibliography

"1984 en informatique", Wikipedia (n.d.), https://fr.wikipedia.org/wiki/1984_en_informatique. Accessed 15 Oct 2022
"1984 in Science", Wikipedia (n.d.), https://en.wikipedia.org/wiki/1984_in_science. Accessed 15 Oct 2022
"Abkhazia", Wikipedia (n.d.), https://en.wikipedia.org/wiki/Abkhazia. Accessed 30 Nov 2022
"About INWES", INWES – International Network of Women Engineers and Scientists (n.d.), https://www.inwes.org/about/. Accessed 15 Mar 2023
"About Us – Earth Day", Earth Day (n.d.), https://www.earthday.org/about-us/. Accessed 10 Nov 2022
"Agenda 21 – Sustainable Development Knowledge Platform" (n.d.), https://sustainabledevelopment.un.org/outcomedocuments/agenda21. Accessed 15 Dec 2022
"Aileen Cavanagh", Engineering and Technology History Wiki (2018), https://ethw.org/Aileen_Cavanagh. Accessed 15 Feb 2021
"Aileen Cavanagh", Society of Women Engineers (n.d.), https://swe.org/membership/swes-past-presidents/aileen-cavanagh. Accessed 15 Feb 2021
"Amour Anna Enrichetta", Scienza a due voci (n.d.), http://scienzaa2voci.unibo.it/biografie/150-amour-anna-enrichetta. Accessed 10 Mar 2021
"Applicon", Wikipedia (n.d.), https://en.wikipedia.org/wiki/Applicon. Accessed 15 Oct 2022
E.d.' Auzac, Nicole Bécarud (1933-2020). Diplômées **278–279**, 73–77 (2021). https://doi.org/10.3406/femdi.2021.10667
"AWIS History", AWIS – Association for Women in Science (n.d.), https://awis.org/about-awis/awis-history/. Accessed 3 Mar 2023
N.C. Baker, 96: Isabel Hardwich, Magnificent Women (2019a), http://www.magnificentwomen.co.uk/engineer-of-the-week/96-isabel-hardwich. Published 19 Oct 2019. Accessed 22 Mar 2023
N.C. Baker, 112: Elizabeth Laverick, Magnificent Women (2019b), http://www.magnificentwomen.co.uk/engineer-of-the-week/112-elizabeth-laverick. Published 25 Nov 2019. Accessed 23 Mar 2023
N.C. Baker, D. Bonfield, *WES President Biographies (1919–1985)*, Women's Engineering Society (2019). https://www.magnificentwomen.co.uk/uploads/4/0/3/7/40379829/wes_presidents_biographies__deceased_.pdf. Accessed 23 Mar 2023
"Beatrice Hicks", EngineerGirl (n.d.), https://www.engineergirl.org/125368/Beatrice-Hicks. Accessed 22 Mar 2023
J. Bindé, *World Knowledge Societies* (UNESCO, 2005) https://unesdoc.unesco.org/ark:/48223/pf0000141843. Accessed 4 Mar 2023
Canadian Committee on Women in Engineering, *More Than just Numbers, Report of the Canadian Committee on Women in Engineering* (NSERC, 1992)
A.M. Cetto et al., *World Conference on Science: Science for the Twenty-First Century. A New Commitment* (Banson, 2000) https://unesdoc.unesco.org/ark:/48223/pf0000120706
"Channel Tunnel", Wikipedia (n.d.), https://en.wikipedia.org/wiki/Channel_Tunnel. Accessed 15 May 2022
"Comecon", Wikipedia (n.d.). https://en.wikipedia.org/wiki/Comecon. Accessed 13 Apr 2022
"Communist Women's International", Wikipedia (n.d.), https://en.wikipedia.org/wiki/Communist_Women's_International. Accessed 24 Feb 2023
"Concorde", Wikipedia (n.d.). https://en.wikipedia.org/wiki/Concorde. Accessed 15 May 2022
"Congrès international des recherches féministes dans la francophonie". Wikipedia (n.d.), https://fr.wikipedia.org/wiki/Congrès_international_des_recherches_féministes_dans_la_francophonie. Accessed 24 Feb 2023
M. Cook, Private School to Open in Old Merrickville Inn, *Ottawa Citizen*, Jan 2, 2013 (2013). https://www.pressreader.com/canada/ottawa-citizen/20130102/282226598063767. Accessed 4 Mar 2023

"DDT", Wikipedia (n.d.), https://en.wikipedia.org/wiki/DDT. Accessed 12 Oct 2021

"Dr. Wanda Wuttunee", Faculty of Arts, Department of Native Studies, University of Manitoba (n.d.), https://umanitoba.ca/faculties/arts/departments/native_studies/people/wuttunee.html. Accessed 25 Dec 2022

"Eureka Prometheus Project", Wikipedia (n.d.), https://en.wikipedia.org/wiki/Eureka_Prometheus_Project. Accessed 11 Nov 2022

"European Graduates – Aalborg Universitet, Ålborg Area, Denmark", European Graduates (n.d.). https://graduates.name/aalborg_universitet-1997-111#id59185838. Accessed 22 Aug 2022

"F. Suzanne Jenniches", Engineering and Technology History Wiki (2010). https://ethw.org/Oral-History:F._Suzanne_Jenniches_(2010). Accessed 23 Mar 2023

R.F. Fellows, A.M.M. Liu, *Research Methods for Construction*, 5th edn. (Wiley Blackwell, 2022)

W. Feuerhahn, P. Rabault-Feuerhahn, Présentation: la science à l'échelle internationale. Rev. Ger. Int. **12** (2010). https://doi.org/10.4000/rgi.246. Accessed 25 Feb 2023

N. Forestell, M. Monaygh (eds.), *Documenting First Wave Feminisms, Volume 2: Canada – National and Transnational Contexts* (University of Toronto Press, 2014)

"Foundation of MWIA", Medical Women's International Association (n.d.), https://mwia.net/history/foundation-of-mwia/. Accessed 28 Feb 2023

M. Frize, *A Woman in Engineering – Memoirs of a Trailblazer. An Autobiography by Monique (Aubry) Frize* (University of Ottawa Press, 2019)

M. Frize, J. Inch, Women in engineering: More than just numbers, in *Proceedings of the 1992 Women in Engineering Conference: Increasing Enrollment and Retention*, (1992)

"Global Village", Wikipedia (n.d.), https://en.wikipedia.org/wiki/Global_village. Accessed 26 Feb 2023

"GWI History", Graduate Women International (n.d.), https://graduatewomen.org/who-we-are/our-story/. Accessed 28 Feb 2023

"History", ATSU (n.d.), https://atsu.edu.ge/index.php/en/teqnologiuris-history-eng. Accessed 30 Nov 2022

"Hommage à Nicole Bécarud", AFFDU (2020), https://www.associationdesfemmesdiplomees.fr/2020/04/08/hommage-a-nicole-becarud/. Published Apr 8 2020. Accessed 22 Mar 2023

M. Hudson et al., *Who's Who of Women in World Politics* (Bowker-Saur, 1991)

"Hydropower Explained: Ocean Thermal Energy Conversion", Independent Statistics and Analysis (2022), U.S. Energy Information Administration. https://www.eia.gov/energyexplained/hydropower/ocean-thermal-energy-conversion.php. Accessed 15 Sept 2022

K. Ighobor, Wangari Maathai, the woman of trees, dies. Africa Renewal (2011), https://www.un.org/africarenewal/web-features/wangari-maathai-woman-trees-dies. Accessed 27 Mar 2023

"Ila Ghose", Wikipedia (n.d.), https://en.wikipedia.org/wiki/Ila_Ghose. Accessed 16 Aug 2022

"International Conference for Women Leaders", Wikipedia (n.d.), https://en.wikipedia.org/wiki/International_Conference_for_Women_Leaders. Accessed 24 Feb 2023

"International Congress of Women", Wikipedia (n.d.), https://en.wikipedia.org/wiki/International_Congress_of_Women. Accessed 24 Feb 2023

"International Congress of Working Women", Wikipedia (n.d.), https://en.wikipedia.org/wiki/1919_International_Congress_of_Working_Women. Accessed 24 Feb 2023

"International Socialist Women's Conferences", Wikipedia (n.d.), https://en.wikipedia.org/wiki/International_Socialist_Women's_Conferences. Accessed 24 Feb 2023

"International Space Station", Wikipedia (n.d.), https://en.wikipedia.org/wiki/International_Space_Station. Accessed 15 Oct 2022

"Izabela Kunińska-Zachwatowicz", Pamięci Architektów Polskich (n.d.), https://www.archimemory.pl/pokaz/izabela_kuninska-zachwatowicz,5100. Accessed 23 Mar 2023

J. Juillard, Advanced ceramics. Impact Sci. Soc. **149**.: New Materials, New Machines, 25–37 (1988)

"Karen Decontie", Indigenous Futures in Engineering at Queen's University (n.d.), https://www.aboriginalaccess.ca/adults/role-models/decontie. Accessed 25 Dec 2022

"L'AFFDU vue de l'intérieur 1920-1995", *Diplômées*, vol 180: Soixante quinze ans de l'histoire de l'AFDU (1997), pp. xliii–liv. https://doi.org/10.3406/femdi.1997.7538. Accessed 25 Feb 2023

"La Fin des certitudes", Wikipedia (n.d.), https://fr.wikipedia.org/wiki/La_Fin_des_certitudes. Accessed 20 Nov 2022

"Landsat Program", Wikipedia (n.d.), https://en.wikipedia.org/wiki/Landsat_program. Accessed 2 Mar 2022

"LaserDisc", Wikipedia (n.d.), https://en.wikipedia.org/wiki/LaserDisc. Accessed 29 Mar 2022

"Lillian Dyck", The Canadian Encyclopedia (n.d.-a), https://www.thecanadianencyclopedia.ca/en/article/lillian-dyck. Accessed 25 Dec 2022

"Lillian Dyck", Wikipedia (n.d.-b), https://en.wikipedia.org/wiki/Lillian_Dyck. Accessed 25 Dec 2022

"Louise Brown", Wikipedia (n.d.), https://en.wikipedia.org/wiki/Louise_Brown. Accessed 29 Mar 2022

"Magnetohydrodynamic Generator", Wikipedia (n.d.), https://en.wikipedia.org/wiki/Magnetohydrodynamic_generator. Accessed 15 Sept 2022

"Marthe Solange Achy Brou", Wikipedia (n.d.), https://fr.wikipedia.org/wiki/Marthe_Solange_Achy_Brou. Accessed 3 Mar 2023

"Mary Okwakol", Wikipedia (n.d.), https://en.wikipedia.org/wiki/Mary_Okwakol. Accessed 13 Mar 2022

K. Mason, Empowering citizen green: Prefigurative politics, autonomous geographies and hoping against Hope, PhD Dissertation, Aberystwyth, 2009

S. Meakin, Le Sommet de la Terre de Rio: Sommaire de la Conférence des Nations Unies sur l'environnement et le développement (1992), https://publications.gc.ca/Pilot/LoPBdP/BP/bp317-f.htm. Accessed Nov 2022

L. Miall, First nations member sought change for 'seven generations': Community mourns loss of a great leader (2014). https://www.concordia.ca/cunews/main/stories/2014/02/19/first-nations-membersoughtchangeforsevengenerations.html. Published 19 Feb 2014. Accessed 25 Dec 2022

M. Moynagh, N. Forestell (eds.), *Documenting First Wave Feminisms, Volume I: Transnational and Collaborations and Crosscurrents* (University of Toronto Press, 2012)

"Naomi McAfee", Engineering and Technology History Wiki (2003), https://ethw.org/Oral-History:Naomi_McAfee_(2003). Accessed 23 Mar 2023

"Organization for Women in Science for the Developing World", Wikipedia (n.d.), https://en.wikipedia.org/wiki/Organization_for_Women_in_Science_for_the_Developing_World. Accessed 3 Mar 2023

G. Orwell, *1984* (Secker & Warburg, 1949)

P.A. Patterson, *Baltimore Sun*, February 8, 2004 (2004), https://www.baltimoresun.com/news/bs-xpm-2004-02-08-0402091382-story.html. Accessed 23 Mar 2023

A.D. Patterson, F.J. Gonzalez, J.R. Idle, Xenobiotic metabolism: A view through the metabolometer. Chem. Res. Toxicol. **23**, 851–860 (2010). https://pubs.acs.org/doi/10.1021/tx100020p.https://doi.org/10.1021/tx100020p. Accessed 23 Mar 2023

"Polish Association of Chemical Engineers SIPTChem", SIPTChem (n.d.), https://sitpchem.org.pl/polish-association-of-chemical-engineers-sitpchem/. Accessed 1 Mar 2023

I. Prigogine, *La fin des certitudes* (Odile Jacob, 1996) (English: The End of Certainty: Time, Chaos, and the New Laws of Nature, Free Press, 1997)

"Priscilla Settee", College of Arts and Science, University of Saskatchewan (n.d.-a), https://artsandscience.usask.ca/profile/PSettee. Accessed 25 Dec 2022

"Priscilla Settee", Wikipedia (n.d.-b), https://en.wikipedia.org/wiki/Priscilla_Settee. Accessed 25 Dec 2022

L.M. Puaca, *Searching for Scientific Womanpower. Technocratic Feminism and the Politics of National Security, 1940–1980* (The University of North Carolina Press, 2014)

"Qui sommes-nous?", Fondation Erna Hamburger (n.d.), https://www.unil.ch/hamburger/buts. Accessed 13 Mar 2022

A. Ricard, Women's history month: Dr. Dormer M. Ellis. Westcoast Women in Engineering, Science and Technology (2018) October 15, 2018. https://www.sfu.ca/wwest/WWEST_blog/women-s-history-month%2D%2Ddr%2D%2Ddormer-m%2D%2Dellis.html. Accessed 21 Mar 2023

"Rohini (satellite)", Wikipedia (n.d.), https://en.wikipedia.org/wiki/Rohini_(satellite). Accessed 29 Mar 2022

S.V. Rosser, *Teaching Science and Health from a Feminist Perspective: A Practical Guide* (Pergamon Press, 1986)

S.V. Rosser, *Feminism within the Science and Health Care Professions: Overcoming Resistance* (Pergamon Press, 1988)

S.V. Rosser, *Female-Friendly Science: Applying Women's Studies Methods and Theories to Attract Students* (Pergamon Press, 1990)

S.V. Rosser, *Biology and Feminism: A Dynamic Interaction* (Twayne Publishers, 1992)

S.V. Rosser, *Women, Science, and Society: The Crucial Union* (Teachers College Press, 2000)

"Ruth Schafer", Wikipedia (n.d.), https://en.wikipedia.org/wiki/Ruth_Shafer. Accessed 12 Oct 2021

Salway Black, S. (1994). Redefining success in community development: A new approach for determining and measuring the impact of development, The 1994 Richard Shramm Paper on Community Development, presented at the Lincoln Filene Centre, Tufts University. https://epub.sub.uni-hamburg.de/epub/volltexte/2009/3321/pdf/tuftreport1.pdf. Accessed 25 Dec 2022

"Satellite Geodesy", Wikipedia (n.d.), https://en.wikipedia.org/wiki/Satellite_geodesy. Accessed 15 May 2022

L. Schiebinger, *Has Feminism Changed Science?* (Harvard University Press, 1999)

"Science for the Twenty-First Century: A New Commitment. Declaration on Science and the Use of Scientific Knowledge. Science Agenda – Framework for Action", UNESCO (2000), https://unesdoc.unesco.org/ark:/48223/pf0000122938. Accessed 15 Dec 2022

"Society of Women Engineer – History", Wikipedia (n.d.), https://en.wikipedia.org/wiki/Society_of_Women_Engineers#History. Accessed 15 Feb 2021

"Somdatta Sinha", Wikipedia (n.d.), https://en.wikipedia.org/wiki/Somdatta_Sinha. Accessed 4 May 2022

J. Stuart, D. Sanders, European Eureka project – PROgraMme for a European traffic of highest efficiency and unprecedented safety (PROMETHEUS) (n.d.). https://researchportal.port.ac.uk/en/projects/european-eureka-project-programme-for-a-european-traffic-of-highe. Accessed 11 Nov 2022

W. Sullivan, Protein from petroleum not ready for use in food. New York Times, May 30, 1976 (1976). https://www.nytimes.com/1976/05/30/archives/protein-from-petroleum-not-ready-for-use-in-food.html. Accessed 13 Oct 2021

"Sumati Bhinde (1932–1999)", Stree Shakti (n.d.), http://www.streeshakti.com/bookS.aspx?author=36. Accessed 22 Mar 2023

"Suzanne Jenniches", Society of Women Engineers (n.d.), https://swe.org/membership/suzanne-jenniches/. Accessed 23 Mar 2023

Z. Szentgyörgyi, A short history of computing in Hungary. IEEE Ann. Hist. Comput. **21**(3), 49–57 (1999)

"The Way We Were in 1981: Technology", BBC News (2010), https://www.bbc.com/news/uk-11789191. Published 19 Nov 2010. Accessed 27 Mar 2023

"Timeline of the Telephone", Wikipedia (n.d.), https://en.wikipedia.org/wiki/Timeline_of_the_telephone. Accessed 29 Mar 2022

"Timeline of Women in Science", Wikipedia (n.d.), https://en.wikipedia.org/wiki/Timeline_of_women_in_science. Accessed 2 Mar 2022

"United Nations Conference on the Human Environment, 5-16 June 1972, Stockholm", United Nations (n.d.), https://www.un.org/en/conferences/environment/stockholm1972. Accessed 15 Oct 2021

"Vous souvenez-vous du 'steak de pétrole'?", France inter (2019), https://www.franceinter.fr/histoire/vous-vous-souvenez-du-steak-de-petrole. Published 26 Nov 2019. Accessed 12 Oct 2021

X. Wang, M. Lu, Microwave power transmission based on retro-reflective beamforming, in *Wireless Power Transfer – Fundamentals and Technologies*, ed. by E. Coca, (2016). https://doi.org/10.5772/62855. Accessed 2 Jan 2022

WEPAN National Conference, *held May 31 – June 2,* 1992*, at The Capital Hilton, Washington, DC*, n.p

“What is IAW –International Alliance of Women”, International Alliance of Women – Alliance international des femmes (n.d.), https://www.womenalliance.org/about-iaw/what-is-iaw/. Accessed 24 Feb 2023

“Women Engineers Plan International Conference”, IEEE/EE **82**(12), 745 (1963). https://doi.org/10.1109/EE.1963.6541838

“Women’s International Democratic Federation”, Wikipedia (n.d.), https://en.wikipedia.org/wiki/Women’s_International_Democratic_Federation. Accessed 24 Feb 2023

“World Conferences on Women”, UN Women (n.d.), https://www.unwomen.org/en/how-we-work/intergovernmental-support/world-conferences-on-women. Accessed 24 Feb 2023

M. Ziolkowski, La Francophonie en Pologne. Hermès, La Revue **40**, 59–61 (2004/2003), https://www.cairn.info/revue-hermes-la-revue-2004-3-page-59.htm. Accessed 24 Jan 2021

Index

A
Aboriginals, 256, 262, 267, 268, 282
Activities, 3, 7, 10, 12, 14, 15, 27, 37, 44, 47, 52, 53, 60, 61, 67, 70, 74, 84, 91, 94, 95, 102, 104–106, 111, 112, 123, 125, 140, 145, 154, 177, 180–182, 186–188, 194, 200, 203, 208, 215, 216, 221, 231, 246, 251, 257–260, 262, 263, 267, 269, 270, 281, 289, 292, 294, 296, 306, 309, 319, 320, 323, 326, 329–331, 333, 334, 349, 351, 353, 357, 365, 372, 373, 383, 398, 401, 405, 407, 411, 412, 423, 430, 441, 442, 446, 448
Applied engineering, 7
Archives of Women in STEM in Canada, 409
Association des Ingénieurs, Techniciens, et Agronomes de la Côte d'Ivoire (AITACI), 157, 158
Association Française des Femmes diplômées des Universités (AFFDU), 100–102, 108, 420

B
Board elections, 375
Board of Directors, 28, 140, 274–276, 279, 280, 297, 309, 318, 330, 342–344, 358, 360, 364, 371, 376, 378, 410
Business meeting, 3, 4, 14, 45–50, 80, 88, 114, 131, 153–154, 158, 159, 190–191, 272, 290, 299, 306, 310, 311, 342, 393, 401, 414, 425

C
Cercle Français des Femmes Ingénieurs, 100
Cercle Ivoirien des Femmes Ingénieurs, Scientifiques, Architectes et Techniciennes Supérieures (CIFISATS), 157, 158, 273, 344, 419
Charity corporations, 365, 411
CIWES-ICFIS, 407–414
Climate change, 198, 224, 256, 258–260, 280, 282, 319, 398, 401, 404, 415
Committees, 14, 33, 66, 79, 99, 119, 158, 211, 256, 289, 310, 398, 408
Communications, 5, 12, 13, 16, 17, 29, 58, 61–64, 66, 75, 76, 84, 87, 96, 117, 121, 122, 125, 144, 148, 150, 157, 159, 161, 163, 166, 171–182, 193, 195, 214, 218, 237, 240, 241, 245, 250, 256, 261, 267, 279, 285, 295, 297, 299, 301, 302, 306, 310–313, 315, 320, 330, 333, 334, 341, 346, 351, 353, 367, 368, 375–377, 380, 381, 383, 387, 389, 395, 398, 399, 402, 403, 411, 415, 416, 423, 430, 433, 435, 436, 446, 447
Computer technologies, 61, 63, 76, 141, 433
Construction, 39, 60, 63, 65, 66, 87, 92, 93, 135, 149, 150, 158, 163, 193, 200, 205, 209, 343, 392, 424
Continuity, 45–47, 51, 82, 103, 114–117, 153, 154, 190, 191, 211, 237, 275, 279, 290, 293, 299, 324, 330, 347, 413–414, 417, 432, 434, 436, 445
Continuous committee, 290

M. Frize et al., *Women's Contribution to Science and Technology through ICWES Conferences*, Women in Engineering and Science,
https://doi.org/10.1007/978-3-031-33800-7

D

Diversity, 4, 26, 186, 224, 257, 258, 281, 302, 349, 398–401, 403, 422

E

Ecology, 5, 66, 215, 219, 220, 222, 224, 227, 228, 231, 249, 250, 261, 399, 400, 415

Education, 7, 12, 13, 19–21, 23, 35, 43, 44, 49, 50, 54, 55, 65, 69, 79–82, 84, 89, 91–93, 103, 104, 107–113, 117, 121, 122, 128–130, 137, 139, 141, 145, 147, 150–152, 158, 162, 166, 171, 172, 179–183, 185, 189, 190, 193, 201, 206–209, 217, 220, 223, 225–227, 231–233, 236, 237, 246, 250, 262, 263, 266–268, 271, 282, 284, 292, 296, 297, 314, 354, 361, 362, 368, 372, 373, 377, 381, 382, 391–393, 399, 404, 407–413, 415–417, 421–423, 431, 433–435, 439, 445, 447

Eleven regions, 362, 382

Environment, 18, 19, 29, 43, 57, 58, 61, 66, 71, 72, 75, 82–84, 88, 100, 105, 106, 120, 129, 148, 149, 161, 167, 174, 179, 181, 185, 194, 198, 200, 217, 221, 223, 224, 227, 229–231, 233, 234, 245, 247, 249, 250, 260, 264, 267, 283, 291, 376, 390, 398, 401, 408, 415–417, 421

Equal opportunities, 68, 69, 71, 109, 111, 173, 185, 207, 217, 238, 271, 429, 436, 448

Ethics, 13, 218, 225, 226, 231, 232, 256, 282, 284, 337, 346, 375, 378, 388, 389, 392, 393, 398, 417

F

Food, 5, 16, 17, 25, 26, 34, 35, 37–41, 48, 52, 54, 104, 106, 121, 123, 126, 132, 133, 163, 164, 169, 174, 229, 231, 250, 267, 283, 285, 395, 404, 415, 420, 430, 431, 434, 436, 448

Founding meeting, 179, 292, 311

Funding, 29, 48, 189, 236, 237, 258, 261, 264, 269–272, 275, 284–286, 293, 297–301, 304, 307, 318, 321, 329, 332, 334, 335, 337, 346, 350, 352, 354, 355, 358, 365, 367, 375, 377, 378, 380, 381, 407, 411, 412, 423, 446

G

Gender Papers, 10, 21–25, 41–45, 68–72, 88–93, 106–113, 126–131, 146–152, 161–162, 182–190, 197, 201–210, 232–233, 262–272, 314, 400

Goals, 4, 12, 13, 35, 37, 43, 44, 69, 107, 108, 110–112, 114, 157, 158, 164, 174, 188, 190, 215, 228, 237, 255, 272, 275, 289, 292, 294, 295, 301–303, 306, 307, 312, 315, 316, 331, 338, 345, 349, 350, 354, 355, 359, 366, 384, 387, 390, 404, 407, 409, 414, 416, 431

H

Health, 7, 19, 35, 37, 39, 40, 65–67, 83, 85, 102, 124, 126, 130, 206, 209, 210, 227–229, 231, 234, 235, 258, 260, 262–264, 284, 417, 431

Human health, 121, 123, 124, 229–230

Human needs, 12, 17, 82, 86, 88, 103, 415, 430

I

ICWES database, 3, 4, 414–417

ICWES guidelines, 190, 272, 276, 277, 279, 445

ICWES I, 3, 8–9, 15, 247

ICWES-II, 39, 40, 114, 432

ICWES III, 10, 63

ICWES-IV, 84, 100, 104, 105, 108, 202

ICWES-V, Rouen, France, 1978, 5, 8, 99–117, 127, 134, 153, 420, 427, 434

ICWES-VI, Bombay, India, 1981, 5, 8, 86, 114, 115, 119–138, 420, 427, 434

ICWES-VII, Washington, DC, USA, 1984, 5, 7, 8, 131, 134, 139–156, 421–423, 427, 435

ICWES-VIII, Abidjan, Ivory Coast, 1987, 4, 153, 154, 157–169, 336, 415, 419, 427, 453

ICWES-IX, Warwick, UK, 5, 8, 171–195

ICWES-X, Budapest, Hungary, 5, 7, 25, 31, 48, 95, 99, 114, 197–218, 221, 222, 224, 289, 290, 424, 427, 453

ICWES-XI, Chiba, Japan, 5, 8, 214, 215, 219–253, 421, 427, 453

ICWES-XII, Ottawa, Canada, 3, 5, 8–10, 84, 179, 244, 248, 252, 253, 255–286, 289–292, 294, 297, 307, 309–311, 317, 325, 326, 342, 344, 353, 354, 363, 373, 409, 413, 427, 451, 453

ICWES-XIII, Seoul, South Korea, 12, 19, 224, 233, 278, 279, 285, 309–311, 317–323, 325–340, 342–347, 353, 357, 373, 393–399, 451

ICWES-XIV, Lille, France, 325, 339, 342, 345, 347–351, 353, 354, 358, 361, 364–390, 397–405
Indian Women Scientists Association (IWSA), 15, 117, 120, 134, 420
International collaboration, 82, 84, 180
International Network of Women Engineers and Scientists (INWES), 6, 46, 48, 179, 263, 272, 274–280, 284–286, 290, 298, 307, 309–324, 326–337, 339–372, 374–390, 392–395, 398, 400, 405, 407, 409–414, 416, 417, 419, 420, 426, 451, 454
International organization, 48, 84, 116, 163, 294, 296, 304, 314–316, 350, 359, 387
INWES Education and Research Institute (INWES-ERI), 377, 382, 407–409, 411–413

K
Korea Woman Scientists and Engineers (KWSE), 326, 339, 344–346, 373, 391, 394, 403

L
Life sciences, 126, 227

M
Medical fields, 87, 225, 226, 229–230, 232
Membership, 13, 47, 115, 120, 137, 188, 279, 281, 297–299, 305, 306, 310, 311, 313, 315, 316, 318, 319, 321, 322, 324, 330, 333–335, 342, 343, 346, 348, 352, 353, 358, 359, 364, 367, 368, 372, 374–377, 381, 382, 384, 386, 389, 390, 394, 424
Membership types, 306
Merrickville, 272, 289–298, 307, 309, 311, 313, 314, 350, 414, 451
Methodology, 200, 269
Michiko Ishii, 5, 220–221, 223, 224, 227, 233, 244, 248, 249, 421
Mission, 14, 60, 221, 281, 294, 300, 301, 319, 365, 397, 400, 416

N
New technologies, 100, 125, 156, 163, 200, 237, 238
Numerical and computer applications, 64, 415
Nutrition, 37, 48, 54, 82, 83, 85, 120, 122–124, 126, 136, 161, 261, 320, 415

P
Peer review, 238, 256, 258, 384

Q
Qualitative documents, 5
Quantitative documents, 5

R
Role of women, 24, 42, 43, 67, 69, 72, 89, 104, 146, 161, 208–210, 257, 339, 345, 351, 398, 409, 417, 429

S
Scientific responsibility, 99, 105, 162, 438
Society of Women Engineers (SWE), 3, 4, 7, 11–31, 33–35, 40, 42, 45, 47, 51, 73, 85, 104, 115, 117, 131, 133, 135–139, 154, 160, 163–165, 191, 195, 200, 211, 212, 214, 215, 217, 244, 245, 263, 273, 280, 281, 285–286, 290, 291, 296, 297, 301, 302, 307, 312, 314, 316, 320, 323–326, 331, 336, 337, 341, 344, 352–355, 357, 358, 365, 366, 374, 377, 380, 387, 402, 409, 410, 413, 419–424, 429, 435, 436, 453
S&T Content, 83–87, 102–105, 142–146, 226–227, 415–416
STEM international conference, 14, 290, 314, 321, 344, 355
Structure, 38, 64, 65, 71, 82, 88, 142, 143, 145, 208, 209, 269–271, 274, 289, 291–293, 296–300, 303, 305, 306, 309–312, 350, 358–360, 365, 375, 378, 385, 411, 417, 440, 441

T
Testimonials, 3, 4, 25–31, 44, 50–55, 75–77, 80, 94–98, 115–117, 131–138, 154–156, 164–166, 191–192, 197, 212–218, 245–247, 280–286, 313, 393–395, 401–405
Trained personnel, 21, 22, 30, 188, 430

W
Women Engineering Society (WES), 4, 11, 13–15, 22, 26, 27, 33, 34, 42, 44, 45, 48, 51, 54, 73, 75, 102, 110, 115, 131, 134–135, 160, 167, 172, 174–175, 191, 193, 212, 217–218, 223, 247–253, 273, 281–286, 290, 291, 312, 314, 326, 329, 344, 393–395, 397, 401, 402, 405, 413, 421–424, 445, 447, 448
Women engineers, 11, 33, 58, 102, 162, 202, 256, 289, 312, 397
Women's employment, 91, 100, 162, 403
Women's role, 68, 70–72, 88, 112–113, 220, 256
Women's studies, 7, 230, 253, 256, 258, 262, 265, 266, 280, 297, 321, 416
Women's training, 189
Workshops, 3, 7, 8, 140, 141, 146, 203, 212, 214–217, 234, 248, 256, 258, 263, 269, 278, 305, 309, 318, 320, 322, 326, 328, 331, 334, 336, 341, 351, 353, 355, 366, 367, 373, 377, 380, 384, 392, 398, 408, 409, 411, 435

SPRINGER NATURE

GPSR Compliance

The European Union's (EU) General Product Safety Regulation (GPSR) is a set of rules that requires consumer products to be safe and our obligations to ensure this.

If you have any concerns about our products, you can contact us on ProductSafety@springernature.com

In case Publisher is established outside the EU, the EU authorized representative is:

Springer Nature Customer Service Center GmbH
Europaplatz 3
69115 Heidelberg, Germany

The manufacturer's authorised representative in the EU is Springer Nature Customer Service Centre GmbH, Europaplatz 3, 69115 Heidelberg, Germany. If you have any concerns regarding our products, please contact ProductSafety@springernature.com

Printed and bound by CPI Group (UK) Ltd, Croydon, CR0 4YY
01/10/2026
02225865-0019